D1090887

Late Quaternary Climate, Tectonism, and Sedimentation in Clear Lake, Northern California Coast Ranges

Edited by
John D. Sims

SPECIAL PAPER

214

Late Quaternary Climate, Tectonism, and Sedimentation in Clear Lake, Northern California Coast Ranges

Edited by

John D. Sims
U.S. Geological Survey
345 Middlefield Road
Menlo Park, California 94025

214

Published by The Geological Society of America, Inc.
3300 Penrose Place, P.O. Box 9140, Boulder, Colorado 80301

GSA Books Science Editor Campbell Craddock

Library of Congress Cataloging-in-Publication Data

Late Quaternary climate, tectonism, and sedimentation in Clear Lake,
Northern California coast ranges / edited by John D. Sims.
p. cm.—(Special paper ; 214)
Includes bibliographies.
Includes index.
ISBN 0-8137-2214-4
1. Paleolimnology—California—Clear Lake (Lake County)
2. Geology, Stratigraphic—Quaternary. I. Sims, John D.
II. Series: Special papers (Geological Society of America) ; 214.
QE39.5.P3L39 1988
551.7'9'0979414—dc19 87-37943
CIP

Contents

Preface

The research papers in this volume were stimulated by core drilling that I did in Clear Lake in 1973 and 1980. The first suite of cores that I collected from Clear Lake in 1973 was prompted by my interest in earthquake-induced sedimentary structures. Although the cores I collected in 1973 did not yield the results I had hoped for, they did show that Clear Lake was a long-lived lake with a complex geologic history, and that one could decipher the record of climate change in a continental record. The results of my research on these early cores with Dave Adam and Mike Rymer was the basis for additional core drilling in 1980. The goal of this latter core drilling was to obtain a longer record of climate change and tectonism than I was able to obtain in 1973. We were confident that Clear Lake contained this longer record because the results of our research on the 1973 suite of cores suggested that the basin had been in existence for perhaps as much as 2 million years. The project to do the coring in 1980 was funded by the Paleoclimate Program of the U.S. Geological Survey. Core drilling was undertaken in July and August of 1980. I was able to interest a number of my colleagues in the Geological Survey and elsewhere to take part in the analysis of the cores to determine just what the paleoclimate signals were and to interpret them. The papers in this volume are the results of that research. The initial results of our work on Clear Lake were presented in a symposium on Clear Lake held at the Geological Society of America, Cordilleran meeting in Salt Lake City in May 1982.

John D. Sims

Geological Society of America
Special Paper 214
1988

Late Quaternary climate, tectonism, and sedimentation in Clear Lake, northern California Coast Ranges

John D. Sims
U.S. Geological Survey
345 Middlefield Road, MS 977
Menlo Park, California 94025

Clear Lake, California, is of interest to geoscientists because it is located in a region characterized by active tectonism, recent volcanism, and plant communities that are sensitive to climatic change. The geologic history of the Clear Lake basin is closely related to the events in the nearby Geysers–Clear Lake geothermal area and to Coast Range tectonism, both of which developed as a result of the northward migration of the northern triple junction of the San Andreas transform system (McLaughlin, 1981). The sedimentary record of Clear Lake itself spans about 450,000 years of nearly continuous deposition. The papers in this volume are concerned with the stratigraphic and sedimentologic record contained in these sediments, as well as the use of this record to determine the tempo of volcanic and tectonic processes, the variation in late Quaternary climate in the northern Coast Ranges, and the evolution of plants and animals that live in the Clear Lake basin.

The present study began in 1973 with a multidisciplinary investigation of the tectonic, stratigraphic, and paleoclimatic history of the Clear Lake basin. Eight cores, ranging in length from 13.9 to 115.2 m, were taken from Clear Lake in 1973 (Fig. 1, Table 1). Two additional longer cores were taken in 1980 to complement and extend the earlier investigations (Fig. 1) (Sims, 1976; Sims and others, 1981). The objectives of the present multidisciplinary study are (1) to describe the late Quaternary paleoclimatic and paleotectonic framework of the region, and (2) to develop a reference section that may be correlated with other continental and marine deposits of the western United States.

Similar multidisciplinary studies of long-lived lakes exist for Lake Biwa, Japan (Horie, 1984), and Searles Lake, California (Smith, 1976, 1979). Both Lake Biwa and Clear Lake are shallow and eutrophic lakes situated in regions of active volcanism and tectonism. Lacustrine sedimentation in the Biwa basin spans as much as 4 m.y. (Nishida and others, 1978). The sediments in Lake Biwa are also thicker than at Clear Lake—about 1,200 m in the former as compared to about 200 in the latter. Studies at Lake Biwa primarily concern geochemical evolution of the lake, as well as age dating of the cores by isotopic and paleomagnetic studies.

Our study at Clear Lake differs from the Lake Biwa study primarily in its emphasis. The research in part concentrates on aspects of sedimentation that contribute to the knowledge of paleoclimatic change; we endeavored to correlate the discoveries at Clear Lake with similar records of paleoclimate in deep ocean cores. Paleomagnetic investigations at Clear Lake reveal that the sediments are all of normal polarity (Liddicoat and others, 1981). The geochemical investigations are primarily concerned with geothermal springs (Sims and Rymer, 1975; Thompson and others, 1981) and with understanding the origin and timing of mercury deposition in the sediments of the lake (Sims and White, 1981). Studies at Searles Lake illustrate the cyclic nature of the deposits, from normal lacustrine to playa and associated geochemical changes, from which paleoclimatic data is inferred. In contrast, Clear Lake has always been a eutorphic, low-salinity shallow-water lake.

PHYSICAL SETTING

Clear Lake is located about 140 km north of San Francisco at an elevation of 404 m. The lake lies in a fault-bounded, seismically active intermontane valley north of The Geysers geothermal field in the northern California Coast Ranges (Fig. 1). The Clear Lake drainage basin has an area of about 1,370 km^2, and elevations range from 404 to 1,475 m above mean sea level. West of the Clear Lake basin the Mayacmas Mountains form the drainage divide that separates the Clear Lake basin from the Russian River basin. Elevations in the Mayacmas Mountains vary between 600 and 1,440 m, whereas ridges north of the Clear Lake basin have maximum elevations generally above 1,200 m. The peaks of Mount Konocti are adjacent to the lake on the south and reach an elevation of 1,310 m.

Clear Lake, the largest natural fresh-water lake wholly within the state of California, is a polymictic, eutrophic lake with a seasonal thermocline. The lake has an average depth of about 6 m, and on calm summer days the bottom waters become depleted in oxygen. The lake has a surface area of approximately 160 km^2 and is composed of three distinct basins: a roughly circular main basin, the east-trending elongate Oaks arm, and the southeast-trending Highlands arm (Fig. 1). The main basin has a surface area of about 110 km^2 and attains a maximum depth of

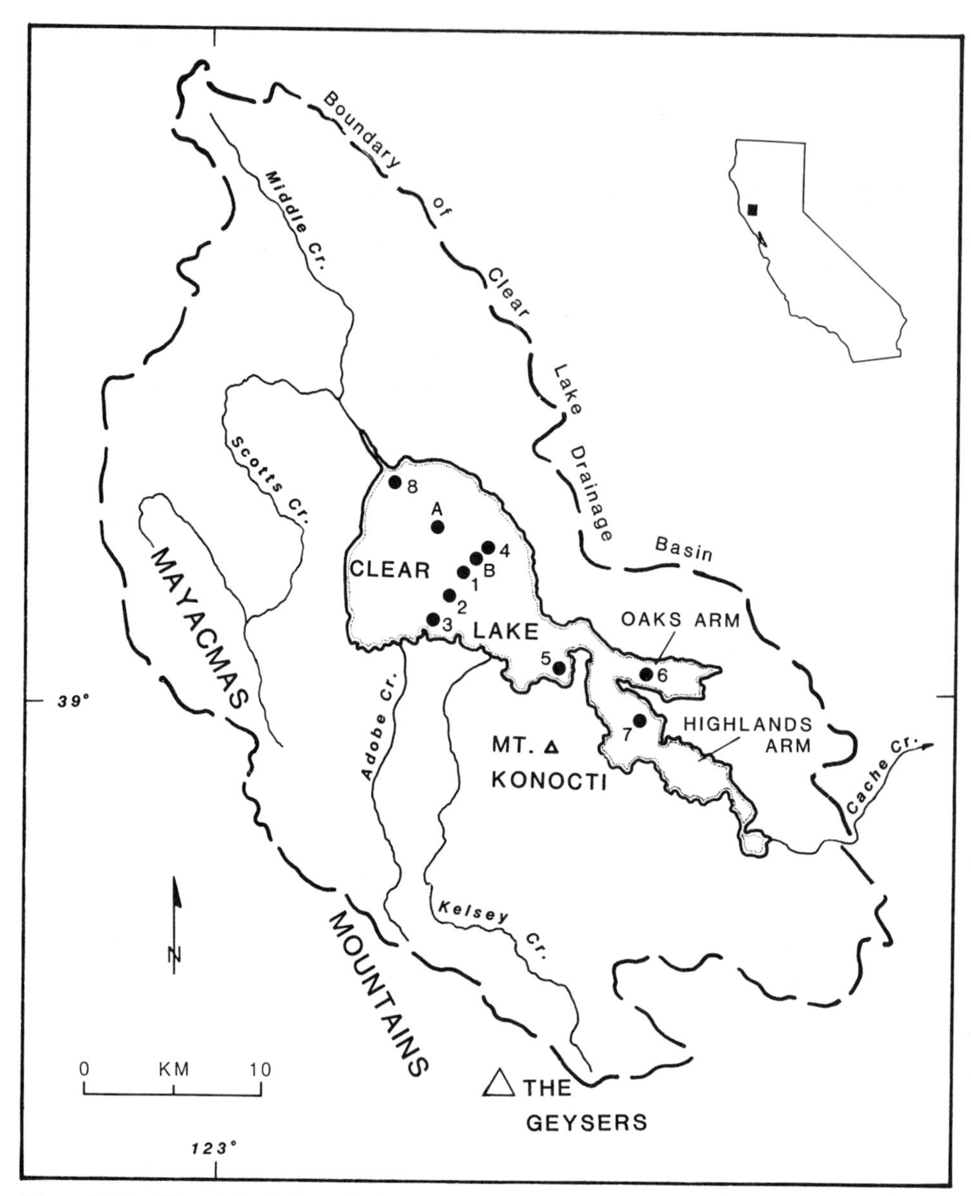

Figure 1. The Clear Lake drainage basin showing geographic features and core locations in the lake. Numbered sites refer to the series of cores taken in 1973 and are designated by the prefix CL-73-. Lettered sites refer to the two cores taken in 1980: A = CL-801, B = CL-80-2.

slightly greater than 8 m. The Oaks Arm has a surface area of 13 km^2 and attains a depth of 13 m, as does the Highlands arm, which has a surface area of about 29 km^2. Middle, Scotts, Adobe, and Kelsey Creeks discharge into the main basin of the lake, which is drained by Cache Creek, flowing out from the southern end of the Highlands arm (Fig. 1).

The Clear Lake region has a Mediterranean climate characterized by wet, cool winters and hot, dry summers. Mean annual temperature is about 13.5°C, with mean January and July temperatures of about 6.5 and 22.0°C, respectively. Winds on Clear Lake are predominantly northwesterly, although major tropically fed storms produce southerly winds. The northwesterly winds have the greatest effect on the lake and its circulation because of the northeast-southeast orientation of the lake, which results in about 32 km of fetch (including the Highlands arm of the lake).

PLEISTOCENE CHRONOSTRATIGRAPHIC TERMINOLOGY

There are many variations in definition of the boundaries of the Pleistocene epoch, because of the broad range of studies such as the variation in ^{18}O in foraminifera in deep-sea cores, the stratigraphy of continental glacial and lacustrine sequences, and magnetostratigraphic studies. Each broad area of research uses datums available to it (Fig. 2). In this volume we use the ages of boundaries proposed by Richmond and Fulerton (in press). The Pliocene-Pleistocene boundary is placed at 1.65 Ma just above the termination of the Olduvai Normal Subchron. The Pleistocene-Holocene boundary is placed at 10,000 yr B.P. just above the initiation of oxygen-isotope stage 1 (Psias and Moore, 1981). The Pleistocene is divided into the following stages: early Pleistocene (1.65 – 0.788 m.y.), middle Pleistocene (0.788 – 0.132 m.y.), and the late Pleistocene (0.132 – 0.01 m.y.) (Fig. 2).

GEOLOGY AND LATE CENOZOIC HISTORY OF THE BASIN

The dominant rocks adjacent to Clear Lake consist of the Franciscan assemblage (Bailey and others, 1964), the Great Valley sequence, the Clear Lake Volcanics (Hearn and others, 1981), and Pliocene and younger sedimentary rocks (Rymer, 1981) (Fig. 3). The Franciscan rocks are Jurassic and Cretaceous in age and crop out in the western, northern, and northeastern portions of the Clear Lake drainage basin. Locally they are found adjacent to the lake. The Franciscan assemblage within the Clear Lake region is composed of chert, graywacke, shale, metasedimentary rocks, metavolcanic rocks of blueschist grade, and ultramafic rocks (Hearn and others, 1976b; McLaughlin, 1978; Hearn and others, this volume). Clastic miogeosynclinal rocks of the Great Valley sequence are exposed in a northwest-trending block that is near the main basin of the lake (McLaughlin, 1981).

The Pliocene and Pleistocene lacustrine and fluvial sedimentary rocks of the Cache Formation crop out south and east of the Oaks and Highlands arms. Pleistocene lacustrine and fluvial sedimentary rocks of the Lower Lake and Kelseyville Formations are found south of the Highlands arm, and in Big Valley south of the main basin of the lake, respectively. The Lower Lake and Kelseyville Formations are in part contemporaneous with the Clear Lake Volcanics and probably accumulated in fault-controlled sub-basins of an ancestral Clear Lake from about 0.64 to about 0.2 m.y. B.P. These deposits dip gently toward Clear Lake; they are also in part correlative with the middle and late Pleistocene and Holocene lacustrine deposits cored beneath the lake (Rymer, 1981; Rymer and others, this volume; Bradbury, this volume).

The late Cenozoic geologic history of the Clear Lake region is dominated by faulting and volcanic activity. Such activity is attributed to east-west extension related to tectonism within the San Andreas transform system (McLaughlin, 1981; Hearn and others, this volume). The Clear Lake basin may have formed as an extensional strike-slip basin of a type that is common within transform systems (McLaughlin, 1981; Crowell, 1974). Vertical movement on basin-margin faults (Sims and Rymer, 1975) maintained the lake basin and allowed the accumulation and preservation of the thick sequence of Pleistocene ahd Holocene lacustrine and fluvial deposits cored for the present study.

Northwest-trending faults of the Coast Ranges are the predominant structural features of this seismically active region and exhibit mainly strike-slip movement (Jenkins and Strand, 1963; Hearn and others, 1976a; McLaughlin, 1981; McLaughlin and Ohlin, 1984; Coffman and von Hake, 1973; Buffe and others, 1981). Local northwest-trending faults are subparallel to the San Andreas fault zone and often are the contact between Mesozoic units and younger deposits in the Clear Lake area (Anderson, 1936; Brice, 1953; Swe and Dickinson, 1970; McLaughlin and Ohlin, 1984). Similar northwest-trending faults cut rocks of the overlying Clear Lake volcanic field and other Cenozoic deposits (Hearn and others, 1976b; Rymer, 1981). The major faults trending north-northwest to northwest, which cut rocks of the Clear Lake volcanic field, are active (Donnelly and others, 1976; Hearn and others, 1981). Late Cenozoic faults are also inferred to control local shoreline orientation in the Oaks and Highlands arms of Clear Lake, as well as the location of the main basin of Clear Lake (Sims and Rymer, 1975; Hearn and others, 1976a).

Numerous volcanic collapse features are present along the shores of the lake, and may also be present within the Highlands arm. The semicircular main basin of the lake may be a product of such volcanic activity, although direct evidence is lacking (Hearn and others, 1981).

TABLE 1. LOCATION, TOTAL LENGTH, AND RECOVERY PERCENT FOR CORES FROM CLEAR LAKE, CALIFORNIA

Core No. (CL-1)	Water Depth (m)	Length of Coring (m)	Recovery (%)
80-1	7.5	177.0	66.7
80-2	8.0	165.8*	65.0**
73-1	8.8	62.6	35.0
73-2	4.3	13.9	88.0
73-3	8.4	69.0	96.0
73-4	8.4	115.2	92.0
73-5	7.6	22.6	94.0
73-6	12.2	21.6	99.0
73-7	12.8	27.4	94.9
73-8	5.2	20.5	99.6

*Core length for core CL-80-2 calculated from top of sediment column to lowest attempted coring depth, even though the intervals from 0 to 53.42 m and 77.20 to 98.68 m were drilled out.

**Recovery percent calculated for cored segments of lake sediment and does not include drilled out segments.

REPORTS OF INVESTIGATIONS—THIS VOLUME

The two groups of papers in this volume discuss aspects of the geology of the area surrounding Clear Lake and utilize evi-

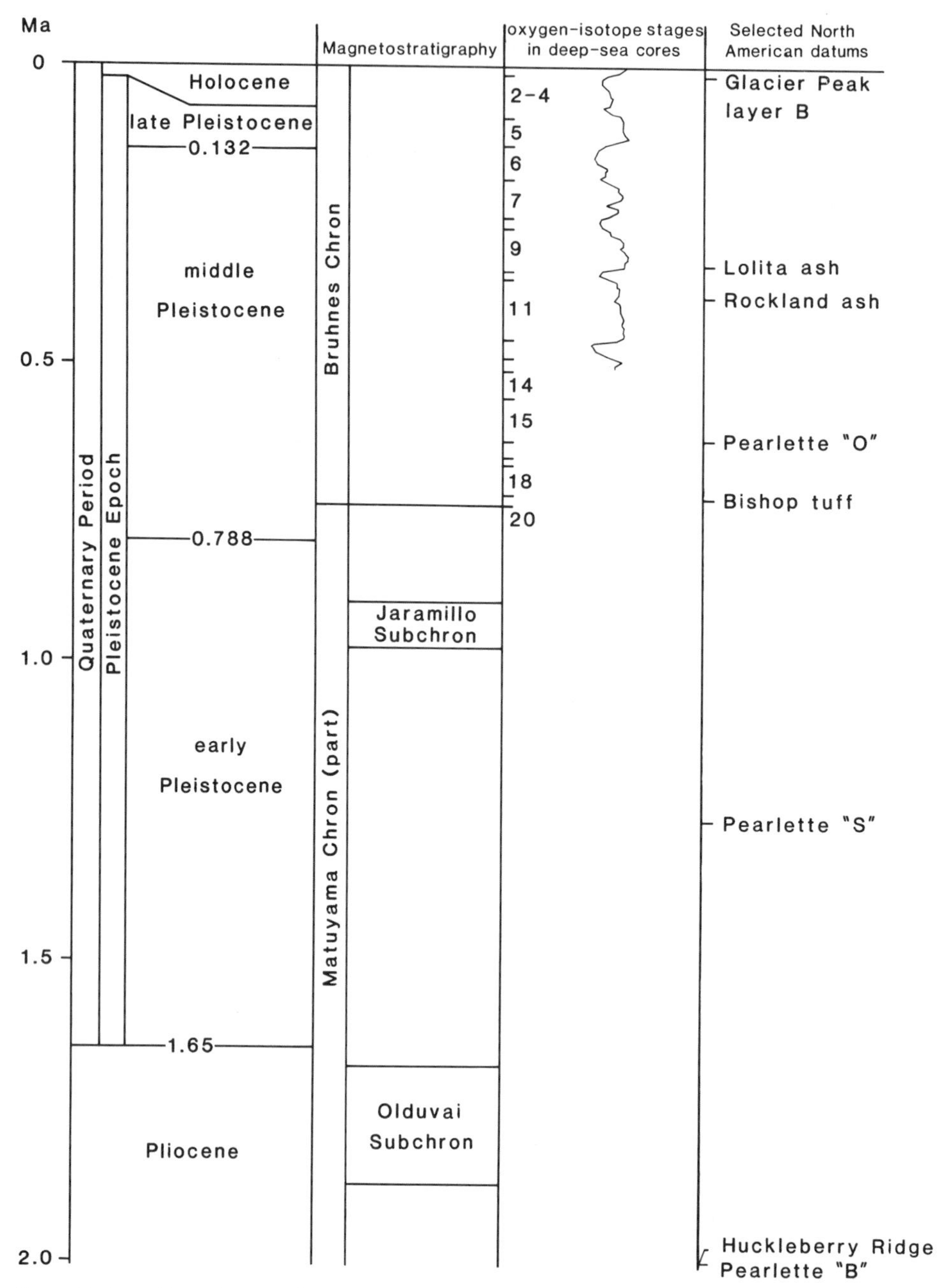

Figure 2. Age of boundaries of Pliocene and Pleistocene, Pleistocene and Holocene, and divisions of the Pleistocene (Richmond and Fullerton, 1986) and their relationship to magnetostratigraphic units (after Mankinen and Dalrymple, 1979), oxygen-isotope stages from deep-sea cores (Psias and Moore, 1981; Shackleton and Opdyke, 1973) and some selected North American Pleistocene datums (Richmond and Fullertin, 1986).

dence in the lake to interpret the lake's history. In the first group, Hearn and others examine the tectonic framework and paleogeography of the region. Rymer and others present interpretations of depositional environments for late Cenozoic basin deposits around Clear Lake. The papers of Eberhardt-Phillips and of Urban and Diment both focus on the magma body believed to lie near the surface just south of Clear Lake. Eberhardt-Phillips approaches the topic through seismic data, whereas Urban and Diment make use of thermal data taken from the core sites.

Hearn and his colleagues show that the Clear Lake basin is shaped by the interaction of faults in the San Andreas system, eruptive activity within the Clear Lake volcanic field, and fluvial and lacustrine erosional and depositional processes. During the past approximately 2 m.y. these processes were most active in

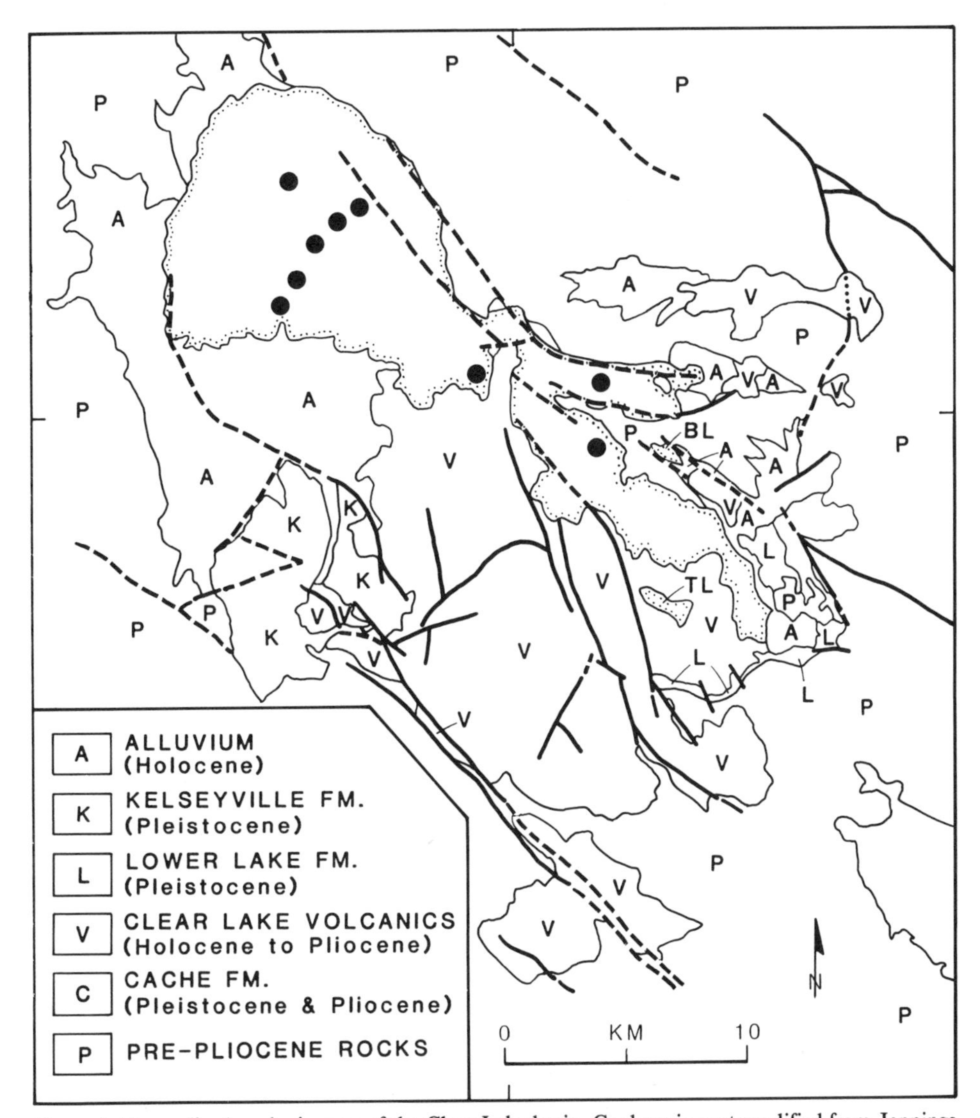

Figure 3. Generalized geologic map of the Clear Lake basin. Geology in part modified from Jennings and Strand (1960), Wagner and Bortugno (1982), Sims and Rymer (1976), Rymer (1981), and Hearn and others (1976). Abbreviations of geographic names and geologic features: BVF = Big Valley Fault; CFZ = Collayomi Fault Zone; BL = Borax Lake; KBF = Konocti Bay Fault; MK = Mount Konocti; TL = Thurston Lake; LP = Lakeport; KV = Kelseyville; RI = Rattlesnake Island.

two major episodes during the late Pliocene and(or) early Pleistocene (up to 1.5 Ma) and the period between about 0.6 Ma and the present. This latter period approximates that covered by the cores from Clear Lake. Faults and volcanic forms with young geomorphic expression, earthquake epicenters, and geodetic measurements all suggest that the Clear Lake basin is continuing to deform.

Eberhardt-Phillips illustrates the localization of current seismicity in the Geysers geothermal area and along the Konocti Bay fault. Little seismicity is associated with the Collayomi fault zone. Focal mechanisms of earthquakes are predominantly strike-slip, with significant dip-slip components. These components are consistent with the San Andreas right-lateral transform boundary and the inferred nearby crustal partial-melt body at the Geysers.

Thermal conductivity measurements by Urban and Diment show that heat flow in Clear Lake, corrected for average sedimentation rates, ranges from 1.7 to 1.8 heat-flow units (HFU) in the main basin to about 2.7 HFU in the Highlands arm. These rates are considerably lower than expected when similar data from the nearby Geysers geothermal field are considered. The authors suggest that these anomalously low heat flows may be due to refraction effects and to water movement along faults and aquifers beneath the lake.

The second group of papers examines the Clear Lake cores

in terms of physical stratigraphy, dating of the sediments, and evidence of the paleolimnology and paleoclimate of Clear Lake. Sims and others illustrate the physical stratigraphy and sedimentology of Clear Lake as a means of determining that chronology and correlation in the cores are a physical paleolimnologic and paleogeographic reconstruction. The related report of Rymer and others details depositional environments in the Kelseyville and Lower Lake Formations. The latter of these formations represents an early stage in the history of Clear Lake.

Four papers in the volume address the problem of the age and correlation of the sediments in Clear Lake and the rate of sediment accumulation. A chronology for core CL-73-4, based on radiocarbon dates, is presented by Robinson and others. This chronology, while not substantially different from earlier interpretations, is more refined and detailed. Correlation of tephra layers from Clear Lake with extrabasinal tephra (Sarna-Wojcicki and others) documents an hiatus in core CL-80-1 between 110 and 123 m. These tephra correlations place the age of base of the core CL-80-1 at 450,000 yr. Correlations by Sarna-Wojcicki and others also provide a direct link between cores CL-80-1 and CL-73-4. Concentrations of total amino acids decrease with depth in core CL-80-1 but not systematically (Blunt and Koenvolden, this volume); the lack of systematic changes may reflect variations in sedimentation and the hiatus suggested by Sarna-Wojcicki and others. Physical correlation of tephra beds between the 10 cores, as well as correlation of other sedimentologic parameters by Sims and others, gives a relative age determination as well as a general timing of paleolimnologic processes in the late Pleistocene and Holocene. Determination of depositional environments in the Pliocene and Pleistocene Cache Formation and the Pleistocene Lower Lake and Kelseyville Formations by Rymer and others provides a basis for constructing and testing correlations with the sediments in Clear Lake.

Biologic indicators of limnology and climate change factors affecting Clear Lake are covered in papers on palynology (Adam), diatoms (Bradbury), ostracods (Forester), and subfossil and modern fish (Hopkirk), respectively.

The Clear Lake palynologic record clearly shows the complex influence of climate change on the composition of the regional flora of the Clear Lake basin. The pollen content of sediments in core CL-73-4 gives a very detailed record because the average sedimentation rate for the core is about 1 m/1,000 yr, a rate that contrasts greatly with sedimentary records from deep-ocean basins that have sedimentation rates on the order of 1 cm/1,000 yr (for example, core V28-238). Adam shows that the record of fluctuation in pollen frequencies for Clear Lake compares closely with those from Grande Pile, France (Woillard, 1979; Woillard and Mook, 1982). Gardner and others show the relationship for the last 20,000 yr between pollen spectra in sediments from cores on the nearby California continental slope and the Clear Lake record of climate change derived from pollen spectra.

The diatom biostratigraphy of Clear Lake indicates the presence of a hiatus in cores CL-80-1 and -2. The absence of a hiatus in core CL-73-4 is also confirmed on the basis of the diatom biostratigraphy. The diatom succession in Clear Lake indicates that the paleoenvironments of the lake were characterized by fresh, moderately deep, nutrient-rich water.

Ostracode remains are uncommon in the sediments of Clear Lake, primarily because the lake bottom has remained too soft and fluid. The generally barren ostracode record of Clear Lake suggests that over the last 450,000 years the lake was similar to the present-day lake and that it was primarily influenced by changes in precipitation rather than by water chemistry or temperature.

The fish population of Clear Lake is of interest because of its endemic modern and subfossil species and their evolutionary trends. The subfossil fish population in Clear Lake is derived from species adapted to stream conditions. Hopkirk concludes that, contrary to inferences by Casteel and others (1977), the frequency of occurrence and the growth rate of Tule perch are controlled by nutrient availability and not simply temperature.

SUMMARY

The geologic, tectonic, and biologic settings of Clear Lake combine to produce a natural laboratory for the study of geologic and ecologic processes during late Pleistocene and Holocene time. Tectonism and volcanism produced the intermontane basin how partly occupied by Clear Lake, and their effects are currently expressed as modern seismicity, active fault movements, The Geysers geothermal field, and numerous thermal gaseous springs in and around the lake. The lake basin lies near the ecotone between forests that are predominantly pine and oak. This ecotone shifted in response to late Quaternary climatic changes, which resulted in differing vegetational cover surrounding Clear Lake. The makeup of the different plant communities during those climatic variations may be read from the pollen content of sediments in Clear Lake. The sediments of Clear Lake also contain the remains of lacustrine plants and animals that reveal conditions and changes of the lake itself through late Quaternary time. Aquatic plants such as algae and diatoms provide insight into stratigraphic succession, age, and environmental history. Animals such as ostracodes give information about water chemistry, turbidity, and lake bottom conditions. The relative age of the sediments of Clear Lake may be determined by several graphic successional relationships. Absolute age-dating is accomplished by radiocarbon dating, amino acid racemization ratios, and tephrachronologic correlation.

SUGGESTIONS FOR FURTHER RESEARCH

Opportunities for fruitful research abound in the Clear Lake basin. The following comments apply only to those areas that may yield data applicable to ambiguities and gaps of knowledge in the cores from the lake. Of critical interest is the development of a detailed absolute chronology for the Kelseyville Formation and a detailed correlation with the longer cores from Clear Lake. Studies of the detailed biostratigraphy of the Kelseyville and Lower Lake Formations, particularly palynologic research, should significantly increase the confidence of correlation, as well as yield paleoclimatic data. Diatom biostratigraphy and tephrachronology may also play major roles in determining the age and correlation of these two units. Additional work on pollen spectra from cores in Clear Lake may further refine and amplify the conclusions of Adam and of Heusser and Sims (1983), as well as enable better correlations among the 10 cores. If better correlations and biostratigraphic detail follow, the late Quaternary section of Clear Lake will become increasingly useful as a reference section of at least provincial importance.

ACKNOWLEDGMENTS

A project of the duration and complexity of the research at Clear Lake brings many people into contact. Many geologists, limnologists, biologists, government officials of Lake County and the State of California, and citizens of Lake County have assisted us with support, information, and encouragement. My thanks and continued gratitude extends to all who have assisted us.

Of particular importance is the foresight of the Lake County Board of Supervisors, who provided financial support for coring the lake sediments during the 1973 drilling program. That invaluable support allowed a vastly more detailed examination of critical sites in the Oaks and Highlands arms of Clear Lake.

REFERENCES CITED

Anderson, C. A., 1936, Volcanic history of Clear Lake area, California: Geological Society of America Bulletin, v. 47, p. 629–664.

Bailey, E., Irwin, W. P., and Jones, D., 1964, Franciscan and related rocks and their significance in the geology of Western California: California Division of Mines and Geology Bulletin 183, 177 p.

Brice, J. C., 1953, Geology of the Lower Lake quadrangle, California: California Division of Mines and Geology Bulletin 166, 71 p.

Buffe, C. E., Marks, S. M., Lester, F. W., Ludwin, R. S., and Stickney, M. C., 1981, Seismicity of the Geysers–Clear Lake region, *in* McLaughlin, R. J., and Donnelly-Nolan, J. M., eds., Research in the Geysers–Clear Lake geothermal area, northern California: U.S. Geological Survey Professional Paper 1141, p. 129–138.

Coffman, J. L., and von Hake, C. A., 1973, Earthquake history of the United States: U.S. Department of Commerce Publication 41-1, 208 p.

Crowell, J. C., 1974, Origin of late Cenozoic basins in southern California, *in* Dickinson, W. R., ed., Tectonics and sedimentation: Society of Economic Paleontologists and Mineralogists Special Publication 22, p. 190–204.

Donnelly, J. M., McLaughlin, R. J., Goff, F. E., and Hearn, B. C., Jr., 1976, Active faulting in the Geysers-Clear Lake area, northern California: Geological Society of America Abstracts with Programs, v. 8, p. 375–376.

Hearn, B. C., Jr., Donnelly, J. M., and Goff, F. E., 1976a, Geology and geochronology of the Clear Lake volcanics, California: United Nations Symposium on the Development and Use of Geothermal Resources, 2d, v. 1, p. 423–428.

—— , 1976b, Preliminary geologic map and cross sections of the Clear Lake volcanic field, Lake County, California: U.S. Geological Survey Open-File Report 76-751, p.

—— , 1981, The Clear Lake volcanics; Tectonic setting and magma sources, *in* McLaughlin, R. J., and Donnelly-Nolan, J. M., eds., Research in the Geysers–Clear Lake geothermal area, northern California: U.S. Geological Survey Professional Paper 1141, p. 25–45.

Heusser, L. E., and Sims, J. D., 1983, Pollen counts for core CL-80-1; Clear Lake, Lake County, California: U.S. Geological Survey Open-File Report 83-384, 28 p.

Horie, S., 1984, Lake Biwa: The Hague, Dr. W. Junk Publishers, 638 p.

Jenkins, C. W., and Strand, R. G., 1963, Geologic map of California; Ukiah sheet: California Division of Mines and Geology, scale 1:250,000.

Liddicoat, J. C., Sims, J. D., and Bridge, W. D., 1981, Paleogeomagnetism of a 177-m-long core from Clear Lake, California: Geological Society of America Abstracts with Programs, v. 13, p. 67.

Mankinen, E. A., and Dalrymple, G. B., 1979, Revised geomagnetic polarity time scale for the interval 0-5 m.y. B.P.: Journal of Geophysical Research, v. 84, p. 615–626.

McLaughlin, R. J., 1978, Preliminary geologic map and structural sections of the central Mayacamas Mountains and the Geysers steam field, Sonoma, Lake, and Mendocino Counties, California: U.S. Geological Survey Open-File Report 78-309.

—— , 1981, Tectonic setting of pre-Tertiary rocks and its relation to geothermal resources in the Geysers–Clear Lake area, *in* McLaughlin, R. J., and Donnelly-Nolan, J. M., eds., Research in the Geysers-Clear Lake geothermal area: U.S. Geological Survey Professional Paper 1141, p. 3–23.

McLaughlin, R. J., and Ohlin, H. N., 1984, Tectonostratigraphic framework of the Geysers–Clear Lake region, California, *in* Blake, M. C., Jr., ed., Franciscan geology of Northern California: Pacific Section Society of Economic Paleontologists and Mineralogists, v. 43, p. 221–254.

Nishida, J., Sasajinean, S., and Aki, E. S., 1978, An estimation of the sediment thickness in Lake Biwa as deduced from Bouguer gravity anomalies, *in* Horie, S., ed., Paleolimnology of Lake Biwa and the Japanese Pleistocene: Kyoto, Japan, Kyoto University, v. 6, p. 146–160.

Psias, N. G., and Moore, T. C., Jr., 1981, The evolution of Pleistocene climate: A time series approach: Earth and Planetary Science Letters, v. 52, p. 450–458.

Richmond, G. M., and Fullerton, D. S., 1986, Summation of quaternary glaciations in the United States of America; Correlation of quaternary glaciations in the Northern Hemisphere: London, Pergamon Press (in press).

Rymer, M. J., 1981, Stratigraphic revision of the Cache Formation (Pliocene and Pleistocene), Lake County, California: U.S. Geological Survey Bulletin 1502-C, 35 p.

Shackleton, N. J., and Opdyke, N. D., 1973, Oxygen isotope and paleomagnetic stratigraphy of Pacific core V28-239; Oxygen isotope temperatures and ice volumes on a 10^5 and 10^6 year scale: Quaternary Research, v. 3, p. 39–55.

Sims, J. D., 1976, Paleolimnology of Clear Lake, California, U.S.A., *in* Horie, S., ed., Paleolimnology of Lake Biwa and the Japanese Pleistocene: Kyoto, Japan, Kyoto University, v. 4, p. 658–702.

Sims, J. D., and Rymer, M. J., 1975, Map of gaseous springs and associated faults, Clear Lake, California: U.S. Geological Survey Miscellaneous Field Investigations Map MF-75-721.

Sims, J. D., and White, D. E., 1981, Mercury in the sediments of Clear Lake, *in* McLaughlin, R. J., and Donnelly-Nolan, J. M., eds., Research in the Geysers–Clear Lake geothermal area: U.S. Geological Survey Professional Paper 114, p. 237–241.

Sims, J. D., Adam, D. P., and Rymer, M. J., 1981, Late Pleistocene stratigraphy and palynology of Clear Lake, Lake County, California, *in* McLaughlin, R. J., and Donnelly-Nolan, J. M., eds., Research in the Geysers–Clear Lake geothermal area: U.S. Geological Survey Professional Paper 1141,

p. 219–230.
Smith, G. I., 1976, Paleoclimatic record in the Upper Quaternary sediments of Searles Lake, California, U.S.A., *in* Horie, S., ed., Paleolimnology of Lake Biwa and the Japanese Pleistocene: Koyto, Japan, University of Kyoto, v. 4, p. 577–644.
——, 1979, Subsurface stratigraphy and geochemistry of Late Quaternary evaporites, Searles Lake, California: U.S. Geological Survey Professional Paper 1043, 130 p.
Swe, W., and Dickensen, W. R., 1970, Sedimentation and thrusting of Late Mesozoic rocks in the Coast Ranges near Clear Lake, California: Geological Society of America Bulletin, v. 81, p. 165–189.
Thompson, J. M., Sims, J. D., Yadav, S., and Rymer, M. J., 1981, Chemical composition of water and gas from five near shore subaqueous springs in Clear Lake, *in* McLaughlin, R. J., and Donnelly-Nolan, J. M., eds., Research in the Geyser–Clear lake geothermal area: U.S. Geological Survey Professional Paper 1141, p. 215–218.
Wagner, D. L., and Bortagno, E. J., compilers, 1982, Geologic map of the Santa Rosa Quadrangle: California Division of Mines and Geology Regional Geologic Map Series Map 2A.
Woillard, G. M., 1979, Grand Pile peat bog; A continuous pollen record for the last 140,000 years: Quaternary Research, v. 9, p. 1–21.
Woillard, G. M., and Mook, W. G., 1982, Carbon-14 dates at Grand Pile; Correlation of land and sea chronologies: Science, v. 215, p. 159–161.

Manuscript Accepted by the Society September 15, 1986

Printed in U.S.A.

Geological Society of America
Special Paper 214
1988

Tectonic framework of the Clear Lake basin, California

B. C. Hearn, Jr.
U.S. Geological Survey
959 National Center
Reston, Virginia 22092

R. J. McLaughlin
J. M. Donnelly-Nolan
U.S. Geological Survey
345 Middlefield Road
Menlo Park, California 94025

ABSTRACT

The actively deforming Clear Lake basin has been shaped primarily by shear and tensional stresses within the broad San Andreas fault system; and it has been modified both by eruption and subsidence of the Clear Lake Volcanics and by depositional processes. Within the San Andreas fault system, shear has been dominantly right-lateral N35°W to N45°W, and maximum tension dominantly east-west. However, the current local maximum tension deduced from focal-plane solutions is N70°W, representing a secular change of unknown, possibly short duration, that could modify tensional basin-forming processes. The Collayomi fault zone (N45°W) partially bounds the basin on the southwest. The seismically active Konocti Bay fault zone (N25°W) across the basin merges into the less well documented N40°W Clover Valley fault zone that includes the faults in Clover Valley and near Lucerne and shorter faults near Clearlake Highlands and Lower Lake. The northeast margin of the main lake and both margins of the Oaks arm are largely fault-controlled; the Highlands arm is partly fault-controlled.

East of the basin, sediments now assigned to the Cache Formation—ranging in age from more than 1.8 to about 1.6 m.y. old—were deposited in an earlier basin that was controlled mainly by N35-45°W (Bartlett Springs) and N20°E (Cross Spring) fault zones. The earliest stage of the present Clear Lake basin is dated by the 0.6-m.y.-old rhyolite of Thurston Creek, which is found at or near the base of lacustrine-fluvial deposits. The extensive rhyolite's thickness and distribution suggest that its eruption triggered subsidence that initiated or accelerated basin formation. Limited data suggest that a dominantly volcanic subsidence feature within the larger Clear Lake Basin could have had dimensions of about 13 by 15 km and either an elliptical shape with major axis oriented west-northwest, or a rectilinear shape. Volcanism occurring between 0.6 and 0.3 m.y. ago partly filled the southern part of the Clear Lake basin with flows, pyroclastic materials, and clastic deposits. Deposition beneath Clear Lake (0.4 mm/yr for the past 0.45 m.y.) has kept pace with subsidence and tilting down to the northeast. Projected maximum sedimentary and volcanic thicknesses could total more than 1 km beneath Clear Lake, suggesting a possible overall subsidence rate of 1.7 mm/yr for the past 0.6 m.y.

INTRODUCTION

This paper outlines the development of the Clear Lake basin and an earlier basin containing the Cache Formation in the tectonic setting of the Coast Ranges during the past 3 m.y., and including the evolution of the Clear Lake Volcanics during the past 2 m.y.

Located in the northern California Coast Ranges approximately 50 km northeast of the main trace of the San Andreas fault, the Clear Lake basin is within the broad zone of deformation associated with the San Andreas fault system along the western edge of the North American plate. The basin is one of several Pliocene and Pleistocene basins (Ukiah, Potter, Little Lake, and Round Valleys; Little Sulphur Creek [McLaughlin and Nilsen, 1982]; and the depositional basin of the Cache Formation [Rymer, 1981]) that are close to known or suspected strike-slip fault zones (Fig. 1). These basins are thought to have developed initially as the Mendocino triple junction migrated northward along the coast, propagating the San Andreas fault system northward along the soft margin of the North American plate during the past 3 to 5 m.y. The Clear Lake basin was shaped by a variety of processes over the past 1 to 2 m.y. The present shape of the basin is a result of three processes: the dominant shear and subsidiary tensional stresses of the San Andreas fault system; the eruption and subsidence of part of the Clear Lake Volcanics; and the erosion and deposition that have influenced the distribution, character, and amount of the basin fill.

FAULTS

Detailed geologic mapping in the past 10 years has extended the limits of recognition of young faults related to the San Andreas fault system in northern California farther eastward than was previously known. Holocene faults of the San Andreas system occur as far east as the western part of the Sacramento Valley (Helley and Herd, 1977; Harwood and Helley, 1982). Earthquake swarms along the Maacama fault zone near Willits in 1977 (events of magnitude $\leqslant 4.9$) and near Ukiah in 1978 (events of magnitude $\leqslant 4.4$), and seismicity along the Bartlett Springs–Green Valley fault zone (Bufe and others, 1981; Eberhart-Phillips, this volume) emphasize that tectonism is continuing in the northern Coast Ranges east of the main San Andreas fault (Cockerham and Herd, 1982). Geodetic measurements reported by Lofgren (1981) indicate that strike-slip strain of at least 12 mm/yr has occurred over the period of 1972 to 1977 between The Geysers steam field and the northeast side of Clear Lake. Additional strain of 1 to 2 mm/yr is occurring across the Maacama fault zone, according to short-term measurements reported by Pampeyan and others (1981).

Herd (1978) outlined evidence for a small separate plate, the Humboldt plate, within the soft margin of the North American plate in the northern Coast Ranges. In the Clear Lake area, the eastern boundary of the Humboldt plate was shown as the Maacama fault zone (Herd, 1978). On the basis of young fault features and seismicity (Bufe and others, 1981; Eberhart-Phillips, this volume) associated with the Bartlett Springs fault zone, and its northward and southward continuations into the Lake Mountain and Green Valley fault zones, respectively, the eastern boundary of the Humboldt plate—or the boundary of another elongate plate—could lie farther east. However, in view of the abundance of Quaternary faults between the Sacramento Valley and the coast, the concept of distinct, brittle small plates may not even be valid for this part of the northern Coast Ranges.

Our understanding of the tectonic framework of the Coast Ranges is limited by uncertainties about the nature of the interaction and boundary at depth between the North American and Pacific plates. There is some evidence from heat flow and from seismic and gravity studies that the North American plate may be decoupled from underlying lithosphere at mid-crustal to lower crustal depths of 12 to 20 km (Lachenbruch and Sass, 1980; Kelleher and Isaacs, 1982; Zandt and Furlong, 1982; Jachens and Griscom, 1983). If the underlying lithosphere is part of, or tied to, the Pacific Plate, and is thus moving northward, that lithosphere might contribute to the broadening of the area of right-lateral shear in the shallow North American plate to the northeast of the main San Andreas fault. The generally incompetent nature of much of the Franciscan assemblage lends further support to the concept of a soft plate margin with a wide zone of deformation rather than a zone of discrete microplates.

Young faulting processes in the Coast Ranges are superimposed on the pre-existing fault boundaries between complex terranes of Franciscan assemblage, Great Valley sequence, and ophiolite-serpentinite slices. Young faulting has reactivated some of these major basement faults (McLaughlin, 1981). The Clear Lake basin is underlain on the southwest by serpentinite, on the west and north by the central terrane of the Franciscan assemblage, on the east by the eastern terrane of the Franciscan assemblage, and on the south and southeast by the Great Valley sequence. Ophiolitic rocks and the Great Valley sequence compose the upper plate of the deformed Coast Range thrust, and are displaced northward along strike-slip faults from coeval rocks that lie above the main trace of the Coast Range thrust on the west side of the Sacramento Valley. A fault contact of the eastern (Franciscan) terrane on top of or against the central (Franciscan) terrane extends approximately south-southeast from the vicinity of Lucerne, and projects beneath the northeastern part of the volcanic field, southwest of eastern terrane exposures on Buckingham Point, Sulphur Bank Point and Baylis Point (Fig. 2). That fault contact between the eastern and central terranes of the Franciscan assemblage approximately coincides with a change in the character of aeromagnetic gradients (Isherwood, 1976, 1981) from a deep southwest-dipping magnetic gradient associated with basement at least 6 km deep on the northeast, to a series of shallow magnetic anomalies that are associated with slabs of ophiolitic rocks near or at the surface on the southwest (R. J. McLaughlin, unpub. data).

The Clear Lake basin and the nearby older basin of deposition of the Cache Formation (Rymer, 1981) (Fig. 2) probably

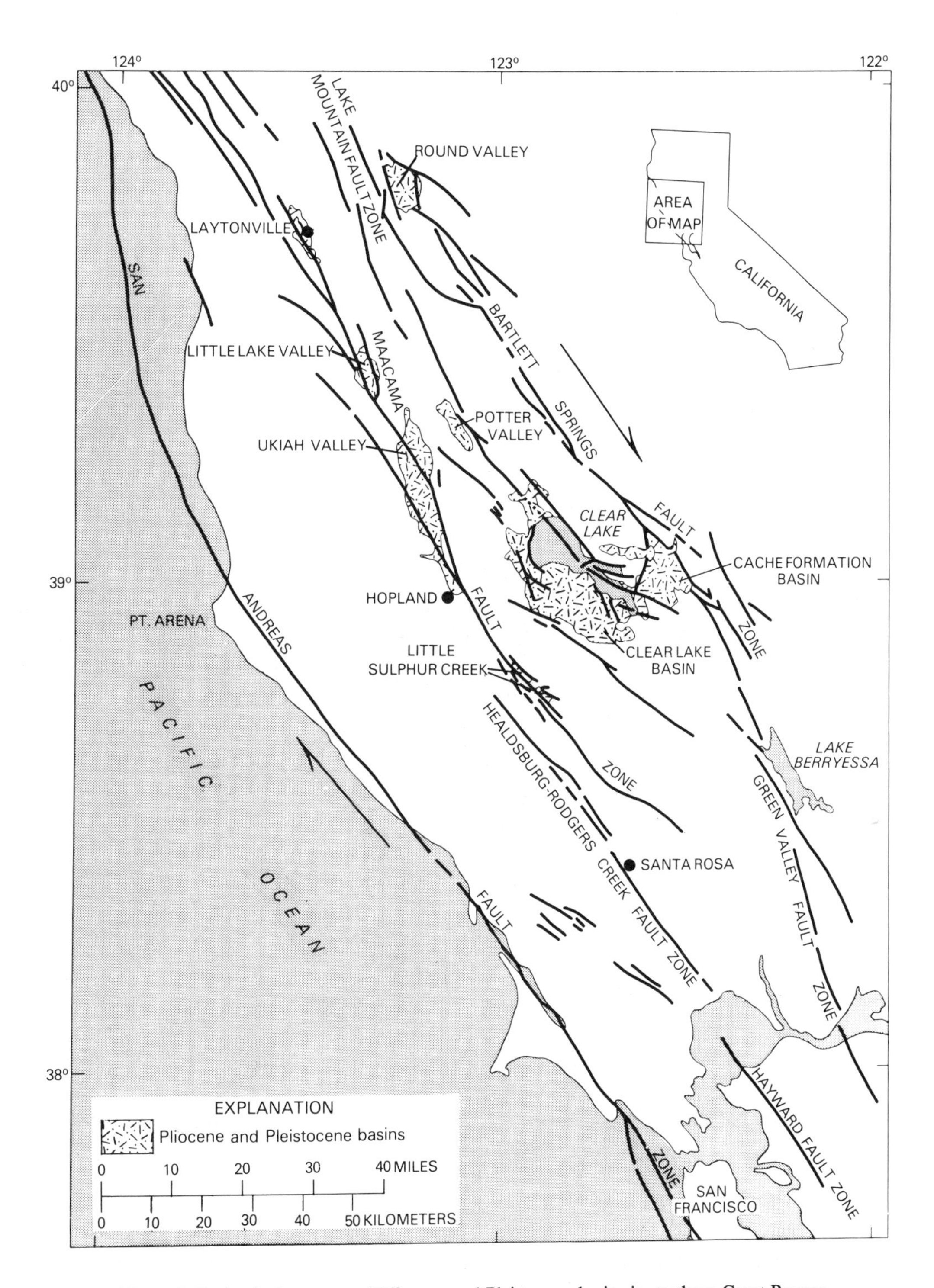

Figure 1. Faults, fault zones, and Pliocene and Pleistocene basins in northern Coast Ranges.

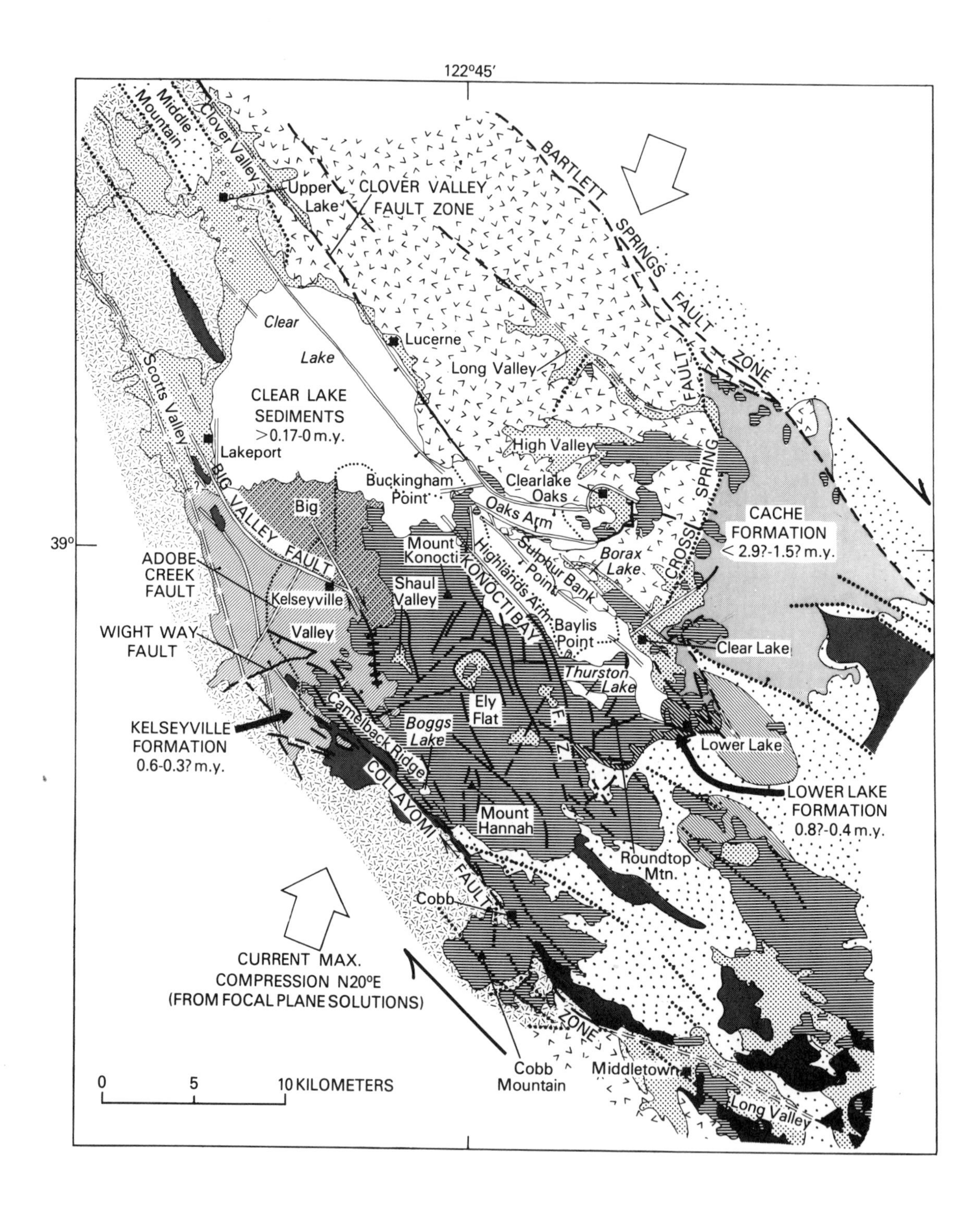

Figure 2. Distribution of basin deposits, the Clear Lake Volcanics, the Cache Formation, and prevolcanic rocks and faults younger than about 3 m.y., classified according to maximum age of youngest known displacement. (Explanation on facing page.)

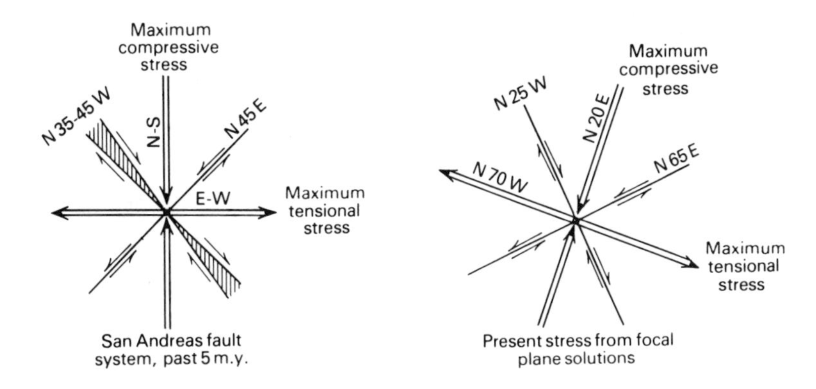

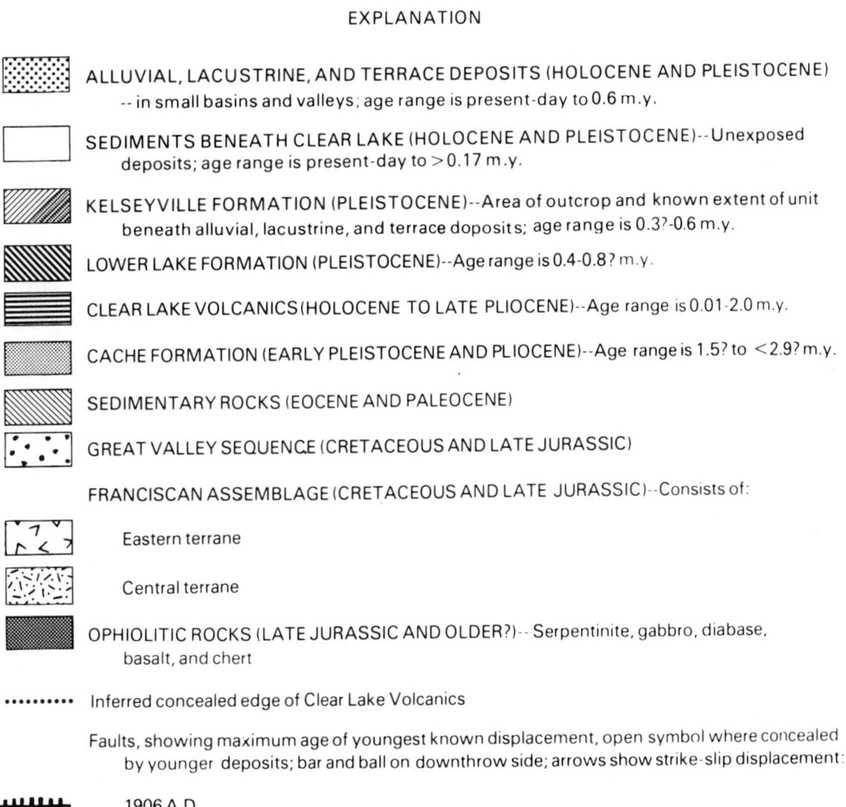

1906 A.D.

0.1 m.y.

0.6 m.y.

0.8 m.y.

1.8 m.y.

represent complex tensional, pull-apart structures that have developed between major strike-slip faults and have evolved by the mechanisms outlined by Crowell (1974). He proposed that, in strike-slip fault systems, many basins form as pull-apart grabens at en-echelon steps between parallel strike-slip faults, or that they form between diverging strike-slip faults. Crowell recognized that complexities such as irregular trends of graben faults, curving of graben faults into the terminal parts of bounding strike-slip faults, and the presence of marginal thrusts and braided strike-slip zones were to be expected. He also recognized that the oldest basin deposits may be absent from the deep central part of an evolved basin that has undergone several stages of extension. Igneous activity from upper mantle depths may occur in the early stages of extensional basin formation, (for example, the Gulf of California), and with continued basin extension, a more shallow magma chamber could form and evolve to a large silicic system. Changes in the orientation of maximum compressive stress could add additional complexities to extensional basin evolution. The Clear Lake basin shows the above general characteristics.

Two major northwest-trending fault zones and several additional faults control the steep topographic margins of the Clear Lake basin and have exerted more subtle control on the basin in the past (Fig. 2). The Collayomi fault zone partially bounds the southwest side of the Clear Lake basin, and probably connects to the faults that control the west side of the basin. The seismically active Konocti Bay fault zone crosses the east-central part of the basin, and merges northward into a fault zone (part of the Clover Valley fault zone) that has shaped the northeast side of the northern Clear Lake basin. The Highlands arm of the lake is partly controlled by the Konocti Bay zone and partly by inferred N45°W faults (Highlands Arm fault zone). The Oaks arm is controlled by curving faults of N70°W to N65°E trends that are somewhat anomalous (Oaks Arm fault zone).

The San Andreas fault system is characterized by relatively long faults in right-stepping arrays. Shear has been dominantly N35°W to N45°W, and the expected or observed direction of tension fractures has been north-south to N10°E. In the Clear Lake Volcanics, the preferred direction of tension fractures is corroborated by two alignments of basaltic and andesitic vents. One group of vents, 0.6 to 0.35 m.y. in age across Mount Konocti, is aligned north-south. A younger group of vents, about 0.1 to 0.01 m.y. in age, is aligned N2°E to N8°E, from Roundtop Mountain to High Valley (Hearn and others, 1981, Figs. 14, 16). Current seismic activity (exclusive of The Geysers steam field where clustered epicenters are largely related to steam production) shows zones of epicenters that are N30°W to north-south (Bufe and others, 1981; Eberhard-Phillips, this volume). These alignments are more northerly than the alignment of the major, older, through-going faults of the San Andreas system, indicating a secular change in orientation of regional stress. To the southwest, epicenters are aligned along the Maacama fault zone. To the southeast and northeast of Clear Lake, respectively, epicenters are clustered near the Bartlett Springs fault zone and the active Green Valley fault zone. Within the Clear Lake basin, epicenters are concentrated along the Konocti Bay fault zone, but are widely scattered in Big Valley and in the vicinity of Lakeport (Bufe and others, 1981; Eberhard-Phillips, this volume).

Focal plane solutions for these recent earthquakes are dominantly strike-slip and indicate (1) that the average maximum compressive stress is oriented N20°E, and (2) that the maximum tensional stress is oriented N70°W (Bufe and others, 1981), which represents a major temporal shift from the dominant stress orientation of the past few million years. Such a shift may be of brief duration in geologic time, but would be expected to produce new strike-slip faults of north-northwest trend or to reactivate old faults of north-northwest trend. In this respect, the concentrations of current seismic activity along the northern Maacama fault zone (N15°W), the Lake Mountain fault zone (N22°W) (Herd, 1978), and the Konocti Bay fault zone (N25°W) are significant. The lack of clustered epicenters along the Collayomi fault zone may indicate that its orientation (N40°W) is not favorable for current movement.

Faults that control the margins of the Clear Lake basin generally seem to fit the San Andreas fault system directions. Most of these faults have not been studied in detail; many are inferred to explain the steep topography or linearity of lake or basin margins. Some faults are inferred by Sims and Rymer (1976) from linear arrays of gaseous springs beneath the lake. Strike-slip displacement has not been established on most of these faults. Descriptions of the individual fault zones are given below.

Collayomi Fault Zone

The Collayomi fault zone is well defined as a S40°E trend from Camelback Ridge southeastward through Cobb Valley and thence along a S60 to 65°E trend to Middletown and Long Valley. From Middletown, one fault extends along the east side of Long Valley, along Butts Canyon southeast of Detert Reservoir (California Department of Water Resources, 1962), and may connect with the active Green Valley fault system (Frizzell and Brown, 1976; Cockerham and Herd, 1982). Another fault extends from Middletown along the west side of Long Valley and may connect with faults in Pope Valley (California Department of Water Resources, 1962). The Collayomi fault zone apparently has limited the southwestern extent of flows of the Clear Lake Volcanics that range in age from 1.5 to 0.6 m.y. old. That apparent limit suggests that a northeast-facing scarp existed along the fault zone. Maximum ages of faulting, based on offsets of dated volcanic units, decrease to the northwest, but young geomorphic features are present along much of the zone. Apparent right-lateral offsets are about 1.1 km for 1.5-m.y.-old andesite and 0.5 km for 0.6-m.y.-old rhyolite, implying average rates of movement of about 0.7 and 0.9 mm/yr, respectively. However, evidence for strike-slip displacement is suggestive rather than definitive, and, as for other examples of faulted units that have low dip angles, offsets can also be explained as the result of purely vertical movement. In addition, apparent displacement rates on single faults may be a small fraction of the total displacement across a fault zone. Geodetic measurements are a better guide to current displacement rates across areas of complex faulting, but the possibility of episodic displacements may limit the validity of extrapolating current displacement rates into the past.

Northwest of Camelback Ridge, the Collayomi fault zone becomes more diffuse where it splays into a fan-shaped array of short, northwest-trending faults (Fig. 2). The zone may connect across the southern part of Big Valley into a major fault, trending N15–25°W, that bounds the west side of the Clear Lake basin. Such a connection is interrupted by the N65°E Wight Way fault that forms a partial boundary for deposition or preservation of coarser basin sediments and has vertical displacement of 60 m or more (Lake County Flood Control and Water Conservation District, 1967). Orientation of that fault suggests parallelism to potential left-lateral shear in the San Andreas fault system, but there is no definitive geologic evidence of lateral displacement. Other faults at the south end of Big Valley are east-west and show small apparent right-lateral displacements (McLaughlin, 1978).

Big Valley Fault

The Big Valley fault (Lake County Flood Control and Water Conservation District, 1967) is a prominent splay of the Collayomi fault zone. In the 1906 San Francisco earthquake, a 2-km segment of the Big Valley fault showed en echelon ground breakage in an area about 3 km southeast of Kelseyville (Lawson, 1908). This N10°W segment shows evidence for maximum right-lateral offset of 0.42 km during the past 0.5 m.y., or about 0.8 mm/yr. The concealed Big Valley fault bends to a N65°W trend near Kelseyville on the basis of water-well data, an eroded topographic scarp, and tilted beds of the Kelseyville Formation (Lake County Flood Control and Water Conservation District, 1967; Rymer, 1981). Vertical displacement of about 70 m (Rymer, 1981, Fig. 13) is based on offset of the Kelsey Tuff Member of the Kelseyville Formation. Other water-well data indicate "a (postulated) branch of the Big Valley fault" that would trend N20-30°W north of Kelseyville (Lake County Flood Control and Water Conservation District, 1967, p. V-3).

Other inferred faults of N30 to 35°W trend in Big Valley show as subtle lineations in alluvium or are probable boundaries of hills of serpentinite that are surrounded by terrace and alluvial deposits near Lakeport. A north-south branch is inferred to control the generally straight shoreline at Lakeport (Sims and Rymer, 1976).

West Margin Fault

A fault of N15 to 25°W trend is inferred along the west margin of Big Valley, and probably continues to the northwest through Scotts Valley. No evidence of strike-slip displacement is known, but the west margin has not been mapped in detail. Although the west margin fault was described as a "dissected fault escarpment" against which the basin deposits were laid down (Lake County Flood Control and Water Conservation District, 1967), movement that postdates the Kelseyville Formation is likely in view of the other young faults and thick sedimentary fill in Big Valley.

Konocti Bay Fault Zone

To the south of the Highlands arm, the seismically active Konocti Bay fault zone (2 to 3 km wide) breaks the Clear Lake Volcanics and trends N25°W. The obvious displacements are mainly normal, and strike-slip offsets are not apparent. However, fault-plane solutions (Bufe and others, 1981; Eberhart-Phillips, this volume) indicate strike-slip mechanisms. A small thrust fault that fits a north-south compression direction is exposed on the east side of Ely Flat. Beneath Konocti Bay, a fault marked by a prominent line of gaseous springs (Sims and Rymer, 1976) has displacement young enough to have produced a 1-m-high scarp in the lake bottom sediments (J. D. Sims, oral communication, 1976) and is in line with an on-shore fault in the Konocti Bay fault zone.

Northeast Margin

The Konocti Bay fault zone continues or merges northwestward discontinuously into a parallel pair of N40°W faults that control the northeast side of the Clear Lake basin. An offshore fault is inferred beneath the lake on the basis of seven aligned gaseous springs (Sims and Rymer, 1976). That fault may continue northwestward and may bound the straight northeast side of Middle Mountain. The second fault (Clover Valley fault zone) extends from the Oaks arm to the shoreline at Lucerne, and continues northwest as the fault that controls the shape of Clover Valley.

Highlands Arm

A N50°W fault that partially controls the northeast shore of the Highlands arm is inferred from steep shoreline topography and localized lake bottom depressions close to shore. Southeastward projection of this fault trend meets northwest-trending faults in the Lower Lake area. The fault is also parallel to faults that control the small structural depression of Borax Lake valley and are loci of springs and a fumarole (Sims and Rymer, 1976; Hearn and others, 1976). The fault that bounds the northeast side of Borax Lake valley shows about 1.5 km of apparent right-lateral offset of Franciscan metabasalt and metachert.

Oaks Arm

The easterly orientations of the faults that control both sides of the Oaks arm seem anomalous. The faults are broadly arcuate, trending N70°W to N65°E. The northern fault connects to the Clover Valley fault zone. The southern fault is marked by at least two gaseous springs (Sims and Rymer, 1976) and probably connects eastward to a fault at the Sulphur Bank Mine that has offset the andesite of Sulphur Bank at least 10 m (White and Roberson, 1961) within the past 45,000 yr (Donnelly-Nolan and others, 1981; Sims and White, 1981). At the east end of the Oaks arm, the subcircular embayment that is partially filled by alluvial and lacustrine sediments has been shaped by fault-controlled subsidence.

BASINS

Depositional Basin of the Cache Formation

An older fault-controlled depositional basin that lies mainly east of the present Clear Lake basin contains the fluvial and lacustrine Cache Formation (Rymer, 1981). The Cache Formation is at least 4,000 m thick, the maximum thickness is unknown, and an unknown amount has been eroded (Rymer, 1981; Rymer and others, this volume). The Cache Formation is Pliocene and early Pleistocene in age, but its maximum age is unknown. Fossil fauna suggest an age of late Pliocene (1.8 to 2.9 Ma) for the lower part of the Cache. Early basaltic rocks of the Clear Lake Volcanics are intercalated with, intruded into, or overlie the Cache Formation, and tend to be localized near faults that bound the basin. One flow of early basaltic andesite, which overlies beds in the upper part of the Cache Formation, has pebbles and cobbles from the Cache scattered on its surface, and is dated at 1.66 ± 0.10 Ma (Donnelly-Nolan and others, 1981). Early basaltic rocks that are interbedded with or intruded into the lower or middle parts of the Cache Formation near the Bartlett Springs fault zone have not been dated. The lack of prominent silicic airfall tuffs in the Cache Formation may indicate that the Cache is younger than the small- to moderate-volume silicic pyroclastic eruptions of the Sonoma Volcanics, for which the youngest date is 2.9 Ma on Mount St. Helena (Mankinen, 1972). Thus a remarkably thick sequence of clastic deposits accumulated in the basin in a period of about 1 to 1.5 m.y., with average sedimentation rates of at least 2.7 to 4 mm/yr.

The Cache basin is bounded on the northeast by the Bartlett Springs fault zone trending N40°W to N50°W, and is bounded on the west by the Cross Spring fault, which bends from N22°E trend in its southern part to N20°W to N30°W trend in its northern part. The southwestern boundary is a fault that trends approximately N65°W. The southeastern limit of the Cache Formation is mainly erosional, but the basin fill is offset by several northwest-trending faults that are parallel to the Bartlett Springs fault zone.

Faults that bound the Cache basin approximately fit the expected directions in the San Andreas system and probably controlled deposition of the Cache Formation. For faults that break the Cache Formation and younger deposits, strike-slip is suspected but not proven, according to recent geologic mapping (McLaughlin and others, 1986; B. C. Hearn, J. M. Donnelly-Nolan, and F. E. Goff, unpublished mapping). The Cache basin was probably formed in the late Pliocene by stress along the major Bartlett Springs fault zone, perhaps at a right-step position between the Bartlett Springs fault zone and faults to the southwest.

Present Clear Lake Basin

There is no evidence of a widespread basin during the period 1.5 to 0.8 Ma in the present position of the Clear Lake basin. Volcanic units 1.5 to 0.8 m.y. in age tend to have northwest-southeast elongation as a result of linear arrangement of vents and control by topography related to northwest-trending faults. Limited data from a few drill holes (Fig. 3) show local interflow volcaniclastic deposits up to 100 m thick, which locally contain clasts of Franciscan rocks. The volcaniclastic material is generally derived from underlying or laterally close units of the Clear Lake Volcanics, and probably was deposited in small local basins produced by faulting or volcanic blockage of drainage. Stream drainage was mainly to the north or northwest and was partly controlled by faulting and by the northwest-trending structural grain of the Franciscan and Great Valley units. Eruption of the rhyolite of Bonanza Springs at 1.02 ± 0.04 Ma (Donnelly-Nolan and others, 1981) blanketed at least 10 to 20 km^2 with rhyolitic ash; this eruption may well have been accompanied by some basin subsidence.

The present Clear Lake basin became a prominent feature about 0.6 m.y. ago, initially as an eastern basin of deposition for the Lower Lake Formation, and a western basin of deposition for the Kelseyville Formation on the western side of the Clear Lake Volcanics. Both formations represent fluvial and lacustrine deposition of silt- to cobble-size, locally derived detritus. Whether these two basins were interconnected initially or at a later time is uncertain, as the intervening area is covered by units of the Clear Lake Volcanics that range in age from 0.6 m.y. to younger.

The Lower Lake Formation is less than 0.9 m.y. old. It is interbedded with the 0.52-m.y.-old dacite of Clearlake Highlands and overlain by the 0.40-m.y.-old dacite of Cache Creek, and contains rhyolitic debris probably derived from the 0.6-m.y.-old rhyolite of Thurston Creek (Rymer, 1981; Donnelly-Nolan and others, 1981). The thickness of the Lower Lake Formation is only 90 m in its western exposures where some material may have been lost by erosion. The eastern part is at least 200 m thick in a steeply tilted section near Lower Lake. To the north, a lesser preserved thickness unconformably overlies an erosional remnant of the Cache Formation.

The main evidence for the time of initiation of the present Clear Lake basin is seen in the Kelseyville Formation (Rymer, 1981) and the interbedded units of the Clear Lake Volcanics. The Kelseyville Formation, exposed mainly in Big Valley, consists of fluvial, deltaic, and lacustrine deposits of conglomerate (containing clasts of Franciscan rocks and the Clear Lake Volcanics), sand, silt, and clay. Near the southern limit of preservation of the Kelseyville Formation, mudflow deposits, airfall tuff, and reworked pumice, all from the rhyolite of Thurston Creek, occur at or within 25 m of the base of the Kelseyville Formation. Conglomerate beds beneath these rhyolitic mudflow deposits are lo-

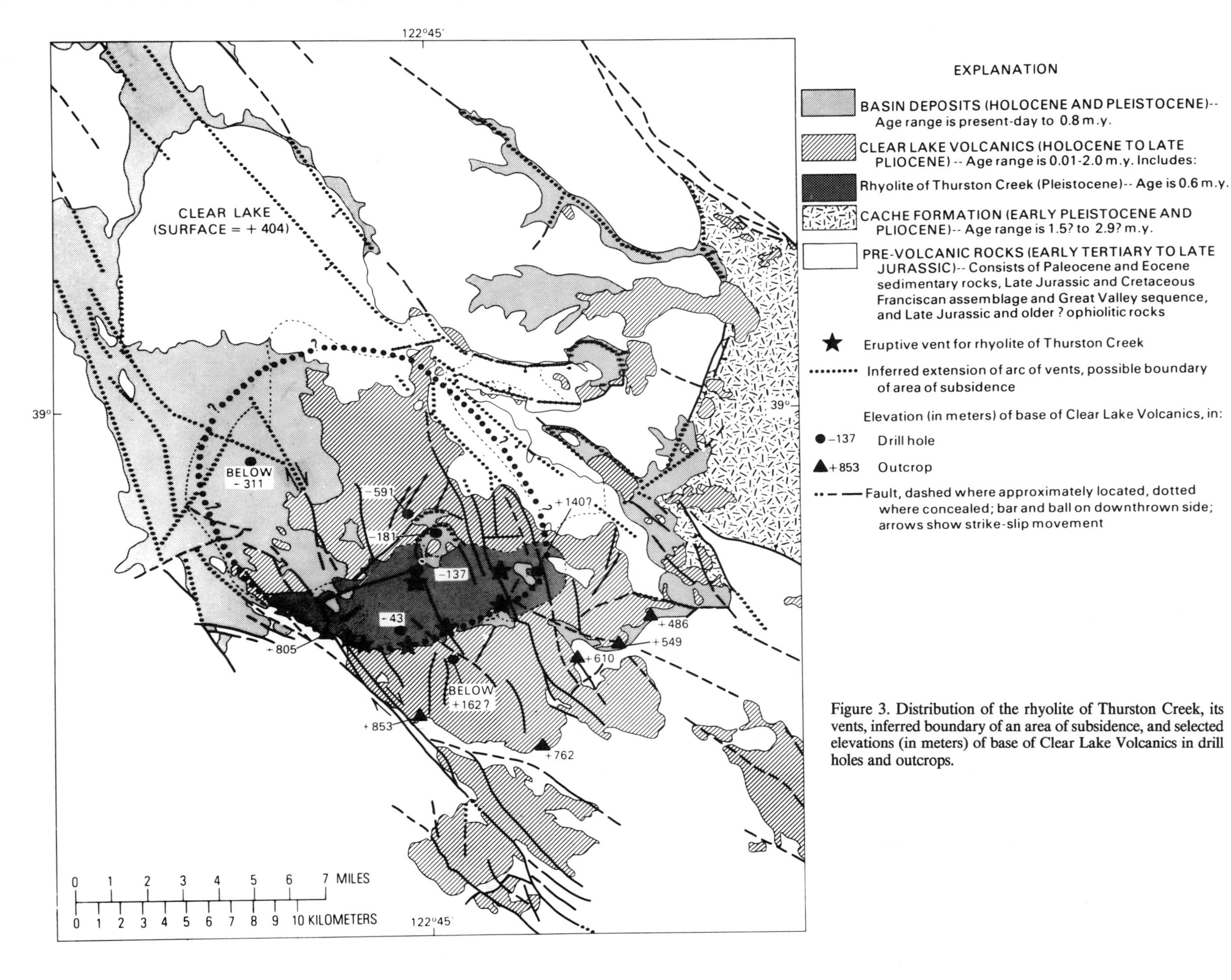

Figure 3. Distribution of the rhyolite of Thurston Creek, its vents, inferred boundary of an area of subsidence, and selected elevations (in meters) of base of Clear Lake Volcanics in drill holes and outcrops.

cally distinctive because of the lack of clasts of the older Clear Lake Volcanics (such as the andesites of Boggs Mountain and Poison Smith Spring, and the rhyolite of Alder Creek) that are typical of conglomerates in the rest of the exposed Kelseyville Formation. Thus, eruption of the rhyolite of Thurston Creek may have initiated the subsidence of the Clear Lake basin and probably changed the existing drainage patterns.

K/Ar ages for the rhyolite of Thurston Creek range from 0.48 ± 0.02 to 0.64 ± 0.03 Ma; the best estimate of its age is about 0.6 Ma, based on the mean of two ages from the western part of the rhyolite and on ages of the overlying units (Donnelly-Nolan and others, 1981). The rhyolite of Thurston Creek erupted from a series of vents that are indicated either by topographic highs or by significant accumulations of pyroclastic deposits (Fig. 3). Most of the vents lie in an arcuate, northward-concave zone close to the southern limit of the rhyolite. The rhyolite flowed mainly to the northwest, north, and northeast, but not southward, suggesting that the southward extent was limited by topography, northward slopes, or contemporaneous subsidence. Although earlier deposited units of the Clear Lake Volcanics may have provided some topographic limits to the rhyolite, the more dominant control appears to have been the probable subsidence of the area north of the arc of vents (Fig. 3). The rhyolite seems to thicken abruptly to the north, as the flow is 300 m thick in a heat-flow test hole northwest of Mount Hannah, only 800 m from the south edge of the flow. In addition, the flow did not fill pre-existing drainages south of the arc, but instead dammed those drainages to produce local basins of fluvial and lacustrine deposition that are approximately contemporaneous with the oldest exposed part of the Kelseyville Formation.

The evidence suggests that the arcuate zone of rhyolite vents lies above an arcuate fault that may be the southern boundary of subsidence that initiated the formation of the basin of deposition of the Kelseyville Formation. The arc is suggestive of a ring-fault segment, but the arc has no exposed silicic vents along its projection to the northeast, and it continues only 2 to 3 km to the northwest as a locus of younger vents for dacites and biotite rhyolite 0.53 to 0.60 m.y. old (Hearn and others, 1976; Donnelly-Nolan and others, 1981; Hearn and others, 1981). The arc is not marked by faults at the surface. It crosses the northern part of the present magma chamber inferred from geophysical data (Isherwood, 1976, 1981) and has curvature opposite to that of the inferred margin of the present magma chamber.

Data on the amount, lateral extent and geometry of subsidence of the Clear Lake Volcanics is quite limited. Several drill holes have penetrated a larger than expected thickness of volcanic rocks (Fig. 3), and certainly some local subsidence around vents is expected. Outcrops and the available data from drill holes for geothermal exploration indicate that the base of the Clear Lake Volcanics descends from about 850 m above sea level near Mount Hannah, to 590 m below sea level (1,260 m thick) on the south side of Mount Konocti (Fig. 3). Such a deep position of the base of the volcanic rocks (if unrelated to local vent collapse) indicates that the Clear Lake Volcanics and interbedded basin deposits younger than 0.60 m.y. could be as thick as 1,000 m at Mount Konocti and beneath the lake. That thickness would represent an average rate of 1.7 mm/yr of subsidence and sedimentation. The deepest drill hole (CL-80-1) beneath the lake has only penetrated 177 m of lake deposits, representing about 0.45 m.y., for an average sedimentation rate of 0.4 mm/yr (Sims and others, 1981b; this volume). Core CL-73-4 from Clear Lake has an average sedimentation rate of 0.9 mm/yr for total depth of 115 m (Sims and others, 1981a). No drill holes have penetrated the base of the Clear Lake Volcanics beneath Big Valley or close to the edge of Clear Lake.

If we assume that much of the volcanic extent in the Mount Konocti area has been limited by fault-bounded subsidence related to volcanic eruptions, then the outer limit of such subsidence should exclude outcrops or shallow subsurface occurrences of Franciscan or Great Valley bedrock. A boundary of possible subsidence could be drawn to exclude Baylis Point, Sulphur Bank Point, and Buckingham Point, and to include Mount Konocti dacites, vent areas of 0.53- to 0.6-m.y.-old dacites and biotite rhyolite at Kelsey Creek gorge, and known subsurface occurrences of the biotite rhyolite of Cole Creek beneath Big Valley. This boundary, if extended from the arc of vents of the rhyolite of Thurston Creek, describes an elliptical area about 13 by 15 km with its major axis oriented approximately west-northwest (Fig. 3). However, in view of the limited amount of data, much of the same area could also be drawn as a rectilinear shape bounded by a combination of northwest- and northeast-trending faults. Confirmation of such a large area of subsidence, and distinction between circular and rectilinear shape, will depend on more detailed drill-hole data or more definitive geophysical surveys.

The Kelseyville Formation and the sediments beneath Clear Lake are keys to the volcanic and structural history of the basin. The total thickness of the Kelseyville Formation is estimated by Rymer (1981) to be at least 500 m. Its maximum thickness is unknown because the base in the deepest part of the basin is unexposed and its top is eroded. The westernmost extent of the Kelseyville Formation is partly limited by a concealed basin-margin fault. The southernmost extent is controlled by initial high topography south of the basin, faulting, and erosion. The easternmost extent is limited by pre-existing and contemporaneous constructional topography of the Clear Lake Volcanics.

The Kelsey Tuff Member, a 1-m-thick bed of lapilli tuff of basaltic andesite, occurs 35 m below the uppermost remaining beds of the Kelseyville Formation. The Kelsey Tuff Member was probably derived from a vent on Mount Konocti or south of it, at about 0.35 to 0.4 Ma, and is preserved over at least 100 km^2. The Kelseyville Formation probably extends northward at depth below the sediments of Clear Lake that have been cored by Sims and Rymer (1981). Rymer (1981) estimated the preserved top of the Kelseyville Formation to be about 0.13 m.y. old, by correlating warmer temperature flora 200 m below the Kelsey Tuff Member, and colder temperature flora 20 m above the Kelsey Tuff Member, with oxygen isotope stages 7 and 6, respectively. However, that glacial-interglacial correlation could correspond to older interglacial to glacial transitions as well. The Kelsey Tuff Member at about 35 m below the present top of the Kelseyville would be about 0.165 m.y. old if the 1-mm/yr sedimentation rate in Clear Lake (Sims and others, 1981a, b) prevailed, an age suggesting that the tuff should have been encountered in the 177-m-deep drill hole in Clear Lake, but it was not. If the Kelsey Tuff Member is really as young as 0.165 m.y., it is more likely derived from one of the lake shore maar-type vents, or from the basaltic andesite of Buckingham Peak. However, recent correlations of diatom successions and other tephra indirectly indicate that the age of the Kelsey Tuff Member is between about 0.28 and 0.32 m.y. (Rymer and others, this volume; Sims and others, this volume). The Kelsey Tuff Member apparently is missing in cores CL-80-1 and CL-80-2 owing to erosional hiatuses.

Tilting of the floor of the Clear Lake basin has been down to the northeast, as shown by tilted volcanic units, dips in the Kelseyville Formation, and the location of the steepest lakeshore topography and deepest water near the northeast edge of the lake. Apparently sedimentation has approximately kept pace with subsidence to keep the water depths generally shallow. Shaul Valley, an alluvial valley on the south side of Mount Konocti, may be an example of the effects of tilting. The lowest part of the valley is the closed north end.

Suggestions for Further Study

The age range of the Kelseyville Formation is not well known. The total depth and maximum age of basin sediments and the shape of the bottom of the basin are all unknown. The maximum thickness and concealed limits of the Clear Lake Volcanics are unknown. Many of the boundary faults are inferred only by reconnaissance mapping, and the detailed studies necessary to establish their periodicity and minimum ages of displacement have not been done.

CONCLUSIONS

The Clear Lake basin has evolved as a result of three processes. (1) The San Andreas fault system has been the dominant control, by initiating the broad subsidence and controlling the margins, probably by the mechanisms outlined by Crowell (1974). (2) Eruption of the Clear Lake Volcanics has caused local subsidence and has filled part of the basin. (3) Erosional and depositional processes have influenced the distribution, character, and amount of basin fill.

An earlier basin that contains the Pliocene and early Pleistocene Cache Formation, which was deposited over a time span of 1 to 1.5 m.y., was controlled by faults in the broad San Andreas fault system. Formation of the Clear Lake basin was initiated or accelerated at about 0.6 m.y., and appears to be related in part to eruption of the rhyolite of Thurston Creek. The basin may contain a poorly defined circular or rectilinear area of subsidence that contains much of the volume of silicic volcanic units 0.6 to 0.3 m.y. in age. Deposition has approximately equalled subsidence; rates of 0.4 to 0.9 mm/yr for the past 0.45 m.y. are suggested by drilled sediments beneath Clear Lake, and a more speculative rate of 1.7 mm/yr for the past 0.6 m.y. is suggested by the deepest known base of the volcanic rocks near Mount Konocti. Faults with young geomorphic features, earthquake epicenters associated with mapped fault zones, and geodetic measurements all show that the Clear Lake basin is continuing to evolve in the broad zone of deformation in the San Andreas fault system along the soft margin of the North American plate.

The Clear Lake basin has outstanding potential for providing answers to the evolution of tectonics, volcanism, sedimentation, and climate in central northern California over the past 600,000 yr, and for projection of that record into the future. Continued research on these problems will be valuable in the exploration and development of geothermal resources, estimation of potential for further volcanic activity, and determination of the periodicity of seismic activity in the Coast Ranges.

ACKNOWLEDGMENTS

We thank R. P. Koeppen and D. G. Herd for helpful reviews and critical insights. We also appreciate the cooperation of numerous landowners in the Geysers–Clear Lake area, and the contributions of industry personnel to the geologic knowledge of the region. This research has been funded by the Geothermal Research Program of the U.S. Geological Survey.

REFERENCES CITED

Bufe, C. G., Marks, S. M., Lester, F. W., Ludwin, R. S., and Stickney, M. C., 1981, Seismicity of The Geysers–Clear Lake region: U.S. Geological Survey Professional Paper 1141, p. 129–137.

California Department of Water Resources, 1962, Reconnaissance report on upper Putah Creek basin investigations: California Department of Water Resources Bulletin 99, 254 p.

Cockerham, R. S., and Herd, D. G., 1982, Seismicity of the north end of the San Andreas fault system: EOS, Transactions of the American Geophysical Union, v. 63, p. 384.

Crowell, J. C., 1974, Origin of late Cenozoic basins in southern California, *in* Dickinson, W. R., ed., Tectonics and sedimentation: Society of Economic Paleontologists and Mineralogists Special Publication 22, p. 190–204.

Donnelly-Nolan, J. M., Hearn, B. C., Jr., Curtis, G. H., and Drake, R. E., 1981, Geochronology and evolution of the Clear Lake Volcanics: U.S. Geological Survey Professional Paper 1141, p. 47–60.

Frizzell, V. A., Jr., and Brown, R. D., Jr., 1976, Map showing recently active breaks along the Green Valley fault, Napa and Solano Counties, California: U.S. Geological Survey Miscellaneous Field Studies Map MF-743, scale 1:24,000.

Harwood, D. S., and Helley, E. J., 1982, Preliminary structure contour map of the Sacramento Valley, California showing major late Cenozoic structural features and depth to basement: U.S. Geological Survey Open-File Report 82-737, 19 p., scale 1:250,000.

Hearn, B. C., Jr., Donnelly, J. M., and Goff, F. E., 1976, Preliminary geologic map and cross-section of the Clear Lake volcanic field, Lake County, California: U.S. Geological Survey Open-File Report 76-751, scale 1:24,000.

——, 1981, The Clear Lake volcanics; Tectonic setting and magma sources: U.S. Geological Survey Professional Paper 1141, p. 25–45.

Helley, E. J., and Herd, D. G., 1977, Map showing faults with Quaternary displacement, northeastern San Francisco Bay, California: U.S. Geological Survey Miscellaneous Field Studies Map MF 881, scale 1:125,000.

Herd, D. G., 1978, Intracontinental plate boundary east of Cape Mendocino, California: Geology, v. 6, p. 721–725.

Isherwood, W. F., 1976, Gravity and magnetic studies of the Geysers–Clear Lake geothermal region, California: United Nations Symposium on Development and Use of Geothermal Resources, 2nd, San Francisco, 1975, Proceedings, v. 2, p. 1065–1073.

——, 1981, Geophysical overview of The Geysers area: U.S. Geological Survey Professional Paper 1141, p. 83–95.

Jachens, R. C., and Griscom, A., 1983, Three-dimensional geometry of the Gorda plate beneath northern California: Journal of Geophysical Research, v. 88, p. 9375–9392.

Kelleher, J., and Isacks, B., 1982, Directional hypothesis; Late stage pre-earthquake loading may often be asymmetric: Journal of Geophysical Research, v. 87, p. 1743–1756.

Lachenbruch, A. H., and Sass, J. H., 1980, Heat flow and energetics of the San Andreas fault zone: Journal of Geophysical Research, v. 85, p. 6185–6222.

Lake County Flood Control and Water Conservation District, 1967, Big Valley ground-water recharge investigation, 63 p.

Lawson, A. C., 1908, The California earthquake of April 18, 1906; Report of the State Earthquake Investigation Committee: Carnegie Institution of Washington Publication 87, v. 1, and atlas, 451 p.

Lofgren, B. E., 1981, Monitoring crustal deformation in The Geysers–Clear Lake region: U.S. Geological Survey Professional Paper 1141, p. 139–148.

Mankinen, E. A., 1972, Paleomagnetism and potassium-argon ages of the Sonoma volcanics, California: Geological Society of America Bulletin, v. 83, p. 2063–2072.

McLaughlin, R. J., 1978, Preliminary geologic map and structural sections of the central Mayacmas Mountains and The Geysers steam field, Sonoma, Lake, and Mendocino Counties, California: U.S. Geological Survey Open-File Report 78-389, scale 1:24,000.

——, 1981, Tectonic setting of pre-Tertiary rocks and its relation to geothermal resources in The Geysers–Clear Lake area: U.S. Geological Survey Professional Paper 1141, p. 3–23.

McLaughlin, R. J., and Nilsen, T. H., 1982, Neogene nonmarine sedimentation and tectonics in small pull-apart basins of the San Andreas fault system, Sonoma County, California: Sedimentology, v. 19, p. 865–876.

McLaughlin, R. J., Ohlin, H. N., and Thormahlen, D. J., 1986, Geologic map and structure sections of the Little Indian Valley–Wilbur Springs geothermal area, northern Coast Ranges, California: U.S. Geological Survey Open-File Report (in press).

Pampeyan, E. H., Harsh, P. W., and Coakley, J. M., 1981, Preliminary map showing recently active breaks along the Maacama fault zone between Hopland and Laytonville, Mendocino County, California: U.S. Geological Survey Miscellaneous Field Studies Map MF-1217, scale 1:24,000.

Rymer, M. J., 1981, Stratigraphic revision of the Cache Formation (Pliocene and Pleistocene), Lake County, California: U.S. Geological Survey Bulletin 1502-C, 35 p.

Sims, J. D., and Rymer, M. J., 1976, Map of gaseous springs and associated faults in Clear Lake, California: U.S. Geological Survey Miscellaneous Field Investigations Map MF-721, scale 1:48,000.

Sims, J. D., and White, D. E., 1981, Mercury in the sediments of Clear Lake: U.S. Geological Survey Professional Paper 1141, p. 237–241.

Sims, J. D., Adam, D. P., and Rymer, M. J., 1981a, Stratigraphy and palynology of Clear Lake, California: U.S. Geological Survey Professional Paper 1141, p. 219–230.

Sims, J. D., Rymer, M. J., and Perkins, J. A., 1981b, Description and preliminary interpretations of core CL-80-1, Clear Lake, Lake County, California: U.S. Geological Survey Open-File Report 81-751, 175 p.

White, D. E., and Roberson, C. E., 1962, Sulphur Bank, California, a major hot-spring quicksilver deposit, *in* Engel, A.E.J., James, H. L., and Leonard, B. F., eds., Petrologic studies; A volume in honor of A. F. Buddington: Geological Society of America, p. 397–428.

Zandt, G., and Furlong, D. P., 1982, Evolution and thickness of the lithosphere beneath coastal California: Geology, v. 10, p. 376–381.

Manuscript Accepted by the Society September 15, 1986

Printed in U.S.A.

Geological Society of America
Special Paper 214
1988

Late Quaternary deposits beneath Clear Lake, California; Physical stratigraphy, age, and paleogeographic implications

John D. Sims
Michael J. Rymer
James A. Perkins
U.S. Geological Survey
345 Middlefield Road
Menlo Park, California 94025

ABSTRACT

Clear Lake, California, lies in a volcano-tectonic depression that received nearly continuous lacustrine deposition for the past 500,000 yr and probably longer. The lake has been shallow (<30 m) and eutrophic throughout its history. Sediments beneath the floor of the lake are fine grained (chiefly >7.0ϕ) and contain fossils of a large lacustrine biota, as well as a pollen record of land plants that lived in the basin. The sediments also contain tephra units of local and regional extent. The ages of the sediments in Clear Lake are determined from radiocarbon dates on the sediments, from correlation of regionally distrbuted tephra units, and from inferred correlation of oak-pollen spectra with the marine oxygen-isotope record. From the chronology of events recorded in the cores from Clear Lake, the late Quaternary history of the lake can be deciphered and the sediments correlated with other basins in northern California.

Comparison of cores from Clear Lake with strata of the Kelseyville Formation, exposed south of the main basin, suggests a general northward migration of lacustrine sedimentation, which in turn suggests a northward tilt of the basin. Migration of the lake was a response to volcanism and tectonism. Volcanic rocks erupted from Mt. Konocti, on the southern margin of the lake, and displaced the shoreline to the west and north. Clear Lake is bounded by faults that are part of the San Andreas fault system. These faults strongly influenced the position, depth, and longevity of Clear Lake. Movement on these boundary faults deepened the Highlands and Oaks arms of the lake about 10 ka. Tectonic movement accompanying faulting was probably also largely responsible for hiatuses in the deposits beneath Clear Lake about 17 and 350 ka that have been inferred from the subbottom stratigraphy of the lake. Climate change, although responsible for large variation in the composition of the terrestrial flora of the Clear Lake drainage basin, has not influenced the areal extent, depth, or position of the lake.

INTRODUCTION

The presence of uplifted late Pleistocene lacustrine deposits bordering the present lake suggests a long period of lacustrine sedimentation in the basin (Sims, 1976; Sims and others, 1981a; Rymer and others, this volume). Thus, 10 cores were taken from Clear Lake to provide the basis for a multidisciplinary study of a long-lived, volcano-tectonic sedimentary basin (Fig. 1). The physical and geologic settings of Clear Lake and its drainage basin are summarized by Sims (this volume).

Tectonic movement played a major role in shaping the Clear Lake basin and in particular in controlling sedimentation (Hearn and others, this volume; Rymer 1981, this volume). Lake boundary faults are well known, but the amount of movement and its timing are not known (Sims and Rymer, 1975a), although the Coloyami, Mayacama, Konocti Bay, and Big Valley faults are presently seismically active (Eberhardt-Phillips, this volume). Rymer (this volume) estimates that the Big Valley fault may have

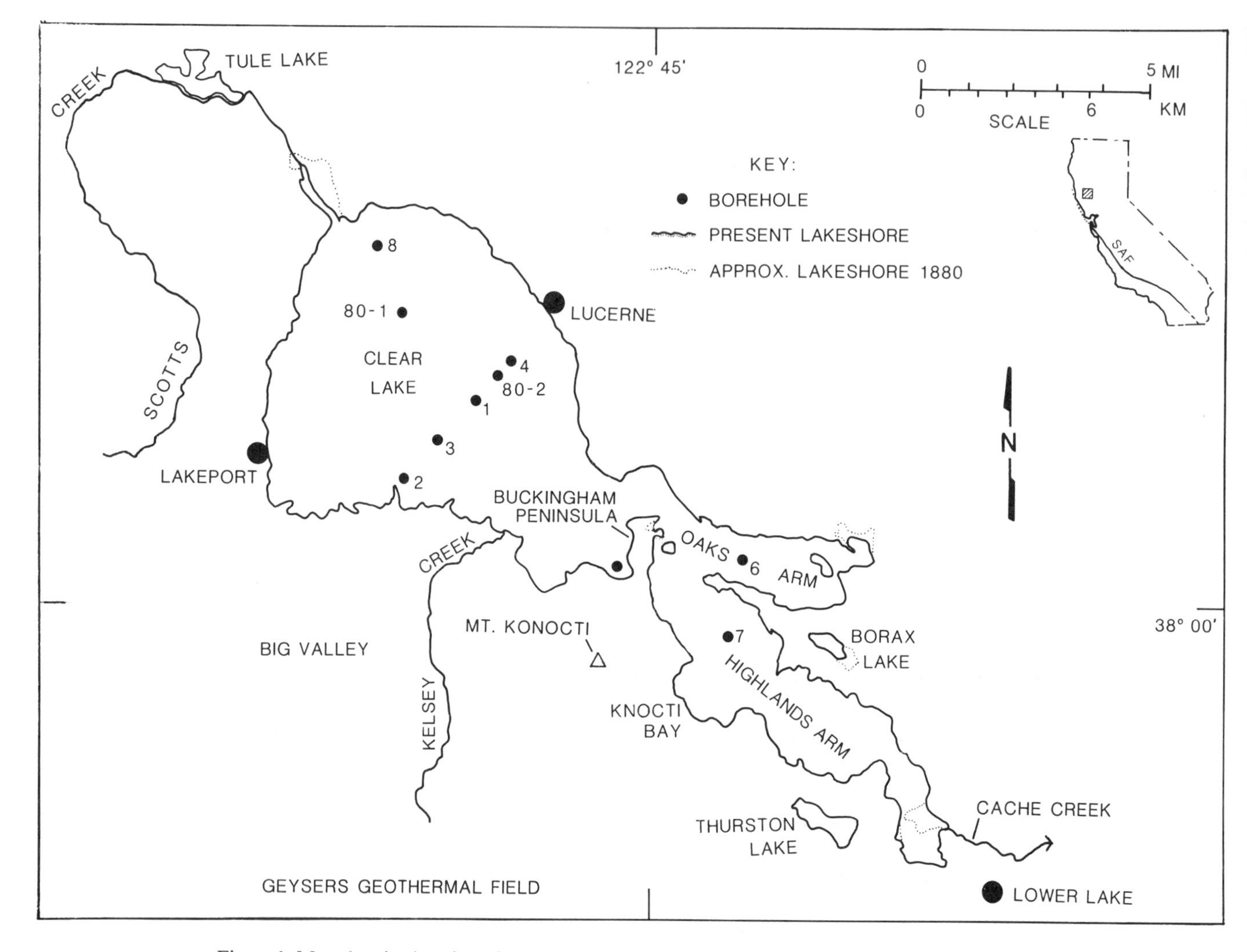

Figure 1. Map showing location of the Clear Lake area in California with respect to the San Andreas fault zone (SAF on inset). The location of cores taken from Clear Lake for this study are shown with two designations: single numbers 1 through 8 represent the locations of the series of cores taken in 1973 that bear the prefix CL-73- in the text; locations 80-1 and 80-2 represent cores taken in 1980 that bear the prefix CL-80- in the text. The approximate 1880 shoreline of Clear Lake is shown for reference (after Becker, 1888).

as much as 0.55 mm yr^{-1} of vertical movement. The average sedimentation rate for the 115.2-m-long core CL-73-4 over the past 127 ka is about 0.9 mm yr^{-1}. Thus vertical displacement on the Big Valley fault could account for a large part of the subsidence that has taken place in the lake basin.

In addition to an extensive record of lacustrine sedimentation, the Clear Lake cores contain diverse assemblages of lacustrine and terrestrial macrofossils and microfossils. Macrofossils include fish scales and bones (Casteel and Rymer, 1981; Casteel and others, 1977, 1979; Cook and others, 1966; Hopkirk, 1974, this volume), molluscs (Rymer and others, this volume), and plant fragments and seeds (D. P. Adam, oral communication). The microfossil assemblages include ostracodes (Forrester, this volume), diatoms (Bradbury, this volume), pollen and spores (Adam, this volume), algal remains and cysts (Mahood and Adam, 1979), and sponge spicules, as well as cladocera and other insects (D. P. Adam, oral communication).

This report first reviews the physical stratigraphy and physical properties of the subbottom sediments. This review is followed by the results of absolute and relative dating studies. The chronology for the cores derived from the dating studies allows correlation between cores and with other Quaternary sections distant from the Clear Lake basin. Sedimentation rates determined from application of the chronology are used to examine generalized patterns of sedimentation in the lake; long-term sedimentation rates in the two most complete of the three longer cores from the lake are also determined. Finally, data from physical stratigraphy and properties of the sediments, sedimentation rates, and the overall chronology for the lake are integrated to produce a series of paleogeographic reconstructions for different stages of development of the lake basin.

PHYSICAL PROPERTIES

Physical properties used here to characterize the sediments of Clear Lake are general lithology, mineralogy, grain size, and bulk density. Detailed descriptions and analytic results for individual cores may be found in Sims and Rymer (1975b–h, 1976), Sims and others (1981b, c).

Lithology

The principal lithology in the 10 cores is uniform sapropelic clay and silt similar to sediment presently accumulating away from the shoreline of the lake (Fig. 2A and B, located in pocket inside back cover). Adjuncts to the sapropelic mud are sand, gravel, peat and peat-rich sediments, and thin volcanic ash beds. Lithologic variability within a core depends on location in the lake. In general, cores near the center of the main basin are dominated by or composed entirely of sapropelic clay and silt. Cores in the arms of the lake or that are near shore in the main basin are more varied, primarily owing to a greater abundance of coarse clastic material derived from streams entering the lake (Fig. 2).

The sapropels are dark olive gray and have texture that is clayey to gritty. Two long cores, CL-73-4 (115.2 m) and CL-80-2 (168.7 m), contain only fine-grained sapropelic sediments; such sediments also make up all of two short cores CL-73-5 (22.6 m) and CL-73-8 (20.5 m). A third long core, CL-80-1 (177.0 m), is composed chiefly of sapropelic mud with intervals of coarse to medium sand and gravel (Fig. 2B). In the remaining cores the uppermost 4.2 (CL-73-2) to 21.6 m (CL-73-6) is composed of sapropelic mud. Examination of core CL-73-4 and analysis at about 1- to 2-m intervals revealed little variation in grain size over the entire 115.2-m length, even though nearby cores (CL-80-1 and CL-73-3) contain abundant coarse clastic material (Sims and Rymer, 1975d; Sims, 1976; Sims, 1982).

Sand and gravel are prominent in cores CL-73-2, CL-73-3, and CL-80-1, and to a lesser degree in core CL-73-1. Coarse clastic beds are thickest and coarest near the present shoreline and become thinner and finer grained basinward (Sims, 1976). Coarse clastic beds in core CL-80-1 occur sparsely between 104 and 119 m, and abundantly between 151 and 168 m. Gravel is common between 168 and 177 m but was not recovered. These coarse-grained deposits occur much deeper than those encountered in cores CL-73-1, -2, and -3, and are thus not correlative with them. Because unconsolidated sand and gravel units are difficult to retrieve in coring operations, the details of grain size and bedding characteristics of the coarser units are not well known.

Peat-rich sediments are found in the two arms of Clear Lake, in cores CL-73-6 and -7 (Fig. 2B). Peat and peat-rich sediment are encountered below 7.6 m in core CL-73-6 and below 6.9 m in core CL-73-7 and are overlain by sapropelic mud (Sims and Rymer, 1978a,c). Peat and peat-rich sediments are more abundant in core CL-73-7 than in core CL-73-6. The peat in core CL-73-7 is predominantly fibrous brown peat of the type most commonly found near the edges of shallow bays and in open marshes (Rigg and Gould, 1957; Cohen, 1973). The peat contains sedge roots as well as seeds of rooted shallow-water plants such as *Zanichellia, Myriophyllum,* and *Nuphar* (Sims and others, 1981a). Based on the plant assemblage, the sediment below 6.9 m in core CL-73-7 was deposited in an open marsh environment having <2 m of water, perhaps much less than 2 m (D. P. Adam, oral communication, 1974). Fish remains are not abundant in the peat, thus the marsh was probably only intermittently flooded by the open lake (Casteel and others, 1975).

Volcanic ash beds are a volumetrically minor component in the cores, but are important for correlation between cores within the lake and with stratigraphic sequences outside the lake basin (Sarna-Wojcicki and others, this volume). The degree of visibility in the tephra units is dependent on the original thickness, the degree of bioturbation, and the degree of disturbance and mixing of the sediments by wind-induced currents in the lake (Sims, 1976, p. 676). Tephra units thicker than a few millimeters and those that escaped intensive bioturbation are often easy to identify megascopically, especially where intercalated within intervals of uniformly fine-grained silt and clay. The tephra layers are distinguished by pale pink, white, or light to medium gray colors and by shard structure of the grains. But many other ash beds from Clear Lake are not obvious except in x-ray radiographs of thin slices from the cores. The number of ash beds ranges from none in core CL-73-2 to as many as 56 in core CL-73-4 (Table 1). The number of ash beds identified within a core depends chiefly on the length and percent of recovery, but also reflects depositional or erosional hiatuses within the cores as well as primary nonuniform distribution of some locally derived tephra units. Several ash beds were deposited from eruptions well outside the Clear Lake basin (Sarna-Wojcicki and others, this volume).

Mineralogy

Sediments introduced to Clear Lake by streams are potentially derived from a variety of rock types present in the Clear Lake basin. About 60 percent of the land area of the Clear Lake drainage basin is underlain by sedimentary and metamorphic rocks of the Franciscan assemblage, 2 percent by sedimentary rocks of the Great Valley sequence, and 13 percent by the Clear Lake Volcanics. Other sources of sediment, the Kelseyville and Lower Lake Formations and Quaternary fluvial deposits, compose 25 percent of the land area. These latter deposits comprise primarily debris from the Franciscan assemblage, Great Valley sequence, and Clear Lake Volcanics.

Using the methods of Schultz (1964), quantitative clay min-

TABLE 1. CORE DATA FROM CLEAR LAKE, CALIFORNIA

Core No.	Water Depth (m)	Core Length (m)	Recovery (%)	No. of Ash Beds	Estimated Age of Core Bottom (yr)
80-1	7.5	177.0	66.7	54	450,000
80-2	8.0	165.8[a]	65.0[b]	23	450,000
73-1	8.8	62.6	35.0	6	40,000[c]
73-2	4.3	13.9	88.0	0	10,000[c]
73-3	8.4	69.0	96.0	43	38,000[c]
73-4	8.4	115.2	92.0	56	130,000[d]
73-5	7.6	22.6	94.0	7	28,000[e]
76-6	12.2	21.6	99.0	8	36,000[e]
73-7	12.8	27.4	94.9	26	40,000[e]
73-8	5.2	20.5	99.6	5	24,000[e]

[a]Core length for core CL-80-2 calculated from top of sediment column to maximum coring depth, even though the intervals from 0 to 53.42 m and 77.20 to 98.68 m were not cored.

[b]Percent recovery calculated for cored segments of lake sediment; does not include drilled out segments.

[c]Estimated age, from Sims and Rymer (1981).

[d]Estimated age, from Sims and others (1981a) and from Adam and others, (1981).

[e]Radiocarbon age of core bottom, from Sims (1976).

eralogy of samples at about 1-m intervals from core CL-80-1 and CL-80-2, and about 1- to 3-m intervals from core CL-73-4, shows little systematic variation with depth (Table 2). The clay mineral fraction of the sediments is dominated by chlorite and illite (~56 percent), common components of Franciscan rocks. Secondary components in the clay fraction are smectite and mixed-layer smectite species (~35 percent) and kaolinite (~9 percent). The smectite and mixed-layer clays probably are mostly derived from the weathering of volcanic rocks but are also present in rocks of the Great Valley sequence and Tertiary sedimentary rocks in the area.

Grain Size

Grain-size analyses of the fine-grained sediment in cores CL-73-4 and CL-80-1 were done to determine if the longest cores contain systematic variations in grain size through late Quaternary time (Table 2). Grain-size analyses were not done on core CL-8-2 because it is similar in lithology to core CL-73-4 and is not as complete as CL-73-4 (Table 1). In core CL-73-4 (Fig. 2) the mean grain-size (X_ϕ) ranges from 7.9 to 9.2ϕ (Fig. 3A, Sims, 1982). Statistical time-series analysis suggests that there is no detectable systematic variation in mean grain size with depth in core CL-73-4 (E. Cranswick, written communication, 1983).

Mean grain size in core CL-80-1 ranges between 5.8 and 8.4ϕ (Fig. 3B), which reflects the relatively greater abundance of coarse-grained material in this core as compared to cores CL-73-4 and CL-80-2. The grain-size analyses for all three cores are biased toward the finer grained samples because core recovery in coarse-grained strata is poorer than in fine-grained strata. This bias particularly affects the grain-size data set from core CL-80-1 because coarse-grained deposits are more abundant in it than in core CL-73-4.

Density and Weight Loss on Ignition

The fact that water content, sediment bulk density, and organic-material content of Clear Lake cores are variable reflects the contrasting depositional environments, sedimentation rates, rates of biologic activity, and depths of burial. Water content in the cores varies from 67 to 87 percent near the tops of the cores to as little as 12 percent near the bottom of the longer cores. Dry bulk density is calculated from the dry weight of a known volume of sediment (Beaver and others, 1976). The increase in dry bulk density with depth results from the expulsion of water and the collapse of void spaces as the sediment is buried. Small departures from a smooth increase in density probably result from variations in mineralogy, grain size, and amount of organic material. Errors in sampling and weighing also account for some minor variation in bulk density. Core CL-73-4 shows a general progressive dewatering of samples at depth that closely fits the density-dewatering curves of Skempton (1970), Yamamoto and others (1974), and Rieke and Chilingarian (1974). It also has a high recovery rate of about 92 percent (Table 1), and the smooth increase in density with depth is well documented (Fig. 4C). Overall in core CL-73-4 there is about a 3½-times increase in dry bulk density in the sediment between the surface and 80 m. Below 80 m, sediment density approaches the average for sedi-

TABLE 2. DATA FROM GRAIN-SIZE ANALYSES AND SEMI-QUANTITATIVE CLAY MINERAL ANALYSES FROM CORES CL-80-1, CL-80-2, AND CL-73-4
(Sims, 1982; Summary statistics are based on moment calculations.)

Core	Sample Number	Depth (cm)	X_ϕ	σ_ϕ	Sk_ϕ	Kt_ϕ	Smectite	Chlorite	Illite	Kaolinite	Mixed Layer	Total
CL-80-1	4	100	7.71	1.01	-0.33	2.80	18	20	25	30	7	100
"	14	650	7.71	1.06	-0.61	4.13						
"	25	950	7.28	1.16	-0.71	4.11	17	29	24	10	20	100
"	26	975	7.55	1.14	-0.67	3.76	18	27	20	8	27	100
"	29	1050	8.25	1.17	-0.61	3.57						
"	38	1300	7.55	1.14	0.69	3.76	22	2	0	0	76	100
"	49	2175	5.79	1.46	0.35	2.53						
"	53	2275	7.40	1.06	-0.41	3.12	20	45	27	14	-5	101
"	59	3054	7.57	1.10	-0.38	2.92	18	35	40	7	0	100
"	83	3825	7.67	1.00	-0.27	2.95	18	22	27	8	25	100
"	96	3915					29	34	18	6	13	100
"	100	4160					27	36	38	0	0	101
"	104	4250					36	32	19	5	8	100
"	131	4800	8.19	1.04	-0.20	3.11	19	33	30	10	7	99
"	134	4875					14	27	23	14	22	100
"	150	5050	8.55	0.52	0.04	2.97	13	27	24	14	22	100
"	168	5325					16	26	25	20	13	100
"	170	5360					21	20	23	16	20	99
"	175	5630					17	16	15	19	32	100
"	185	6035	7.66	1.03	-0.42	3.00						
"	204	6360					17	18	18	0	47	100
"	215	6725					13	23	20	17	26	99
"	230	6850	7.28	1.24	-0.43	2.95	19	29	23	12	17	100
"	236	7560	6.89	1.29	-0.21	2.81	25	36	32	6	1	100
"	244	7760	7.22	1.25	-0.37	2.75	14	34	32	11	10	101
"	254	8075	7.09	1.38	-0.367	2.51	15	24	33	26	3	101
"	273	8550	6.95	1.42	-0.41	2.55	16	32	27	18	7	100
"	282	8850	6.52	1.46	-0.25	2.17	16	23	38	24	0	101
"	298	9075	7.22	1.10	-0.25	3.24	13	27	17	11	32	100
"	319	9585	6.85	1.28	-0.05	2.52	16	17	15	18	34	100
"	348	9975	7.43	1.41	-0.69	3.57	13	24	28	-5	41	100
"	352	10075	7.17	1.10	-0.12	2.98	14	42	26	11	7	100
"	377	10375					7	19	29	17	28	100
"	425						16	21	13	9	42	100
"	443	11450					12	17	13	10	48	100
"	453	11750	8.38	1.26	-0.73	3.32						
"	457	11850	7.33	1.45	0.09	2.32						
"	460	11930					11	22	26	12	30	101
"	480	12375	7.65	1.16	-0.70	3.17	9	25	25	13	29	101
"	484	12475					9	16	17	12	46	100
"	495	12650					25	8	27	24	15	99
"	499	12750					5	24	29	14	28	100
"	503	12850	7.86	1.23	0.01	2.32						
"	509	13050	7.75	1.00	-0.03	4.02						
"	514	13150	8.06	1.25	-0.60	3.18						
"	524	13250					14	25	19	15	27	100
"	528	13350	7.31	1.27	-0.36	2.53	11	27	23	14	25	100
"	540	14200	7.71	1.48	-0.35	2.73						
"	564	13650										
"	572	14300	7.51	1.06	0.15	3.19						
"	575	14430					4	30	20	12	23	100
"	591	14825	8.55	0.79	-0.24	3.62						
"	595	14925	8.00	1.12	-0.02	2.97	7	35	23	18	17	100
"	599	15025	7.66	0.99	-0.35	2.96						
"	604	15150					5	33	39	8	17	100
"	609	15329					14	55	21	16	27	99
"	611	15550					12	20	32	18	18	100
"	617	15850					2	41	21	0	35	99
"	621	15950					2	32	24	11	31	100
"	625	16150					5	49	29	0	17	100
"	629	16250					2	44	17	15	21	99
"	636	16425					4	27	28	12	29	100
"	645	16525					2	43	42	11	2	100
"	649	16625	7.25	1.00	0.12	2.61	7	31	36	18	8	100
"	656	16750					8	29	30	6	27	100
Cl-80-2	004	5607					14	34	29	1	21	99
"	008	5707					20	30	22	9	19	100
"	012	5942					21	25	27	14	14	102
"	016	6284					17	31	24	16	12	100
"	020	6540					21	32	20	8	19	100
"	023	6857					21	24	29	14	12	100
"	028	7132					21	30	23	0	26	100

TABLE 2. (CONTINUED)

Core	Sample Number	Depth (cm)	X_ϕ	σ_ϕ	Sk_ϕ	Kt_ϕ	Smectite	Chlorite	Illite	Kaolinite	Mixed Layer	Total
CL-80-2	044	7551					23	30	23	9	15	100
"	048	9874					18	30	24	5	22	99
"	052	9974					21	30	29	7	13	100
"	056	10074					17	30	30	6	16	99
"	060	10243					15	39	28	2	16	100
"	064	10343					16	34	26	6	17	99
"	068	10443					20	35	30	0	17	102
"	072	10538					16	25	38	16	5	100
"	080	10929					17	31	29	8	16	100
"	084	11029					16	39	33	2	10	100
"	092	11230					17	29	20	6	27	99
"	106	11780					16	39	30	0	17	102
"	110	12090					27	45	31	2	0	105
"	114	12190					21	32	27	13	5	99
"	118	12290					29	19	25	13	14	100
"	122	12390					25	0	22	0	54	101
"	126	12505					16	41	37	6	0	100
"	134	12823					16	30	25	7	22	100
"	138	12931					18	33	29	12	8	100
"	142	13276					17	32	30	12	9	100
"	146	13376					10	34	29	15	13	101
"	150	13476					13	31	30	9	17	100
"	154	13580					20	32	29	2	17	100
"	158	13680					15	36	25	7	17	100
"	162	13837					18	37	36	8	2	101
"	166	13937					25	35	28	2	9	99
"	169	14057					12	31	25	11	22	100
"	170	14447					15	25	30	16	14	100
"	174	14547					12	27	24	15	22	100
"	178	14647					17	31	32	10	11	100
"	180	14725					16	28	33	10	13	100
"	184	14825					11	26	11	6	46	100
"	188	15378					24	41	39	0	0	104
"	192	15503					12	27	25	3	32	100
"	196	15603					22	34	34	0	12	102
"	200	15703					15	26	25	13	20	99
"	208	16043					19	24	231	5	31	100
"	212	16143					18	31	36	11	4	100
"	216	16243					17	28	26	6	23	
"	220	16354					11	40	35	6	8	100
CL-73-4	2046	100	8.15	1.07	-0.80	4.70						
"	2062	220	8.75	0.84	-0.45	3.73						
"	2070	310	8.16	1.38	-1.06	3.88						
"	1849	409	8.61	0.75	+0.05	3.57						
"	1837	595	8.62	0.96	-0.70	4.27						
"	1831	710	8.72	0.81	-0.47	4.13						
"	1823	792	8.78	0.77	-0.22	3.12						
"	1817	890	8.74	0.87	-0.65	4.13						
"	1809	990	8.38	1.12	-0.77	3.47						
"	1787	1100	9.08	0.57	-0.52	4.88						
"	1255	1196	7.93	1.19	-0.53	2.98						
"	1287	1301	8.72	0.92	-0.60	4.00						
"	1889	1409	8.72	0.75	-0.02	3.13						
"	1897	1500	8.51	0.86	-0.25	3.26						
"	1903	1590	8.39	0.93	-0.60	4.21						
"	1911	1700	8.41	0.90	-0.42	3.56						
"	1919	1800	8.32	0.84	-0.30	3.73						
"	1855	1895	8.39	0.94	-0.99	5.61						
"	1869	2000	7.73	1.07	-0.53	3.47						
"	1877	2090	8.67	0.73	-0.34	4.34						
"	1685	2210	8.40	0.88	-0.72	5.27						
"	1675	2515	8.90	0.76	-0.13	0.89						
"	1669	2600	8.91	0.81	-0.71	4.31						
"	1663	1680	8.83	0.72	-0.21	3.61						
"	1655	2990	8.02	1.39	-1.23	4.36						
"	1649	3090	8.63	0.77	0.00	2.64						
"	1641	3170	8.58	0.77	-0.48	4.32						
"	1623	3300	8.95	0.77	-0.75	5.14						
"	1617	3408	8.80	0.74	0.06	2.39						
"	1745	3495	8.74	0.75	-0.32	3.76						
"	1731	3690	8.62	0.77	-0.38	4.10						
"	1723	3604	8.57	0.88	-0.68	4.29						
"	1719	3800	8.30	0.90	-0.46	4.02						
"	1713	3910	8.70	0.70	-0.20	3.85						

TABLE 2. (CONTINUED)

Core	Sample Number	Depth (cm)	X_{ϕ}	σ_{ϕ}	Sk_{ϕ}	Kt_{ϕ}	Smectite	Chlorite	Illite	Kaolinite	Mixed Layer	Total
CL-73-4	1706	4010	8.85	0.82	-0.37	3.31						
"	1697	4085	8.83	0.75	-0.07	3.08						
"	1779	4219	8.44	0.86	-0.21	3.17						
"	1737	4320	9.21	0.66	-0.47	3.94						
"	1743	4430	8.30	1.25	-1.40	5.25						
"	1753	4600	8.63	0.89	-0.51	3.61						
"	1763	4710	8.59	0.77	-0.30	3.94						
"	1769	4804	8.32	0.90	-0.24	3.61						
"	1775	4900	8.47	-0.75	4.76							
"	1239	4999	8.86	0.72	-0.33	3.43						
"	1269	5100	8.60	0.88	-0.45	3.45						
"	1541	5199	8.77	0.71	-0.19	4.20						
"	1527	5315	8.66	0.77	-0.41	4.26						
"	1521	5435	8.68	0.84	-0.32	3.20						
"	1511	5505	8.26	0.83	-0.33	4.12						
"	1299	5595	8.57	0.80	-0.38	3.88						
"	1559	5700	8.45	0.76	-0.46	5.07						
"	1553	5810	8.55	0.85	0.46	3.95						
"	1245	5903	8.78	0.74	-0.20	3.41						
"	1275	6001	8.41	0.86	-0.36	3.69						
"	1253	6097	8.48	0.80	-0.45	4.26						
"	1583	6190	8.51	0.56	-0.36	3.73						
"	1591	6289	8.55	0.73	0.09	2.89						
"	1571	6410	8.62	0.78	-0.15	3.14						
"	1601	6510	8.78	0.88	-0.72	4.21						
"	1593	6600	8.23	0.95	-0.32	3.36						
"	1609	6700	8.53	0.83	-0.37	3.74						
"	648	6796	8.94	0.72	-0.13	3.32						
"	657	6903	8.81	0.81	-0.61	3.97						
"	1427	6991	8.53	0.82	-0.43	4.18						
"	1423	7108	8.79	0.81	-0.42	3.49						
"	1415	7194	8.72	0.87	-0.41	3.45						
"	1409	7300	8.65	0.86	-0.59	4.24						
"	1401	7387	8.50	0.88	-0.84	5.19						
"	1389	7520	8.64	0.85	-0.45	3.75						
"	1381	7607	8.64	0.74	-0.43	4.37						
"	1373	7693	8.58	0.77	-0.17	3.87						
"	1367	7799	8.37	0.97	-0.66	3.98						
"	1361	7896	8.47	0.94	-0.47	3.17						
"	1355	8002	8.60	0.77	-0.14	3.52						
"	1347	8109	8.60	0.80	-0.47	4.42						
"	1341	8195	8.63	0.82	-0.52	4.37						
"	1335	8291	8.61	0.78	-0.08	3.80						
"	1465	8387	8.55	0.86	-0.94	6.05						
"	1315	8521	8.50	0.81	-0.39	4.06						
"	1309	8617	8.70	0.81	-0.20	3.28						
"	1445	8704	8.25	0.94	-1.01	6.27						
"	1453	8799	8.55	0.71	0.15	3.06						
"	1431	8896	8.46	0.78	-0.29	3.67						
"	1439	8993	8.39	0.82	-0.70	5.15						
"	1305	9110	8.56	0.89	-0.31	3.38						
"	1483	9195	8.67	0.90	-0.75	4.43						
"	1299	9312	8.62	0.81	-0.55	4.25						
"	1479	9394	8.15	0.85	-0.19	3.90						
"	1501	9491	8.36	0.82	-0.50	4.41						
"	1493	9572	8.56	0.81	-0.18	3.88						
"	1503	9668	8.91	0.78	-0.38	3.82						
"	1509	9947	8.49	0.95	-0.78	4.43						
"	1487	10014	8.65	0.75	0.08	2.40						
"	1131	10099	8.05	1.32	-1.13	4.20						
"	1139	10201	8.55	0.78	-0.33	4.57						
"	1161	10297	8.71	0.79	-0.02	2.45						
"	1155	10400	8.61	0.91	-1.02	6.00						
"	1449	10496	7.96	1.30	-1.04	3.93						
"	1177	10578	8.78	0.73	-0.62	4.93						
"	1179	10710	8.12	1.14	-0.74	3.48						
"	1187	10798	8.38	0.89	-0.62	4.52						
"	1193	10587	8.51	0.84	0.41	4.46						
"	1125	11000	8.47	1.27	-1.42	5.37						
"	1199	11111	8.30	1.36	-1.61	5.98						
"	1231	11213	8.45	0.86	-0.46	4.18						
"	1215	11294	8.29	0.89	-0.15	3.67						
"	1223	11391	8.21	0.86	-0.12	3.61						
"	1209	11496	8.43	0.82	-0.47	4.02						

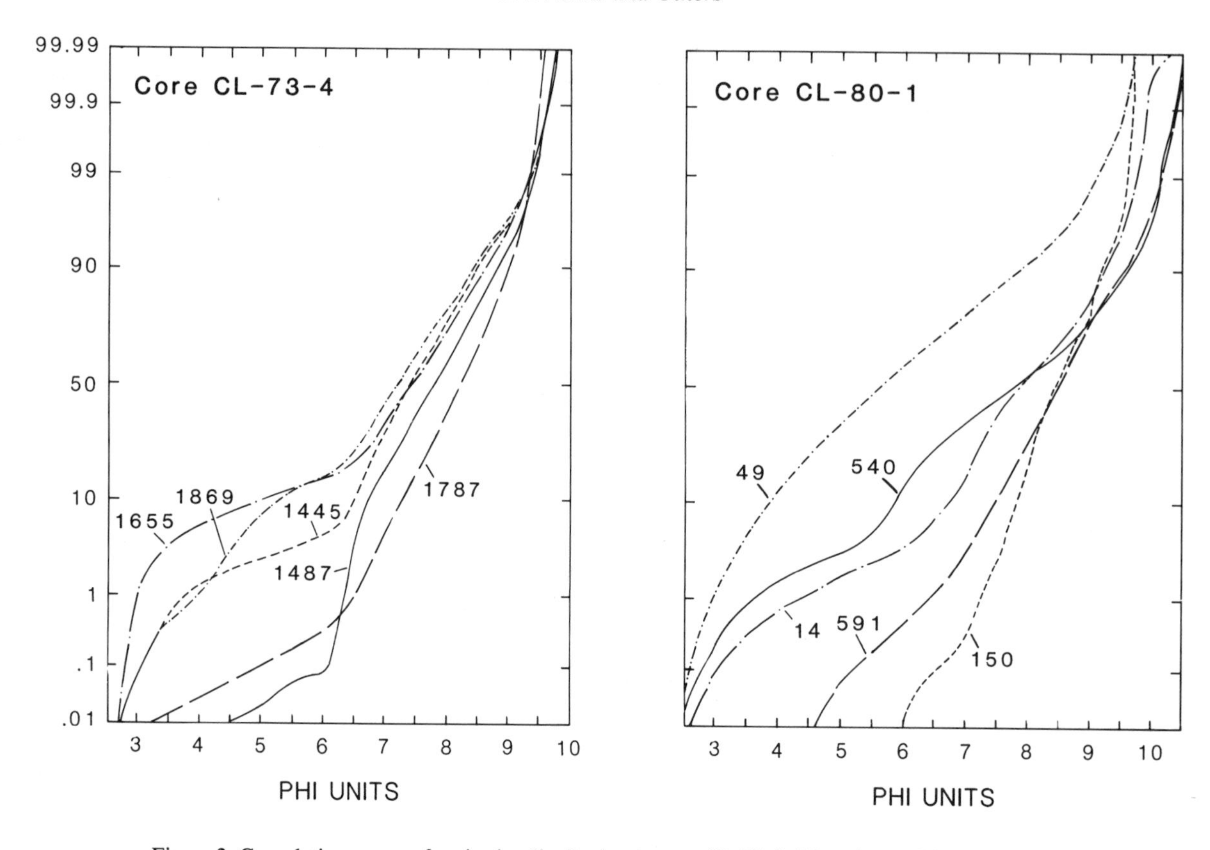

Figure 3. Cumulative curves of grain-size distribution in core CL-73-4 (A) and core CL-80-1 (B). The number beside each curve refers to sample in Table 2. The samples illustrated represent the maximum range of grain-size summary statistical parameters from the two cores. Thus the maximum and minimum mean grain size (X_ϕ), standard deviation (σ_ϕ), skewness (Sk_ϕ), and kurtosis (Kt_ϕ) are represented in the diagrams.

mentary rock at 115 m, the maximum depth reached by the core. Density in the upper 26 m more than doubles, from about 0.6 g cm^{-3} at 0 m to about 1.4 g cm^{-3} at 26 m, with an average density for the interval of about 1.0 g cm^{-3}.

Dry bulk density (*d*) is calculated for the remaining cores except CL-73-7 and CL-80-2. Dry density is highly variable in these cores as compared to core CL-73-4, but generally shows a gradual increase with depth, ranging from 0.1 to 0.6 near the top of cores to 0.5 to 1.4 near the bottom of short cores (Fig. 4A) and to 1.4 to 2.65 near the bottom of the long cores (Fig. 4B through D). Exceptions to the gradual increase with depth are cores CL-73-2, CL-73-3, and CL-80-1, which show abrupt, large-scale changes in density. In core CL-73-2, an abrupt increase in dry density occurs at about 4.2 m where the density increases from about 0.6 to an average of about 1.65 below 4.2 m (Fig. 4A). Between about 7 and 10 m, the density is erratic, owing to the presence of compact clay and gravel (Sims and Rymer, 1975c). Sediment density in core CL-73-3 shows an abrupt increase at about 8.5 m (from ~0.35 to ~1.4); below about 8.5 m, sediment density averages about 1.4 throughout the rest of the core (Fig. 4B). In core CL-73-4, values for bulk density are similar to values below ~40 m (Robinson and others, this volume). Because core CL-73-3 is 69 m long, bulk density is expected to increase with depth similar to the depth-density relationship found in core CL-73-4. Density variation in core CL-80-1 is of a different character, increasing gradually from about 1.0 near the top of the core toa bout 1.3 at 150 m and then increasing sharply below 150 m to about 2.65 at 170 m (Fig. 4D). The abrupt change in density suggests that sediments were subjected to greater overburden pressures than they are at present. In the case of both cores CL-73-2 and CL-73-3, these pressures are equivalent to about 25 to 35 m of sediment when compared to equivalent density in core CL-73-4. The depth-density relationship in core CL-80-1 is difficult to interpret because there is little variation in density between 0 and 150 m. Density at the top of the core is 1.0 compared to 0.6 in core CL-73-4, but does not increase above about 1.35 at depths of less than about 150 m (Fig. 4D). Changes in lithology or in water and organic material content do not explain the lower than expected density in core CL-80-1.

Weight loss on ignition (LOI) was determined for core CL-80-1 (Fig. 5). Losses at 550° and 950°C nominally correspond to the amount of combustible organic carbon and inorganic carbon

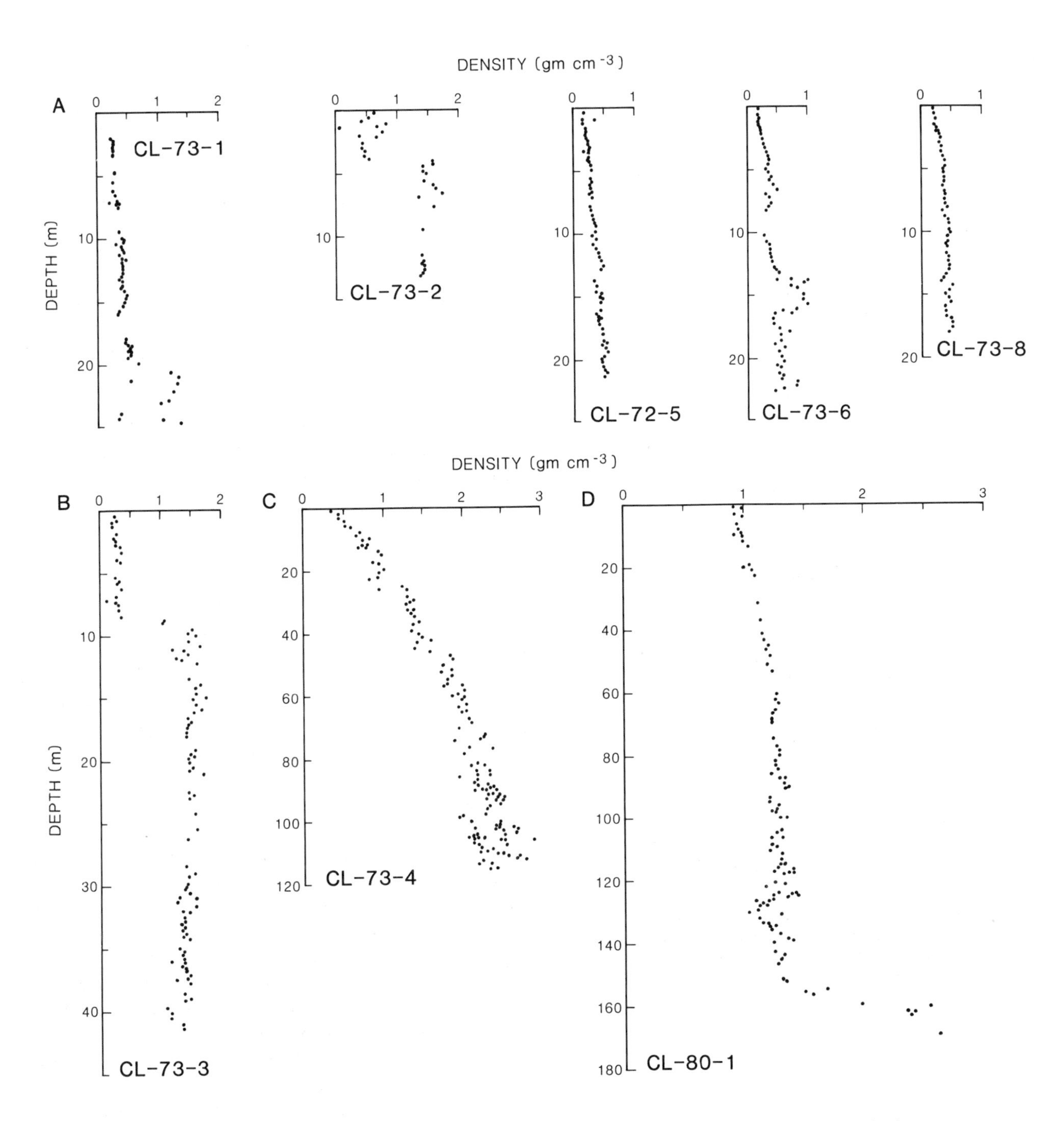

Figure 4. Sediment bulk density for cores from Clear Lake. Data for shorter cores are shown in A. Density depth curves for the longer cores (CL-73-3, -4, and CL-80-1) are shown individually in B through D, respectively.

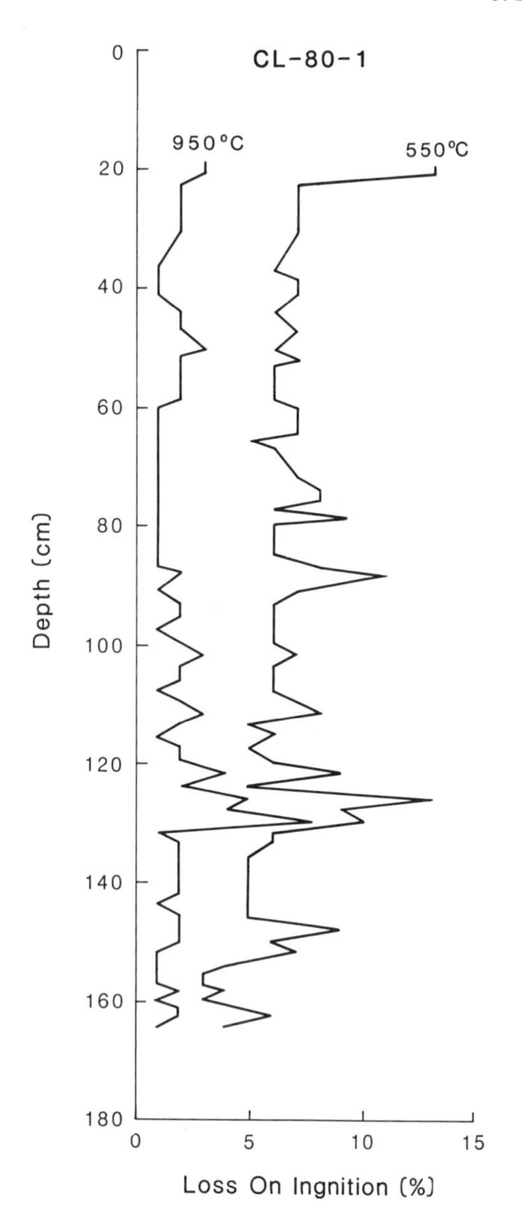

Figure 5. Weight loss on ignition (LOI) for sediment in Clear Lake cores CL-80-1. A, LOI at 550°C represents combustible carbonaceous material of organic origin. B, LOI at 950°C represents combustible carbonaceous material of inorganic origin (may also contain small amount of LOI due to dewatering of opaline silica and dehydroxylization of clay minerals).

species, respectively. Weight loss at 950°C also includes the loss due to the dehydration of opaline silica and the combustion of biogenic carbonate (both of which are probably small). Weight loss at 550°C for core CL-80-1 varies from about 3 to 13 percent, with an average of about 6 percent; weight loss on ignition at 950°C varies from about 1 to 5 percent, and averages about 2 percent (Fig. 5). The greatest variability in LOI is between about 112 and 132 m. Organic and inorganic combustibles reach the highest values at about 126 and 128 m, respectively.

CHRONOLOGY AND CORRELATION

Introduction

In this section we establish the chronostratigraphic framework for the cores from Clear Lake, primarily for core CL-80-1. This framework allows us to interpret sedimentation rates and paleogeographic reconstructions in later sections. Because the study of the cores from Clear Lake is a multidisciplinary one, multiple lines of evidence are used to establish the chronostratigraphic framework for the cores. Radiocarbon analysis is used for the shorter cores. Tephra correlation provides critical age dates mostly for the longer cores and is the primary evidence for the age-depth relationships in core CL-80-1. Oak-pollen frequency variation is the primary evidence used for correlation between deep-sea cores and cores in Clear Lake. We demonstrate that at least two major hiatuses exist in the lower third of core CL-80-1. Evidence from tephra and oak-pollen correlations allows us to interpret the length of the hiatuses and to suggest the amount of section missing because of erosion and nondeposition.

The ages of the sediments and primary correlation data between the cores in Clear Lake come from three sources: (1) radiocarbon dates such as the well-dated sequence of peat and tephra units in core CL-73-7, and correlation of tephra between cores CL-73-7, -6, and -3; (2) correlation of tephra in core CL-80-1 and CL-73-4 with well-dated tephra outside the Clear Lake basin; and (3) oak-pollen frequency diagrams for cores CL-73-4, CL-80-1, and CL-73-7, and their presumed correlation with ^{18}O records from deep-sea cores. These correlations are augmented by data on bulk density, general lithology, and heavy mineral components of tephra.

Radiocarbon dates for Clear Lake were determined on a variety of materials (Fig. 2C). In cores CL-73-6 and -7, dates were determined primarily on peat and peat-rich sediment. In the remaining cores, dates were determined primarily on disseminated carbonaceous material extracted from bulk sediment samples. A few samples yielded enough wood fragments for conventional radiocarbon dating. In Clear Lake, and generally in lacustrine sediments of the western United States, a practical older limit for ^{14}C dates is about 30 ka (Robinson, 1984). The reasons for this limit are poorly known but are generally thought to be due to contamination, bioturbation, mobilization of soluble humic components, and "hard water" effects (Olsson, 1984).

Tephra units are deposited virtually instantaneously over a large area and provide—where laterally traceable—an excellent basis for correlation and dating. Age control by tephra in Clear Lake was achieved by matching trace-element chemical composition of tephra layers in the cores with dated counterparts inside the lake basin, with distant basins in Northern California, and adjacent continental shelf core sites. The trace element analysis and correlation with dated tephra units outside the Clear Lake basin are reported by Sarna-Wojcicki and others (this volume).

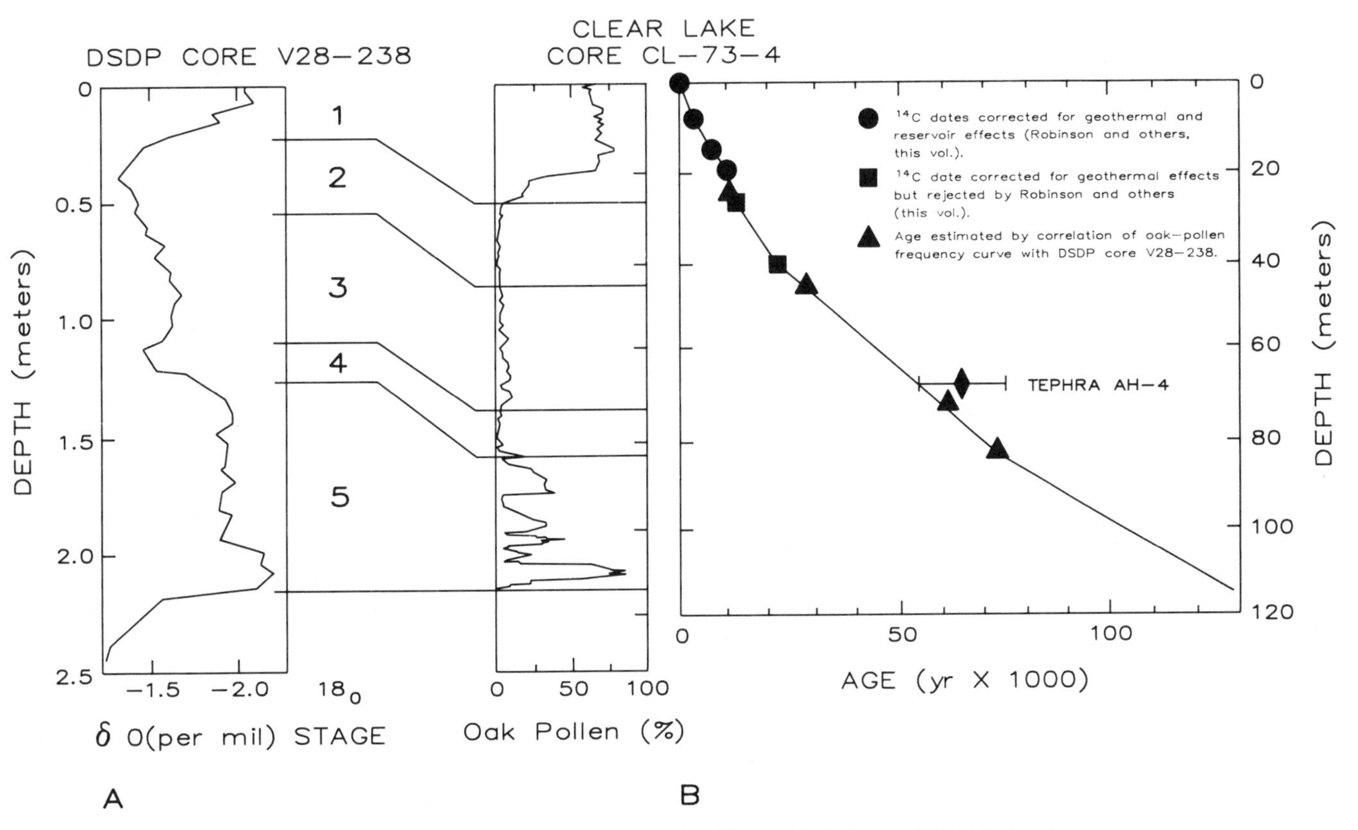

Figure 6. A, Correlation between Clear Lake core CL-73-4 and DSDP core V28-238, showing correspondence between oak-pollen frequency variation and ^{18}O-isotope variation (after Adam, this volume). The age of oxygen isotope stage boundaries is given in parentheses (Pisias and Moore, 1981). Age-depth relationship in core CL-73-4 determined from ^{14}C dates; age of oxygen isotope stage boundaries correlated with oak-pollen zones; and tephra AH-4, which has an estimated age of 55 to 75 ka and is correlated with tephra in Tule Lake, California, by Sarna-Wojcicki and others (this volume).

Pollen analysis is useful for biostratigraphic zonation of the cores and for correlation between cores in Clear Lake. Most of the pollen present in Clear Lake sediments falls into three classes: oak, pine, and TCT (*Taxodiaceae-Cupresaceae-Taxaceae*) (Adam, this volume). This simple and easily recognizable assemblage makes core-to-core correlation possible at Clear Lake, as well as correlations between Clear Lake and the deep sea (Adam and others, 1981; Sims and others, 1981a; Adam, this volume). Correlation of oak-pollen frequency in Clear Lake with ^{18}O abundance in foraminifera from deep-sea sediments suggests a time scale for sedimentation in Clear Lake. The correlation between the curves is strengthened by radiocarbon dates from the upper 41.0 m of core CL-73-4 (Fig. 6), which have been corrected for both reservoir and "old carbon" effects (Robinson and others, this volume). Age estimates from radiocarbon, tephra, and amino acid diagenesis further support the pollen correlations between Clear Lake and deep-sea cores (Sarna-Wojcicki and others, this volume; Blunt and Kvenvolden, this volume).

Amino acid racemization analyses were performed on whole sediment samples from cores CL-73-4 and CL-80-1 (Blunt and others, 1981; Blunt and Kvenvolden, this volume). Amino acid concentration decreases with depth in core CL-80-1, but not uniformly. The D/L ratios of individual amino acids also increase nonuniformly with depth. Because inference of age from D/L ratios is dependent on calibration by radiocarbon dates, we tend to rely on tephra correlations for the parts of the cores older than about 40 ka.

Correlation Based on Radiocarbon Dating

The age of sediment in the upper part of individual cores is generally established by radiocarbon dating. The ^{14}C age data fall into two patterns of age-depth relationships. Age-depth lines from cores CL-73-5, -6, and -7 form one family of line segments that intersect the coordinates of the graph at or near the origin. Age-depth lines from cores CL-73-1, -3, -4, and CL-80-1 form another family of line segments that intersect the abscissa at values of between 4 and 15 ka (Fig. 7). There are insufficient data from core CL-73-8, and no data from cores CL-73-2 and CL-80-2, to determine reliable age-depth relationships. The radiometric age data from each core are evaluated by the use of tephrochronology, statistical analysis of the relationships between dates within a core and type of material dated (Ward and Wilson, 1978; Wilson and Ward, 1981; International Study Group, 1982). Radiocarbon dates on lacustrine sediments commonly contain large unexplainable errors (Robinson, 1984); such errors are most common for large (250 to 500 gm) whole sediment samples. Dates on disseminated carbonaceous material in whole sediment samples from depths greater than about 40 to 60 m in Clear Lake are commonly younger than the samples from lesser depths (Fig. 7). When a series of dates on whole sediment samples are available, they often fail to increase with depth and show stratigraphic reversals.

Core CL-73-7 yields a consistent set of ^{14}C dates (Fig. 2C, 7). The ^{14}C dates are determined on peat, which accumulates rapidly from plant material that is autochthonous, receives CO_2 from the atmosphere, and is commonly derived from single-season plant growth. Furthermore, the data for CL-73-7 has a "zero-year" intercept near the present sediment-water interface. We use this core, therefore, as a standard of comparison for age-depth relationships in Clear Lake. A linear regression on the 12 samples of peat or peaty sediments in core CL-73-7 (Figs. 2C, 7) yields an equation of $Y = (-1{,}184.54 + 15.06X)$ with $r^2 = 0.99$. The linear relationship applies also because compaction is also approximately linear for sediment thicknesses of less than about 30 m.

Core CL-73-7 also contains a distinctive tephra unit, tephra A, at 12.3 m that is radiometrically dated by sediment samples at 12.25 and 12.76 m (Fig. 2A, analyses I-7756 and W-3064). Dates for these two samples are not significantly different and may be combined to yield an age of 17,550 ± 380 yr for the tephra. This tephra unit is correlated to lithologically similar tephra in core CL-73-6 at 12.93 m (Sims, 1976, p. 675) and in core CL-73-3 at 11.3 m.

Radiocarbon dates from core CL-73-6 show an age-depth trend similar to the series of dates from core CL-73-7 (Fig. 7). Four dates between 8.18 and 10.23 m in CL-73-6 range in age from 13,090 ± 210 to 13,890 ± 500 yr (Fig. 8). Because the dates are not statistically different, they are considered to date the interval and are combined by the method of Ward and Wilson (1978, Case I situation) to yield a composite date of 13,280 ± 270 yr. The depth assigned to the date is the midpoint of the interval from which the samples were taken (9.21 m). Because the two lowest dates in core CL-73-6, W-3225 at 19.30 m and W-3226 at 19.91 m, are not statistically different and are from a thin stratigraphic interval, they too are combined to yield a date of 30,780 ± 710 yr (Fig. 7). Tephra A, present at 12.93 m in core CL-73-6, is correlated with the tephra at 12.29 m in core CL-73-7 (Sims and White, 1981), which is dated at 17,550 ± 380 (discussed above). This correlation is confirmed by trace element analysis of the tephra in cores CL-73-7 and CL-73-3, which shows them to have identical trace element compositions (Sarna-Wojcicki and others, this volume), as well as by radiocarbon dates from each core (Figs. 2A, 7). An age of 19,770 yr for the tephra in core CL-73-6 is indicated by linear regression of age-depth data that excludes date W-3199 (Fig. 8), and that is in reasonably close agreement with the radiocarbon age of the tephra in cores CL-73-3 and -7, considering the errors inherent in the dates.

Age-depth relationships determined from radiocarbon dates in core CL-73-5 are limited because only two dates are available. A line segment that passes through the two dates intersects the age axis at about -6,000 yr, which suggests that both dates are at least 6 ka too young if the line segment belongs to the family of lines containing CL-73-6 and -7 (Fig. 7). Thus, an age-depth relationship may be suggested by translating the age-depth line that connects the two dates to pass through the origin of the graph. Such a translation yields alternate age dates of ~16 and ~30 ka. Such age estimates are in agreement with age-depth relationships in core CL-73-6 and -7. The alternate hypotheses are: (1) to accept the age dates and suggest that a line segment between date W-3295 at 11.5 m passes through the origin and that the sedimentation rate increased between about 10 ka to the present; (2) to conclude that the younger date is too old and that the intercept of the line segment is at the origin; or (3) to conclude that the older date is too young and that the intercept of the line segment passes through the origin. There is no evidence in the core to support the suggestion of increased sedimentation rates. However, such a rate (~1.2 mm yr^{-1}) is not out of line for sediment <10 ka in the main basin of Clear Lake (discussed below; see also Table 4). There are no correlatable tephra layers in this core. Decisive interpretation of the age-depth relationships in core CL-73-5 awaits additional data.

Radiocarbon dates in core CL-73-1 are from a 2.16-m interval and are all close in age to one another. The dates as a group do not meet the criteria of the test statistic T of Ward and Wilson (1978) for combination ($\boldsymbol{T} = 12.66 > 9.49 = \chi^2$). However, four of the dates were made on wood; one is bulk sediment. Because bulk sediment dates can have large errors and this date presents a stratigraphic reversal, we exclude it. Exclusion of the date on bulk sediment (I-7754) results in $\boldsymbol{T} = 1.35$; the test for combination is met. Combination of the two dates yields a composite date on wood of 14,140 ± 220 yr for the interval whose midpoint is 22.34 m, and the age-depth relationships for the core are ambiguous without additional data. However, the age-depth relationship for this core is probably similar to that for core CL-73-4, because the two sites are about 1 km apart and are likely to have had similar rates of sediment accumulation.

The upper control point on the age-depth relationships in core CL-73-3 is based primarily on the correlation of the tephra at 11.3 m with the tephra at 12.3 m in core CL-73-7 (Fig. 7) which is well dated at 17,550 ± 380 (see above). The two tephra have identical trace element compositions (Sarna-Wojcicki and others, this volume) and similar lithology. Thus, the age of the tephra at 11.3 m in core CL-73-3 is 17,550 yr. The radiocarbon age of a 17-cm-thick unit of peaty mud between 12.14 and 12.31 m in core CL-73-3 is 17,600 ± 450 yr (Fig. 2C; sample USGS-

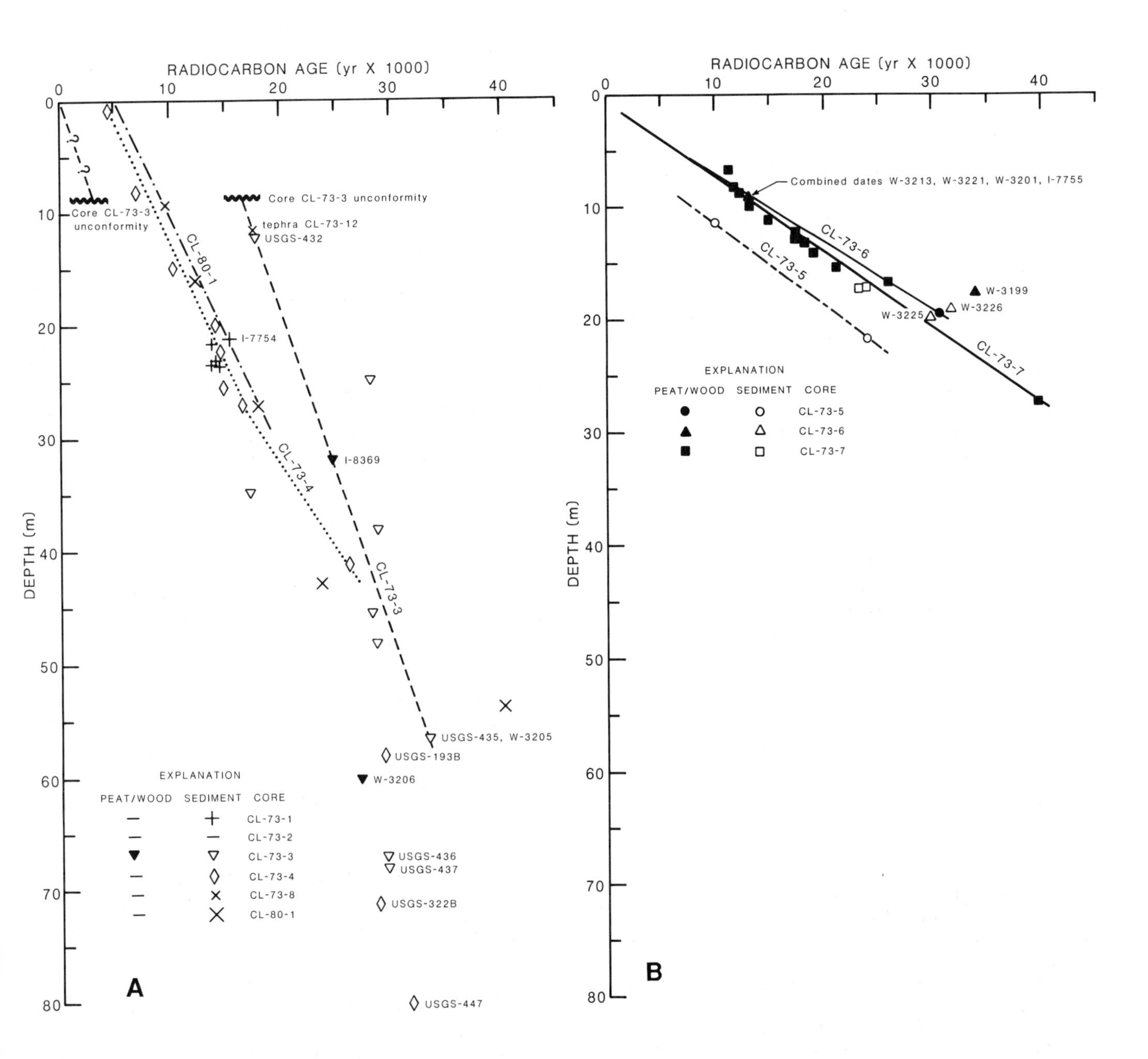

Figure 7. A, Radiocarbon age vs. depth for the cores in Clear Lake. Solid symbols represent analyses performed on peat or wood specimens. Open symbols represent analyses performed on whole-sediment samples. Age-depth lines are based on the interpretation of the dates from single cores. Locally, dates were combined (indicated by arrows) or rejected before determination of the age-depth lines. See text for the discussion of specific dates and cores. Dates identified by their number (Table 4) are mentioned in the text discussion or are identified in this figure for clarity. The age-depth line for core CL-73-3 above ~10 is estimated to show the maximum hiatus on the unconformity that is interpreted to be at ~10 m. B, Age-depth relationship of ^{14}C dates in core CL-73-6 and their bearing on the age estimate of the ash at 12.93 m, which is correlated to tephra in cores CL-73-7 and CL-73-3 (Fig. 2A).

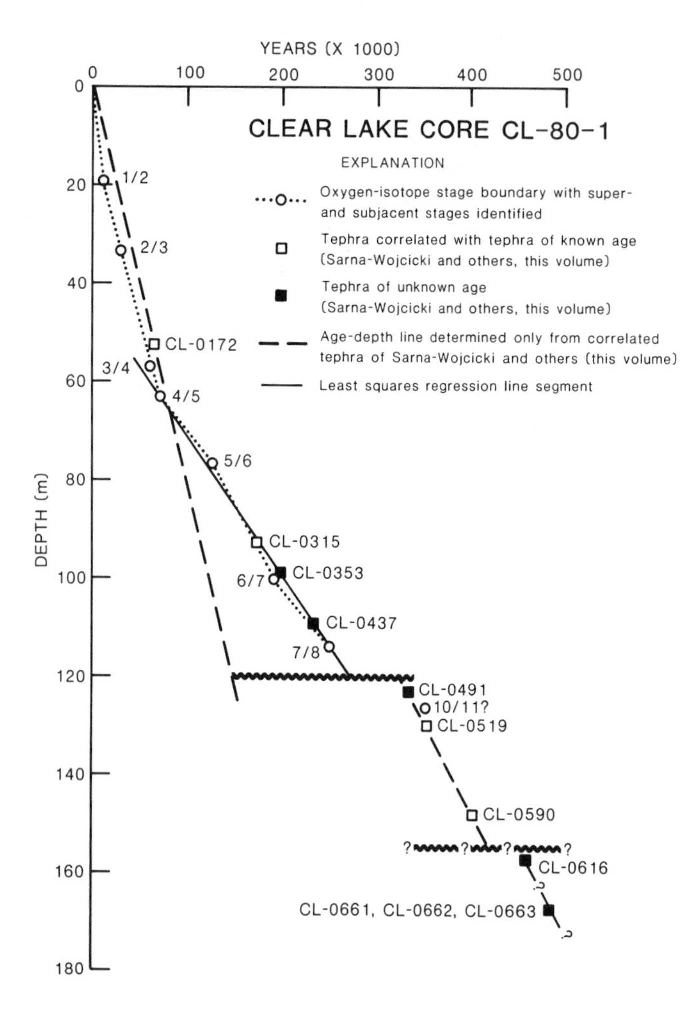

Figure 8. Age-depth relationships in core CL-80-1 showing postulated hiatus at 120 m. Dashed line represents age vs. depth as interpreted using only correlations of tephra in Clear Lake to tephra of known age (open squares) according to Sarna-Wojcicki and others (this volume). Tephras analyzed by Sarna-Wojcicki and others but not correlated to tephra of known ages are also shown (closed squares). Oxygen isotope stage boundaries determined from the correlation of core CL-80-1 with deep-sea core V28-238 (Shackleton and Opdyke, 1973) are plotted (open circles) with ages from Pisias and Moore (1981). Oxygen isotope stage 8/9 boundary is not plotted because it may lie with the hiatus associated with the postulated unconformity at 120 m (see text). A visual "best fit" of the age-depth relationship from the boundary between the oxygen isotope stages 3 and 4 to the postulated hiatus at 120 m, and from the hiatus to oxygen-isotope stage boundary 12/13, is shown by the light solid line segments. The offset between the two line segments represents an estimate of the length of time represented by the hiatus (~40 ka).

432). The remaining dates in core CL-73-3 may be interpreted on the basis of the age of the ash at 11.3 m and on the abrupt increase in bulk density at about 8.5 m, which suggest that the sediment below about 8.5 m has been overlain by a greater thickness of sediment that created greater compaction pressure than is now present. The complete set of dates contains ^{14}C data from bulk sediment analyses and from wood. The scatter in the age dates from core CL-73-3 is large and stratigraphic reversals are present (Fig. 7). We believe that ^{14}C date USGS-432 (17,600 $\pm$ 450) is reasonable because it is supported by a tephra correlation and because the density of the sediment is high.

Because ^{14}C dates on wood are generally considered to be more reliable than dates on whole sediment, only the dates on wood are utilized in the following discussion, with the exception of date W-3206, which deviates greatly from stratigraphically higher dates (USGS-435 and W-3205) and is thus excluded from the discussion of the age-depth relationship in core CL-73-3. Dates USGS-435 and W-3205 are combined, using the criteria of Ward and Wilson (1978) and Wilson and Ward (1981) to yield a date of 33,830 $\pm$ 1170 yr for the interval. This date is considered acceptable even though it is beyond the general practical limit of ^{14}C because the material is wood and the date conforms to the criteria established by Ward and Wilson (1978) and Wilson and Ward (1980).

The dates from core CL-73-4 are critical to the assignment of ages to the pollen zonation of that core (Adam, this volume) and thus have received closer scrutiny. Robinson and others (this volume) aplied reservoir and geothermal corrections to conventional radiocarbon dates at and above 22.1 m in core CL-73-4. Dates below 22.1 m were discarded by them as unreliable owing to possible contamination by bacterial growth during a 4- to 6-yr period of unrefrigerated storage of the cores. However, comparison of dates below 22.1 m, to which the 4-ka geothermal correction is added, yields two dates that are in general agreement with age estimates derived from oak-pollen correlations (Fig. 6). Thus, we accept all radiocarbon data from depths between 22.1 and 41 m that are corrected for reservoir and geothermal effects (Robinson and others, this volume). Below 41 m, however, radiocarbon dates fail to increase in age (Fig. 7). Although poorly understood, such a phenomenon is common in lacustrine sediments, and is generally attributed to bioturbation and contamination by bacterial growth during storage.

Radiocarbon dates from CL-80-1 show an increase in age with depth. However, the upper two dates (USGS-1577 and USGS-1579) are of questionable value because of the probability that the recovered core segments are affected by drilling disturbance and may have been contaminated (Sims and others, 1981b). Correlation of the oak-pollen frequency zonation to the deep-sea cores suggests that the dates are too old for their depth in the core (Figs. 2A, 9). The two lower dates (USGS-1578 and USGS-1580), although in proper sequence, may be too young for their depth of recovery. Sample USGS-1580 at 53.4 m is 0.2 m above tephra CL-0173 at 53.6 m. The tephra is correlated to tephra AH-4 in core CL-73-4 and with tephra from Tule Lake,

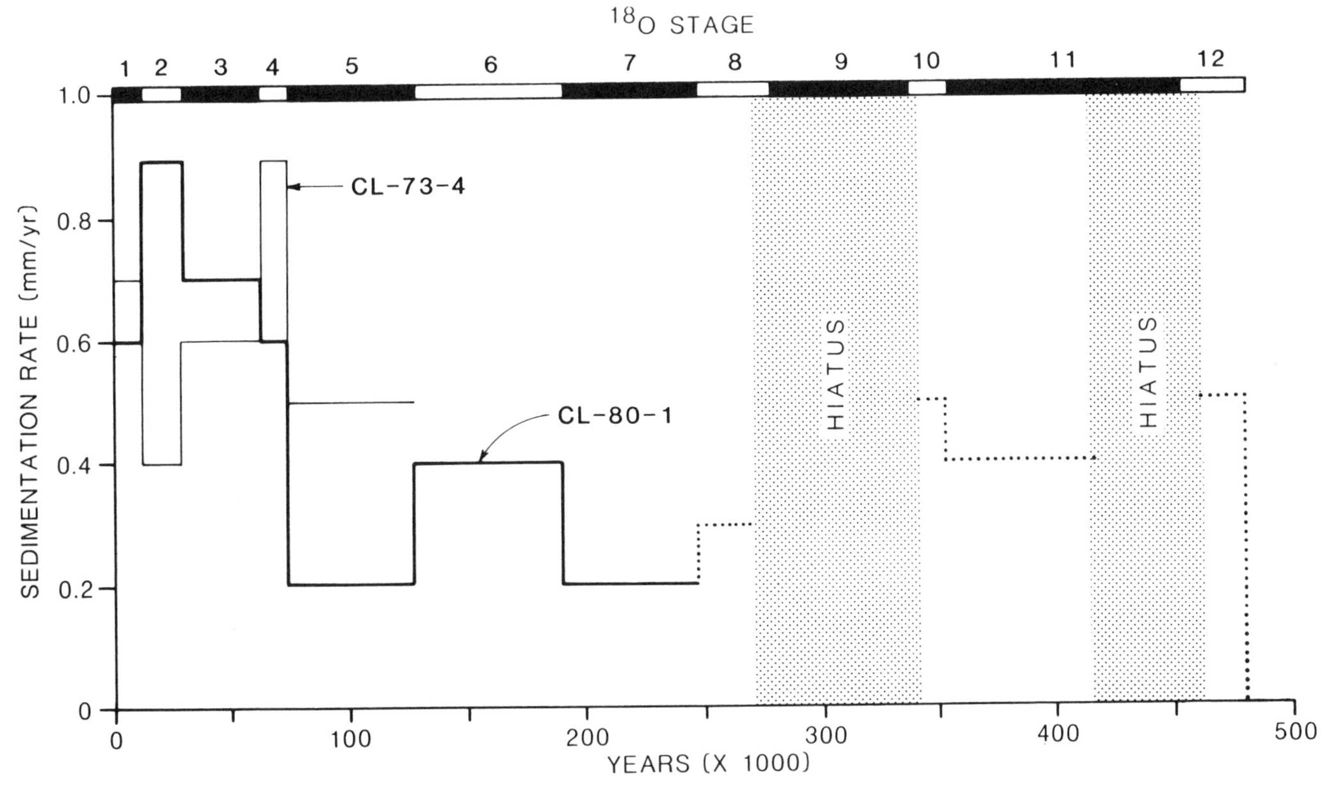

Figure 9. Sedimentation rates for cores CL-73-4 (dotted-dashed line) and CL-80-1 (light line). The rates for oxygen isotope zones 8, 10, 11, and 12 are reconstructed based on data in Figure 8 (see also Table 3 and Fig. 2E). Shaded area indicate the length of postulated hiatuses in the lower part of core CL-80-1. The rates for zones 8 through 12 have large uncertainties. See text for discussion of correlations based on tephra ages and uncertainty in length of hiatuses.

California core T-1, by Sarna-Wojcicki and others (this volume). The age of the tephra is between 55 and 75 ka (Fig. 2A, and discussed below). Dates for the ash in core CL-73-4 suggested by amino acid D/L ratios and by correlation of pollen zonation to deep-sea cores suggest an age of 55 ka (Blunt and others, 1981; Fig. 6). Correlation of oak-pollen frequency in core CL-73-4 (discussed below) suggests an age of 61 ka at 73 m and an interpolated age of ~55 ka at 68.5 m, the position of tephra AH-4, but we reject all ^{14}C dates for core CL-80-1 as estimators of age at the various depths.

Correlation by Tephra

The cores from Clear Lake contain numerous tephra units of which only a few can be confidently correlated between cores and with tephra in other northern California basins and nearby offshore sites. Correlation of tephra layers from one core to another can be hampered by bioturbation (Sims, 1976, p. 676), current-induced sediment mixing, nonhomogeneous distribution of ash (Sarna-Wojcicki and others, 1981), and postdepositional foundering of tephra layers (Anderson and others, 1984). Current-induced sediment mixing is not easily demonstrated by physical data from the sediments. However, present-day wind-induced currents in Clear Lake resuspend sediments almost daily. Phreatomagmatic eruptions were important in the development of the Clear Lake volcanic field but account for only minor volumes of deposits in Clear Lake (Hearn and others, 1981). The small volume of phreatomagmatic eruptions, coupled with the often highly direction nature of such eruptions, may account for the localized deposition of some of the ash beds in Clear Lake.

The correlation of three tephra units from the 177-m-long core CL-80-1 is critical to the interpretation of the chronology and the correlation of the oak-pollen frequency record for that core. The three tephra units lie at about 53, 130, and 148 m, and are designated CL-0172, -0519, and -0590, respectively (Fig. 2A-B). Tephra sample CL-0172 was initially correlated on the basis of lithologic similarity with the tephra at 68.5 m in core CL-73-4 (sample AH-4), 2.5 km to the southeast in the lake (Sims and Rymer, 1981). This correlation is confirmed by Sarna-Wojcicki and others (this volume) who have correlated samples CL-0-0172 and AH-4 with a tephra present in Tule Lake, California (core T-1, sample T-199), and Summer Lake, Oregon (sample DR-24). Sarna-Wojcicki and others have estimated the age of this tephra as between 55 and 75 ka. We use the median age of 65 ka, pending more precise age data.

The other two tephra units in core CL-80-1 at 130 and 148 m are correlated with dated tephra in the Humbolt basin, Tule Lake core T-1, and DSDP cores 36 and 173. Tephra CL-0519 at 130 m is correlated with the Loleta ash bed whose age is estimated at 340 to 350 ka (Sarna-Wojcicki, and others, this volume). Tephra CL-0590 at 148 m is correlated with an ash bed in Tule Lake core T-2 (sample T-296) that lies 11 cm above the Rockland ash bed (sample T-291). Because sample T-296 is stratigraphically near the Rockland ash and there is no evidence of a hiatus, the age of the Rockland ash (400 ka) is used as the age of sample T-296 in the Tule Lake core and by correlation for sample CL-0590 (Sarna-Wojcicki and others, this volume). Three other tephra units, CL-0315, CL-0437, and CL-0441, were analyzed by Sarna-Wojcicki and others (this volume) but could not be correlated with tephra of known age.

A plot of chronologic data in core CL-80-1 using the tephra correlations suggests a major hiatus in the lower half of the core (Fig. 8). This hiatus is postulated on the basis of correlation of tephra in Clear Lake with dated tephra elsewhere in northern California, as established by trace-element composition and by the presence of etched hornblende grains in tephra units at and below 123.7 m in core CL-80-1 (Sarna-Wojcicki and others, this volume). The hiatus lies somewhere between 110.6 and 123.7 m, the levels from which the lowest unetched and highest etched hornblende grains occur. No other physical expression of a break in sedimentary record is seen at this level in the core. On the basis of only the tephra correlations of Sarna-Wojcicki and others (this volume), the hiatus is from about 150 to 320 ka, a period of about 170,000 yr. On the basis of the correlation of oak-pollen frequency variation to deep-sea cores (Fig. 9, and discussion below), the hiatus is interpreted to span a 40,000-yr interval between about 280 and 320 ka.

Correlation of the remaining tephra among cores CL-73-4, CL-80-1, and CL-80-2 is hampered by the large number of tephra in the upper part of CL-73-4, poor recovery in the upper 75 m of CL-80-1, and incomplete coring of CL-80-2 (Fig. 2A). In core 80-2, for example, the ash bed equivalent of CL-0172 in core CL-80-1 and AH-4 in CL-73-4 was sought between 53.4 and 78.2 m. The ash bed was not recovered, nor were any ash beds in core CL-80-2 megascopically identifiable. Thus, tephra correlation of CL-73-4 with CL-80-1 relies on a single tephra, AH-4 in core CL-73-4 and CL-0172 in core CL-80-1.

A number of tephra correlations shown in Figure 2A-B are yet to be tested using the trace element fingerprinting method. A tentative correlation is made using the four megascopically visible tephra between 34.6 and 35.2 m in core CL-73-3 with the tephra at 43.7 m in core CL-73-4 (Fig. 2A-B). The tephra in core CL-73-3 have an aggregate thickness of 6 cm; the tephra in CL-73-4, a thickness of 3 cm. The correlation is suggested because of the age-depth relationships discussed above. Thus, all of core CL-73-3 probably correlates with the upper 45-50 m of core CL-73-4 (Fig. 2A). The tephra at 12.95 m in core CL-73-6 is correlated with a tephra in cores CL-73-7 and CL-73-3 by lithologic similarity, stratigraphic position, and ^{14}C dates (Fig. 2A). Additional work is required to confirm these two correlations.

Correlation by Oak-Pollen Frequency

Cores CL-73-4 and -7, and CL-80-1 were analyzed for their pollen content (Sims and others, 1981a; Adam, this volume; Heusser and Sims, 1983). Biostratigraphic zonation of these cores is based primarily on oak-pollen frequency variation with depth. This variable is useful for biostratigraphic zonation because Clear Lake is located near the ecotone separating oak and pine forests. Climatic variation causes the ecotone to shift, with oak forests being more widespread during warm and dry climatic periods but being largely replaced by pine forests during cool and wet climatic periods (Adam and West, 1983). Because of this sensitivity to paleoclimatic conditions, oak-pollen frequency is also useful for correlation with other paleoclimatic indicators. The similarity between oak-pollen frequency variation in core CL-73-4 and the variation of ^{18}O-isotope abundance in foraminifera from deep-sea sediments (Fig. 6) is remarkable and can be used as an uncalibrated paleoclimate curve (Sims and others, 1981a; Adam and others, 1982; Adam, this volume). Cores CL-73-4 and CL-73-7 have the most detailed records of pollen in Clear Lake, but core CL-80-1 is longer and covers a longer time span than CL-73-4 and -7. Core recovery in core CL-80-1 is less than in CL-73-4, and the pollen analysis is less detailed (Heusser and Sims, 1983). Cores CL-73-4 and -7 span about 130 and 40 ka, respectively, based on radiocarbon analysis, and, in the case of CL-73-4, correlation between pollen and $\delta^{18}O$ curves (Fig. 2E). Core CL-80-1 extends back to about 450 to 475 ka, based on correlation with core CL-73-4, Tule Lake core T-1 (Sarna-Wojcicki and others, this volume), and the correlation of pollen curves from Clear Lake and ^{18}O curves (Fig. 2E). Correlation between cores CL-73-4 and -7 is treated in detail by Adam (this volume).

The pollen analysis of core CL-80-1 is less detailed because the interpretation of the preliminary pollen analysis was impeded by the lack of age data for the core. Early suggestions of the maximum age of the core at about 175 to 200 ka (Sims and others, 1981b; Heusser and Sims, 1983) are now believed to be in error because the chronostratigraphic framework provided by tephra correlations was not available earlier (see above and Sarna-Wojcicki and others, this volume). The correlation of pollen zones in core CL-73-4 with core CL-80-1 is supported by the correlation of tephra C in the two cores (Fig. 2A), radiocarbon dates in the upper part of CL-73-4 and CL-80-1 (Fig. 2C; Robinson and others, this volume), and the correlation of pollen zones from CL-73-4 with the well-dated pollen zones in core CL-73-7 (Fig. 2E). Thus, the pollen zones in the upper 76 m of core CL-80-1 (Heusser and Sims, 1983) are correlated with the 115-m-long sequence of pollen data from core CL-73-4 (Sims and others, 1981a; Adam and others, 1981; Blunt and others,

1981), implying further that ^{18}O stages 1 through 5 are represented in the same interval.

Below 76 m in core CL-80-1, in sediments believed to be older than the base of oxygen-isotope stage 5, variation in oak-pollen frequency is interpreted in a manner similar to that of Sims and others (1981a) and Adam (this volume): high oak-pollen frequency indicates warm and dry climatic conditions and low oak-pollen frequency represents cold and wet climatic conditions. Oak-pollen frequency between 76 and 100 m in core CL-80-1 is low and generally analogous to the frequency of between 20 and 63 m in CL-80-1, and possibly analogous to the frequency found between 20 and 84 m in CL-73-4, which is correlated to stages 2, 3, and 4. Thus, the zone between 76 and 100 m is correlated with ^{18}O stage 6.

Correlation of oak-pollen frequency with oxygen isotope stages below 100 m, the inferred base of stage 6 in core CL-80-1, is complicated by an unconformity postulated to lie somewhere within the interval 110.7 to 123.7 m (Sarna-Wojcicki and others, this volume). The unconformity is postulated on the basis of etched detrital hornblende grains that occur at and below 123.7 m but not in samples at and above 110.7 m, and that are interpreted to represent the effect of subaerial weathering. Age relationships based on tephra at 53, 130, and 148 m, correlated with tephra of known ages of about 65, 350, and 400 ka, respectively, also support the interpretation of a hiatus. Age-depth relationships based solely on the correlated tephra indicate a hiatus but do not define its position or duration (Fig. 9). We tentatively place the unconformity near 120 m because of the change in bulk dry density that occurs between about 118 and 130 m (Fig. 4), and because of a layer of coarse sand that occurs between 118.7 and 119.4 m.

The high oak-pollen frequency between about 100 and 113 m (Fig. 2E) may or may not be part of a continuous depositional sequence owing to uncertainty in the position of the hiatus, although there is no physical evidence for an unconformity in the interval. However, the interval is assigned to oxygen isotope stage 7 because the oak-pollen frequency is generally high but variable, suggesting alternating warm and cold intervals similar to the interpretation of ^{18}O values in deep-sea core V28-238. The time span suggested by the thickness of the interval also has a duration similar to stage 7 in core V28-238 (Fig. 2E).

Placement of the unconformity at about 120 m and assignment of the interval between 100 and 113 to oxygen-isotope stage 7 suggests that the interval between 113 and about 120 m may be assigned to oxygen-isotope stage 8, a stadial. Such an assignment is supported by the low oak-pollen frequency for the interval (Fig. 2E) and age-depth relationships for younger ^{18}O stages. Exact correlation of sediments between 100 and 120 m awaits further investigation of the position and nature of the unconformity within the interval.

Oak-pollen frequency of the interval between 120 m and the bottom of the core (176 m) is variable. It is high from 120 to 131 m, and 147 to 152 m, and low from 131 to 147 m and 152 to 167 m (Fig. 2D). Two correlated tephra layers, CL-0519 and CL-0590, are present at 130 and 148 m. The upper tephra is correlated with the Loleta ash bed whose age is estimated to be between 300 and 400 ka (Sarna-Wojcicki and others, 1987, this volume; A. M. Sarna-Wojcicki, oral communication, 1984, 1987). The lower tephra, at 148 m, is correlated with a tephra closely associated with the Rockland ash in a core from Tule Lake, California, and is assigned an age of 400 ka by Sarna-Wojcicki and others (this volume). The lithology of the interval between 120 and 167 m is uniform fine-grained sediments except between 152 and 158 m, which contains sand and gravelly mud as well as being an interval of poor recovery. The coarse-grained sediments of this interval suggest the presence of an unconformity that we tentatively place at the base of a gravelly mud layer at 155.4 m (Sims and others, 1981b) and just above the interval of low oak-pollen frequency recorded at 156.25 m and below (Heusser and Sims, 1983).

The age of the Loleta ash bed is not well constrained in deep-sea cores or in its type locality; its source area correlation of this part of core CL-80-1 is also poorly constrained. Tephra CL-0590, dated at 400 ka and thought to be well constrained, suggests that at least part of this section of the core is equivalent to oxygen-isotope stage 11 (Fig. 2D). However, the oak-pollen frequency between about 131 and 147 m is anomalously low when compared to other interstadials in the core. If the correlation of tephra CL-0590 is correct, then one may estimate the age of the Loleta ash in Clear Lake based on sedimentation rates of comparable zones (discussed in next section) higher in the core where compaction of the sediment has occurred, and the 400-ka age of tephra CL-0590.

The sedimentation rates of interstadials, such as oxygen isotope stage 11, are generally lower than those of stadials, for example, stages 5 and 7 (Fig. 9, Table 3). The sedimentation rate for interstadial oxygen isotope stages 5 and 7 in Clear Lake is 0.2 mm yr^{-1}. If one applies this rate to the 18-m separation of the Loleta and tephra CL-0590, a time span of 90,000 yr results, which suggests an age of about 310 ka for the Loleta. However, oak-pollen frequency of this interval suggests conditions similar to stage 6 with a rate of 0.4 mm/yr (Table 3). If this rate is applied, the time span between the two tephra is about 45,000 yr and the estimated age of the Loleta would be about 355 ka. Thus, depending on the assumptions, one may correlate the section that includes the Loleta ash and tephra CL-0590 in Clear Lake with the lower part of oxygen isotope stage 9 to the upper part of stage 11, or, alternatively, with only the upper part of stage 11. The data in Clear Lake, however, do not support an age of 380 to 400 ka, as suggested by Sarna Wojcicki and others (1987).

Our preference is to take the age of the Loleta at 355 ka, as suggested by the sedimentation rate for a low oak-pollen frequency. We then correlate the section from 120 to 126 m with isotope stage 10, 126 to 155 m, with about the upper three-fourths of stage 11, and 155 to 167 m with most of stage 12 (Fig. 2E). The oak-pollen frequencies do not show the transition from stage 12 to stage 11, and thus we suggest that part of stage 12 lies in the hiatus associated with the unconformity at 155 m. Better

TABLE 3. SEDIMENTATION RATES FOR CLEAR LAKE CORES CL-73-4 AND CL-80-1

^{18}O Stage	Median Density	Thickness (m)[a] CL-73-4	CL-80-1	Length of Stage (yr x 1000)	Age of Base[b] (ka)	Rate (mm/yr)[c] CL-73-4	Cl-80-1	Climate
1	0.95	20 (7)	19 (7)	11	11	1.8 (0.7)	1.8 (0.6)	Warm
2	1.60	26 (16)	10 (6)	18	29	1.4 (0.9)	0.6 (0.4)[d]	Cold
3	2.10	26 (21)	26 (21)	32	61	0.8 (0.6)	0.8 (0.7)[d]	Cool
4	2.40	12 (11)	8 (7)	12	73	1.0 (0.9)	0.7 (0.6)[d]	Cold
5	2.65	29 (29)	13 (13)	54	127	0.5[e]	0.2 (0.2)[e]	Warm
6		g	24	63	190	---	0.4	Cold
7		g	13	57	247	---	0.2	Warm
8		g	7 (9)[f]	29	276	---	0.2 (0.3)[f]	Cold
9		g	NR[f]	60	336	---	NR[f]	Warm
10		g	6 (8)[f]	16	352	---	0.4 (0.5)[f]	Cold
11		g	28 (47)[f]	101	453	---	0.3 (0.4)[f]	Warm
12		g	14 (16)[f]	27	480	---	0.5 (0.6)[f]	Cold

[a]Thickness of deposits assigned to oxygen isotope stage in core CL-73-4 from Adam (this volume) and for core CL-80-1 modified from Heusser and Sims (1983); numbers in parentheses refer to calculated compacted thickness calculated from a reference bulk dry density of 2.65.

[b]Age of base of stage boundary (after Pisias and Moore, 1981).

[c]First figure is rate if compaction not considered; figure in parentheses is sedimentation rate on compacted basis (see text).

[d]Average for stadial, which is composed of stages 2 through 4, is 0.55 mm yr^{-1}.

[e]Dry bulk density of this and lower intervals is equal to reference compacted density; thus apparent rate is compacted rate.

[f]Deposits assigned to stage in core CL-80-1, but stage may not be completely represented; rate calculated for existing section. Number in parentheses is estimated true thickness or rate for time interval. NR indicates stage not present in this core. See text for discussion of determination of sedimentation rate.

[g]Not penetrated.

constrained correlations, particularly that with stage 12, depend on the better constrained age of the Loleta ash bed, on confirmation of the correlation by Sarna-Wojcicki and others (this volume) of tephra CL-0590 with an age of 400 ka, and on better constraints on the age and correlation of tephra below 155 m in Clear Lake with tephra of known age.

The length of the hiatuses at 120 and 155 m can be estimated from age-depth relationships above and below the unconformity at 120 m. Our preferred correlations below the unconformity at 120 m indicate that stage 9 and probably the topmost part of stage 10 are absent. Thus, the hiatus is approximately 85,000 yr (Fig. 2E). The length of the hiatus associated with the unconformity at 155 m is more poorly constrained. An estimate of its length is largely dependent on the correlation of strata below the unconformity with stage 12. Such a correlation is very uncertain. However, if the correlation is correct, then at least the uppermost part of stage 12 is missing and the length of the hiatus is about 40,000 to 50,000 yr. Placement of the unconformity at about 120 m and assignment of the interval between 100 and 113 to oxygen-isotope stage 7 suggest that the interval between 113 and about 120 m should be assigned to oxygen-isotope stage 8, a full glacial period. Such an assignment is supported by the low oak-pollen frequency for the interval (Fig. 2E) and age-depth relationships for younger ^{18}O stages. Exact correlation of sediments between 100 and 120 m awaits further investigation of the position and nature of the unconformity within the interval.

SEDIMENTATION RATES

Sedimentation rates were calculated for the two most complete long cores, CL-80-1 and CL-73-4 (Table 3). The basis for subdividing the cores is the composite dating scheme discussed earlier and the correlation of pollen zones with deep-sea oxygen isotope stages (Fig. 2E). In both cores the highest apparent sedimentation rate is in oxygen isotope stage 1; in general, both sites have similar sedimentation trends (Table 3). These high sedimentation rates in stage 1, and possibly in stages 2 through 4, as compared with strata in the lower parts of the cores, result from lack of compaction.

Because compaction, and thus sediment density, increases with depth (Fig. 4), sedimentation rates in the upper part of the cores are recalculated to reflect compaction to allow comparison of sedimentation rates for various periods of time. The compacted thickness is determined by the ratio of the average measured density for an interval (d_{int}) to the reference compacted density

(d_c = 2.65) times the thickness of the interval (T_{int}) to yield an estimated compacted thickness (T_c)

$$(d_{int}/d_c) \times T_{int} = T_c.$$

For example, deposits assigned to oxygen isotope stage 1 in core CL-73-4 have a thickness of 20 m. Complete compaction of the upper 20 m yields a thickness of 7.2 m. The time span represented by the interval is 11 ka. Therefore, a corrected sedimentation rate for the interval is 0.7 mm yr^{-1} (Table 3). Similar estimates of compacted sediment thickness are applied to the remaining subdivisions of core CL-73-4, as well as to core CL-80-1. In both cases a density of 2.65 g cm^{-3} is used as the compacted density of the sediments (Table 3). The change in density with depth for core CL-80-1 (Fig. 4) is not a gradual and even relationship (Fig. 4D), which complicates comparison with core CL-73-4. Density in core CL-73-4 increases uniformly from ~0.5 g cm^3 at the surface to ~2 g cm^3 at 60 m. Beyond 60 m, density increases at a lower rate and reaches a maximum of ~2.65 g cm^3 at about 100 m. The density-depth relationship in core CL-80-1 is different in that initial density of surface samples is greater and increases at a lower rate; maximum values are reached only about 150 m in an abrupt increase in density from ~1.4 g cm^3 to ~2.5 g cm^3. The reason for the different density depth relationships is unknown. Therefore, we calculate the compacted thickness from the d_{int}/d_c values used in corresponding intervals in core CL-73-4, and complete compaction is assumed below ^{18}O stage 5 in core CL-80-1.

Uncertainty in the calculated sedimentation rates for stages 8 through 12 arises from uncertainties in the position of unconformities and the span of the hiatuses associated with them, as well as from uncertainty in the ages of tephra used to correlate the oak-pollen curve with oxygen isotope stages. Interpretation of the span of the hiatus associated with the postulated unconformity at 120 m in core CL-80-1 affects the sedimentation rates during oxygen isotope stages 8 and 9. Our interpretation of age-depth relationships (Fig. 9) and correlation of the oak-pollen frequency with deep-sea cores (Fig. 2E) suggests that the lower part of oxygen isotope stage 8, all of stage 9, and the uppermost part of stage 10 are missing because of erosion or nondeposition. The estimated time span represented by strata assigned to stage 8 (113 to 120 m) is about 19,000 yr. Thus the corrected sedimentation rate for stage 8 is about 0.3 mm yr^{-1}. The sedimentation rate for stage 10, a short-lived stadial, is dependent on the assumptions we used to correlate the oak-pollen frequency between 120 and 126 m with the oxygen isotope record. We calculate a rate of 0.5 mm yr^{-1} corrected for the span of the hiatus associated with the unconformity at 120 m and for compaction of sediment.

The stratigraphic interval in core CL-80-1 assigned to oxygen isotope stage 11 is 28 m thick (126 to 154 m). Stage 11 was 101 ka long, which yields a corrected sedimentation rate for the 101-ka-long stage of 0.4 mm yr^{-1}. However, the lower boundary of the stage falls in an interval of poor recovery (Sims and others, 1981b) and thus few pollen analyses (Heusser and Sims, 1983). Gravel and pebbly mud occur between about 153 and 155 m. Thus stage 11 extends from 146.5 to as deep as 155.4 m, but cannot be confidently placed below 151.2 m (Fig. 2E). We place the lower boundary at 155.4 m, the location of a proposed unconformity, below which lie strata tentatively assigned to oxygen-isotope stage 12. This assignment yields a sedimentation rate of 0.4 mm yr^{-1} when corrected for probably missing strata (Table 3).

We calculate a sedimentation rate for strata assigned to stage 12 (Table 3). However, the poorly constrained correlations and ages of tephra suggest that this rate is suspect.

Sedimentation rates in core CL-80-1 (Table 3) vary between 0.9 and 0.06 mm yr^{-1} and may fall into a pattern of alternating increased and decreased rates. Increased rates occur in oxygen isotope stages 1, 3, 4, 6, 8, 10, and 12; a decreased pattern of alternating increased, and decreased sedimentation rates exist (Fig. 9). Increased sedimentation, in even-numbered stages, is associated with cold, wet climates, and decreased sedimentation, in the odd-numbered stages, is associated with warm, dry climatic conditions. The apparent pattern does not hold for stages 2 through 5.

PALEOLIMNOLOGY AND PALEOGEOGRAPHY

Data from the cores in Clear Lake may be used to reconstruct the paleogeography of the lake at various times, and, because most of the sediment in cores from Clear Lake is similar in grain size and composition to that presently accumulating, the depositional regime of the modern lake may be used to infer past conditions. Important to these reconstructions are unconformities, lithology, correlation, and chronology. Clear Lake is presently accumulating fine-grained sapropelic detritus throughout most of the basin. The bottom of modern Clear Lake is soft and fluid to semifluid, which we observed in situ during SCUBA observations. At present, coarse-grained rock detritus and coarse-grained organic-rich detritus are minor constituents. Sand and gravel deposits are commonly found near the mouths of inflowing streams and in the littoral zone of the lake. Coarse-grained, organic-rich deposits may be found along the southern margin of Clear Lake west of the mouth of Kelsey Creek, and in the Highlands arm of the lake near the Cache Creek outlet.

The cores from Clear Lake contain a record of lacustrine sedimentation that encompasses late and much of middle Quaternary time. An uninterrupted record of the past 130,000 yr is interpreted to be present in core CL-73-4 on the basis of correlation with oxygen isotope stages 1 through 5 of deep-sea cores (Fig. 6A; Sims and others, 1981a; Adam, this volume; Heusser and Sims, 1983). Evidence of the hiatus (Fig. 2A-B) present at about 10 m in core CL-73-3, at 12 m in core CL-73-2, and possibly present in core CL-73-1 (Fig. 2A) is not identifiable in core CL-73-4 (Sims and others, 1981a; Adam and others, 1981; Adam, this volume). The sediments in Clear Lake are characterized by organic-rich, bioturbated clay and silt with adjuncts of sand and gravel. Sand and gravel, present in cores CL-73-1, -2, -3, and CL-80-1, represent fluvial-deltaic incursions into the lake basin primarily from the southern part of the drainage basin through the ancestral Kelsey and Adobe Creek drainages. All the Clear Lake deposits accumulated in a shallow, well-mixed lake that remained eutrophic through much of the last 450,000 yr (J. E. Sanger, written communication, 1982; Hopkirk, this volume; Bradbury, this volume). Ash bed correlations coupled with interpretation of oak-pollen frequency variation suggest that the oldest sediments cored in Clear Lake are from core CL-80-1 and are dated about 450 to 475 ka.

Because data used in reconstruction of ancient Clear Lake are more abundant for younger deposits, those reconstructions are discussed from youngest to oldest. Historic shorelines of the lake (ca. 1880) show a slightly larger lake (Fig. 10A). The 1880 shoreline moved lakeward primarily because of landfill, reclamation, and settlement along the lakeshore. The lake reached its present general configuration about 3,500 yr ago with the flooding of the southwest half of the historic main basin. Between 3,500 and 8,000 yr ago the southwest half of the main basin was exposed to erosion (Fig. 10B). Core CL-73-3 records this episode well with a prominent unconformity at about 8 to 10 m. Core CL-80-1 has a less well-expressed unconformity at about 13–15 m (Fig. 2A), interpreted on the basis of sandy cuttings returned during drilling operations. Also at about this time, an arm of the main basin extended northwest into the area now occupied by the shallow intermittent Tule Lake (Fig. 10B). Tectonic deformation is presumed to be the cause of the exposure of the southwest part of the main basin, although such an origin cannot be detailed. Tectonic movements along the Main Basin and Konocti Bay faults (Fig. 10B) may also have altered the northeast shoreline of the main basin and the Oaks and Highlands arms. Sedimentation patterns in cores from these locations show that fine-grained detritus similar to modern sediments accumulated (Fig. 2A). Thus, between about 3,500 and 8,000 yr ago, Clear Lake was a narrow linear body extending from Tule Lake in the northwest into the Highlands and Oaks arms to the southeast (Fig. 10B). This configuration is parallel to the regional tectonic grain and suggests fault control of the shorelines. Analysis of plant pigments and their degradation products in the sediments suggests that the lake was shallow and eutrophic (Jon Sanger, written communication, 1982). There is no evidence to indicate whether Cache Creek or Cold Creek (west-northwest of Tule Lake) was then the outlet of Clear Lake.

Between about 17,500 and 40,000 yr ago the main basin of the lake was more extensive than in the preceding interval (Fig. 10C). Evidence for this comes mainly from lacustrine sediments in core CL-73-3 (Fig. 2A). A delta-like wedge of coarse sand and gravel was present along the southern margin of the main basin (Fig. 2B). The wedge of sediments is interpreted as a coarse-grained facies of the finer grained basinward deposits of cores CL-73-1 and -4 and CL-80-2. The sand and gravel is greenish-gray, which suggests a reducing environment; however, the color may have developed after burial. Evidence that suggests subaerial deposition of the coarse-grained deposits is lacking. If the deposits were subaerially deposited, then the southern margin of the lake for the interval 17.5 to 40 ka would probably have been somewhat more basinward. The Highlands and possibly the Oaks arms of the lake were occupied by marshes during this interval. The lower 20 m of core CL-73-7 consists of peat and peat-rich sediments that were deposited between about 40 and 9.5 ka (Fig. 6). Core CL-73-6 contains fewer peat and peat-rich sediments, which suggests overall deeper water conditions in the Oaks arm than in the Highlands arm between about 36 and 11 ka (Figs. 2A, 8). Plants that make up the bulk of the peat deposits require water depths of 2 m (Sims, 1976). The uppermost peat deposits in core CL-73-7 are about 20 m below the present lake surface, which suggests a rate of subsidence for the past 10,000 yr of about 0.5 mm yr^{-1}. Although tectonic activity may have played a major role in the position of the shoreline in Clear Lake, particularly in the two arms of the lake (Sims and Rymer, 1975a; Rymer, this volume), the effects at any one time are poorly known. Thus the areal extent and the position of the northeast shoreline in this period is conjectural (Fig. 10C).

The data for shoreline positions in Clear Lake that are older than 40,000 yr is derived from cores CL-80-1 and -2 and CL-73-4, the distribution of volcanic rocks (Hearn and others, 1981), and from data on the Kelseyville Formation (Rymer, 1981; Rymer and others, this volume). In the period from about 125 to 250 ka, fine-grained sediment similar to present-day sediment accumulated in the lake (Fig. 10D). Nothing is known of the paleogeography of the two southeast arms of the lake because cores do not penetrate past about 40 ka. However, the areas on either side of the Buckingham peninsula (Fig. 10D) were subjected to numerous maar eruptions (Donnelly-Nolan and others, 1981). The shoreline for this time was drawn to the west of the maar eruptive areas and tentatively south of the Big Valley fault, which may have been intermittently active at this time (Lake County Flood Control and Water Conservation District, 1967; Rymer, 1983, this volume). The length of this period in which a large lake was present in the Clear Lake basin may be attributed to the general quiescence in the Clear Lake volcanic field and uplift that can accompany volcanic activity.

Paleogeographic reconstruction of Clear Lake between about 260 and 340 ka shows that the lake consisted of a narrow northwest-trending area of the present main basin and a southwest-trending arm that extended into the present northern part of Big Valley (Fig. 10E). The western third of the main basin and possibly an elongate area basinward of the present northeast shoreline were exposed (Fig. 10E). Primary evidence for restriction of the lake in the main basin and the extension into Big Valley is the unconformity at 120 m in core CL-80-1 (Fig. 2) and lacustrine sediments interpreted to have been deposited in this time interval (Rymer, 1981; this volume). The presence of the unconformity of this age in core CL-80-2 or in the Kelseyville Formation is unconfirmed. In the absence of such confirmation the lake is interpreted as a single basin elongate parallel to the major structural grain of the region (Fig. 10B). The unconformity at about 120 m lies within a sequence of coarse sand in core CL-80-1. Coarse sand between 118.7 and 119.4 m is interpreted

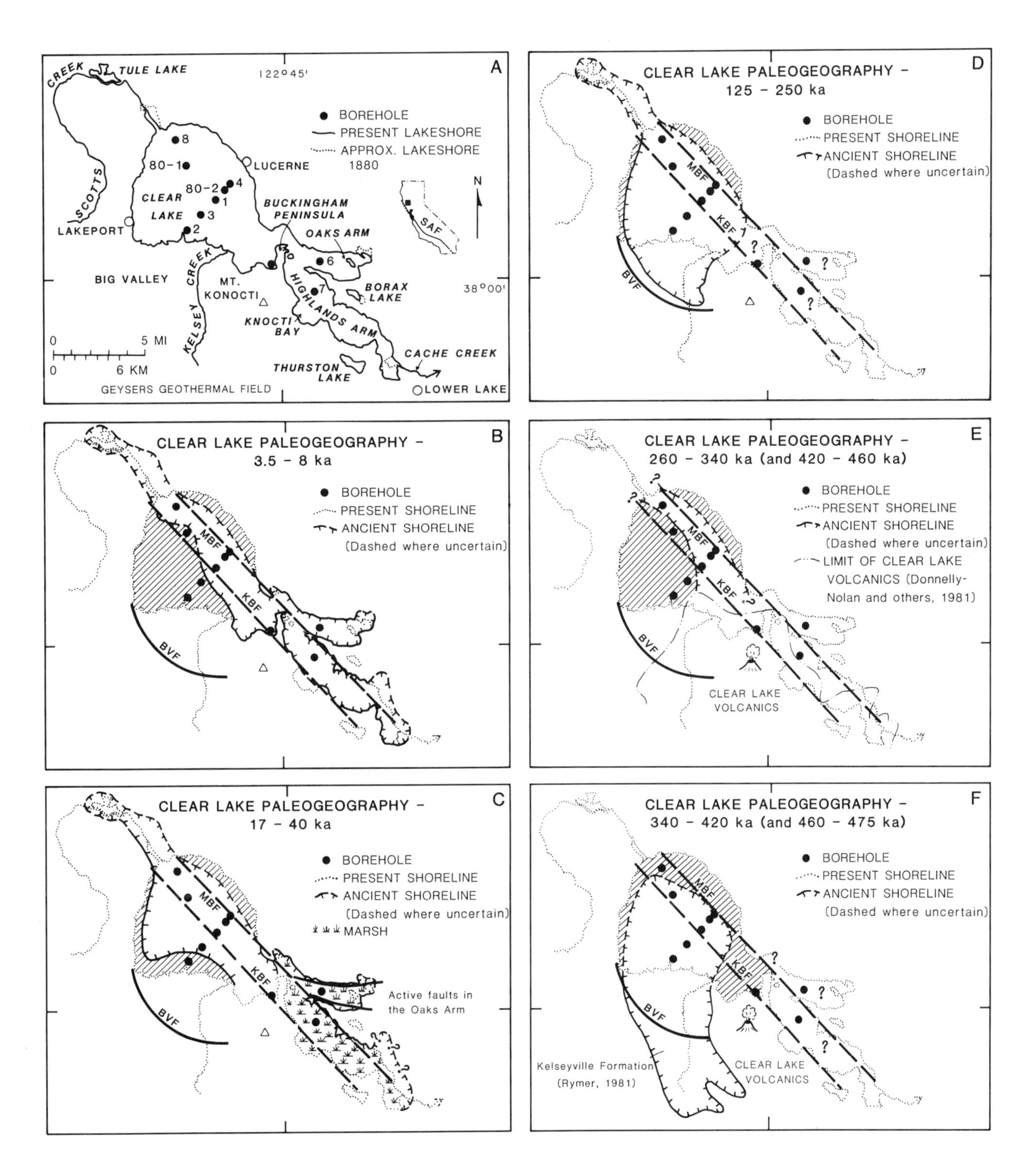

Figure 10. Paleogeographic reconstruction of Clear Lake at present and five earlier times in the Holocene and late Pleistocene. Hatchured areas indicate subaerial exposure of previously deposited lake sediments; dashed areas, questionable. Stippled areas represent prelacustrine bedrock areas now occupied by Clear Lake. Faults that affected the position of the lake shoreline are shown by the following: KBF, Konocti Bay fault and its extension into the main basin; MBF, Main Basin fault from Sims and Rymer (1975a) and its extension into the Highlands Arm of the lake; BVF, Big Valley fault. Unnamed faults in the Oaks Arm of Clear Lake are also shown.

to represent lag deposits on the erosional surface developed on the unconformity (Sims and others, 1981b). A similar unit of sand and gravel occurs in subsurface deposits in Big Valley (Lake County Flood Control and Water Conservation District, 1967). A possible unconformity also exists in the type section of the Kelseyville Formation below its Kelsey Tuff Member, which is estimated to bet ween 247 and 273 ka (Rymer and others, this volume). The Kelsey Tuff Member is absent in cores CL-80-1 and -2.

Between about 340 and 420 ka was another long period in which the basin was occupied by a large and apparently eutrophic lake. Evidence for the size of the lake comes primarily from cores CL-80-1 and -2, and from data derived from the Kelseyville Formation. The sedimentation rate in core CL-80-1 may have been higher than average during the time represented by oxygen isotope stage 10 (Fig. 2E). This increased rate of sedimentation may be a response to increased activity in the Clear Lake volcanic field (Donnelly-Nolan and others, 1981). Such increased volcanic activity was not accompanied by increased tectonic movements that would have displaced the shorelines or caused uplift and erosion of the lake bed. The shoreline for this period is drawn to encompass all of the area underlain by the Kelseyville Formation and most of the main basin of the present lake.

Paleogeographic reconstruction of the lake for the period from about 420 to 460 ka, the time of development of the unconformity at 155 m, is interpreted to be similar to the reconstruction for 260 to 340 ka (Fig. 10E). The extent of the unconformity is unknown because its presence is not confirmed in core CL-80-2 or in the Kelseyville Formation.

For a paleogeographic reconstruction at 460 to 475 ka, only cores CL-80-1 and -2 have long enough records. At this time the Clear Lake volcanic field had few eruptions (Donnell-Nolan and others, 1981), and much of the lacustrine Kelseyville Formation was also deposited around this time (Rymer and others, this volume). Because lacustrine sediments were deposited in Big Valley and the volcanic field was generally quiescent, we suggest an expansion of the lake to encompass the Kelseyville Formation and use the northwest bounding fault of the main basin to restrict the shoreline, similar to the reconstruction for the period between 260 to 340 ka (Fig. 10F). Most of the eruptive rocks in the Clear Lake volcanic field at this time accumulated in the Mt. Konocti and Highlands arm areas, and thus the lake was probably restricted to the area of the main basin and Big Valley.

A search for a pattern in the fluctuation of the position of the shoreline of Clear Lake suggests that there was a gradual shift in the depositional basin. Early in the history of Clear Lake, sedimentation was centered in the area of Big Valley with the formation of the deposits of the Kelseyville Formation. The oldest lacustrine deposits of the Clear Lake basin, excluding those of the Lower Lake Formation, are those of the Kelseyville Formation (Rymer, 1981). Stratigraphic relationships between the Lower Lake Formation and the deposits of Clear Lake are poorly known. Deposition of the Kelseyville Formation in Big Valley probably ended before about 125 to 150 ka. Precise determination of the end of deposition in the Big Valley area awaits palynologic study of the Kelseyville Formation and correlation between it and the younger deposits of Clear Lake. Since about 125 to 150 ka, the main basin underwent pronounced shoreline fluctuation and general enlargement. Since about 40 ka the two arms of the lake were probably formed and were deepened about 10 ka.

The lake was restricted at three different times, about 3.5 to 8, 260 to 340, and about 420 to 450 ka. All these periods were at least partly during warm and dry climatic periods, oxygen isotope stages 1, 9, and 11, respectively. However, restriction of the lake is not seen during other warm and dry climatic periods. On this basis we suggest that tectonism rather than climate change was the major influence in the determination of the size of the lake and position of the shoreline. Sedimentation rates, however, were generally more influenced by climate change than by tectonism and volcanism. This pattern however is not well established owing to uncertainty in the length of hiatuses associated with unconformities, and to decreased precision in the chronology beyond about 260 ka.

CONCLUSIONS

Sediments in Clear Lake consist predominantly of fine-grained, organic-rich clay and silt. Adjuncts to the clay and silt are peat and peat-rich sediments, and sand and gravel. The latter sediment types are temporally and spatially restricted (Table 2, Fig. 4). The lake was shallow and eutrophic throughout its history, as illustrated by diatom assemblages and by algal chlorophyll degradation products (J. E. Sanger, written communication, 1982).

Dating and correlation of the sediments in Clear Lake by ^{14}C tephra and pollen indicates that the lake has existed for at least 450,000 yr. Although Clear Lake remained a shallow eutrophic lake throughout its life, the configuration of its shoreline has varied considerably (Fig. 10). The primary influence on shoreline position is tectonism. The faults that bound the lake are subparallel to the grain of the Coast Range structure and the San Andreas fault system. The faults were active over the life of the lake, but details of their activity are poorly known. The best documented fault activity is from the Highlands and Oaks arms of the lake about 10,000 yr ago, and from the Big Valley fault that cuts the Kelseyville Formation.

Volcanic activity in the Clear Lake volcanic field contributed volumetrically minor amounts of detritus to the Clear Lake sediments. However, volcanic detritus as tephra units are of major importance to intrabasin and interbasin correlation. The control of shoreline position by volcanic activity is limited to areas on the west side of the Highlands arm and on the north and west sides of Mt. Konocti, where the volcanic field bounds Clear Lake and the Kelseyville Formation.

Climate change, although it was pronounced throughout the history of Clear Lake, produced little effect on the paleogeography of the lake. The major hiatuses present in the sedimentary record of the lake do not coincide with changes in climatic conditions. Thus we conclude that climate change affected the lake more subtly than did tectonism and volcanism. Rymer and others (this volume) have similarly concluded that a fluvial incursion corresponding to a hiatus in the Kelseyville Formation was due to tectonism, and not to climate change.

REFERENCES CITED

Adam, D. P., Sims, J. D., and Throckmorton, C. K., 1981, A 130,000-year continuous pollen record from Clear Lake, Lake County, California: Geology, v. 9, p. 373–377.

Adam, D. P., and West, G. J., 1983, Temperature and precipitation estimates through the last glacial cycle from Clear Lake, California, pollen data: Science, v. 219, p. 168–170.

Anderson, R. Y., Nuhfer, E. B., and Dean, W. E., 1984, Sinking of volcanic ash in uncompacted sediment in Williams Lake, Washington: Science, v. 225, p. 505–508.

Becker, G. F., 1888, Geology of the quicksilver deposits of the Pacific slope with an atlas: U.S. Geological Survey Monographs, v. 13, p. 233–270.

Blunt, D. J., Kvenvolden, K. A., and Sims, J. D., 1981, Amino acid dating of sediments from Clear Lake, California: Geology, v. 9, p. 378–382.

Casteel, R. W., and Rymer, M. J., 1981, Pliocene and Pleistocene fishes from the Clear Lake area, *in* McLaughlin, R. J., and Donnelly-Nolan, J. M., eds., Research in the Geysers–Clear Lake geothermal area, northern California: U.S. Geological Survey Professional Paper 1141, p. 231–235.

Casteel, R. W., Adam, D. P., and Sims, J. D., 1975, Fish remains from core 7, Clear Lake, Lake County, California: U.S. Geological Survey Open-File Report 75-173, 67 p.

Casteel, R. W., Beaver, C. K., Adam, D. P., and Sims, J. D., 1977, Fish remains from core 6, Clear Lake, Lake County, California: U.S. Geological Survey Open-File Report 77-639, 154 p.

Casteel, R. W., Williams, J. H., Throckmorton, C. K., Sims, J. D., and Adam, D. P., 1979, Fish remains from core 8, Clear Lake, Lake County, California: U.S. Geological Survey Open-File Report 79-1148, 98 p.

Cohen, A. D., 1973, Petrology of some Holocene peat sediments from the Okefenokee swamp-marsh complex of southern Georgia: Geological Society of America Bulletin, v. 84, p. 3867–3878.

Cook, S. F., Jr., Moore, R. L., and Conners, J. D., 1966, The status of the native fishes of Clear Lake, Lake county, California: Washington Journal of Biology, v. 27, p. 141–160.

Donnelly-Nolan, J. M., Hearn, B. C., Jr., Curtis, G. H., and Drake, R. E., 1981, Geochronology and evolution of the Clear Lake volcanics, *in* McLaughlin, R. J., and Donnelly-Nolan, J. M., Research in the Geysers–Clear Lake geothermal area: U.S. Geological Survey Professional Paper 1141, p. 47–60.

Hearn, B. C., Jr., Donnelly-Nolan, J. M., and Goff, F. E., 1981, The Clear Lake volcanics; Tectonic setting and magma sources, *in* McLaughlin, R. J., and Donnelly–Nolan, J. M., eds., Research in the Geysers–Clear Lake geothermal area: U.S. Geological Survey Professional Paper 1141, p. 25–45.

Heusser, L. E., and Sims, J. D., 1983, Pollen counts for core CL-80-1, Clear Lake, Lake County, California: U.S. Geological Survey Open-File Report 83-384, 28 p.

Hopkirk, J. D., 1974, Endemism in fishes of the Clear Lake region of central California: University of California Publications in Zoology, v. 96, 135 p.

International Study Group, 1982, An inter-laboratory comparison of radiocarbon measurements in tree rings: Nature, v. 298, p. 619–623.

Lake County Flood Control and Water Conservation District, 1967, Big Valley ground-water recharge investigation: Unpublished report to Lake County Flood Control and Water Conservation District, 78 p.

Mahood, A. M., and Adam, D. P., 1979, Late Pleistocene Chrysomonad cysts from core [CL-73-7] Clear Lake, California: U.S. Geological Survey Open-File Report 79-971, 11 p.

Olsson, I. U., 1984, Dating non-terrestrial materials, *in* Mook, W. G., and Waterbolk, H. T., eds., ^{14}C and archaeology, First International Symposium Proceedings, Groningen, 1981: European Study Group on Physical, Chemical, Biological, and Mathematical Techniques Applied to Archaeology, p. 277–294.

Pisias, N. G., and Moore, T. C., Jr., 1981, The evolution of Pleistocene climate; A time series approach: Earth and Planetary Science Letters, v. 52, p. 450–458.

Rieke, H. H., III, and Chilingarian, G. V., 1974, Compaction of argillaceous sediments; Developments in sedimentology, vol. 16: Amsterdam, Elsevier, 424 p.

Rigg, G. B., and Gould, H. R., 1957, Age of Glacier Peak eruption and chronology of post-glacial deposits in Washington and surrounding areas: American Journal of Science, v. 225, p. 341–363.

Robinson, S. W., 1984, Radiocarbon dating of lacustrine deposits: Geological Society of America Abstracts with Program, v. 16, p. 637.

Rymer, M. J., 1981, Stratigraphic revision of the Cache Formation (Pliocene and Pleistocene), Lake County, California: U.S. Geological Survey Bulletin 1502-C, 35 p.

Sarna-Wojcicki, A. M., Wait, R. B., Jr., Woodward, M. J., Shipley, S., and Wood, S. H., 1981, Premagmatic ash erupted from March 27 through May 14, 1980; Extent, mass, volume, and composition, *in* Lipman, P. W., and Mullineaux, D. M., eds., The 1980 eruptions of Mount St. Helens, Washington: U.S. Geological Survey Professional Paper 1250, p. 509–576.

Sarna-Wojcicki, A. M., Morrison, S. D., Meyer, C. E., and Hillhouse, J. W., 1987, Correlation of upper Cenozoic tephra layers between sediments of the western United States and eastern Pacific Ocean and comparison with biostratigraphic and magnetostratigraphic age data: Geological Society of America Bulletin, v. 98, p. 207–223.

Schultz, L. G., 1964, Quantitative interpretation of mineralogical composition from x-ray and chemical data for the Pierre Shale: U.S. Geological Survey Professional Paper 391-C, 31 p.

Shackleton, N. J., and Opdyke, N. D., 1973, Oxygen isotope and paleomagnetic stratigraphy of equatorial Pacific core V28-238; Oxygen isotope temperatures and ice volumes on a 10^5 year and 10^6 year scale: Quaternary Research, v. 3, p. 39–55.

Sims, J. D., 1976, Paleolimnology of Clear Lake, California, U.S.A., *in* Horie, S., ed., Paleolimnology of Lake Biwa and the Japanese Pleistocene: Kyoto, Japan, Kyoto University, v. 4, p. 658–702.

—— , 1982, Granulometry of core CL-73-4, Clear Lake, California: U.S. Geological Survey Open-File Report 82-70, 5 p.

Sims, J. D., and Rymer, M. J., 1975a, Map of gaseous springs and associated faults, Clear Lake, California: U.S. Geological Survey Miscellaneous Field Investigations Map MF-721.

—— , 1975b, Preliminary description and interpretation of cores and radiographs from Clear Lake, Lake County, Claifornia; Core 1: U.S. Geological Survey Open-File Report 75-665, 55 p.

—— , 1975c, Preliminary description and interpretation of cores and radiographs from Clear Lake, Lake County, California; Core 2: U.S. Geological Survey Open-File Report 75-266, 40 p.

—— , 1975d, Preliminary description and interpretation of cores and radiographs from Clear Lake, Lake County, California; Core 4: U.S. Geological Survey Open-File Report 75-660, 178 p.

—— , 1975e, Preliminary description and interpretation of cores and radiographs from Clear Lake, Lake County, California; Core 5: U.S. Geological Survey Open-File Report 75-381, 46 p.

—— , 1975f, Preliminary description and interpretation of cores and radiographs from Clear Lake, Lake County, California; Core 6: U.S. Geological Survey Open-File Report 75-569, 53 p.

—— , 1975g, Preliminary description and interpretation of cores and radiographs from Clear Lake, Lake County, California; Core 7: U.S. Geological Survey Open-File Report 75-144, 58 p.

—— , 1975h, Preliminary description and interpretation of cores and radiographs from Clear Lake, Lake County, California; Core 8: U.S. Geological Survey Open-File Report 75-306, 47 p.

—— , 1976, Preliminary description and interpretation of cores and radiographs from Clear Lake, Lake County, California; Core 3: U.S. Geological Survey Open-File Report 76-208, 123 p.

——, 1981, Deep coring of Quaternary sediment in Clear Lake, California: Geological Society of America Abstracts with Programs, v. 13, p. 106.

Sims, J. D., and White, D. E., 1981, Mercury in the sediments of Clear Lake, *in* McLaughlin, R. J., and Donnelly-Nolan, J. M., eds., Research in the Geysers–Clear Lake geothermal area: U.S. Geological Survey Professional Paper 1141, p. 237–241.

Sims, J. D., Adam, D. P., and Rymer, M. J., 1981a, Late Pleistocene stratigraphy and palynology of Clear Lake, Lake County, California, *in* McLaughlin, R. J., and Donnelly-Nolan, J. M., eds., Research in the Geysers–Clear Lake geothermal area: U.S. Geological Survey Professional Paper 1141, p. 219–230.

Sims, J. D., Rymer, M. J., and Perkins, J. A., 1981b, Description and preliminary interpretation of core CL-80-1, Clear Lake, Lake County, California: U.S. Geological Survey Open-File Report 81-751, 175 p.

Sims, J. D., Rymer, M. J., Perkins, J. A., and Flora, L. A., 1981c, Description and preliminary interpretation of core CL-80-2, Clear Lake, Lake County, California: U.S. Geological Survey Open-File Report 81-7323, 112 p.

Skempton, A. W., 1970, The consolidation of clays by gravitational compaction: Geological Society of London Quarterly Journal, v. 125, p. 373–411.

Ward, G. K., and Wilson, S. R., 1978, Procedures for comparing and combining radiocarbon age determinations; A critique: Archaeometry, v. 20, p. 19–31.

Wilson, S. R., and Ward, G. K., 1981, Evaluation and clustering of radiocarbon determinations; Procedures and paradigms: Archaeometry, v. 23, p. 19–39.

Yamamoto, A., Kanari, S., Fukuo, Y., and Horie, S., 1974, Consolidation and dating of the sediments in core samples from Lake Biwa, *in* Horie, S., ed., Paleolimnology of Lake Biwa and the Japanese Pleistocene: Kyoto, Japan, Kyoto University, v. 2, p. 135–144.

Manuscript Accepted by the Society September 15, 1986

Printed in U.S.A.

Geological Society of America
Special Paper 214
1988

Depositional environments of the Cache, Lower Lake, and Kelseyville Formations, Lake County, California

Michael J. Rymer
U.S. Geological Survey
345 Middlefield Road
Menlo Park, California 94025

Barry Roth
University of California, Berkeley
Berkeley, California 94720

J. Platt Bradbury
Richard M. Forester
U.S. Geological Survey
Denver Federal Center
Denver, Colorado 80225

ABSTRACT

We describe the depositional environments of the Cache, Lower Lake, and Kelseyville Formations in light of habitat preferences of recovered mollusks, ostracodes, and diatoms. Our reconstruction of paleoenvironments for these late Cenozoic deposits provides a framework for an understanding of basin evolution and deposition in the Clear Lake region. The Pliocene and Pleistocene Cache Formation was deposited primarily in stream and debris flow environments; fossils from fine-grained deposits indicate shallow, fresh-water environments with locally abundant aquatic vegetation. The fine-grained sediments (mudstone and siltstone) were probably deposited in ponds in abandoned channels or shallow basins behind natural levees. The abandoned channels and shallow basins were associated with the fluvial systems responsible for deposition of the bulk of the tectonically controlled Cache Formation. The Pleistocene Lower Lake Formation was deposited in a water mass large enough to contain a variety of local environments and current regimes. The recovered fossils imply a lake with water depths of 1 to 5 m. However, there is strong support from habitat preferences of the recovered fossils for inferring a wide range of water depths during deposition of the Lower Lake Formation; they indicate a progressively shallowing system and the culmination of a desiccating lacustrine system. The Pleistocene Kelseyville Formation represents primarily lacustrine deposition with only minor fluvial deposits around the margins of the basin. Local conglomerate beds and fossil tree stumps in growth position within the basin indicate occasional widespread fluvial incursions and depositional hiatuses. The Kelseyville strata represent a large water mass with a muddy and especially fluid substrate having permanent or sporadic periods of anoxia. Central-lake anoxia, whether permanent or at irregular intervals, is the simplest way to account for the low numbers of benthic organisms recovered from the Kelseyville Formation. Similar low-oxygen conditions for benthic life are represented throughout the sedimentary history of Clear Lake. Water depths for the Kelseyville Formation of 10 to 30 m and 12 m near the margins of the basin are inferred both before and after fluvial incursions. These water-depth fluctuations cannot be correlated with major climatic changes as indicated by pollen and fossil leaves and cones; they may be due to faulting in this tectonically active region.

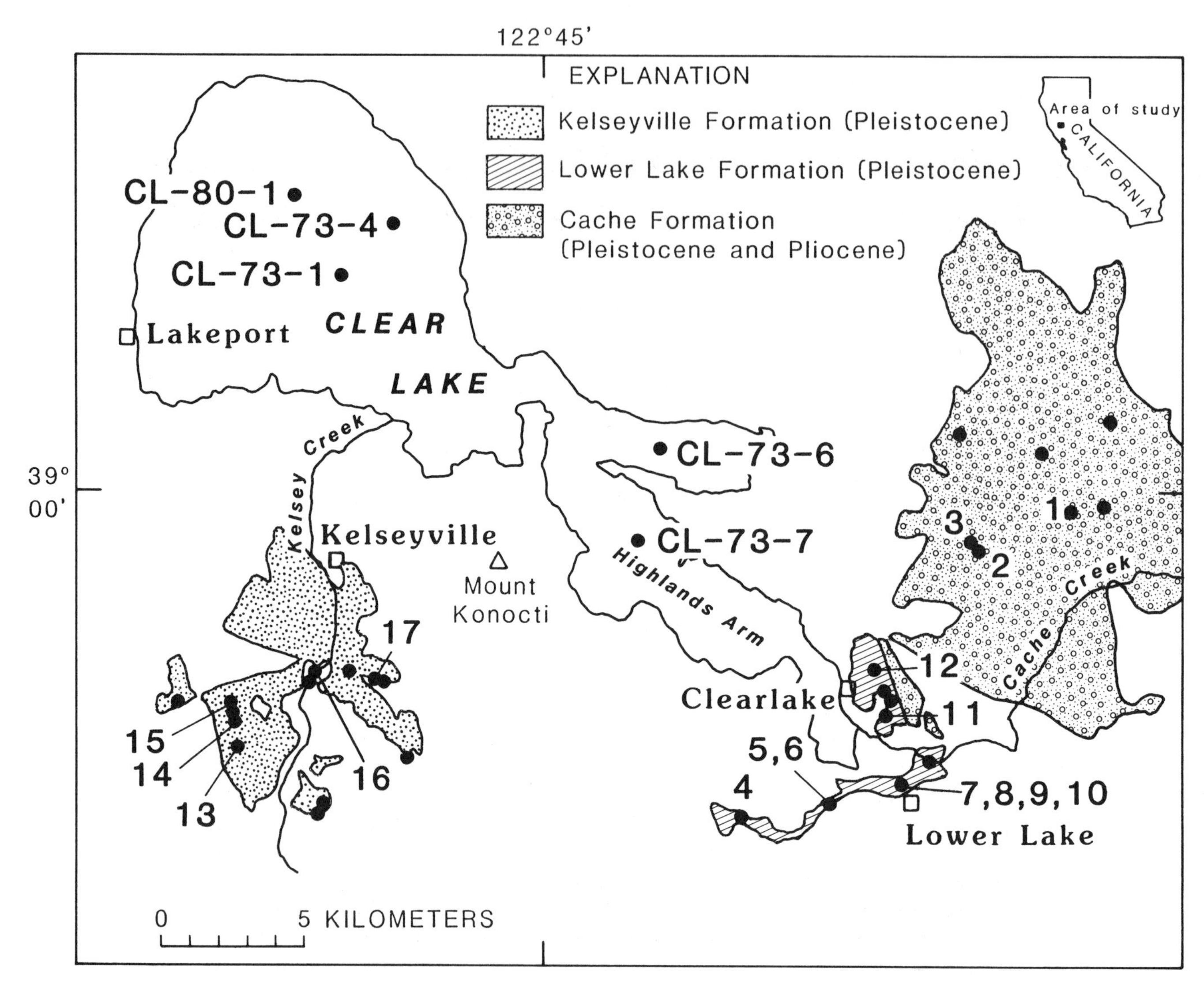

Figure 1. Fossil mollusk, ostracode, and diatom localities in Cache, Lower Lake, and Kelseyville Formations, Lake County, California. Labeled fossil localities represent samples used in this report. Unlabeled sites represent barren samples. Locality descriptions are given in Appendix A.

INTRODUCTION

The Cache, Lower Lake, and Kelseyville Formations consist of late Cenozoic nonmarine sedimentary rocks located east and south of Clear Lake (Fig. 1). These formations comprise a late Cenozoic record of basin evolution and variations in depositional environments for the region. Faulting, erosion, and lack of continuous outcrops have prevented a complete analysis of these environments through space and time.

Previous geologic investigations (Becker, 1888; Anderson, 1936; Brice, 1953; Lake County Flood Control and Water Conservation District, 1967; Rymer, 1978, 1981) concentrated on the stratigraphy and age of the formations and made generalized interpretations on their depositional environments, as summarized below. The first geologic study of the area (Becker, 1888) suggested that the Cache and Lower Lake Formations (usage of Rymer, 1981, and this report) were lacustrine in origin. A more detailed study by Anderson (1936) stated that deposits in the Cache Formation were primarily formed in fluvial environments, whereas those in the Lower Lake Formation were predominantly of lacustrine origin. According to Brice (1953), the Cache Formation was deposited in a subsiding tectonic basin with major fluvial deposition and only minor lacustrine intervals. The Lower Lake Formation was inferred by Brice to represent deposition in a large lake in which chemical, organic, and volcaniclastic sediments accumulated. The Kelseyville Formation was referred to as representing lacustrine and flood-plain environments (Lake County Flood Control and Water Conservation District, 1967). According to Rymer (1978, 1981), the Kelseyville was deposited in a large lake, and is probably correlative with lacustrine sediments

TABLE 1. LOCALITIES, IDENTIFICATIONS, AND QUANTITIES OF MOLLUSK REMAINS RECOVERED IN THE CACHE, LOWER LAKE, AND KELSEYVILLE FORMATIONS

Locality	Genus and Species	Material Recovered
Cache Formation		
1	Valvata humeralis Say	9 partial shells and fragments
	"Hydrobia" cf. "H." andersoni (Arnold)	8 intact, 18 partial shells
	Gyraulus parvus (Say)	1 partial shell
	Genus indeterminate	1 fragment, 2 mm across
2	"Hydrobia" cf. "H." andersoni (Arnold)	8 intact shells, 24 fragments
	Gyraulus parvus (Say)	1 partial shell
Lower Lake Formation		
7	Anodonta cf. A. wahlamatensis (Lea)	4 fragments
	Pisidium compressum Prime	17 valves, 3 fragments
	Valvata humeralis Say	45 shells
	"Hydrobia" cf. "H." andersoni (Arnold)	1 immature shell
	Lymnaea, species indeterminate	2 immature shells--1 intact and 1 partial
	Gyraulus parvus (Say)	4 intact, 1 partial shell
	Planorbella cf. P. tenuis (Dunker)	8 fragments
	Physa cf. P. virgata Gould	4 fragments
11	Valvata humeralis Say	1 shell
Kelseyville Formation		
13	Valvata virens Tryon	5 partial shells and fragments
16	Valvata virens Tryon	12 intact, 7 partial shells
	"Hydrobia" cf. "H." andersoni (Arnold)	6 intact, 3 partial shells
	Helisoma newberryi (Lea)	8 partial shells and fragments
17	Valvata virens Tryon	2 nuclear tips, 2 partial shells

beneath Clear Lake. Casteel and Rymer (1975, 1981) concluded that fossil fish from the Cache, Lower Lake, and Kelseyville Formations represent growth and deposition in mostly lacustrine environments.

This report describes the depositional environments for the Cache, Lower Lake, and Kelseyville Formations in light of new evidence. Our paleoenvironmental interpretations are based on sedimentologic and stratigraphic studies as well as known habitat preferences of fossil mollusks, ostracodes, and diatoms. Major emphasis is placed on fine-grained, fossiliferous deposits, because these are the most helpful in determining the nature of ancient lacustrine basins. Also, the possibility of succession or correlation of these three formations with Clear Lake deposits is discussed. (More complete descriptions of Clear Lake and its depositional environments are given in Sims and others, this volume; Goldman and Wetzel, 1963; Horne and Goldman, 1972; Bradbury, this volume; and Forester, this volume.)

METHODS AND DEFINITIONS

The fossils used in this study are listed in Tables 1, 2, and 3. The fossils were collected during an investigation of the stratigraphy of the three formations (Rymer, 1981). Other small sets of samples were obtained in subsequent visits to collect diatom and ostracode samples from the Lower Lake and Kelseyville Formations and Clear Lake. Mollusk and ostracode materials were collected sporadically at uneven stratigraphic intervals where these organisms were visible in the field; thus they do not represent a comprehensive stratigraphic sample. The diatom samples are from rocks collected for the stratigraphic analysis, complemented by the later collections. Figure 2 shows the stratigraphic distribution of samples used in this report and also indicates the type of organism found at each site. For each site where either ostracodes or diatoms are indicated, both types were sought; the mollusk sample set was not necessarily examined for ostracodes or diatoms. Our paleoenvironmental interpretations may not be applicable to an entire formation.

The inferred and known habitats of the identified diatoms and ostracodes are summarized in Tables 3 and 4, respectively; habitats of mollusks are given in the text. The information on habitat characteristics is based on the available literature (mollusks: Burch, 1975; Hannibal, 1912; Taylor, 1966, 1981, 1986; ostracodes: Delorme, 1970a, b, 1971; Furtos, 1933; diatoms: Cholnoky, 1968; Rosen and others, 1981; and references cited in Bradbury, this volume), and on our observations of modern and fossil species collections. The habitat and zonal subdivisions discussed in the text are defined as follows: fresh-water environments include all waters with total dissolved solids (TDS) of less than or equal to 3,000 parts per million (ppm) (Forester, 1983). Marginal lacustrine environments include marshes, swamps, and

TABLE 2. LOCALITIES, IDENTIFICATIONS, AND QUANTITIES OF OSTRACODES RECOVERED IN THE CACHE, LOWER LAKE, AND KELSEYVILLE FORMATIONS*

Taxon	Cache Formation		Lower Lake Formation					Kelseyville Formation		
	1	2	5	8	9	10	11	13	14	15
Cypridopsis vidua (Mueller)	V	P	P	V	R					
Candona n. sp. 1	C		C	A	V	C	C	R	R	P
Candona n. sp. 2			P	R	P	V				
Cyclocypris serena (Koch)			A	V	C	R				
Limnocythere itasca Cole			P	C	C	V			j	
Physocypria globula Furtos			P							
Ilyocypris gibba (Ramdohr)				P		V				
Ilyocypris bradyi Sars					V					
Heterocypris incongruens? (Ramdohr), complex					R	V				
Limnocythere paraornata? Delorme					R	C				
Cypricercus deltoidea (Delorme)						C				
Darwinula sp.						C				
Candona sp., indet. juveniles		j								
Physocypria sp., indet. juveniles								j		

*Presence and relative abundance of a species at individual sample sites indicated by letters. Abundance data based on total counts of whole adult valves in equal volumes of sample and follows the scheme:

Abundant ≥ 56 Common = 9-20 Rare ≤ 3
Very common = 21-55 Present = 4-8 "j" = juveniles only.

wetlands, which are often active groundwater recharge or discharge sites. These marginal lacustrine environments often occur in basins where the boundary between land and water is poorly defined, and where water depths are shallow (usually less than 3 m) and vary seasonally. If water does flow in these environments, it moves only slowly. Extensive emergent as well as submergent vegetation is often found in such lacustrine environments. The upper littoral zone of some lakes possesses some of the same physical, and to a lesser extent, chemical, characteristics as the marginal lacustrine environments. The subdivisions of the littoral zone are based on the discussion and subdivision of Wetzel (1975, p. 358). The fluvial habitat includes both channelized flow environments and assorted aquatic environments that border the channels. The aquatic environments that border channels resemble the marginal lacustrine category, but have a greater flow component and are subject to periodic flooding. Both flowing water and flooding tend to destroy mollusk and ostracode habitats, and contain fewer taxa than other environments.

CACHE FORMATION

The Cache Formation, which crops out east of Clear Lake (Fig. 1), is composed of sandstone, siltstone, and conspicuous, resistant conglomerate (Rymer, 1981). The Cache sediments were deposited in a basin bounded by the Bartlett Springs fault zone on the northeast, the Cross Spring fault on the west, and various smaller faults within the San Andreas fault system on the southwest and southeast (Brice, 1953; B. C. Hearn, Jr., J. M. Donnelly-Nolan, and F. E. Goff, unpublished data; M. J. Rymer, unpublished data; McLaughlin and others, 1985). The origin of the basin is probably the result of movements along these faults. Fossil horse remains from the lower part of the Cache Formation, identified by C. A. Repenning, suggest an age range for the Cache of approximately 1.8 to 3.0 Ma (Rymer, 1981). A local basalt flow conformably overlying the Cache has a K-Ar age of 1.66 ± 0.10 Ma (Donnelly-Nolan and others, 1981).

In general, the Cache Formation is folded into a northwest-plunging anticline (Brice, 1953). Strata along the southwest limb of the anticline were used to show that the formation is at least 1,600 m thick (Rymer, 1981). However, a more recent study of the area indicates a much greater thickness for the Cache. We used an average dip of 30° for strata along the axis of the anticline to estimate an approximate thickness of 4,000 m for the southern two-thirds of the formation. The stratigraphic thickness of the northern one-third of the Cache is unknown and thus is not included; this figure represents only a minimum for the total thickness of the formation. The newly calculated thickness may include beds repeated by faulting, although such faults are unknown in this area.

Bedding characteristics within the lower part of the Cache Formation, present in the southeast part of the exposed Cache and at the type locality along the North Fork of Cache Creek, are suggestive of fluvial, predominantly alluvial fan and debris flow deposition. Conglomerates in this section commonly contain

TABLE 3. PERCENTAGES AND HABITAT PREFERENCES OF DOMINANT AND COMMONLY OCCURRING DIATOMS FROM OUTCROP SAMPLES OF THE CACHE AND LOWER LAKE FORMATIONS*

	Habitat Preference	Cache Formation		Lower Lake Formation							
		2	3	4	5	6	8	9	10	11	12
Fragilaria construens v. *venter* (Ehr.) Grun.	P	68			33	40	20	36		1	
Cocconeis placentula (Ehr.)	P	6	31	2	+	+	+	3	1	+	
Epithemia adnata (Kutz.) Breb	P	5	1	2	4	+	+	+		9	
Synedra ulna (Nitz.) Ehr.	P, T	5	11	4	+		+	1	1	1	
Melosira italica (Ehr.) Kutz.	T	4			10	+	1	+			
Gomphonema subclavatum (Grun.) Grun.	P	3									
Fragilaria construens v. *subsalina* Hust.	P, T	3			1	+		1			
Rhoicosphenia curvata (Kutz.) Grun.	P		23		1			+			
Achnanthes lanceolata Breb.	P	+	6					+	13		
Achnanthes hauckiana Grun.	P		3								
Amphora ovlis (Kutz.) Kutz.	P, M	+	3	4	1		+	4	1	9	
Gomphonema parvulum Kutz.	P		3				+	1	1		
Nitzschia frustulum (Kutz.) Grun.	T, P, M		3	2	1		+	1	1	2	
Fragilaria construens (Ehr.) Grun.	T, P				4	57	60	23		1	
Fragilaria brevistriata Grun.	T, P	1		30	39	1	15	10	+	3	
Melosira distans (Ehr.) Kutz.	T					+		5	51		+
Navicula pseudolanceolata	M							4	1		
Gomphonema angustatum (Kutz.) Rabh.	P						+		12		2
Nitzschia communis Rabh.	P, M, T								8		2
Rhopalodia gibba (Ehr.) O.Mull.	P	+			2		+	+		2	
Navicula oblonga (Kutz.) Kutz.	M			26	1	+	+	+	+	15	
Rhopalodia gibberula (Ehr.) O.Mull.	P		1	2	+	+	1	2		6	
Cymbella mexicana (Ehr.) Cl.	P			12				+		1	
Anomoeoneis sphaerophora (Ehr.) Phitz.	M	+		10	+	+	+	+	+	8	
Navicula peregrina (Ehr.) Kutz.	M	+	2		+	+	+	+		14	
Hantzschia amphioxys (Ehr.) Grun.	A	+									46
Naviculamutica Kutz.	A										38
Navicula cuspidata (Kutz.) Kutz.	A, M				1	+	+			1	8
Diatom Concentration (valves/mm of 150-mm-wide traverse)		72	6	1	91	1000	318	135	69	14	6

*Habitat preferences are indicated for the following categories: Tychoplanktonic, benthic Periphytic, benthic Motile, and Aerophilic.

+Indicates that species is present in sample, but outside of area counted on slide.

cobbles, pebbles, and locally boulders and are characterized by massive to thick-bedded, horizontally stratified beds up to 6 m thick. Basal contacts are sharp and commonly erosional. Sandstone and siltstone beds in the same area are characterized by poorly sorted, dominantly massive to weakly horizontally stratified beds up to 2 m thick.

Bedding characteristics within the upper part of the Cache Formation, especially on the southwest limb of the anticline, are suggestive of predominantly low-energy braided stream deposition. Conglomerates high in the section generally contain pebble- and, locally, cobble-sized clasts and beds up to 2 m thick that are thick-bedded, moderately sorted, and horizontal to trough cross-stratified. Fining-upward sequences and local channels are common; basal contacts are distinct, rarely transitional, and locally erosional. Sandstone and siltstone units in this part of the section are characterized by moderately sorted beds up to 0.5 m thick that are horizontal to cross-stratified.

Quiet-water deposits occur locally within the inferred fluvial deposits of the Cache Formation. These quiet-water deposits are extremely limited stratigraphically and areally. They vary widely in lithology, which includes peat, diatomite, and mudstone. Fossils, including mollusks, ostracodes, diatoms, and abundant plant remains, were recovered from three sites in the middle and upper parts of the Cache Formation (Figs. 1, 2).

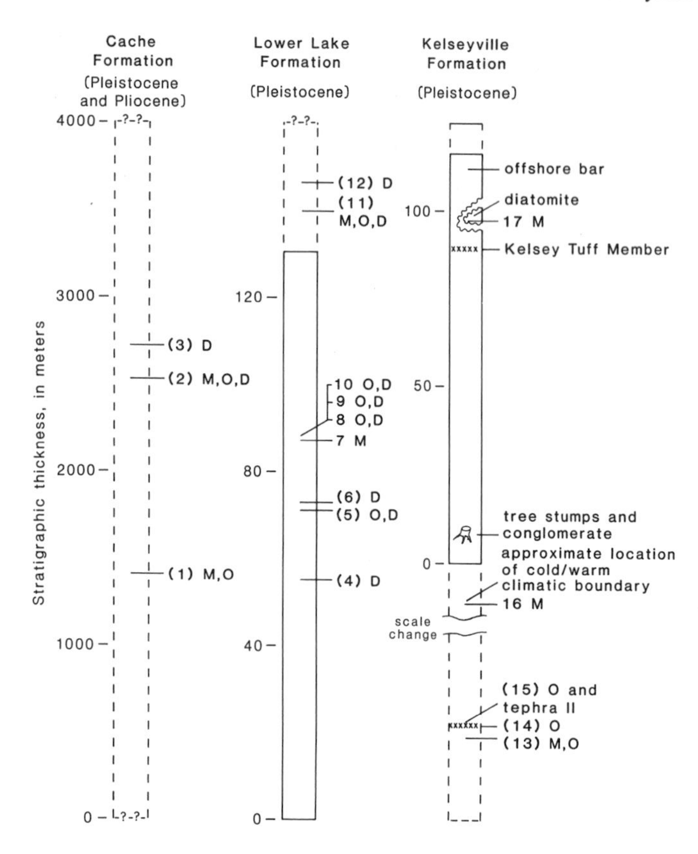

Figure 2. Schematic stratigraphic columns showing relative stratigraphic location of sample sites and amount of stratigraphic coverage. Solid lines along columns represent measured sections (Rymer, 1981). Parentheses around sample number means sample is projected onto measured section or is placed in approximate stratigraphic position. Type(s) of fossils found at each site are indicated with letters: M = mollusk, O = ostracode, D = diatom. Diatoms from the Kelseyville Formation are reported in Bradbury (this volume). Note different thickness scales for the formations.

Mollusk fossils from the Cache Formation occur at localities 1 and 2. Three identifiable species of fresh-water gastropods were recovered from the formation (Table 1): *Valvata humeralis, Gyraulus parvus,* and a hybrobioid prosobranch tentatively called *"Hydrobia" andersoni;* the hydrobioid is the most common species in both samples. A finely ribbed fragment, 2 mm across, from a larger planorbid gastropod is present from locality 1 but is not identifiable.

Valvata humeralis and *Gyraulus parvus* are extant species with wide biogeographic ranges in North America. *V. humeralis* lives in lakes, ponds, marshes, and slow perennial streams on mud bottom, commonly in dense vegetation (Taylor, 1981). *Gyraulus parvus* is found in perennial waters of lakes, ponds, reservoirs, rivers, creeks, and ditches, characteristically among dense aquatic vegetation (Taylor, 1981); it is primarily a quiet-water species (Taylor, 1960). *"Hydrobia" andersoni* is known from the basal part of the Tulare Formation associated with faunas of Blancan age (Taylor, 1960).

Mollusks reported from the Cache Formation by several authors (Becker, 1888; Hannibal, 1912; Anderson, 1936) are presently unavailable for examination (Taylor, 1966). Taylor (1966, p. 37) recorded the following bivalves and gastropods from USGS Cenozoic locality 222, in the Cache Formation and are probably from or near Sec. 2 or 3, T.14N.,R.6W., approximately 10 km north of the localities reported herein: Bivalvia: *Anodonta wahlamatensis* Lea and *Pisidium compressum* Prime; Gastropoda: *Valvata humeralis* Say, *Hydrobia*? cf. *H.*? *andersoni* Arnold, Planorbidae (fragment), and *Physa.* The two bivalve species are extant. The habitat of *Anodonta wahlamatensis* includes both lakes and slow rivers; *Pisidium compressum* occurs in perennial creeks or rivers (Taylor, 1981) or lakes with some current action (Taylor, 1960). *Physa* is not diagnostic of a particular habitat, although it may occur in perennial rather than seasonal water bodies.

Ostracodes recovered from the Cache Formation are also from sites 1 and 2 and are listed in Table 2. *Cypridopsis vidua,* the predominant ostracode taxa, is a short-lived, parthenogenic species found in most low-energy, shallow, fresh-water to slightly saline environments, and especially in ponds, ditches, and in the upper to middle littoral areas of lakes. The low ostracode diversity, the dominance of *C. vidua,* and the absence of numerous ostracode species that live in lakes suggest a pond or low-energy stream. Salinity probably varied seasonally within the limited range of 500 to 1,000 ppm, but it could fall within the broad range of 200 to 2,500 ppm. Few autecologic data are available for *Candona* n.sp. 1, but it lives in marshes and ponds near modern-day Clear Lake. The charophytes *Nitellopsis (Tectochara)* n.sp. and *Chara* spp. support the ostracode paleoenvironmental interpretation and further suggest generally low turbidity and moderate to low turbulence.

Megafossil plant remains of *Nymphaeites nevadensis* (Knowlt.) R. W. Brown (water lily) were recovered from both sites (Rymer, 1978).

Diatoms from the Cache Formation are from sample sites 2 and 3 (Fig. 2, Table 3). *Fragilaria construens* v. *venter* is the dominant diatom in sample 2. This common and widespread diatom characterizes fresh to slightly saline shallow water. It lives either on the bottom of ponds, pools, and shallow regions of larger lacustrine systems or is loosely attached to the stems of aquatic macrophytes. The other common species in this sample, *Cocconeis placentula, Ephithemia adnata, Synedra ulna,* and *Melosira italica,* are from similar habitats and are commonly found in association with each other in modern habitats of oxygen-rich, circumneutral, or moderately alkaline water. The stratigraphically higher sample (3) contains many of the same species, although in different percentages (Table 3). The presence of large numbers of *Rhoicosphenia curvata,* a diatom that also lives attached to substrates in shallow-water environments, implies somewhat elevated salinities for this aquatic environment. *R. curvata* is able to tolerate great fluctuations in salinity-related

TABLE 4. GENERALIZED SPECIES HABITAT CHARACTERISTICS FOR OSTRACODE SPECIES FROM CACHE, LOWER LAKE, AND KELSEYVILLE FORMATIONS

	Lacustrine				Water Temperature[1]				Salinity[2]		Climatic					Precipitation/ Evaporation Ratio[3]		
Ostracodes	Littoral	Profundal	Marginal Lacustrine	Fluvial and Marginal Lacustrine	Cold	Cool	Warm	Hot	Fresh Water	Saline	Arctic	North Temperate	Temperate	Warm Temperate	Tropical	>1.0	≅1.0	<1.0
Cypridopsis vidua (Mueller)	*		*	x		x	*	x	*	x		x	*	*	x	x	x	
Candona n. sp. 1	*		x			x	*	*	x			x	*	*	x	x	x	
Candona n. sp. 2	*		?	?	?	?	?	?	?	?	?	?	?	?	?	?	?	?
Cyclocypris serena (Kock)	*		*			*	?		*	x		x	*	?			*	x
Limnocythere itasca Cole	*		x			*	*		*	x		x	*	x		x	*	
Physocypria globula Furtos	*		x			x	*	*	*				x	*	x	x	*	
Ilyocypris gibba (Ramdohr)	x		*	*	*	*	*	?	*	x	x	*	*	*	?	*	*	?
Ilyocypris bradyi Sars	x		*	*	x	x	x	?	x	x	x	*	*	x		x	x	x
Hetorocypris incongruens? (Ramdohr), complex			*	x		*	*	x	*	x		x	*	*	x			
Limnocythere paraornata? Delorme	x		x	x		x	x		x	x		*	*	x				
Cypricercus deltoidea (Delorme)			*	x		*	x		*			x	*					
Darwinula sp.	x	?	x	x	x	x	x	x	x		x	*	*	*	x	x	x	

x = Occurrence

* = Preference

[1]Cold = always less than 15°C; cool = seasonally cold, seasonal highs to 25°C; warm = seasonally cool to cold, but mostly greater than 20°C; hot = always greater than 20°C.

[2]Fresh water = salinity always less than 3 ppt; saline = salinity at least seasonally greater than 3 ppt.

[3]P/E ratio >1.0 (e.g., forests); ≅1.0 (e.g., forest-prairie transition); <1.0 (e.g., semi-arid grassland to desert).

changes in osmotic pressure (Cholnoky, 1968). Other taxa associated with *R. curvata* in sample 3, however, do not indicate excessive salinities.

In summary, mollusks, ostracodes, and diatoms from the few fine-grained sample sites in the Cache Formation indicate shallow, fresh water to possibly slightly saline aquatic environments with locally abundant aquatic vegetation. Stratigraphic relations indicate that environments suitable for mollusks and ostracodes were of limited extent. Certain species, such as the mollusk *Pisidium compressum* and ostracode *Cypridopsis vidua,* suggest locally greater current activity. Such environments were probably associated with the fluvial systems responsible for the deposition of the bulk of the Cache Formation, perhaps as ponds in abandoned channels or in shallow basins behind natural levees. The reported fauna and flora can also exist in marginal or littoral lacustrine habitats of larger lakes, but there is no stratigraphic or paleontologic evidence for the existence of fully developed lacustrine environments in the Cache Formation (Rymer, 1981).

Although units of quiet-water origin within the Cache Formation can be stratigraphically correlated, no meaningful inferences can be made about fluctuations or changes in the aquatic environment through time. The differences between faunal and floral assemblages could be the result of climatic and/or hydrologic influences of short duration.

LOWER LAKE FORMATION

The Lower Lake Formation crops out southeast of Clear Lake (Fig. 1) and is locally at least 130 m thick (Rymer, 1981). In general, the Lower Lake is folded and dips to the northwest.

Lithology near the base of the Lower Lake Formation is dominated by poorly sorted, poorly stratified sandstone. Higher in the section the Lower Lake is composed of sandstone, siltstone, conglomerate, tuff, diatomite, marly siltstone, and marlstone. The Lower Lake is considered middle Pleistocene in age in this report, based on K-Ar ages of 0.92 ± 0.3 Ma for the locally underlying dacite of Diener Drive (Donnelly-Nolan and others, 1981) and 0.40 ± 0.04 Ma for the locally overlying dacite of Cache Creek (Donnelly-Nolan and others, 1981).

Bedding characteristics and lithologies within the Lower Lake Formation suggest diverse lacustrine environments with minor fluvial incursions. A general fining-upward sequence is present from the base of the formation in the type section to about 95 m above its base (Rymer, 1981). The sequence varies from local pebbly sandstone to limestone. Above this is another fining-upward sequence, ranging in lithology from pebbly sandstone and local pebble conglomerate to diatomite. The upper 40 m of the lower fining-upward sequence is composed of fine-grained rocks rich with fossil fauna and flora (samples 7 through 10, described below). Southwest of its type section the Lower Lake is dominantly siltstone and sandstone with local marlstone, tuff, and diatomite. North of the type section, near the town of Clearlake, formerly called Clearlake Highlands, the Lower Lake is predominantly marly siltstone, marlstone, tuff, and local diatomite. Macrofossils are more common in the type section than to the southwest and north, although near the town of Clearlake, tule root casts are abundant.

Mollusks from the Lower Lake Formation (localities 7 and 11) coincide with USGS vertebrate localities M-1239 and M-1349, which contain fossil fish (Casteel and Rymer, 1981). Two species of fresh-water bivalves and six species of fresh-water gastropods were recovered from the formation (Table 1), whereas all species are present at locality 7. All of the identified species in the Lower Lake are known from faunas at least as old as Blancan, and all but *"Hydrobia" andersoni* are still extant. *"Hydrobia" andersoni* has not been reported from faunas younger than Blancan age (1.7 to 4.8 Ma). However, age control for the Lower Lake Formation (see above and Rymer, 1981) indicates that the present occurrence of *"H." andersoni* extends the chronologic age of the species into late Irvingtonian or possibly early Rancholabrean time (Berggren and Van Couvering, 1974).

The mollusks are indicative of a lacustrine environment with some current action. *Pisidium compressum* is found in perennial creeks, rivers, or lakes with some current action, but it is not found in ponds, swamps, or bogs (Taylor, 1960). Fast current is ruled out by the presence of quiet-water species *Anodonta* cf. *A. wahlamatensis, Valvata humeralis,* and *Gyraulus parvus. Valvata humeralis* and *Gyraulus parvus* indicate the presence of dense aquatic vegetation; *V. humeralis* suggests a muddy substrate.

Ostracodes from five sample sites in the Lower Lake Formation are listed in Table 2. Samples 8, 9, and 10 constitute an ascending stratigraphic order and were collected from the middle part of the type section of the formation. (Sample 8 approximately coincides with the stratigraphic position of sample 7, which was analyzed only for mollusks.) Samples 5 and 11 are from stratigraphically lower and higher positions, respectively (Fig. 2).

The ostracodes from sample 5 typically live in lakes or large permanent ponds, but not in streams or ephemeral marginal lacustrine environments. Moreover, these species are common members of the littoral zone, especially the upper and middle littoral, but do not live in deeper water. Collectively these taxa indicate fresh water, with seasonal salinity ranges of about 350 to 800 ppm and a solute composition that is Ca-Mg-HCO_3-CO_3-(SO_4)-dominated. *Cyclocypris serena* and *Limnocythere itasca* indicate a seasonal, probably summer, drought that maintains an Mg/Ca ratio of 1.0 or greater. *Physocypria globula* indicates water temperatures at least seasonally above 20°C, and the absence of numerous cold temperate taxa indicates mild winters with water temperatures probably remaining above 5° to 10°C. Although these taxa indicate a generally mild climate, they imply conditions both cooler and drier than the modern climate in the region.

Ostracode sample 8 is similar to 5, although *Physocypria globula* is not present and *Ilyocypris gibba* appears for the first time. *Ilyocypris* spp. tolerate moderate current activity and hence are found in streams, small ponds in windy areas, or occasionally in the upper littoral zone of lakes. The appearance of *I. gibba,* and the disappearance of *P. globula,* suggest an environment less littoral than the one represented by 5 and very probably near a stream entering the lake.

Samples 9 and 10 contain the ostracodes present in sample 8, suggesting the same littoral zone-stream environment. *Heterocypris incongruens* and *Cypricercus deltoidea,* species that commonly live in groundwater discharge or recharge zones, suggest nearby groundwater activity. Probably the shore and edge of the lake marsh zone were closer to sample sites 9 and 10, which implies that the lake receded. *Darwinula* sp. may also reflect groundwater activity, but as with *Limnocythere paraornata, Darwinula* spp. also live in streams and littoral areas of lakes, so its presence may not indicate groundwater activity. Sample 11 contains only *Candona* n.sp. 1 and thus a precise depositional environment cannot be interpreted for this sample site.

Eight samples from the Lower Lake Formation were analyzed for diatoms (Table 3, Fig. 2). These samples from the Lower Lake are similar to those from the Cache Formation (Table 3) and indicate comparable environments.

The shallow-water diatoms of both the Cache and Lower Lake Formations represent four partially distinctive habitats, although there is some overlap in species distribution among them. Most distinctive is the terrestrial diatom habitat, which is composed of aerophilic diatoms that can tolerate complete desiccation and thus can prosper in moist meadows or in the upper few centimeters of damp soils. *Hantzschia amphioxys* and *Navicula mutica* are common, widely distributed inhabitants of such environments today, and their presence implies similar conditions at sample site 12.

Shallow-water diatoms live either obligately attached to

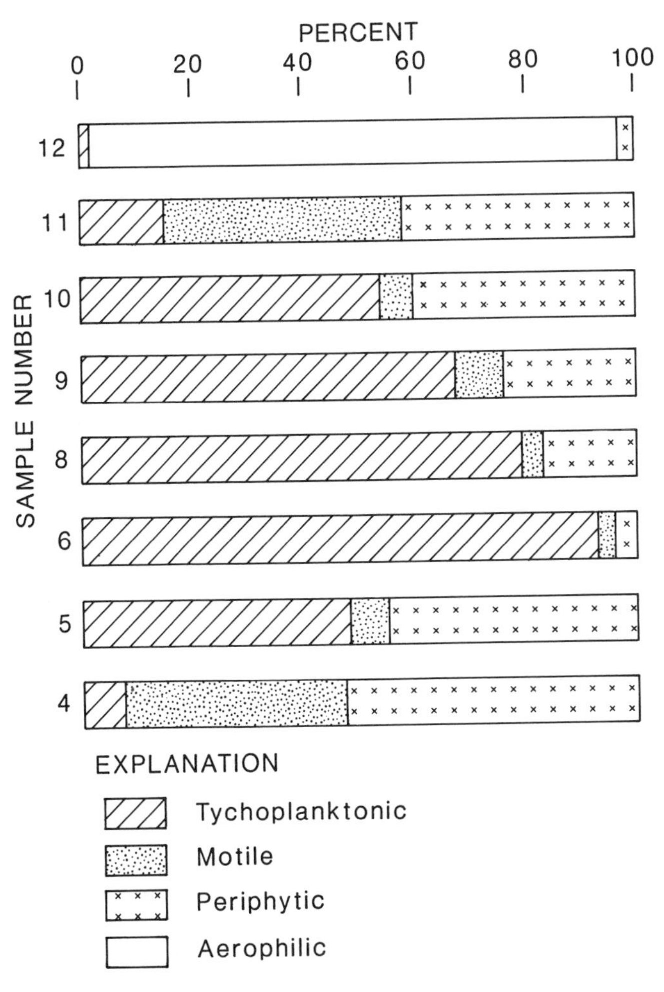

Figure 3. Proportion of tychoplanktonic, motile, periphytic, and aerophilic diatoms from tentative stratigraphic succession of samples from the Pleistocene Lower Lake Formation.

stems of higher aquatic plants (periphytic diatoms, the second habitat group) or loosely resting on bottom sediments or on other surfaces including plants, rocks, and pieces of wood (motile, the third habitat group). These diatoms can become planktonic if turbulence is sufficient to carry them into open water and keep them suspended. Thus, this group is known as tychoplanktonic (opportunistically planktonic). Tychoplanktonic diatoms commonly live lakeward (in somewhat deeper water) from the zone of rooted aquatic vegetation where the periphytic and motile diatoms dominate.

These environmentally controlled groups allow interpretation of water depth in a general way for littoral diatom floras (Fig. 3). Samples from or near the type section just west of the town of Lower Lake are dominated by tychoplanktonic species, principally varieties of *Fragilaria construens* and *Melosira* species. These assemblages imply somewhat deeper water, perhaps in the range of 1 to 5 m, than the assemblages to the north and west that contain greater percentages of periphytic diatoms. Samples collected in known stratigraphic order (8, 9, and 10) illustrate a progressive increase in the proportion of periphytic diatoms from bottom to top. This suggests that the lake responsible for depositing this sediment may have become shallower with time, or at least that rooted aquatic vegetation became reestablished nearer to the sample site.

Diatom concentration, as determined by a count of the number of diatom valves along a microslide transect (Batterbee, 1973), also characterizes the depositional environment. Samples with high proportions of tychoplanktonic diatoms also have the highest concentrations of diatoms (Table 3). In some cases samples, such as sample 6 with approximately 1,000 diatom valves per millimeter of transect, are almost pure diatomite. Such high concentrations of diatoms occur in productive waters that receive little or no dilution by clastic or calcareous materials, and are logically associated with deeper water or areas of shallow-water systems that are surrounded by aquatic vegetation that traps and holds sediment.

Samples collected both west and north of the type section (stratigraphically equivalent and higher, respectively) generally show large percentages of periphytic diatoms and low concentrations of diatom valves, implying shallower water environments nearer shore. Sample 12, east of the town of Clearlake (Fig. 1) has more than 90 percent aerophilic diatoms, indicating deposition at the very margin of the lake or altogether beyond the ancestral lake.

Structural and large-scale stratigraphic relationships in the Lower Lake Formation suggest a tentative stratigraphic arrangement for the diatom sample suite as shown in Figure 2. Only samples 8, 9, and 10 were collected in known ascending stratigraphic order over a distance of about 2 m, but the changing percentages of tychoplanktonic and periphytic diatoms in the Lower Lake samples substantiates the evolution of a deepening, then shallowing, and finally desiccating lacustrine system. In this tentative sequence (Fig. 3) the basal sample with low numbers of tychoplanktonic diatoms suggests shallow-water conditions that rapidly deepen, as indicated by samples 5 and 6. Samples thereafter (samples 8 through 11) indicate a progressively shallower environment, and the final sample (sample 12) indicates a seasonally moist terrestrial environment (Fig. 3). To prove this speculative paleolimnologic sequence, stratigraphic relations and time control are needed along with a larger, denser sampling.

The mollusk, ostracode, and diatom assemblages indicate a body of water large enough to present a variety of local environments and current regimes. Proportions of periphytic, tychoplanktonic, and motile diatoms from the type section of the Lower Lake Formation suggest a water depth of 1 to 5 m. However, diatom species, percentages, and abundances from other localities indicate a range in environments and inferred water depths from sublittoral to marginal lacustrine, and even terrestrial. This environmental range may represent deepening and then progressively shallowing of the lacustrine system. Further support of the deepening and shallowing sequence inferred from diatom habitats is given by the two fining-upward sequences in the type section, which range in lithology from sandstone and pebble conglomerate near the bases of the sequences to diatomite and marlstone at the tops (Rymer, 1981). The faunal and floral assemblages reported herein indicate dense aquatic vegetation with local muddy substrates. Specific ostracode species also suggest diverse lacustrine environments that shallow through time.

KELSEYVILLE FORMATION

The Kelseyville Formation is exposed south of the main basin of Clear Lake (Fig. 1) and consists predominantly of sandstone, siltstone, and mudstone, and lesser amounts of conglomerate (Rymer, 1981). The thickness of this formation is about 500 m. Exposures of the Kelseyville are on the axis and eastern limb of a gently folded north-plunging syncline.

The Kelseyville is considered middle Pleistocene in age in this report, but its lower and upper age limits are not known precisely. A tuff near the base of the formation correlates mineralogically with the rhyolite of Thurston Creek (Rymer, 1981). This rhyolite has yielded K-Ar ages of 0.48 ± 0.02 to 0.64 ± 0.03 Ma, and the best estimate for its age is about 0.6 Ma (Donnelly-Noland and others, 1981). In addition, normal magnetic polarity determinations by J. C. Liddicoat (written communication, 1982) for the lowest exposed beds suggest that these strata belong to the Brunhes Normal Chronozone and are thus younger than 0.72 Ma, the age of the Brunhes Normal-Matuyama Reversed Chronozone boundary (Mankinen and Dalrymple, 1979).

The age of the uppermost part of the Kelseyville is more difficult to determine. To date, the best estimate for the age of the exposed uppermost part of the Kelseyville comes from the comparison of Kelseyville and Clear Lake strata. The upper and lower parts of the Kelseyville are distinguished by the presence of fossil leaves and cones indicative of growth during cold and warm climates, respectively (Rymer, 1981). Within the inferred warm (lower) and cold (upper) sections are local volcanic tuff beds, two of which are important here. The Kelsey Tuff Member of the Kelseyville lies about 200 m above the paleoclimatic boundary, and a thin andesitic tuff, informally referred to as "tephra II," is located an unknown distance below the boundary (Fig. 2). Diatom biostratigraphy shows a very strong correlation of diatom species, abundance, and succession immediately below tephra II and an ash bed at a depth of 123.7 m in Clear Lake core CL-80-1 (Bradbury, this volume). These two ashes are mineralogically similar but have a very weak chemical similarity coefficient (Sarna-Wojcicki and others, this volume). The ash at 123.7 m in core CL-80-1 is near the end of a climatically warm period, as indicated by oak pollen frequencies, and can be correlated with deep-sea oxygen-isotope stage 9 (Heusser and Sims, 1983; Sims and others, this volume). If these correlations hold, the succeeding cold period represented in core CL-80-1 and in the Kelseyville strata represents deposition during ^{18}O-stage 8. The stratigraphically higher Kelsey Tuff Member was not found in core CL-80-1, but a hiatus is suggested in the inferred ^{18}O-stage 8 stratigraphic section (Sims and others, Bradbury, and Sarna-Wojcicki and others, this volume). In conclusion, the age of the uppermost part of the Kelseyville Formation is tentatively placed at 250,000 to 310,000 yr, the inferred extreme limits of ^{18}O-stage 8 (Kominz and others, 1979).

Bedding characteristics for strata in the Kelseyville Formation suggest general lacustrine deposition (Lake County Flood Control and Water Conservation District, 1967; Rymer, 1981). Except at the margins of the basin, the formation is characterized by structureless mudstone to thin and moderately well bedded, laterally continuous siltstone and sandstone beds generally less than 1 cm thick. These deposits represent deposition in a large mud- and sand-bottomed lake. Many bedding features visible in outcrops of the Kelseyville are identical with features in cores of Clear Lake sediment described by Sims, Rymer, and Perkins (this volume).

Conglomerate beds away from the margins of the depositional basin in the Kelseyville Formation are paleoenvironmentally significant. Pebble conglomerate beds, as much as 40 cm thick, are locally present near the center of the basin. These beds are interbedded with laterally continuous sandstone and siltstone beds that contain a rich assemblage of well-preserved fossil leaves and cones, indicative of quiet-water lacustrine deposition. The conglomerates and local stumps of fossil trees, in growth position, indicate that the paleolake either receded or disappeared during this period of deposition.

An offshore bar is present, approximately 2.5 km south of Kelseyville and near the top of the section, which reflects on the water depth during its formation. The bar is approximately 5 m high, assuming the bar was relatively stable; water depth above the bar would have been about 7 m according to the formula of Keulegan (1948).

A diatomite (sample 17) included in the Kelseyville Formation by Rymer (1981) warrants further discussion. The diatomite is located along state highway 29 approximately 4.5 km southeast of Kelseyville and has always been problematic in the reconstruction of depositional environments for the Kelseyville Formation. The diatomite is of limited stratigraphic extent and has no correlative units within nearby clastic rocks occupying similar stratigraphic positions relative to the underlying Kelsey Tuff Member. The diatomite is at the same stratigraphic horizon relative to the tuff, as are medium-grained sandstones that have local wavy parallel lamination and local pebbly sandstone and pebble conglomerate beds. These clastic rocks and bedforms indicate significantly higher energy environments than does the diatomite. We retain the diatomite in the Kelseyville Formation, but note that its limited stratigraphic extent and anomalous lithology indicate a significantly younger age than the surrounding Kelseyville strata (Bradbury, this volume).

Fresh-water gastropods were recovered from three sample sites in the Kelseyville (Table 1); the three species are *Valvata virens, Helisoma newberryi,* and *"Hydrobia" andersoni. Valvata virens* is the most common species in sample 16 and is the only species in samples 13 and 17. As a modern species, *Valvata virens* is restricted to Clear Lake and a pond near Watsonville (Taylor,

1981). The early whorls of some of the shells in the present samples are coiled nearly in one plane, as in *V. v. platyceps* Pilsbry known from the Blancan fauna of the lower part of the Tulare Formation (Taylor, 1966). Others are high-spired, as in typical modern *V. virens,* suggestive of a transitional state in endemism of *V. virens. "Hydrobia" andersoni* represents a chronologic extension into the middle Pleistocene, as in the Lower Lake Formation. The indicated environment is lacustrine. *Valvata virens* is known only from lake and pond habitats (Taylor, 1981). The inferred environment of mollusks from the Tulare Formation (including *V. v. platyceps*) and *"Hydrobia" andersoni* is a large, shallow lake (Pilsbry, 1935). *Helisoma newberryi* is a burrower of mud substrate, and is found in larger lakes and slow rivers, including larger spring sources and spring-fed creeks (Taylor, 1981). The environments of all three localities in the Kelseyville Formation were probably similar. *H. newberryi* possibly indicates the proximity of a spring-fed tributary near locality 16.

Three species of ostracodes have been identified in three samples from the Kelseyville Formation (Table 2), whereas eight other samples were barren. *Candona* n.sp. 1 is the most common species present in the samples, as is the case in the Lower Lake Formation. This species is indicative of fresh-water lacustrine environments (Table 4) that are marginal to littoral. The other two species, *Physocypria* sp. and *Limnocythere itasca,* are represented only by juvenile carapaces, which suggests that they were transported to the site of deposition. The low ostracode diversity in the three samples, and the total absence of ostracodes in other samples, indicates that the local Kelseyville lacustrine environment was unfavorable for ostracodes. Forester (this volume) discusses a similar situation in Clear Lake. The preferred explanation for the generally barren ostracode record in Clear Lake is that the turbulent, shallow, high-productivity conditions present now, which produce anoxic bottom waters, have always existed (Forester, this volume).

Diatoms from the Kelseyville Formation are described by Bradbury (this volume). The general depositional environment inferred by diatoms is a shallow to deep (about 30 m) mesotrophic to oligotrophic lake.

Collectively, mollusks, ostracodes, and diatoms recovered from the Kelseyville Formation indicate a large, marginal to littoral lacustrine environment. Water depth varied during deposition of the Kelseyville in both space and time. Diatoms from a diatomite located west of Highland Springs Reservoir indicate a water depth of 10 to 30 m (J. P. Bradbury, oral communication, 1984); the presence of cross-bedded conglomerates and nearby tree stumps in growth positions exposed in Kelsey Creek indicates shallow to subaerial conditions; and the offshore bar south of Kelseyville indicates a water depth near the margins of the inferred lake of about 12 m. Timings of water depth fluctuations in the so-called Kelseyville lake are poorly known, but on the basis of paleobotanic data and their inferred climatically warm and cold growth conditions at the various deep-water sites, it is apparent that major climatic changes did not significantly affect the lake's depth. The presence of the gastropod *Helisoma newberryi* suggests proximity to a spring-fed tributary and muddy substrates; Adobe, Kelsey, and Cole Creeks, which presently flow through the area, are all spring-fed. The lack of diversity and general absence in the Kelseyville Formation of mollusks and especially ostracodes, both of which are nearly ubiquitous in lacustrine environments, indicate adverse conditions. Similar adverse conditions for ostracodes have been present through the interpreted history of nearby Clear Lake (Forester, this volume). Therefore, we infer that similar conditions existed during deposition of the Kelseyville Formation as they presently exist in Clear Lake. This includes turbulent, turbid, high oxygen-consuming productivity that results in at least intermittent bottom-water sediment anoxia and unstable littoral substrates.

The diatomite southeast of Kelseyville, which we infer to be stratigraphically anomalous and probably younger than enclosing strata, also has a different depositional environment. Diatoms recovered from the diatomite, especially *Melosira ambigua* and *Stephanodiscus niagarae,* indicate a mesotrophic to eutrophic lake of moderate to large size and depth with a littoral zone some distance away from the site of deposition (Bradbury, this volume). Fossil fish recovered from this deposit likewise indicate an anomalous environment relative to the rest of the Kelseyville Formation (Casteel and Rymer, 1981). First, the fish from this deposit (site M-1351) are much more diverse than from other sample sites from the Kelseyville Formation. Second, the presence of *Mylophardon* and *Catostomus* in the fish assemblage indicates large sluggish creeks and small rivers draining into the lake represented by this deposit (Casteel and Rymer, 1981); this suggests a significantly higher precipitation/evaporation ratio during deposition of the diatomite than during the rest of the formation. We do not know the precise age of this deposit, nor do we know the geographic limits of the inferred lake. How this lacustrine system fits into the history of the Kelseyville Formation and/or the Clear Lake lacustrine deposits is unknown. Bradbury (this volume) correlates planktonic diatoms from this deposit with similar species at a depth of 30 to 40 m in Clear Lake core CL-80-1, a depth range that has an inferred age of about 35,000 to 50,000 yr (J. D. Sims, oral communication, 1984).

DISCUSSION AND CONCLUSIONS

Many questions as to the possible succession of depositional environments between the Cache, Lower Lake, and Kelseyville Formations and the Clear Lake deposits can now be answered in light of new data presented in this paper. The possible continuity of deposition between the Cache and Lower Lake Formations appears to be unlikely. Although there is partial commonality of mollusk, ostracode, diatom, and fish species, including endemic species, between the two formations, the percentage, diversity, and specific habitats of species recovered indicate dissimilar environments. Furthermore, there is no indication of continuity on the basis of stratigraphy or age. The youngest exposed Cache beds are about 1 m.y. older than the oldest Lower Lake beds where the Lower Lake Formation overlies the Cache Formation. Also, this

TABLE 5. DISTRIBUTION OF MOLLUSKS AND OSTRACODES IN CACHE, LOWER LAKE, AND KELSEYVILLE FORMATIONS AND IN CORES FROM CLEAR LAKE AND IN PRESENT-DAY CLEAR LAKE

Taxon	Cache Fm.	Lower Lake Fm.	Kelseyville Fm.	Clear Lake Cores CL-73-1	CL-73-4	CL-73-6	CL-73-7	CL-80-1	Present
Mollusca									
Ostrea cf. *O. lurida* (adventitious)	-	-	-	x	-	-	-		-
Anodonta wahlamatensis	x	cf	-	-	-	-	-		x
Pisidium compressum	x	x	-	-	-	x	cf		-
Pisidium insigne	-	-	-	-	-	-	x		-
Valvata humeralis	x	x	-	-	-	x	x		-
Valvata virens	-	-	x	-	x	x	x		x
"*Hydrobia*" cf. "*H.* *andersoni*"	x	x	x	-	-	-	-		-
Lymnaea n. sp.	-	x	-	-	-	-	-		?
Gyraulus parvus	x	x	-	-	-	x	x		x
Helisoma minus	-	-	-	-	-	x	-		x
Helisoma newberryi	-	-	x	-	-	-	-		-
Planorbella tenuis	-	cf	-	-	-	-	-		x
Physa costata	-	-	-	-	-	x	-		x
Physa virgata	-	cf	-	-	-	-	cf		-
Physa sp. (of Taylor, 1966)	x	-	-	-	-	-	-		-
Ostracoda									
*Cypridopsis vidua**	x	x	-		-		x	-	-
Candona n. sp. 1*	x	x	x		x		x	x	-
Candona n. sp. 2*	x	x	-		-		x	-	-
Cyclocypris serena	-	x	-		-		-	-	-
Limnocythere itasca	-	x	x		-		-	-	-
Physocypria globula	-	x	-		-		-	x	-
Ilyocypris gibba	-	x	-		-		-	-	-
Heterocypris incongruens	-	x	-		-		-	-	-
Limnocythere paraornata?	-	x	-		-		-	-	-
Limnocythere n. sp.*	-	-	-		-		x	x	-
Cypericercus deltoidea	-	x	-		-		-	-	-
Darwinula sp.*	-	x	-		-		-	-	-
Cypria ophtalmica	-	-	-		-		x	x	x

Note: Clear Lake core data for mollusks are listed in Appendix B. Present-day records based on California Academy of Sciences collection and Taylor (1981) for mollusks and local sampling for ostracodes (Forester, this volume).

*The margins of Clear Lake have ostracode taxa, marked with asterisk, plus *Physocypria pustulosa* (Sharpe) and *Candona albicans* Brady.

contact is represented by an angular unconformity of about 15° (Rymer, 1981), indicating an intervening period of uplift, warpage, and erosion of Cache Formation strata before deposition of the Lower Lake strata. Apparently the faults that controlled deposition of the Cache Formation continued to move after deposition of Cache sediments. However, the partial similarity of the recovered fauna and flora in the Cache and Lower Lake Formations suggests at least some continuity within ancestral drainages.

Correlation of the Lower Lake strata and the majority of the Clear Lake deposits is unlikely because of contrasting lithologies and inferred depositional environments. The general littoral lacustrine environment and locally dense aquatic vegetation in diverse substrates for the Lower Lake Formation contrast strongly with the general open-water, anoxic, highly fluid sapropelic mud substrates of Clear Lake. However, the environment at the southeast end of the Highlands or "Lower" arm of Clear Lake, adjacent to the Lower Lake Formation, is similar to that inferred for the Lower Lake Formation. The southeast end of the arm supports dense aquatic vegetation standing in less than 1 to 3 m of water. Preliminary analysis of ostracode and diatom samples from this locality shows diverse assemblages suggestive of a turbulent, nutrient-rich setting. Other indicators of possible correlation between the Lower Lake Formation and the Clear Lake deposits come from Clear Lake core CL-73-7, located in the middle of the Highlands arm (Sims and others, this volume). Below approximately 6.8 m of sapropelic mud, similar to that present elsewhere in Clear Lake, there are peat, peat-rich mud, mud, and thin diatomite and ash beds (Sims and Rymer, 1975d). Spores and pollen from these deposits indicate deposition in less than 2 m of water (Sims and others, 1981). Furthermore, mollusks and ostracodes recovered from this section of core CL-73-7 (Table 5) are diverse and more similar to a general Lower Lake Formation assemblage than the Clear Lake assemblage, although the former assemblage suggests a colder and drier climate than the one in Clear Lake. Possibly the active tectonics of the region (Hearn and others, this volume) caused the southward migration of Lower Lake–type depositional environments in the Highlands arm. In summary, succession of Lower Lake Formation environments to Clear Lake is not proven, but remains a strong possibility. Further proof may lie deeper below Clear Lake than core CL-73-7 penetrated.

Partial correlation of the Kelseyville Formation and the Clear Lake deposits is strongly suggested (Rymer, 1981; Bradbury, this volume). Except for a general coarser grain size in the Kelseyville Formation, which is nearer source streams and bordering steep mountains, the two deposits are lithologically similar. Bedding features visible in outcrops of the Kelseyville Formation are identical to features in cores of Clear Lake sediment as described by Sims and others (this volume). Distinctive planktonic diatom species are identical in the Kelseyville Formation and Clear Lake long core CL-80-1, and the taxa occur in the same stratigraphic order in both deposits. Mollusks and ostracodes identified from the Kelseyville Formation and the main basin of Clear Lake are consistent with correlation of the two deposits. The lack of diversity, disproportionate number of juvenile vs. adult ostracodes, and general absence of benthic fauna in the Kelseyville Formation are similar to features observed in Clear Lake sediment (Forester, this volume). Further support for correlation is afforded by new age control on Clear Lake sediment. Sarna-Wojcicki and others (this volume) show correlation of ash beds in core CL-80-1 with dated ash beds exposed elsewhere in northern California, and conclude that sediment beneath about 120 m in core CL-80-1 is in the range of about 0.35 to 0.40 m.y. old, making this section in core CL-80-1 coeval with the upper part of the Kelseyville Formation. The direct correlation of Kelseyville and Clear Lake strata is dependent on the weak chemical similarity of tephra II and the ash bed at 123.7 m in core CL-80-1 and the diatom correlations for sediment immediately beneath these two ash beds (Bradbury, this volume).

The likelihood of correlation of the Lower Lake and Kelseyville Formations is quite low. The lithologies, fossils, and inferred depositional environments are all dissimilar. However, in light of the potential correlation of depositional environments in the Lower Lake strata, with the Clear Lake deposits at the southeast end of the Highlands arm and with sediment deeper than 6.7 m in Clear Lake core CL-73-7, a possible setting can be envisioned. Hypothetically, partly coeval Lower Lake and Kelseyville lacustrine deposits might have been deposited in different subbasins of the same lake, with different substrates, amounts of aquatic vegetation, nutrients, and circulation. This model, although currently unsupported, could be substantiated by extensive coring in the Highlands arm.

APPENDIX A: LOCALITY DESCRIPTIONS

Locality	Latitude	Longitude	Description
1	38°59.70'N	122°32.58'W	Lower Lake 7.5-minute Quadrangle, 1975 edition. Massive mudstone from streambank exposure of Cache Formation of Pliocene and Pleistocene age. Collected 1975, field number 5R101A.
2	38°58.84'N	122°34.66'W	Lower Lake 7.5-minute Quadrangle, 1975 edition. Diatomaceous mudstone from roadcut exposure of Cache Formation of Pliocene and Pleistocene age. Collected 1975, field number 5R041B.
3	38°58.97'N	122°33.44'W	Lower Lake 7.5-minute Quadrangle, 1975 edition. Silty mudstone from roadcut exposure of Cache Formation of Pliocene and Pleistocene age. Collected 1975, field number 5R0038A.
4	38°54.31'N	122°40.41'W	Clearlake Highlands 7.5-minute Quadrangle, 1975 edition. Mudstone from from roadcut exposure of Lower Lake Formation of Pleistocene age Collected 1975, field number 5R003A.
5	38°54.63'N	122°38.16'W	Clearlake Highlands 7.5-minute Quadrangle, 1975 edition. Diatomaceous mudstone from roadcut exposure of Lower Lake Formation of Pleistocene age. Collected 1981, field number Ost-10.
6	38°54.63'N	122°38.14'W	Clearlake Highlands 7.5-minute Quadrangle, 1975 edition. Diatomaceous mudstone from roadcut exposure of Lower Lake Formation of Pleistocene age. Collected 1981, field number Ost-9.
7	38°54.75'N	122°36.69'W	Lower Lake 7.5-minute Quadrangle, 1975 edition. Diatomaceous marly mudstone from roadcut exposure of Lower Lake Formation of Pleistocene age. Collected 1974, field number 4R007A
8	38°54.75'N	122°36.69'W	Lower Lake 7.5-minute Quadrangle, 1975 edition. Diatomaceous marly mudstone from roadcut exposure of Lower Lake Formation of Pleistocene age. Collected by J. P. Bradbury, 1978, field number 25IX78-2A.
9	38°54.75'N	122°36.69'W	Lower Lake 7.5-minute Quadrangle, 1975 edition. Diatomaceous marly mudstone from roadcut exposure of Lower Lake Formation of Pleistocene age. Collected by J. P. Bradbury, 1978, field number 25IX78-2B.
10	38°54.75'N	122°36.69'W	Lower Lake 7.5-minute Quadrangle, 1975 edition. Diatomaceous marly mudstone from roadcut exposure of Lower Lake Formation of Pleistocene age. Collected by J. P. Bradbury, 1978, field number 25IX78-2C.
11	38°56.21'N	122°36.87'W	Lower Lake 7.5-minute Quadrangle, 1975 edition. Diatomaceous marly mudstone from roadcut exposure of Lower Lake Formation of Pleistocene age. Collected 1975, field number 5R005A.
12	38°57.00'N	122°37.34'W	Lower Lake 7.5-minute Quadrangle, 1975 edition. Diatomaceous mudstone from roadcut exposure of Lower Lake Formation of Pleistocene age. Collected by J. P. Bradbury, 1978, field number 25IX78-3A.
13	38°55.50'N	122°52.27'W	Kelseyville 7.5-minute Quadrangle, 1975 edition. Mudstone from gully exposure of Kelseyville Formation of Pleistocene age. Collected 1977, field number 7R030A.
14	38°56.01'N	122°52.37'W	Kelseyville 7.5-minute Quadrangle, 1975 edition. Mudstone from gully exposure of Kelseyville Formation of Pleistocene age. Collected 1976, field number 6R009A.
15	38°56.20'N	122°52.50'W	Kelseyville 7.5-minute Quadrangle, 1975 edition. Mudstone from gully exposure of Kelseyville Formation of Pleistocene age. Collected 1976, field number 6R007A.
16	38°56.78'N	122°50.38'W	Kelseyville 7.5-minute Quadrangle, 1975 edition. Silty sandstone from streambank exposure of Kelseyville Formation of Pleistocene age. Collected 1977, field number 7R002A.
17	38°56.73'N	122°48.92'W	Kelseyville 7.5-minute Quadrangle, 1975 edition. Muddy diatomite from roadcut exposure of Kelseyville Formation of Pleistocene age. Collected 1976, field number 6R001A.

Note: Specimens collected by M. J. Rymer, except as noted.

APPENDIX B: SAMPLES, IDENTIFICATIONS, AND QUANTITIES OF MOLLUSK REMAINS RECOVERED IN CLEAR LAKE CORES

Sample No.*	Depth (cm)	Genus and Species	Material Recovered
Core CL-73-1			
2176	730	*Ostrea* cf. *O. lurida* Carpenter	1 valve, slightly abraded
Core CL-73-4			
1526	10890	*Valvata virens* Tryon	1 partial shell
Core CL-73-6			
351	750	*Valvata virens* Tryon	2 intact, 39 partial shells
353	721	*Pisidium*, species indeterminite	2 valves
		Valvata virens Tryon	Ca. 250 shells, about 1/3 of them intact
386	1178	*Valvata virens* Tryon	Ca. 70 partial shells
393	1262	*Valvata virens* Tryon	1 partial shell
444	2035	*Pisidium compressum* Prime	1 valve
		Valvata humeralis Say	1 intact shell
		Valvata virens Tryon	Ca. 600 shells, most intact
		Gyraulus parvus (Say)	9 intact, 2 partial shells
		Helisoma minus (Cooper)	5 intact, 17 partial shells
		Physa costata Newcomb	2 fragments
446	2057	*Valvata virens* Tryon	1 intact shell
463	1883	*Pisidium compressum* Prime	1 valve
		Valvata humeralis Say	23 intact shells, 3 fragments
		Gyraulus parvus (Say)	11 intact shells
Core CL-73-7			
	(see next page)		

*Sample numbers from reports on Clear Lake cores (Sims and Rymer, 1975a-d).

APPENDIX B: SAMPLES, IDENTIFICATIONS, AND QUANTITIES OF MOLLUSK REMAINS RECOVERED IN CLEAR LAKE CORES
(continued from previous page)

Sample No.*	Depth (cm)	Genus and Species	Material Recovered
Core CL-73-7			
48	1304	*Valvata virens* Tryon	19 intact, 64 partial shells
54	1360	*Valvata virens* Tryon	5 fragments
67	1500	*Valvata virens* Tryon	1 partial shell
78	1610	*Gyraulus parvus* (Say)	3 intact, 3 partial shells
		Lymnaea, species indeterminite	2 partial juvenile shells, 1 fragment
84	1690	*Valvata virens* Tryon	1 intact shell, 4 fragments
167	680	*Valvata virens* Tryon	Ca. 350 shells, about 2/3 of them intact
		Pisidium cf. *P. compressum* Prime	3 valves, 3 fragments
169	700	*Pisidium* cf. *P. compressum* Prime	2 valves
		Valvata virens Tryon	29 intact, 15 partial shells
175	770	*Valvata virens* Tryon	3 partial shells
179	810	*Valvata virens* Tryon	9 fragments
		Gyraulus parvus (Say)	1 fragment
189	924	*Valvata humeralis* Say	1 partial shell
		Gyraulus parvus (Say)	1 intact shell
219	2020	*Pisidium* cf. *P. compressum* Prime	5 valves
		Pisidium insigne Gabb	8 paired and 1 single valve
		Valvata humeralis Say	59 intact shells, 3 fragments
		Lymnaea, species indeterminate	4 fragments
		Gyraulus parvus (Say)	6 intact shells
226	2070	*Valvata humeralis* Say	2 intact shells
229	2100	*Valvata virens* Tryon	1 intact, 2 partial shells
234	2140	*Pisidium insigne* Gabb	1 valve
		Valvata virens Tryon	2 partial shells
		Gyraulus parvus (Say)	10 intact shells, 4 fragments
260	2420	*Gyraulus parvus* (Say)	2 intact shells, 2 fragments
		Physa cf. *P. virgata* Gould	1 intact shell, 1 fragment
271	2540	*Lymnaea*, species indeterminate	3 fragments
		Gyraulus parvus (Say)	1 intact shell

*Sample numbers from reports on Clear Lake cores (Sims and Rymer, 1975a-d).

REFERENCES CITED

Andserson, C. A., 1936, Volcanic history of the Clear Lake area, California: Geological Society of America Bulletin, v. 47, p. 629–664.

Batterbee, R. W., 1973, A new method for the evaluation of absolute microfossil numbers with reference especially to diatoms: Limnology and Oceanography, v. 18, p. 647–653.

Becker, G. F., 1888, Geology of the quicksilver deposits of the Pacific slope: U.S. Geological Survey Monograph 13, 486 p.

Berggren, W. A., and Van Couvering, J. A., 1974, The late Neogene; Biostratigraphy, geochronology, and paleoclimatology of the last 15 million years in marine and continental sequences: Amsterdam, Elsevier, 216 p.

Brice, J. C., 1953, Geology of the Lower Lake quadrangle, California: California Division of Mines and Geology Bulletin 166, 72 p.

Burch, J. B., 1975, Freshwater sphaeriacean clams (Mollusca: Pelecypoda) of North America: Hamburg, Michigan, Malacological Publications, 204 p.

Casteel, R. W., and Rymer, M. J., 1975, Fossil fish from the Pliocene or Pleistocene Cache Formation, Lake County, California: U.S. Geological Survey Journal of Research, v. 3, p. 619–622.

—— , 1981, Pliocene and Pleistocene fishes from the Clear Lake area, Lake County, California, *in* McLaughlin, R. J., and Donnelly-Nolan, J. M., eds., Research in the Geysers–Clear Lake geothermal area, northern California: U.S. Geological Survey Professional Paper 1141, p. 231–235.

Cholnoky, B. J., 1968, Die Okologie der Diatomeen in Binnengewassern: J. Cramer, Lehre, 699 p.

Delorme, L. D., 1970a, Freshwater ostracodes of Canada, Pt. I, subfamily Cypridinae: Canadian Journal of Zoology, v. 48, p. 153–168.

—— , 1970b, Freshwater ostracodes of Canada, Pt. III, family Candonidae: Canadian Journal of Zoology, v. 48, p. 1099–1127.

—— , 1971, Freshwater ostracodes of Canada, Pt. V, families Limnocytheridae, Loxoconchidae: Canadian Journal of Zoology, v. 49, p. 43–64.

Donnelly-Nolan, J. M., Hearn, B. C., Jr., Curtis, G. H., and Drake, R. E., 1981, Geochronology and evolution of the Clear Lake volcanics, *in* McLaughlin, R. J., and Donnelly-Nolan, J. M., eds., Research in the Geysers–Clear Lake geothermal area, northern California: U.S. Geological Survey Professional Paper 1141, p. 47–60.

Forester, R. M., 1983, Relationship of two lacustrine ostracode species to solute composition and salinity; Implications for paleohydrochemistry: Geology, v. 11, p. 435–438.

Furtos, N. C., 1933, The ostracoda of Ohio: Ohio Biological Survey Bulletin 29, v. 5, p. 413–524.

Goldman, C. R., and Wetzel, R. G., 1963, A study of the primary productivity of Clear Lake, Lake County, California: Ecology, v. 44, p. 283–294.

Hannibal, H. B., 1912, A synopsis of the recent and Tertiary freshwater Mollusca of the Californian Province, based upon an ontogenetic classification: Malacological Society of London Proceedings, v. 10, p. 112–211.

Heusser, L. E., and Sims, J. D., 1983, Pollen counts for core CL-80-1, Clear Lake, Lake County, California: U.S. Geological Survey Open-File Report 83-384, 28 p.

Horne, A. J., and Goldman, C. R., 1972, Nitrogen fixation in Clear Lake, California; I. Seasonal variations and the roles of heterocysts: Limnology and Oceanography, v. 17, p. 678–692.

Keulegan, G. H., 1948, An experimental study of submarine sandbars: U.S. Army Corps of Engineers Beach Erosion Board Technical Report 3, 40 p.

Kominz, M. A., Heath, G. R., Ku, T.-L., and Pisias, N. G., 1979, Brunhes time scales and the interpretation of climate change: Earth and Planetary Science Letters, v. 45, p. 394–410.

Lake County Flood Control and Water Conservation District, 1967, Big Valley ground-water recharge investigation: unpublished report to Lake County Flood Control and Water Conservation District, 78 p.

Mankinen, E. A., and Dalrymple, G. B., 1979, Revised geomagnetic polarity time scale for the interval 0-5 m.y. B.P.: Journal of Geophysical Research, v. 84, p. 615–626.

McLaughlin, R. J., Ohlin, H. N., Thormahlen, D. J., Jones, D. L., Miller, J. W., and Blome, C. D., 1985, Geologic map and structure sections of the Little Indian Valley–Wilbur Springs geothermal area, northern Coast Ranges, California: U.S. Geological Survey Open-File Report 85-285, 24 p., scale 1:24,000.

Pilsbry, H. A., 1935, Mollusks of the freshwater Pliocene beds of the Kettleman Hills and neighboring oil fields, California: Philadelphia Academy of Natural Sciences Proceedings, v. 86, p. 541–570.

Rosen, B. H., Kingston, J. C., and Lowe, R. L., 1981, Observations of differential epiphytism on *Cladophora glomerata* and *Bangia atropurpurea* from Grand Traverse Bay, Lake Michigan: *Micron,* v. 12, p. 219–220.

Rymer, M. J., 1978, Stratigraphy of the Cache Formation (Pliocene and Pleistocene) in Clear Lake basin, Lake County, California: U.S. Geological Survey Open-File Report 78-924, 102 p.

—— , 1981, Stratigraphic revision of the Cache Formation (Pliocene and Pleistocene), Lake County, California: U.S. Geological Survey Bulletin 1502-C, 35 p.

Sims, J. D., and Rymer, M. J., 1975a, Preliminary description and interpretation of cores and radiographs from Clear Lake, Lake County, California; Core 1: U.S. Geological Survey Open-File Report 75-665, 55 p.

—— , 1975b, Preliminary description and interpretation of cores and radiographs from Clear Lake, Lake County, California; Core 4: U.S. Geological Survey Open-File Report 75-666, 178 p.

—— , 1975c, Preliminary description and interpretation of cores and radiographs from Clear Lake, Lake County, California; Core 6: U.S. Geological Survey Open-File Report 75-569, 53 p.

—— , 1975d, Preliminary description and interpretation of cores and radiographs from Clear Lake, Lake County, California; Core 7: U.S. Geological Survey Open-File Report 75-144, 68 p.

Sims, J. D., Adam, D. P., and Rymer, M. J., 1981, Late Pleistocene stratigraphy and palynology of Clear Lake, Lake County, California, *in* McLaughlin, R. J., and Donnelly-Nolan, J. M., eds., Research in the Geysers–Clear Lake geothermal area, northern California: U.S. Geological Survey Professional Paper 1141, p. 219–230.

Taylor, D. W., 1960, Late Cenozoic molluscan faunas from the High Plains: U.S. Geological Survey Professional Paper 337, 94 p.

—— , 1966, Summary of North American Blancan nonmarine mollusks: Malacologia, v. 4, p. 1–172.

—— , 1981, Freshwater mollusks of California; A distributional checklist: California Fish and Game, v. 67, p. 140–163.

—— , 1986, Evolution of freshwater drainages and molluscs in western North America, *in* Leviton, A. E., and Smiley, C. J., eds., Late Cenozoic history of the Pacific Northwest: American Association for the Advancement of Science, Pacific Division, Symposium (in press).

Wetzel, R. G., 1975, Limnology: Philadelphia, W. B. Saunders Co., 743 p.

Printed in U.S.A.

Geological Society of America
Special Paper 214
1988

Pollen zonation and proposed informal climatic units for Clear Lake, California, cores CL-73-4 and CL-73-7

David P. Adam
U.S. Geological Survey
345 Middlefield Road
Menlo Park, California 94025

ABSTRACT

Clear Lake occupies a structural depression in the northern California Coast Ranges at an elevation of 404 m. Eight sediment cores were taken from the lake in 1973. This paper reports the palynology of cores CL-73-4 and CL-73-7. The former is 115 m long, and is interpreted to cover the entire last glacial cycle; the latter is 27.5 m long and covers at least the last 40,000 radiocarbon yr.

The pollen record of core CL-73-4 is dominated by three pollen types (oak, pine, and TCT [Taxodiaceae, Cupressaceae, and Taxaceae]) that together account for between 75 and 99 percent of the pollen in each sample. Core CL-73-7 is similarly dominated by these pollen types, but aquatic and riparian pollen types are also locally important.

The present vegetation around Clear Lake consists of oak woodland; mixed coniferous forest is found at higher elevations in the surrounding mountains. The present pollen rain into Clear Lake is dominated by oak pollen. During the cooler parts of the last glacial cycle, oak pollen influx to the sediments of Clear Lake was largely and at some times entirely replaced by coniferous pollen (mostly pine and TCT) in response to vertical migration of vegetation belts caused by climatic changes.

Pollen data were reduced using a Q-mode factor analysis. Five factors were defined that account for more than 98 percent of the variance. Three of the factors summarize aspects of the behavior of the regional forest vegetation around Clear Lake, and two summarize the behavior of the aquatic and swamp vegetation in the lake itself.

Zoning of the pollen diagrams was accomplished using an iterative program that minimized the total sums of squares of the factor loadings within zones. Twenty-one pollen zones are defined for core CL-73-4, four for core CL-73-7.

The pollen zones of core CL-73-4 are used to propose a series of informal climatic units that include the time interval from the penultimate glaciation to the present. The major units proposed, from oldest to youngest, are: (1) Tsabal cryomer, (2) Konocti thermomer, (3) Pomo cryomer, and (4) Tuleyome thermomer (Holocene).

The Pomo cryomer is divided into early, middle, and late phases. The early Pomo includes a series of five cold/warm oscillations that are designated the Tsiwi cryomers and the Boomli thermomers, numbered from Tsiwi 1 (oldest) to Boomli 5 (youngest). Middle Pomo time includes the Cigom 1 cryomer and the Halika thermomers, a series of three minor warm intervals. Late Pomo time includes the Cigom 2 cryomer and a transitional interval following it and preceding the Holocene.

The climatic oscillations of the Tsiwi cryomers and Boomli thermomers were often quite abrupt; both sudden warmings and sudden coolings occurred. The most severe of these changes was the cooling that occurred at the end of the Konocti thermomer, when

oak pollen frequencies dropped from more than 60 percent to about 25 percent within a stratigraphic interval of only 23 cm. These sudden changes were climatic catastrophes for the ecosystems that experienced them.

The record in the sediments of algae with acid-resistant remains indicates that lake productivity was relatively high during warm intervals in the past, and that overall productivity increased as the lake became shallower and its thermal inertia decreased. The lake waters were probably transparent during the cooler parts of the last glacial cycle, but Clear Lake has probably not been as clear a lake during the Holocene.

INTRODUCTION

Clear Lake, California, occupies a structural depression of complex origin in the northern Coast Range of California (Sims and Rymer, 1976; McLaughlin, 1981). In the fall of 1973, eight sediment cores were recovered from the lake (Fig. 1; Sims, 1976). Two of these cores, cores CL-73-4 and CL-73-7, have been analyzed for pollen. Core CL-73-4 comprises one of the longest continuous sedimentary records of upper Pleistocene events recovered within the United States.

This paper presents the pollen records of the two cores, defines and describes pollen zones for each of the cores, and proposes a set of informal climatic units to facilitate discussion of the record. These informal units are used both in the description of the local environmental history inferred from the pollen record (this paper) and in a companion paper that attempts to correlate the Clear Lake pollen record with long climatic records elsewhere (Adam, this volume). A third, closely related paper describes the derivation of the time scale used for the core CL-73-4 sequence (Robinson and others, this volume). This paper and the two mentioned above are extracted from a longer report (Adam, 1987) that contains a complete discussion of the pollen record from the two cores. This paper is restricted to a discussion of pollen percentage data, and refers only briefly to pollen concentration data and algal counts that are described in the longer work.

The data described herein are independent of later work on cores collected in 1980. One of those cores, core CL-80-1, has been examined for diatoms by Bradbury (this volume) and for pollen by Heusser (Heusser and Sims, 1981, 1983). Bradbury (this volume) concluded that the lower part of the 177-m-long core CL-80-1 is probably correlative with the Kelseyville Formation, which is exposed south of Clear Lake (Rymer, 1981). Examination of tephra from cores CL-80-1 and CL-73-4 yielded only one match: the tephra found at 67 m in core CL-73-4 is the same as that found at a depth of 54 m in core CL-80-1. Other tephra layers found deeper in core CL-80-1 have not been found in core CL-73-4, and one tephra at a depth of 155 m in core CL-80-1 has been identified as the Loleta ash bed, with an estimated age of about 360,000 yr (Sarna-Wojcicki and others, this volume). The limited tephra work available from the Clear Lake cores thus supports Bradbury's suggestion that there is a hiatus in core CL-80-1.

I believe that no such hiatus is present in core CL-73-4. Continuity of the core is suggested by grain size that is uniform and very fine throughout (Sims, 1982) and by a lack of weathering zones visible in either cross sections or radiographs of the cores (Sims and Rymer, 1975). The climatic sequence inferred from the pollen data appears to correlate remarkably well with numerous long climatic sequences elsewhere (Adam, this volume), and the time scale suggested by such correlations agrees well with age estimates based on extrapolation of radiocarbon-derived sedimentation rates (Robinson and others, this volume) and amino acid racemization rates (Blunt and others, 1981). Also, etching of pyroxene and hornblende phenocrysts in the tephra layers in core CL-80-1, which suggests subaerial exposure of the deposits, is observed only below depths of about 115 to 120 m (A. Sarna-Wojcicki, written communication, 1983), deeper than the base of core CL-73-4. Because the tephra layer and Bradbury's correlative diatom floras found in both cores occurs at a shallower depth in core CL-80-1, it is reasonable to infer that if correlative hiatuses are present at both sites, any resulting subaerially exposed deposits in core CL-73-4 would also occur deeper than in core CL-80-1. If that is true, the hiatus inferred in core CL-80-1 from such deposits lies below the bottom of core CL-73-4. This paper assumes that no significant hiatus exists in core CL-80-4, although Bradbury (this volume) believes that this assumption has not been rigorously tested.

STATISTICAL METHODS

Pollen samples from cores CL-73-4 and CL-73-7 are divided into a series of pollen zones using a procedure that involves (1) reducing the dimensionality of the data to five orthogonal factors, using the CABFAC program of Klovan and Imbrie (1971), as modified by Adam (1976); and (2) using an iterative boundary-fitting process to group the samples according to their factor loadings. Pollen counts used are presented in two open-file reports (Adam, 1979a, b).

Factor loadings are used as input to an interactive zoning algorithm that employs a least-squares criterion to find the optimum position for a single zone boundary within a given series of samples, subject to the constraint that every zone should contain at least two samples. Within a block of samples being zoned, all possible ways of dividing the block into two smaller blocks are considered, and the function:

$$S = \sum_{k=1}^{2} \sum_{j=1}^{5} \sum_{i=\mathrm{top}_k}^{\mathrm{bottom}_k} (x_{ij} - \bar{x}_{jk})^2 \qquad (1)$$

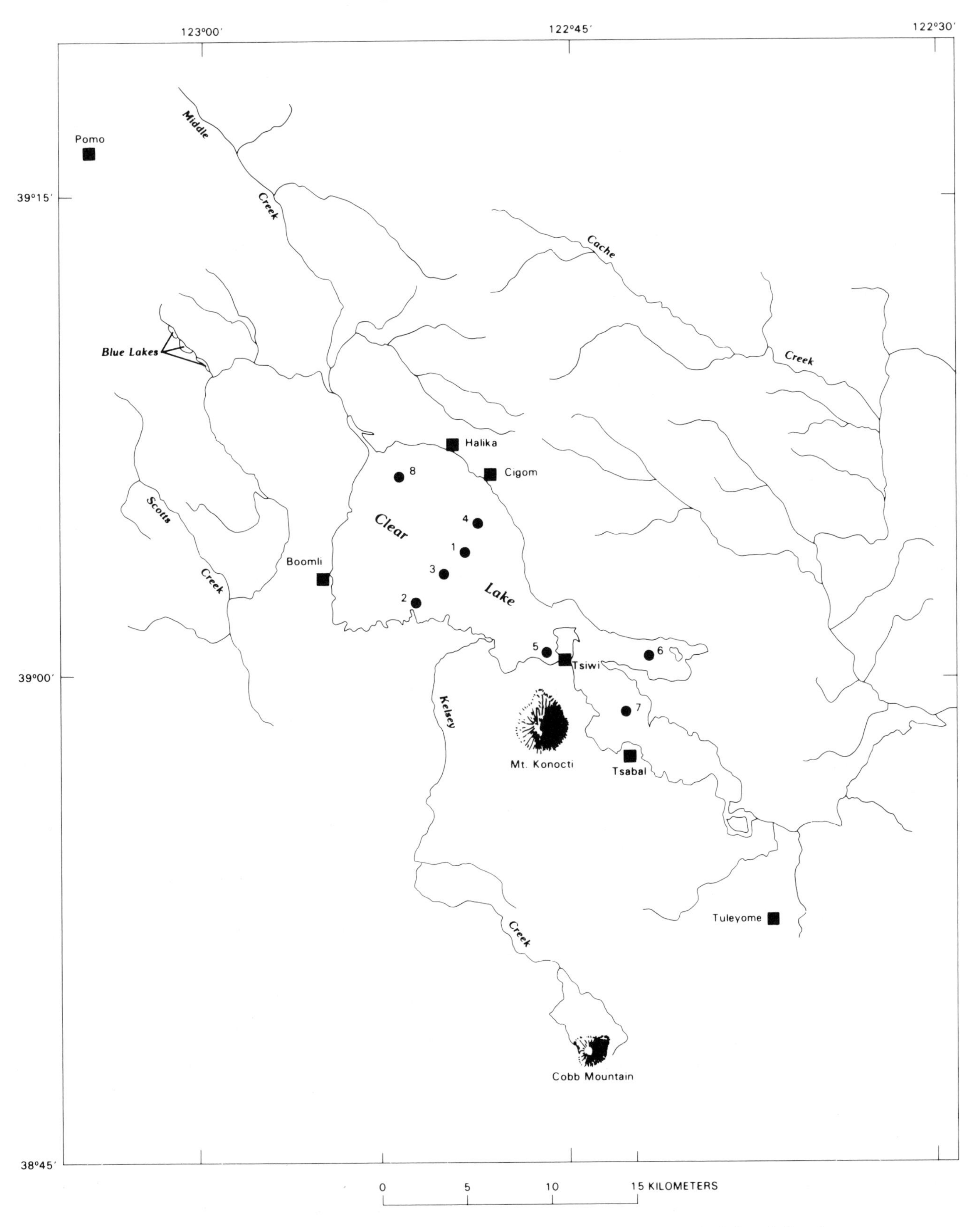

Figure 1. Map showing locations of core sites, former Indian villages used as names for informal climatic units, and other features mentioned in text.

is evaluated for each possibility, where x_{ij} is the value of the jth factor score for sample i, $\bar{x}_{jk}$ is the mean value of the observations of factor j in zone k, and top_k and $bottom_k$ are the top and bottom samples included in zone k. The boundary chosen is the one that produces the minimum value of S.

The choice of the upper and lower bounds of the block of samples to be subdivided can significantly affect the position of the boundary selected, especially if the block being divided contains more than two acceptable zones. The initial choice of blocks to be subdivided is made by subjective inspection of the factor score curves; the resulting zone boundaries then define new blocks to be divided. After several iterations, this process should produce a series of boundaries that are stable and consistent; that is, each zone boundary is selected as the best dividing point for the two zones that is separates.

PALYNOLOGY

All fossil pollen types are included in the pollen sum, as are microspores of *Isoetes* (quillwort). The inclusion of nonarboreal pollen and aquatic pollen in the sum has very little effect on the core CL-73-4 percentage data because those types are scarce in that core. In core CL-73-7, however, aquatic pollen types are abundant in the lower part of the section. The best estimates of the regional arboreal pollen (AP) rain therefore come from the core CL-73-4 record.

The pollen sum consists of at least 200 grains of fossil pollen. Pollen is well preserved throughout the core; the main hindrance to easy counting is the abundance of acid-resistant algal remains found near the top of both cores. The pollen identification key of Kapp (1969) and a U.S. Geological Survey reference collection of California pollen types were the principal resources used to identify unknown pollen grains.

The basic percentage data for the major pollen types are shown in graphic form in Plate 1. The data used are the first 38 variables of the summary data set of Adam (1987), which include all fossil pollen and also *Isoetes* (quillwort) spores and Nymphaeaceae (water lily) leaf hairs. The latter two variables are included because they represent vascular plants. Also shown in Plate 1 is a curve showing sediment density versus depth. Data for that curve were obtained by drying samples of known volume overnight at 110°C. (Plates 1 and 2 in back cover pocket.)

The Clear Lake arboreal pollen record is dominated by three pollen types—oak (*Quercus*), pine (*Pinus*), and TCT (an acronym for the plant families Taxodiaceae, Cupressaceae, and Taxaceae). TCT pollen in the Clear Lake record probably represents mostly incense cedar (*Calocedrus decurrens*), but juniper (*Juniperus*), cypress (*Cupressus* spp.), and redwood (*Sequoia sempervirens*) are probably also represented. The oak and pine pollen types also represent several species that cannot be reliably distinguished on the basis of their pollen grains.

The three major pollen curves are interpreted in terms of a vegetation model in which oak woodland such as now grows at elevations near and below that of Clear Lake produces the highest percentages of oak pollen. Above the oak woodland is a belt of mixed coniferous forest, including incense cedar, that produces the highest percentages of TCT pollen. Above the mixed coniferous forest is montane forest that produces the highest percentages of pine pollen. High percentages of oak pollen in the Clear Lake cores are thus interpreted as indicators of warm conditions, whereas high frequencies of pine and TCT pollen indicate cooler conditions. This model is the same as that of Adam and others (1981) and Sims and others (1981), and is consistent with the simpler model used by Adam and West (1983) and the modern surface sample data of Heusser (1983). A more complete summary is presented by Adam (1987).

Factor Analysis Results

The reduced set of 38 variables and 237 samples (165 from core CL-73-4 and 72 from core CL-73-7) was subjected to a factor analysis as described above, which produced five VARIMAX factors that together account for more than 98 percent of the variance in the data set. Each factor is dominated by only a few variables. These factors are shown in Plate 2, and the scaled factor scores are shown in Table 1. The five pollen factors are described here in order to present the data used to define the zone boundaries below, but interpretation of the pollen record is based on individual pollen types, primarily oak.

Factor 1 accounts for nearly half the variance, and responds primarily to TCT (factor score, 0.88) and *Pinus* (0.42), with smaller negative scores for Cyperaceae (–0.17), *Quercus* (–0.11), and Nymphaeaceae leaf hairs (–0.07) (Table 1). High factor loadings for factor 1 identify the most common type of sample in the data set. The negative factor scores for factor 1 are for variables that have high scores on subsequent factors that identify various other kinds of samples.

Factor 1 produces high factor loadings (>0.6) at depths of 26.00 to 108.67 m in core CL-73-4, as well as below 112.13 m, and low loadings above 26.00 m and between 108.00 and 112.00 m. In core CL-73-7, loadings above 7.20 m are generally comparable to those at the top of core CL-73-4. From 7.70 to 9.00 m, slightly negative values occur. Below 9.00 m are a series of oscillations between low and high values unlike any found in core CL-73-4 (Plate 2).

Factor 2 accounts for about one-fourth of the total variance, and is dominated by *Quercus* (factor score, 0.98), with lesser contributions from *Pinus* (0.10), TCT (0.08), and Rhamnaceae (0.07) (Table 1). Factor 2 produces high factor loadings in the top 20.00 m of core CL-73-4 and the top 6.8 m of core CL-73-7, and also between 108.90 and 112.03 m in core CL-73-4. Generally low values occur between 27.00 and 84.00 m in core CL-73-4, although some systematic fluctuations that affect groups of adjacent samples are found in this interval, and also below 112.00 m. Low values are also found below 7.20 m in core CL-73-7. Between 84.00 and 108.00 m in core CL-73-4, factor 2 shows a series of wide oscillations that change gradually with depth (Plate 2).

TABLE 1. SCALED POLLEN FACTOR SCORES FOR PERCENTAGE DATA, CORES CL-73-4 AND CL-73-7

Variable Name	VARIMAX Factors				
	1	2	3	4	5
Quercus	-0.1083	0.9806	0.0004	-0.0510	-0.0563
Pinus	.4159	.0975	.0749	.1767	.8766
Abies	.0179	-.0037	.0058	.0135	.0071
Picea	.0020	.0007	.0018	.0054	.0035
Tsuga	.0063	-.0008	-.0007	.0063	.0039
Pseudotsuga	.0106	.0040	-.0005	-.0020	.0043
TCT	.8821	.0772	.0515	.0917	-.4500
Alnus	-.0117	.0530	.0018	-.0009	.0202
Salix	.0046	.0217	.0004	-.0025	.0029
Brasenia	-.0018	.0009	.0130	-.0008	-.0018
Cyperaceae	-.1683	.0192	.0139	.9707	-.1211
Typha tetrads	-.0088	.0044	.0022	.0309	.0005
Typha monads	-.0050	.0014	.0004	.0421	-.0006
Cruciferae	-.0007	.0119	.0014	.0008	.0013
Potamogeton	-.0033	.0070	.0121	.0236	-.0089
Isoetes	.0286	-.0096	.0034	-.0179	.0816
Unknown A	.0009	.0112	.0071	-.0026	.0075
Artemisia	.0209	-.0063	.0129	.0614	.0018
Rhamnaceae	-.0208	.0737	-.0015	.0024	.0233
Cheno-ams	-.0005	.0087	.0039	.0161	.0085
High-spine Compositae	.0140	.0474	-.0024	.0023	.0253
Gramineae	.0073	.0261	-.0001	.0070	.0102
Polygonaceae	.0031	.0028	-.0008	.0017	.0017
Caryophyllaceae	.0030	-.0007	-.0002	.0022	-.0009
Arceuthobium	.0002	.0015	.0005	.0012	.0005
Portulacaceae	.0071	-.0009	-.0010	-.0029	-.0019
Other trees and shrubs	-.0006	.0030	.0003	.0000	.0031
Other aquatics	-.0023	.0005	.0170	.0137	-.0033
Umbelliferae	-.0091	.0021	.0006	.0502	-.0094
Chrysolepis	-.0065	.0237	-.0006	.0008	-.0088
Corylus	-.0002	.0060	.0001	-.0008	.0010
Juglans	.0005	.0021	.0001	-.0004	.0008
cf. Tilia	.0020	.0020	.0002	-.0007	.0105
Fraxinus	.0036	.0051	-.0007	-.0022	.0047
Nymphaeaceae leaf hairs	-.0749	-.0124	.9953	-.0333	-.0412
Lithocarpus	-.0022	.0027	-.0003	.0053	.0012
Other Compositae	.0010	.0052	.0006	-.0005	.0019
Unknowns	.0177	.0981	.0042	.0602	.0237

Factor 3 responds primarily to Nymphaeaceae leaf hairs (factor score, 0.99), with minor contributions from *Pinus* (0.07) and TCT (0.05) (Table 1). Loadings are low throughout core CL-73-4 except between 21.10 and 27.00 m. The top 7.00 m of core CL-73-7 show similar low loadings, as does the interval between 9.80 and 16.40 m; high loadings occur between 7.20 and 9.40 m. Below a depth of 16.60 m is a series of oscillations between high and low loadings (Plate 2).

Factor 4 responds mostly to Cyperaceae (factor score, 0.97), and also to *Pinus* (0.18) and TCT (0.09) (Table 1). *Typha* tetrads and monads also have low positive factor scores, whereas Nymphaeaceae leaf hairs and *Isoetes* spores show low negative scores. Factor loadings are low throughout core CL-73-4 and for the top 9.40 m of core CL-73-7. Between 9.80 and 15.80 m in core CL-73-7, the factor 4 loadings are consistently high; below 15.80 m, a series of oscillations occur that appear to be inversely related to the oscillations of factor 3 within this interval (Plate 2).

Factor 5, which accounts for only 4 percent of the variance in the data, is more complex than the first four factors. The variables that contribute positively to factor 5 are dominated by *Pinus* (factor score, 0.88) with a smaller contribution by *Isoetes* (0.08) (Table 1). Negative-effect variables include TCT (–0.45) and Cyperaceae (–0.12). High factor loadings for factor 5 occur for samples that have a high frequency of *Pinus* pollen relative to TCT pollen.

Pollen Zones

The factor loadings were used to select boundaries between pollen assemblage zones using the procedure described above. The zones selected by this procedure are listed in Table 2 and are shown on the pollen and factor diagrams (Plates 1, 2). These zone boundaries account for 92.43 percent of the variance of the core CL-73-4 factor loadings, and for 67.04 percent of the core CL-73-7 variance. Table 2 also presents means and standard deviations within each zone for selected variables that are important in the interpretation of the fossil record; descriptions of the pollen zones are presented in the Appendix.

One zone (F) has been further subdivided empirically. When the zone boundary between zones Fa and Fb was included

TABLE 2. POLLEN ASSEMBLAGE ZONES, ZONE DEPTHS, NUMBERS OF SAMPLES, AND MEANS AND STANDARD DEVIATIONS* FOR EACH ZONE FOR SELECTED VARIABLES

Pollen Zone	Sample Depths Top	Sample Depths Bottom	No. of Samples	Quercus	Pinus	TCT	Alnus	Chrysolepis	Rhamnaceae	High-spine Compositae
Core CL-73-4										
A	0.10	20.00	22	59.17± 5.01	15.34± 3.67	13.66± 3.66	5.16±1.40	1.95±1.31	6.17±3.91	3.22±1.61
B	21.10	22.95	3	23.16± 7.69	52.52±12.78	14.36± 2.86	4.03±2.29	0.19±0.33	0.39±0.34	2.09±1.46
C	25.05	25.43	2	15.15± 0.81	45.43± 3.35	32.20± 5.52	2.02±1.07	0.00±0.00	0.00±0.00	3.00±0.67
D	26.00	28.00	3	4.97± 4.56	51.40± 4.45	34.52± 5.74	0.18±0.31	0.19±0.32	0.00±0.00	0.54±0.02
E	29.40	44.30	18	1.22± 0.73	42.53± 6.76	51.87± 6.69	0.18±0.32	0.03±0.12	0.12±0.30	1.59±1.07
F	46.00	83.88	41	4.06± 3.25	29.46± 6.15	62.28± 6.11	0.28±0.32	0.01±0.07	0.14±0.26	1.82±1.10
Fa	46.00	70.98	26	4.72± 3.10	28.91± 5.73	61.83± 5.35	0.39±0.33	0.02±0.09	0.12±0.27	1.75±1.01
Fb	72.04	83.88	15	2.91± 3.27	30.41± 6.93	63.07± 7.38	0.09±0.20	0.00±0.00	0.17±0.25	1.94±1.27
G	84.25	84.55	2	18.30± 0.23	37.67±10.48	40.95± 9.54	0.00±0.00	0.26±0.37	0.00±0.00	1.55±0.01
H	84.85	86.07	3	5.10± 2.83	32.70± 2.51	58.30± 1.89	0.00±0.00	0.00±0.00	0.00±0.00	2.58±1.05
I	87.04	92.62	8	28.13± 4.86	22.68± 5.92	44.00± 4.34	0.62±0.82	0.06±0.18	0.27±0.50	1.12±0.74
J	92.82	94.90	4	4.03± 2.20	40.64± 3.34	50.97± 2.31	0.20±0.39	0.00±0.00	0.00±0.00	0.93±0.43
K	95.82	96.78	2	8.75± 6.07	19.68± 5.08	68.57± 1.05	0.00±0.00	0.00±0.00	0.25±0.35	1.14±0.22
L	98.51	101.30	5	27.51± 6.43	28.79± 7.97	39.08± 6.17	0.62±0.63	0.00±0.00	0.07±0.15	1.22±1.20
M	101.62	102.12	4	7.13± 1.57	43.66± 7.17	46.02± 6.22	0.63±0.47	0.00±0.00	0.00±0.00	1.68±0.80
N	102.22	104.36	11	31.65± 5.71	23.68± 6.40	40.48± 5.96	0.98±0.77	0.21±0.25	0.18±0.34	1.46±1.09
O	104.56	105.68	6	7.22± 3.31	55.85± 5.29	33.04± 4.58	0.29±0.23	0.00±0.00	0.07±0.18	1.90±1.16
P	105.88	107.10	4	15.88± 4.74	37.09± 5.14	42.65± 2.92	0.41±0.53	0.38±0.26	0.52±0.46	2.30±1.31
Q	107.60	108.07	2	5.35± 1.03	52.75±16.79	36.60±16.22	0.00±0.00	0.00±0.00	0.52±0.73	4.12±5.11
R	108.17	108.67	6	18.69± 4.18	42.23± 4.56	33.27± 2.31	0.41±0.38	1.60±1.65	3.45±1.00	7.55±2.58
S	108.90	112.13	12	66.87±12.76	8.35± 5.42	20.70± 7.18	0.63±0.65	0.40±0.66	0.04±0.15	3.35±1.95
T	112.23	114.01	5	11.10± 8.64	32.79± 2.50	48.32± 8.78	0.55±0.51	0.00±0.00	0.19±0.26	1.15±0.65
U	114.57	115.07	2	0.23± 0.33	54.46± 1.63	39.47± 0.31	0.00±0.00	0.00±0.00	0.00±0.00	0.90±0.71
Core CL-73-7										
A7	0.00	6.80	15	41.48± 5.28	15.46± 3.90	9.67± 2.18	2.13±0.77	1.33±1.16	4.95±3.67	2.58±1.36
B7	7.20	9.40	9	10.12± 5.67	27.70±11.13	15.55± 4.12	0.54±0.66	0.06±0.17	0.60±0.48	0.71±0.61
C7	9.80	16.40	20	0.25± 0.37	20.27± 7.34	20.10± 7.95	0.04±0.12	0.05±0.14	0.13±0.25	0.70±0.82
D7	16.60	27.10	28	2.02± 1.73	28.40± 8.51	32.65± 8.05	0.12±0.25	0.02±0.09	0.12±0.21	0.59±0.56

in the zonation scheme the boundary between zones F and G became unstable. Although the Fa/Fb boundary is useful for interpretation, it is included as a different class of boundary because of this stability problem.

CLIMATIC UNITS

The Clear Lake pollen record is divided here for convenience of discussion (Adam, this volume; Robinson and others, this volume) into a series of informal climatic units, which are based on the pollen zones summarized in Tables 1 and 2 and their interpretation in terms of the vegetation model described above. These units are intended to permit unequivocal reference to specific parts of the Clear Lake pollen sequence, rather than to establish a regional system for other work. The approach used follows Menke (1976), who recognized cold periods as cryomers and warm periods as thermomers in the early Weichselian climatic record of West Holstein (Germany).

As used herein, the term cryomer refers to an interval that is interpreted to be cooler than the intervals that precede and follow it; a thermomer is warmer. Either unit may contain lesser cryomers and thermomers within it. Names have been selected from the ethnogeography of the Pomo and neighboring Indians (Barrett, 1908). Locations of the former Indian villages and landmarks whose names are used in this report are shown in Figure 1. Spelling follows the Barrett versions with the exception of Mt. Konocti (Barrett: Kanaktai), for which the current spelling is used. The correspondence between cryomers, thermomers, and pollen zones is shown in Table 3.

Reference Section

The reference section for all of the climatic units described below is core CL-73-4, taken from beneath 8.4 m of water in the main basin of Clear Lake, about 2.5 km southwest of the town of Lucerne (Fig. 1). The length of the core is 115.2 m; coring was done between 25 September and 17 October, 1973. The core, described in detail by Sims and Rymer (1975), is presently stored at the U.S. Geological Survey in Menlo Park, California, under the curation of J. D. Sims.

Definition of Climatic Units

Tsabal cryomer. The oldest unit, the Tsabal cryomer, is named after the Southeastern Pomo Indian camp of Tsabal on the south shore of the Highlands Arm of Clear Lake (Fig. 1). The unit, which includes pollen zones T and U, is characterized in the reference section by *Quercus* pollen frequencies of less than 20

TABLE 2. POLLEN ASSEMBLAGE ZONES, ZONE DEPTHS, NUMBERS OF SAMPLES, AND MEANS AND STANDARD DEVIATIONS* FOR EACH ZONE FOR SELECTED VARIABLES (continued)

Pollen Zone	Sample Depths Top	Sample Depths Bottom	No. of Samples	Artemisia	Gramineae	Cyperaceae	Isoetes	Nymphaeaceae leaf hairs
Core CL-73-4								
A	0.10	20.00	22	0.86±0.64	2.58±1.38	3.34± 1.89	0.47±0.55	0.55± 0.94
B	21.10	22.95	3	0.75±0.34	0.73±0.63	3.55± 0.42	11.11±3.14	8.23± 4.39
C	25.05	25.43	2	0.63±0.89	0.32±0.45	15.70± 5.30	9.05±3.90	53.66± 5.62
D	26.00	28.00	3	1.06±0.88	1.59±1.02	3.44± 0.74	3.39±2.97	12.55± 7.47
E	29.40	44.30	18	3.93±2.03	2.25±1.15	2.52± 1.46	8.87±6.09	1.63± 1.10
F	46.00	83.88	41	2.50±1.69	1.15±0.94	1.99± 1.33	1.77±7.65	0.46± 0.88
Fa	46.00	70.98	26	2.83±1.93	0.95±0.82	1.99± 1.26	2.73±9.54	0.67± 1.05
Fb	72.04	83.88	15	1.94±0.97	1.51±1.04	2.00± 1.50	0.10±0.21	0.10± 0.22
G	84.25	84.55	2	1.03±0.72	2.06±0.71	2.31± 1.80	0.26±0.37	0.00± 0.00
H	84.85	86.07	3	0.91±0.84	2.04±1.58	1.49± 0.83	0.00±0.00	0.00± 0.00
I	87.04	92.62	8	0.46±0.34	0.27±0.57	1.28± 0.74	0.06±0.18	0.00± 0.00
J	92.82	94.90	4	0.59±0.47	0.29±0.36	1.59± 0.66	0.09±0.19	0.09± 0.19
K	95.82	96.78	2	1.70±1.80	0.71±0.40	2.60± 1.22	0.25±0.35	0.00± 0.00
L	98.51	101.30	5	0.39±0.41	0.56±0.58	0.94± 1.21	0.00±0.00	0.00± 0.00
M	101.62	102.12	4	0.89±0.26	0.13±0.27	1.15± 0.51	0.00±0.00	0.00± 0.00
N	102.22	104.36	11	0.58±0.44	0.34±0.41	1.00± 0.82	0.00±0.00	0.00± 0.00
O	104.56	105.68	6	1.18±0.48	0.32±0.39	1.56± 1.45	0.00±0.00	0.00± 0.00
P	105.88	107.10	4	1.24±0.91	0.13±0.25	3.43± 0.81	0.00±0.00	0.13± 0.25
Q	107.60	108.07	2	0.27±0.39	0.27±0.39	2.37± 1.00	0.00±0.00	0.00± 0.00
R	108.17	108.67	6	0.47±0.57	1.31±1.06	3.31± 1.26	0.26±0.43	0.07± 0.17
S	108.90	112.13	12	0.34±0.53	0.30±0.27	1.83± 0.78	0.19±0.38	0.04± 0.12
T	112.23	114.01	5	0.29±0.42	0.88±1.10	1.04± 0.60	0.16±0.35	0.00± 0.00
U	114.57	115.07	2	0.66±0.38	1.68±0.39	0.46± 0.66	0.00±0.00	0.23± 0.33
Core CL-73-7								
A7	0.00	6.80	15	0.39±0.33	2.20±1.27	4.17± 3.39	0.09±0.27	1.27± 1.69
B7	7.20	9.40	9	0.49±0.70	0.60±0.77	12.09± 7.41	0.00±0.00	506.76±396.25
C7	9.80	16.40	20	3.15±1.79	0.68±1.11	38.11±12.86	0.00±0.00	1.40± 2.72
D7	16.60	27.10	28	4.24±2.39	0.47±0.44	17.12±14.49	0.07±0.29	96.83±106.23

*Standard deviations are estimated values for each zone, rather than calculated values for a population consisting of the samples in the zone. Depths given are for the uppermost and lowermost samples in each zone.

percent, *Pinus* frequencies of more than 30 percent, and TCT frequencies of more than 40 percent, as well as minor amounts of *Abies, Picea, Tsuga,* and *Pseudotsuga.* The upper boundary is defined on the basis of an abrupt increase in the frequency of *Quercus* pollen between the depths of 112.23 and 112.13 m to frequencies in excess of 50 percent. The nature of the lower boundary is unknown. The presently known thickness of the deposits is less than 3 m but the unit is probably much thicker.

Konocti thermomer. The Konocti thermomer includes the interval represented by the deposits of pollen zone S, which are characterized by *Quercus* pollen frequencies in excess of 50 percent. Both the upper and lower boundaries are placed at abrupt changes in oak pollen frequency. The thickness of the deposits of the Konocti thermomer in core CL-73-4 is about 3.5 m. These deposits are the only deposits below those of the Holocene containing *Quercus* pollen frequencies that equal or exceed those of the Holocene; on that basis, the Konocti thermomer is correlated with the last pre-Holocene interglaciation. The unit is named after Mt. Konocti, a large volcano of middle and upper Pleistocene age that dominates the Clear Lake basin.

Pomo cryomer. The Pomo cryomer includes pollen zones B through R in core CL-73-4, and is characterized by *Quercus* pollen frequencies of less than 50 percent; on that basis, the Pomo cryomer is correlated with the last (=Wisconsinan) glaciation. Pollen zones B7, C7, and D7 are also included in the Pomo cryomer. The unit is named after the Pomo Indian village of Pomo in Potter Valley (Fig. 1). The deposits of the Pomo cryomer extend from 21.10 to 108.67 m in core CL-73-4, and numerous climatic fluctuations are recorded within the unit. Both the lower and upper boundaries are defined by abrupt changes in *Quercus* pollen frequency (Plate 1). The upper boundary is somewhat more gradual on the pollen diagrams, probably because the sediments at the upper boundary are not as compact as those at the lower boundary. The Pomo cryomer is divided into early, middle, and late phases, which are defined below.

Boomli thermomers. The climate during the early part of the Pomo cryomer oscillated several times between relatively warm conditions, characterized by *Quercus* pollen frequencies of more than 10 percent, and relatively cool conditions, with *Quercus* pollen frequencies less than 10 percent. Five local maxima of *Quercus* pollen, corresponding to pollen zones G, I, L, N, and P, are designated here as the Boomli thermomers. They are named after the Pomo Indian village of Boomli, whose site was within the present town of Lakeport (Fig. 1). The thermomers are numbered from Boomli 1 (oldest) to Boomli 5 (youngest).

Tsiwi cryomers. The deposits of the Boomli thermomers are separated from each other and from the underlying deposits of the Konocti thermomer by a series of beds that contain local

TABLE 3. CORRESPONDENCE BETWEEN INFORMAL CLIMATIC UNITS AND POLLEN ZONES RECOGNIZED IN THIS STUDY*

Informal Climate Unit	Pollen Zones		Boundary Age (yr)	Estimated Duration (yr)
	Core CL-73-4	Core CL-73-7		
Tuleyome thermomer	A	A7		10,400
			10,400	
Pomo cryomer				
Late	B,C,D	B7		5,200
			15,600	
Cigom 2 cryomer	E	C7		13,550
Middle			29,150	
Halika thermomers	Fa (in part)	D7		29,750
			58,900	
Cigom 1 cryomer	Fb			14,150
Early			73,050	
Boomli 5 thermomer	G			1,150
			74,200	
Tsiwi 5 cryomer	H			3,250
			77,450	
Boomli 4 thermomer	I			12,000
			89,450	
Tsiwi 4 cryomer	J			5,200
			94,650	
(Unnamed transitional interval)	K			4,250
			98,900	
Boomli 3 thermomer	L			6,750
			105,650	
Tsiwi 3 cryomer	M			1,500
			107,150	
Boomli 2 thermomer	N			4,700
			111,850	
Tsiwi 2 cryomer	O			2,400
			114,250	
Boomli 1 thermomer	P			3,100
			117,350	
Tsiwi 1 cryomer	Q			1,400
			118,750	
(Unnamed transitional interval)	R			1,250
			120,000	
Konocti thermomer	S			7,000
			127,000	
Tsabal cryomer	T,U			

*Boundary age and interval duration estimates follow Robinson and others (this volume).

stratigraphic minima for *Quercus* pollen, usually associated with local stratigraphic maxima for *Pinus* as well. *Quercus* pollen frequencies within these deposits are less than 10 percent. These *Quercus* minima are attributed to relatively cold conditions that are here designated the Tsiwi cryomers, named after the Pomo Indian village of Tsiwi at Little Borax Lake (Fig. 1). The cryomers are numbered from Tsiwi 1 cryomer (oldest) to Tsiwi 5 cryomer (youngest); each Tsiwi unit underlies the Boomli unit with the same number.

Cigom cryomers. The sediments overlying the Tsiwi cryomers and Boomli thermomers, pollen zones E and Fb, contain the lowest frequencies of *Quercus* pollen encountered in the Pomo cryomer. They are interpreted to represent cold conditions in the Clear Lake basin that correspond to the two major glacial maxima of the last glacial cycle. These intervals are designated here as the Cigom 1 and Cigom 2 cryomers, named after the Pomo Indian village of Cigom on the northeastern shore of Clear Lake (Fig. 1). *Quercus* pollen frequencies within the deposits of the Cigom cryomers are less than 5 percent, with the exception of a single sample in the Cigom 1 deposits. Cigom 1 corresponds to pollen zone Fb; Cigom 2, to pollen zone E. The boundaries of the Cigom cryomers are less well defined than those of the units defined above. Pollen zone C7 is also correlated with the Cigom 2 cryomer; the Cigom 1 cryomer is not represented in core CL-73-7.

Halika thermomers. The interval between the Cigom 1 and Cigom 2 cryomers was characterized by somewhat warmer climatic conditions. *Quercus* pollen frequencies rose above 5 percent during three separate intervals, each of which is represented by at least two samples in core CL-73-4. These intervals are herein designated as the Halika thermomers after the Pomo Indian village of Halika, which was located near Cigom on the northeast shore of Clear Lake (Fig. 1). The three intervals are designated as Halika 1, 2, and 3, from oldest to youngest; the samples included in the Halika thermomers are listed in Table 4. They are separated from each other by sediments with *Quercus* pollen frequencies of less than 5 percent. All three thermomers are found within pollen zone Fa.

Tuleyome thermomer (=Holocene). The Tuleyome thermomer is herein named after the Northern Moquelumnan Indian village of Tuleyome, which was located about 3.2 km south of the town of Lower Lake (Fig. 1). Pollen zone A is designated as the reference section, and zone A7 is also correlated with the Tuleyome thermomer. The reference section is characterized by *Quercus* pollen frequencies of more than 40 percent, and almost always more than 50 percent. Both *Pinus* and TCT pollen frequencies are generally less than 20 percent. Remains of the algae *Pediastrum* spp., *Botrycoccus,* and *Coelastrum* are abundant.

TABLE 4. SAMPLE DATA FOR THE HALIKA THERMOMERS

Halika Thermomer	Depth (m)		No. of Samples	*Quercus* (max. %)
	Top Sample	Bottom Sample		
1	57.00	58.00	2	9.5
2	63.20	67.00	5	9.7
3	70.01	72.04	3	10.4

ENVIRONMENTAL HISTORY

Tsabal Cryomer

The few samples from the Tsabal cryomer suggest that the climate was quite cold; oak pollen is almost completely absent from the lowest samples in the core. The major pollen types are TCT and pine, indicating that a mixed conifer forest surrounded Clear Lake at that time. The highest observed frequencies of spruce, hemlock, and Douglas fir are found in the deposits of the Tsabal cryomer; these suggest that the character of the forest may have been somewhat different than that of the forests that occupied the area during the Pomo cryomer. Minor amounts of the algae *Botryococcus* and *Pediastrum* were growing in the lake, especially during the transitional conditions of pollen zone T. The Gramineae (grass) pollen curve also shows a slight local maximum in the bottom three samples, which may indicate either more open conditions in some areas or the presence of some riparian grasses.

Konocti Thermomer

The Konocti thermomer began with a sudden doubling of oak pollen frequencies at the end of the transitional pollen zone T. This probably represents a very sudden climatic amelioration. Small abrupt jumps in the frequencies of Cyperaceae and high-spine Compositae pollen occurred at the beginning of the Konocti, and pine and TCT frequencies dropped suddenly.

The marked shift in vegetational types at the onset of the Konocti thermomer was accompanied by an abrupt increase of more than 25 percent in sediment density (Plate 1). This increase was probably the result of increased soil erosion in the Clear Lake watershed at a time when warmth-loving plant communities had not yet become well established. After the initial peak in sediment density at the onset of the Konocti, sediment density decreased throughout the interval, probably in response to a decreasing input of clastic sediment from the watershed and a somewhat increased productivity of the lake as it reached higher summer temperatures.

The middle of the Konocti thermomer was significantly warmer than the early and late phases, judging from the frequency of oak pollen. During the middle part of the Konocti, oak frequencies exceeded 80 percent, and pine pollen frequencies dropped to well below 10 percent. These conditions are not matched anywhere else in the record. It seems likely that the middle part of the Konocti thermomer was significantly warmer and/or drier than the warmest part of the Tuleyome thermomer (Holocene) at Clear Lake, and that the role of *Pinus* in the vegetation was less important than it is today. Adam and West (1983) estimated that temperatures were as much as 1.5°C warmer than at present during the Konocti thermomer. The amount of grass pollen reaching the lake sediments was very low, which suggests that the high oak pollen frequencies do not represent oak grassland in the Clear Lake basin.

Pomo Cryomer

The Pomo cryomer is here divided into three main phases that differed significantly in their climatic responses. The proposed divisions are shown in Plate 1.

Early Pomo cryomer. *Onset of the Pomo cryomer.* The shift from about 60 percent *Quercus* pollen at the end of the Konocti thermomer to about 25 percent at the onset of the Pomo cryomer probably occurred over a stratigraphic interval of no more than 23 cm. This represents an abrupt change, particularly since the sediments are not laminated and the sharpness of the shift has probably been blurred through bioturbation. However, the sudden end of the Konocti thermomer did not immediately lead to maximum cold conditions. The *Quercus* frequencies of pollen zone R drop from 25 percent at the base of the zone to values near 15 percent in the middle and top of the zone. Not until the onset of the Tsiwi 1 cryomer some 60 cm above the top of the Konocti deposits do *Quercus* pollen frequencies drop below 5 percent.

A slight reversal of the decline in sediment density occurred at the end of the Konocti thermomer (Plate 1), but was not nearly so large as the earlier change at the start of that interval. The climatic change at the end of the thermomer was in the direction of more moisture and cooler conditions, so that the stabilizing effect of plant cover on the soil probably increased.

The response of the vegetation of the Coast Ranges to the

climatic change at the end of the Konocti thermomer was dramatic. The oaks clearly responded suddenly; probably many were killed by the change, whereas others perhaps stayed alive for a time but were unable to reproduce. The proportions of different species of *Quercus* contributing pollen to Clear Lake sediments also probably changed significantly at the onset of the Pomo cryomer.

The decrease in *Quercus* pollen frequencies was matched by dramatic increases in the frequencies of TCT and *Pinus* pollen. *Pinus* frequencies increased more rapidly than those of TCT. This may represent a more rapid migration of *Pinus* than TCT species into habitats suddenly made suitable for them. It may also represent the ability of *Pinus* pollen to disperse great distances; if local pollen production were suppressed by a sudden climatic shift, then wind-transported pollen blown in from distant forests could become an important part of the pollen rain until the vegetation was able to adjust to the new climate.

The Tsiwi cryomers and Boomli thermomers. The Tsiwi cryomers and Boomli thermomers represent a series of broad oscillations of the regional climate of the Clear Lake area during an interval of 40,000 to 50,000 yr. The total thickness of the deposits is about 25 m; sediment density (Plate 1) shows both regular and erratic changes. Based on changes in oak pollen percentages, the climatic fluctuations of the early Pomo cryomer appear to have an amplitude of about half that observed between the warmest thermomer and the coolest cryomer. Adam and West (1983) estimated that the Boomli 2, 3, and 4 temperature maxima were about 1–2°C cooler than at present, and that minimum temperatures during the Tsiwi cryomers were about 7°C cooler.

The Tsiwi 1 cryomer was of short duration, and less than 1 m of sediment was deposited. *Quercus* pollen frequencies dropped to about 5 percent, whereas *Pinus* and TCT frequencies increased abruptly. A small but systematic peak of *Fraxinus* (ash) began during the Tsiwi 1 cryomer and persisted through the first half of the Boomli 1 thermomer that followed; a similar peak occurred for the high-spine Compositae.

Tsiwi 1 through Tsiwi 3. The interval from the beginning of the Tsiwi 1 cryomer through the end of the Tsiwi 3 cryomer was a period of marked climatic and environmental instability in the Clear Lake basin. The sediment density curve (Plate 1) shows many rapid oscillations that relate in a general way to three cycles of pollen concentration changes (Adam, 1987) during the same interval (101.62 to 108.07 m). During the initial phases of both the Tsiwi 1 and Tsiwi 2 cryomers, a rise in pollen concentration in the sediment accompanied a decrease in sediment density, suggesting in each case that a shift to cooler, moister conditions led to an increase in vegetative cover and a decrease in the erosion rate. These conditions were reversed during the warming conditions that led to the Boomli 1 and 2 thermomers.

Tsiwi 1 and 2 are both characterized by high frequencies of pine pollen and low frequencies of oak. Small amounts of *Botryococcus* are found in Tsiwi 1 deposits, and in the Tsiwi 2 deposits minor amounts of both *Botryococcus* and *Pediastrum* occur.

Boomli 1 thermomer. The Boomli 1 thermomer is a minor warm interval during the Early Pomo cryomer. Oak pollen frequencies increase from 5 percent to almost 20 percent and the ratio of pine to TCT pollen drops (Plates 1, 2). Pollen concentration data (Adam, 1987) indicate that the percentage figures shown in Plate 1 do not provide an adequate view of the Boomli 1 oak pollen record. The Boomli 1 oak pollen percentage peak occurs before the oak pollen concentration peak. Between depths of 107.60 and 106.55 m, pollen concentrations of both TCT and pine pollen drop by more than two-thirds, whereas oak concentrations increase by almost 30 percent. This suggests that the climatic warming that produced the Boomli 1 thermomer may have been too sudden for the oaks to adjust, and that the sudden drop in pine and TCT frequencies represents a sudden restriction in their range. The highest oak concentration did not occur until the top sample of the interval, so apparently the oaks were still migrating into the area and becoming established when the Boomli 1 thermomer came to an end.

Tsiwi 3 cryomer. The Tsiwi 3 cryomer is perhaps the most remarkable interval during the early Pomo cryomer because of its short duration and its abrupt start and end. Tsiwi 3 deposits include a stratigraphic interval of at least 50 cm but no more than 92 cm, and both the lower and upper boundaries are sharply defined in the percentage data (Plate 1). The upper boundary is somewhat less sharp in the pollen concentration data than in the percentage data (Adam, 1987), but still represents a rapid change.

A small but distinct increase in sediment density at the start of Tsiwi 3 time was the last of the rapid density fluctuations observed during the early part of early Pomo time (Plate 1). Total pollen concentration in the sediment dropped to the lowest values observed in the core. Although sediment density increased by only 8.5 percent, however, total pollen concentration dropped by 59 percent. Concentrations of oak, pine, and TCT pollen all dropped abruptly. The drop was proportionally the greatest for oak and the least for pine, so that constraint effects produced a percentage peak for pine. The changes in pollen concentration observed during the Tsiwi 3 cryomer result from some combination of changes in sediment density, sedimentation rate, and pollen influx. Changes in sediment density are of sufficient magnitude to account for only a small part of the changes, so pollen influx and sedimentation rate changes must have been the primary causes.

Because the decreases in the concentrations of pine, oak, and TCT pollen are not proportional, it seems unlikely that a simple increase in sedimentation rate could alone account for the changes. However, both the higher sediment density and the absence of an increase in pine pollen to accompany the decrease in oak suggest that the sedimentation rate during the Tsiwi 3 cryomer is probably higher than usual (pine and oak show an inverse relationship in the samples below the Tsiwi 3 deposits). If the sedimentation rate is in fact higher than normal, then the Tsiwi 3 cryomer is even shorter and more abrupt than its appearance in the pollen diagrams would indicate.

Boomli 3 thermomer. The onset of the Boomli 3 thermomer is nearly as sudden as the start of the Tsiwi 3 cryomer that precedes it. Dry sediment density decreases by nearly 20 percent and total pollen concentration more than doubles relative to the extreme conditions of the Tsiwi 3 cryomer, which suggest that sedimentation and erosion rates are relatively low. The Boomli 3 thermomer as defined here includes only the samples between 98.51 and 101.30 m (pollen zone L). The interval is unusual among the events of the Early Pomo cryomer in that it does not end suddenly; the transition from the oak maximum of Boomli 3 to the oak minimum of Tsiwi 4 is spread over a stratigraphic interval of 6.2 m. The upper part of the transition (pollen zone K) is marked by a great increase in TCT pollen at a time when oak pollen is decreasing. Pine pollen frequencies change little within the two zones. The upper part of the Boomli 3 thermomer is also characterized by low peaks of Douglas fir (*Pseudotsuga*), ash (*Fraxinus*), and high-spine Compositae pollen (Plate 1).

Lake vegetation varied during the Boomli 3 thermomer and the succeeding Boomli 3/Tsiwi 4 transition (Adam, 1987). The overall impression is that the lake became shallower during the transition from Boomli 3 warmth toward the colder conditions of the Tsiwi 4 cryomer. Because the shallow-water indicators (*Typha/Sparganium* and *Potamogeton*; Adam, 1987) do not persist, the low lake levels during the Boomli 3/Tsiwi 4 transition are probably the result of climatic shifts, rather than irreversible changes such as faulting.

Tsiwi 4 cryomer. The Tsiwi 4 cryomer is the longest and most severe cold interval of the early Pomo cryomer. Sediment density is near normal, but shows a tendency to increase with time. Oak pollen is quite scarce and pine pollen increases somewhat, whereas TCT frequencies decrease greatly from their peak in pollen zone K. Fir (*Abies*) pollen is also found, which suggests distinctly cool conditions in the basin. The pronounced pine pollen percentage peak (Plate 1) during the Tsiwi 4 cryomer is apparently largely an artifact of the percentage constraint (Adam, 1987).

Boomli 4 thermomer. The Boomli 4 thermomer is the longest of the warm intervals during the early Pomo cryomer. Warm conditions appear suddenly; the transition from maximum cold conditions to the maximum in oak pollen occurs over a stratigraphic interval of 76 cm. Total pollen densities are relatively low at the start of the interval but increase with time. Sediment density is higher than normal at the beginning and decreases with time.

A gradual cooling during the latter half of the Boomli 4 thermomer is implied by increases in pine and TCT concentrations. There is also a small rise in oak concentrations, but not enough to keep pace with the increases in pine and TCT, so oak percentages fall.

During the first half of the Boomli 4 thermomer there are low but persistent peaks in alder (*Alnus*) and willow (*Salix,* not shown on Plate 1) pollen, which indicate that these plants grew around the margin of the lake. The amount of fir pollen in the Boomli 4 samples is also somewhat higher than during the previous warm intervals; this suggests that stands of fir that became established in the mountains around Clear Lake during the Tsiwi 4 cryomer may have persisted as relict stands in protected habitats.

Tsiwi 5 cryomer. The Tsiwi 5 cryomer is marked by a drop in oak pollen and a rise in fir. High-spine Compositae and grasses also increase. Sediment density and total pollen concentration are relatively low.

Boomli 5 thermomer. The Boomli 5 thermomer is a relatively minor event, comparable in magnitude to the earlier Boomli 1 thermomer, but of shorter duration and lower amplitude than the other Boomli intervals. Concentrations of oak, pine, and TCT pollen in the sediment increase; sediment density is about normal.

Middle Pomo cryomer. In contrast with the early Pomo cryomer, the middle Pomo cryomer was a time of relative climatic stability. The climate was much colder than at present, but there is no record of repeated sudden changes from one climatic regime to another. In part this may be a result of the low frequencies of oak pollen during middle Pomo time. Oak is the most sensitive recorder of climatic changes in the lower part of the core, but when oak pollen is scarce the sensitivity to climatic changes is damped. Both pine and TCT pollen show changes in frequency, but the general impression is that most of those changes represent noise, rather than systematic climatic signals.

Cigom 1 cryomer. The Cigom 1 cryomer is marked by low pollen concentrations for pine, oak, and TCT, and by oak pollen frequencies at or near zero. This must represent the total or near-total elimination of oak from the Clear Lake basin and the establishment of a mixed coniferous forest down to elevations of 400 m (the elevation of the lake) or below. The ratio of pine to TCT pollen during Cigom 1 time is relatively low, suggesting that incense cedar was able to compete successfully against the pines around Clear Lake, and that the cooling did not depress the incense cedar distribution below 400 m.

Pollen Zone Fa and the Halika thermomers. The environmental complacency established during the Cigom 1 cryomer persists throughout the Middle Pomo cryomer with only a few perturbations. Three brief intervals with oak pollen percentages in excess of 5 percent are designated the Halika thermomers. These thermomers are all cooler than even the coolest of the Boomli thermomers, and are marginally warmer than the warmest of the Tsiwi cryomers, based on their oak pollen percentages. TCT pollen percentages and concentrations are high throughout middle Pomo time, whereas pine pollen concentrations show little change. The sharp peak in *Isoetes* microspores in pollen zone Fa marks the beginning of a period during which the quillwort is an important aquatic plant in the lake; this suggests that the lake water was clear and the lake bottom was within the photic zone.

The earliest part of the core CL-73-7 pollen record corresponds to the upper part of the middle Pomo cryomer. Pollen zone D7 is correlated with the upper part of pollen zone Fa using the resemblance of the *Quercus,* TCT, and *Artemisia* curves. Unfortunately, the lack of both sensitivity and detail preclude a precise correlation of the base of core CL-73-7 with the core

CL-73-4 sequence. The most likely guess is that the core CL-73-7 oak peak at 21.03 m corresponds to the core CL-73-4 peak at 58.00 m, and the pair of peaks at 24.70 and 25.50 m in core CL-73-7 correspond to the peaks at 64.10 and 66.00 m in core CL-73-4.

The Highlands Arm of Clear Lake was a shallow swamp during the last part of the middle Pomo cryomer. Sedge (Cyperaceae) pollen is abundant; standing water is indicated by relatively high percentages of *Potamogeton* (pondweed) and other aquatic pollen and by Nymphaeaceae leaf hairs. *Typha* (cattail) pollen is also found in a few levels.

The factor analysis results (Plate 2) indicate that shallow-water conditions fluctuated considerably during middle Pomo time, as is shown by the curves for factors 3 and 4. Shallow submerged vegetation is displaced several times by emergent vegetation, mostly sedges. Whether these water-level fluctuations are the result of climatic changes or variations in the overflow (sill) level for the swamp is not clear.

Late Pomo cryomer. **Cigom 2 cryomer.** The Cigom 2 cryomer is the time of maximum development of the cold climatic conditions of the last glacial cycle. This inference is based on the very low incidence of oak pollen, the high ratio of pine to TCT pollen, and the relatively high frequencies of *Abies* (fir), *Artemisia* (sagebrush), and Gramineae (grass) pollen. The high ratio of pine to TCT pollen suggests that incense cedar was not able to compete as successfully with the pines in the Clear Lake basin as it did earlier, and that the main part of the distribution of incense cedars was below the elevation of Clear Lake (404 m). This is consistent with the occurrence of incense cedar foliage below present sea level at Mountain View, along the western margin of San Francisco Bay, in deposits dated at between 21,000 and 24,000 radiocarbon yr B.P. (Helley and others, 1972).

A prominent feature of the pollen record during the Cigom 2 cryomer is the consistent presence of *Isoetes* (quillwort) spores. These plants probably grew on the lake bottom in a fairly wide belt around the edge of the lake. The spore record is interpreted here to mean that the lake water was probably much more transparent than during the Holocene; Clear Lake was actually a clear lake. (The lake may have been clear at earlier times as well, but too deep to support a bottom flora of quillworts.)

An increase in the algal flora at the start of late Pomo time marks a fundamental change in the limnology of Clear Lake. During and prior to early Pomo time, the ratio of algae with acid-resistant remains to pollen grains in the sediment sometimes reaches values of as high as 4, and during middle Pomo time the ratio does not exceed 2 (Plate 1). At the start of the Cigom 2 cryomer the ratio jumps to 7, and algae are more common than pollen in all but one of the samples of Cigom 2 and younger ages. It is possible that the increase in algal remains is the result of a shift in lake productivity from species that do not survive pollen extraction techniques to species that do, but it is more likely that the increase in algal remains is the result of a major increase in lake productivity.

The inferred increase in productivity can best be explained as the result of three interacting factors whose joint effect was to increase the supply of nutrients in the lake. The gradual shallowing of the lake as sediment accumulates in the basin has at least two effects: the thermal inertia of the lake decreases, and wind-generated turbulence becomes more effective in mixing the lake water. The climate also became cooler at the onset of the Cigom 2 cryomer, and may have caused a shift from monomixis to dimixis in the water column if the surface water temperature dropped below 4°C in the winter.

In the Highlands Arm, the shallow standing water of the preceding late-middle Pomo time was replaced by emergent swampy vegetation during the Cigom 2 cryomer. Pondweed pollen, Nymphaeaceae leaf hairs, and other aquatic pollen are absent from Cigom 2 samples, whereas sedge pollen is abundant.

End of the Pomo cryomer. The final stages of the Pomo cryomer occur after the end of the Cigom 2 cryomer. In core CL-73-4, three pollen zones (B, C, D) are recognized during the transition from the Cigom 2 cryomer to the Tuleyome thermomer, but only one zone (B7) is recognized in core CL-73-7. Oak pollen begins to increase in frequency at the start of pollen zone D, and by the end of zone D time it is more common than at any time since the end of the Boomli 5 thermomer.

Pollen zone D has much lower frequencies of *Artemisia* pollen than pollen zone E. This decrease, together with the decrease in TCT percentages and the beginning of the increase in oak pollen, marks the end of maximum cold conditions in the Clear Lake basin at about 13,000 radiocarbon yr ago.

However, not all of the pollen types diagnostic of full glacial conditions disappear immediately at the start of pollen zone D. In particular, relatively high frequencies of fir (*Abies*) and Douglas fir (*Pseudotsuga*) persist until the end of pollen zone D. Two pollen types that are common in pollen zones A through C—*Alnus* and the high-spine Compositae—are very scarce in pollen zone D.

Successive rises in the concentrations of pine, TCT, and oak pollen in pollen zone D (Fig. 2) are interpreted here as the release of the pines from climatic constraints on their growth and pollen production, followed by the arrival of competitors (incense cedars and oaks) from lower elevations and the eventual displacement of the pines and incense cedars by the oaks. The TCT increase is more than offset by the increase in pine pollen concentration, so that there is no prominent peak in the TCT percentage curve. The incense cedars apparently did not become a well-established part of the forest around Clear Lake during the transition from the Pomo cryomer to the Tuleyome thermomer.

Pollen zone C is characterized by pollen frequencies unlike those found anywhere else in core CL-73-4. Pine pollen frequencies drop below the levels found in the adjoining pollen zones B and D, whereas oak and TCT frequencies do not change markedly from those found at the top of pollen zone D. *Alnus* (alder) and high-spine Compositae pollen appear in significant numbers as arriving members of the developing Holocene vegetation.

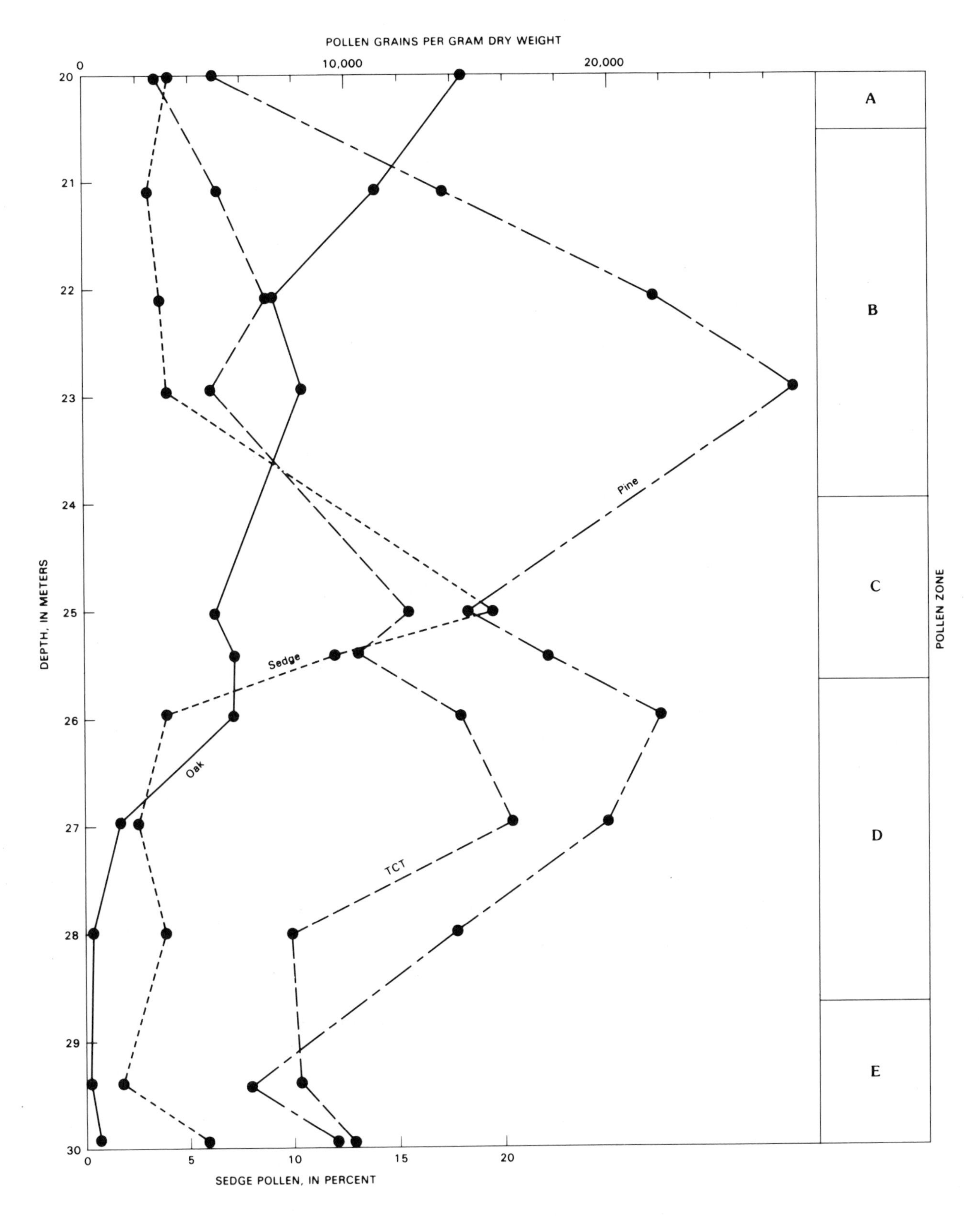

Figure 2. Concentrations of pine, TCT, and oak pollen plotted against depth for part of core CL-73-4 corresponding to transition from Pomo cryomer to Tuleyome thermomer.

The most remarkable feature of pollen zone C is the record of aquatic and riparian plants. Sedge pollen attains a frequency of nearly 20 percent, and there are peaks of cattails (*Typha* and *Typha/Sparganium*), Cruciferae, other aquatic pollen, and Nymphaeaceae leaf hairs as well. Taken together, these types suggest that the main basin of Clear Lake was shallower during pollen zone C time than it was before or after. *Isoetes* spores are common, which suggests that the water was clear; Nymphaeaceae (water lily) leaf hairs attain frequencies of up to 57 percent of total pollen, indicating standing water no more than a few meters deep.

The sediment density curve for core CL-73-4 (Plate 1) does not show any sign of desiccation at the core CL-73-4 site, nor does the core lithology (Sims and Rymer, 1975). Therefore, the explanation preferred here is a decrease in water depth, so that exposed mud flats around the margin of the lake supported a heavy growth of sedges and water lilies. The deposits of pollen zone B7 in core CL-73-7 also reveal changes in water level, but in an opposite direction from those observed in core CL-73-4. At the start of pollen zone B7, sedge pollen drops abruptly in frequency, and is replaced by abundant pollen of *Brasenia* (the water-shield), and also by pollen of *Potamogeton* (pondweed) and other aquatic pollen. Leaf hairs of Nymphaeaceae, probably from the *Brasenia* plants, are up to 10 times as common as pollen.

The opposite changes in water level between the two core sites suggest regional tilting or faulting, rather than climatic change, as the cause of the shifts in aquatic vegetation. Shallower conditions in the main body of the lake (core CL-73-4) are only temporary; the aquatic pollen peaks disappear at the end of pollen zone C. The shallow water plants also diminish at the end of pollen zone B7 in core CL-73-7, but at that site the water becomes still deeper. The peats and shallow-water muds of the lower part of the core are replaced by open-water lake muds containing the bones of large fish (Casteel and others, 1975, 1977). Lithologic and faunal changes similar to those observed in core CL-73-7 were also observed in core CL-73-6 (Sims, 1976; Casteel and others, 1977; Casteel and Beaver, 1978), so whatever mechanism caused the hydrographic changes must have acted on the Highlands and Oaks Arms of the lake as a unit.

An alternate explanation is that the increase in shallow-water indicators in core CL-73-4 could be the result of a drainage shift that caused the flooding of the Oaks and Highlands Arms and diverted their outflow from Cache Creek into the main body of Clear Lake, and thence into the Russian River through Blue Lakes and Cold Creek. According to this hypothesis, the landslide that presently blocks the head of Cold Creek would not have been formed at the time represented by pollen zone C. Some local tilting or faulting would still be required to initiate the drainage change.

The final phase of the Pomo cryomer, during pollen zone B, is characterized by very high frequencies of pine pollen and low frequencies of both TCT and oak pollen. Toward the end of the interval, the high pine frequencies diminish, and oak pollen resumes its rapid increase toward postglacial levels. Alder pollen increases in frequency, and by the start of the Tuleyome thermomer it is fully established as a member of the vegetation.

Tuleyome thermomer

The Tuleyome, or present, thermomer was fully established in the Clear Lake basin by about 10,000 radiocarbon yr ago, based on interpolation between the top radiocarbon date in core CL-73-7 and the top of the core. This is in agreement with the age of 9,990 ± 800 B.P. reported by Adam (1967) for the establishment of postglacial vegetation at Osgood Swamp in the Sierra Nevada.

There are no apparent systematic changes in the frequencies of oak, pine, or TCT during the Tuleyome thermomer in core CL-73-4. The area that contributes pollen to the main basin of Clear Lake is so large that the core CL-73-4 pollen record did not act as a sensitive recorder of minor climatic fluctuations during the Holocene. The core CL-73-7 pollen record, however, shows a distinct maximum in the oak percentage curve during the middle of the Tuleyome, and suggests a warmer and/or dryer interval during the mid-Holocene in the northern Coast Ranges.

A warm interval during the mid-Holocene is also supported by variations in growth annulus measurements of tule perch scales from cores CL-73-6 and CL-73-7 (Casteel and others, 1977; Casteel and Beaver, 1978). In those studies, fish scales from the middle part of the Holocene are found to have wider growth annuli than scales from the early or late Holocene. Wider growth rings are interpreted to mean warmer water temperatures, although no absolute temperatures are given.

Two pollen types have distributions within the Tuleyome thermomer that may be useful stratigraphically. *Chrysolepis (Castanopsis)*, the chinquapin, does not appear until well after the start of the Holocene, but the timing of its initial appearance is not consistent between cores CL-73-4 and CL-73-7. The highest frequencies of chinquapin are found in the top half of the Holocene deposits. Pollen of the Rhamnaceae (buckthorn family) appears irregularly throughout the pollen record of both cores, but its first appearance in substantial frequency appears to coincide with the oak pollen maximum of core CL-73-7. Rhamnaceae pollen may thus be a useful stratigraphic indicator for samples of mid-Holocene ("altithermal") or younger age.

The ratio of algae to pollen is high throughout the Holocene. Acid-resistant algal remains range from twice as abundant as pollen in parts of core CL-73-7 to more than 25 times as abundant at a depth of 20 cm in core CL-73-4 (Plate 1). *Isoetes* spores are present only in low frequencies, which suggests that the plants are restricted to very shallow submerged and emergent habitats at the edge of the lake.

Although the acid-resistant algal forms recorded in this study are not the major planktonic forms found in Clear Lake at present (Sandusky and Horne, 1978), they are found in large numbers in modern sediment, and their presence in older sediments is taken as an index of high lake productivity not only for themselves, but probably also for other more abundant but non-acid-resistant forms. High frequencies of algae throughout the Holocene are taken to mean that the productivity of the lake has been very high throughout the last 10,000 yr, and that the transparency of the waters of Clear Lake has probably not been high during that interval.

ACKNOWLEDGMENTS

J. D. Sims initiated and supervised drilling operations and the Clear Lake project. Core sampling and record keeping by C. K. Throckmorton and technical assistance by M. J. Rymer, R. O. Oscarson, R. Carr, M. Jackson, and J. Olsen are acknowledged. Pollen extractions were done under the supervision of P. J. Mehringer, Jr., at the Laboratory of Anthropology at Washington State University, Pullman.

APPENDIX 1. DESCRIPTIONS OF POLLEN ZONES

CORE CL-73-4

Zone A

Pollen zone A extends from the mud/water interface to a depth of 20.00 m, and includes 22 samples. Loadings are high (>0.9) for factor 2 and low for the other factors (Plate 2). The dominant pollen type is *Quercus; Pinus* and TCT are also common. Other relatively common pollen types are Rhamnaceae, *Alnus,* high-spine Compositae, Gramineae, and Cyperaceae (Plate 1, Table 2). Two types show systematic variations within zone A. Rhamnaceae pollen is present in only minor amounts in the bottom six samples of zone A and rises in frequency above a depth of 13.00 m. *Chrysolepis* is absent from the bottom three samples, but is present in low frequencies above a depth of 16.00 m.

Zone B

Zone B includes three samples that range in depth from 21.10 to 22.95 m. Loadings for factor 2 are lower than in zone A, and decrease with increasing depth from 0.66 at 21.10 m to 0.39 at 22.95 m (Plate 2). The factor loadings for the other factors are higher than in zone A, and the factor loadings for factor 5 are the highest observed in either core. *Pinus* is the most common pollen type in zone B, followed by *Quercus* and TCT (Plate 1, Table 2). *Isoetes* spores, Nymphaeaceae leaf hairs, and Cyperaceae pollen are the most common aquatic elements. Significant minor types are high-spine Compositae and *Alnus,* which triples in frequency from the bottom to the top of zone B.

Zone C

Zone C includes only two samples, at 25.05 m and 25.43 m. Loadings for factors 2 and 5 are lower than in zone B, factor 1 loadings are about the same, and factors 3 and 4 have the highest values observed in core CL-73-4 (Plate 2). *Pinus* and TCT are the most common AP types, followed by *Quercus* (Plate 1, Table 2). Aquatic types are important in zone C, and include Nymphaeaceae leaf hairs, Cyperaceae pollen, and *Isoetes* spores. Minor types include high-spine Compositae and *Alnus.* Within and above zone C, *Alnus* is consistently present in core CL-73-4; below zone C, it occurs only sporadically.

Zone D

The three samples in zone D are from depths of 26.00 to 28.00 m. Factor loadings for factor 1 are higher than in zones A through C (0.70 to 0.84), and loadings for factor 2 are lower (<0.28) (Plate 2). Loadings for factor 3 are fairly high (0.17 to 0.40) and increase upward toward the peak in zone C. Factor 4 loadings are similar to those in zones B and C. Factor 5 loadings are higher than in zone C (0.42 to 0.54), but not as high as in zone B.

Pinus is the most common pollen type, followed by TCT and *Quercus* (Plate 1, Table 2). *Abies* (not shown) is present as a minor element, along with Gramineae and *Salix.* Both *Abies* and *Salix* are missing from the top sample. Nymphaeaceae leaf hairs, Cyperaceae pollen, and *Isoetes* spores are important, although much less so than in zone C.

Zone E

Zone E includes 18 samples ranging in depth from 29.40 to 44.30 m. Loadings are high for factor 1 (mostly >0.9) and low for factors 2 and 3 (<0.17) (Plate 2). Factor 4 loadings are fairly high compared to the rest of core CL-73-4 (0.17 to 0.25), but not as high as in parts of core CL-73-7. Loadings for factor 5 are systematically positive within zone E, but vary considerably (0.07 to 0.61). The most common pollen types are TCT and *Pinus* (Plate 1, Table 2). Minor types include *Artemisia,* Gramineae, *Abies,* and high-spine Compositae. *Quercus* is rare. Aquatic types include *Isoetes,* Cyperaceae, and Nymphaeaceae leaf hairs. Nymphaeaceae leaf hairs are less abundant than Cyperaceae or *Isoetes,* in contrast to zones C and D.

Zone F

Zone F is the thickest zone recognized in the data set, and includes 41 samples ranging in depth from 46.00 to 83.88 m. The factor loadings for factors 1, 3, and 4 show very little variability within the zone, except for a single-sample local minimum for factor 1 at a depth of 59.04 m (Plate 2). Factor 2 shows a series of low-amplitude oscillations above a depth of 70.98 m, and factor 5 fluctuates in an apparently random way throughout the zone. The dominant pollen type is TCT, and the other major type is *Pinus* (Plate 1, Table 2). *Quercus* is a minor type in zone F; other minor types include *Artemisia,* Cyperaceae, high-spine Compositae, *Abies,* and Gramineae. *Artemisia* is more common in the top third of the zone, as is *Abies* to a lesser extent. Three local maxima of *Isoetes* spores occur in the top half of the zone, including the prominent peak (48.9 percent) at 59.04 m (Plate 1).

As noted above, pollen zone F is further subdivided empirically to facilitate discussion. The upper part of zone F shows several systematic fluctuations in the *Quercus* pollen curve, but the zoning algorithm did not produce useful and stable boundaries within zone F. The zone is divided into an upper part (Fa) and a lower part (Fb) using a criterion of 5 percent *Quercus* pollen, which places the boundary between the two subzones between 73.01 and 72.04 m. The same criterion is used later to delimit subunits within subzone Fa.

Zone G

Zone G includes two samples at 84.25 and 84.55 m. The primary feature setting the zone apart from those on either side is relatively high factor loadings for factor 2 (about 0.43) and slightly lower loadings (<0.90) for factor 1 (Plate 2). The lower sample also shows a sharp peak for factor 5 (0.40). Factor loadings for factors 3 and 4 are similar to those of adjoining zones. TCT and *Pinus* are the most common pollen types, but *Quercus* is also important (Plate 1, Table 2). Minor types include Cyperaceae, Gramineae, high-spine Compositae, *Abies, Artemisia,* and other Compositae.

Zone H

Zone H includes three samples with depths between 84.85 and 86.07 m. Loadings are similar to those in zone F. TCT and *Pinus* are the most common pollen types, followed by *Quercus* (Plate 1, Table 2). Minor types include high-spine Compositae, *Abies,* Gramineae, and Cyperaceae.

Zone I

Zone I includes eight samples that range in depth from 87.04 to 92.62 m. Loadings for factor 1 are <0.80, and loadings for factor 2 are >0.40 (Plate 2). These joint criteria distinguish zone I samples from samples in the adjacent zones. TCT is the most common pollen type, followed by *Quercus* and *Pinus* (Plate 1, Table 2). *Quercus* frequencies decrease upward within zone I, whereas *Pinus* and TCT increase. Small amounts of *Alnus* and *Salix* pollen occur in the bottom half of the zone but not in the top half. Minor pollen types include Cyperaceae, high-spine Compositae and *Abies.*

Zone J

Zone J includes four samples ranging in depth from 92.82 to 94.90 m. Factor loadings are similar to those of zone E. TCT and *Pinus* account for most of the arboreal pollen; *Quercus* is scarce (Plate 1, Table 2). Minor types include *Abies* and Cyperaceae.

Zone K

Zone K includes two samples, at depths of 95.82 and 96.78 m. Factor loadings are generally similar to those of zone J, except that loadings for factor 5 are negative instead of positive (Plate 2). TCT pollen is very common, much more so than *Pinus* or *Quercus* (Plate 1, Table 2). Minor types include Cyperaceae, *Artemisia,* and high-spine Compositae (Plate 1, Table 2).

Zone L

Zone L includes five samples at depths between 98.51 and 101.30 m. Factor loadings are similar to those of zone I, but factor 5 has more positive loadings in zone L, notably in the lowest two samples (Plate 2). TCT, *Pinus,* and *Quercus* are all important pollen types; minor types include *Fraxinus* and high-spine Compositae.

Zone M

Zone M consists of four samples from the 50-cm interval between 101.62 and 102.12 m. Zone M is set apart from adjacent zones by low loadings for factor 2 (<0.27), which contrast with values >0.45 throughout zones L and N (Plate 2). TCT and *Pinus* are nearly equally common, whereas *Quercus* is much reduced from its frequency in the adjoining zones (Plate 1, Table 2). Minor types include high-spine Compositae and Cyperaceae.

Zone N

Zone N includes 11 samples between 102.22 and 104.36 m. The most common type is TCT, and *Quercus* is higher in frequency than *Pinus* (Plate 1, Table 2). Minor types include high-spine Compositae, Cyperaceae, and *Alnus.*

Zone O

Zone O includes six samples at depths between 104.56 and 105.68 m. Loadings for factors 1, 3, and 4 are slightly higher than in Zone N (Plate 2). Zone O has distinctive patterns for factor 2, which has much lower loadings than in the adjacent zones, and factor 5, which has much higher loadings. The most common pollen type is *Pinus,* followed by TCT (Plate 1, Table 2). *Quercus* is relatively scarce. Minor types include high-spine Compositae, Cyperaceae, *Artemisia,* and *Abies.*

Zone P

Zone P includes four samples that range in depth from 105.88 to 107.10 m. Loadings for factor 2 are in the range 0.33 to 0.52, higher than in the adjoining zones O and Q but lower than in zones I, L, N, and S (Plate 2). Loadings for factor 5 are low (<0.32), in contrast to zone O. Loadings for factors 3 and 4 differ little from those of adjacent zones. The most common pollen types are TCT and *Pinus* (Plate 1, Table 2). *Quercus* is present in frequencies higher than those of adjacent zones O and Q. Less common types include Cyperaceae, high-spine Compositae, *Artemisia, Fraxinus,* and Cheno-Ams.

Zone Q

Zone Q consists of two samples, at depths of 107.60 and 108.07 m. The zone is separated from the adjoining zones on the basis of low (<0.25) loadings for factor 2 (Plate 2). There is also a peak value of 0.64 (Plate 2) for factor 5 in the bottom sample of zone Q, but that value appears to be more closely related to the values in zone R than to the other zone Q value. The dominant pollen type is *Pinus,* and TCT pollen is also common (Plate 1, Table 2). Minor types include high-spine Compositae, *Fraxinus,* and Cyperaceae.

Zone R

Zone R consists of six samples at 10-cm intervals between the depths of 108.17 and 108.67 m. These samples plus the bottom sample of zone Q came from a single section of the core. Loadings for factor 1 are in the range 0.70 to 0.79, somewhat lower than in the two overlying zones but much higher than in zone S (Plate 2). Factor 2 loadings range from 0.38 to 0.58, again intermediate between the overlying and underlying zones. Loadings for factor 5 reach a local maximum if the lower sample of zone Q is included, and factors 3 and 4 both show increases above the values found in zone S. *Pinus* and TCT are the major pollen types, but *Quercus* is also fairly common (Plate 1, Table 2). This zone also includes the highest frequencies of high-spine Compositae found in either core. Minor types include Rhamnaceae, Cyperaceae, *Chrysolepis,* and Gramineae.

Zone S

Zone S consists of 12 samples that range in depth from 108.90 to 112.13 m. Loadings for factor 1 are similar to those for zone A, but attain even smaller values in the middle of this zone (Plate 2). Loadings for factor 2 rise from a low of 0.80 for the bottom sample to values in excess of 0.95 for the top eight samples. The dominant pollen type is *Quercus,* which in the middle of zone S reaches the highest frequencies found in either core, whereas *Pinus* is less frequent than in any other part of the core. Pollen of the high-spine Compositae and Cyperaceae are found throughout the zone. Several minor types are found primarily in the top half of zone S, including Cruciferae, *Chrysolepis, Fraxinus,* and *Juglans. Salix* appears in higher frequencies at the bottom and top of zone S than in the middle.

Zone T

Zone T includes five samples ranging in depth from 112.23 to 114.01 m. Loadings for factor 1 are high, but decline from a value of 0.971 at the base of the zone to 0.85 at the top (Plate 2). Factor 2 loadings are quite low at the base of the zone (0.12), but rise to value of >0.45 at the top. Loadings for factors 3 and 4 are comparable to those in zones O through R. Factor 5 has low positive values. The most common pollen types are TCT and *Pinus* (Plate 1, Table 2). TCT values reach a local maximum in zone T, whereas *Pinus* values are steady within the zone and *Quercus* values rise almost steadily through time. Zone T is notable for the low but steady frequencies of a number of conifers in addition to *Pinus* and TCT; these include *Pseudotsuga* (2.7 ± 0.7 percent), *Tsuga* (1.8 ± 0.9 percent), *Picea* (0.9 ± 0.8 percent), and *Abies* (0.9 ± 0.4 percent). Other minor types include high-spine Compositae and Cyperaceae.

Zone U

Zone U includes the bottom two samples of core CL-73-4, at depths of 114.57 and 115.07 m. Loadings for factors 1, 3, and 4 are similar to those found in zone T (Plate 2), factor 2 loadings are lower than in zone T (and among the lowest found in the core), and factor 5 loadings are high (>0.40). The zone is dominated by *Pinus* and TCT pollen, followed by *Tsuga* (1.9 ± 0.7 percent), *Pseudotsuga* and Gramineae (both 1.7 ± 0.4 percent), and *Abies* (1.0 ± 0.8 percent). Only a single grain of *Quercus* was found in the two samples.

CORE CL-73-7

Core CL-73-7 pollen zones are distinguished from core CL-73-4 zones by a "7" appended to the zone letter. Corresponding letters in the two cores do not necessarily designate correlative units.

Zone A7

Zone A7 includes the top 15 samples in core CL-73-7; these range in depth from 0.00 to 6.80 m. The dominant factor pattern is one of high loadings for factor 2 (Plate 2). The other factors all show higher loadings at the top and bottom of the zone than in the middle. The most common pollen type is *Quercus*, followed by *Pinus* and TCT (Plate 1, Table 2). Rhamnaceae pollen is common above a depth of 4.40 m, and *Chrysolepis* pollen is consistently present in the top half of the zone. *Alnus* is present throughout the zone, and high-spine Compositae pollen is most common below a depth of 2.00 m. Gramineae pollen shows two peaks, a broad one in the middle of the zone and a sharper one in the topmost sample. Cyperaceae pollen is most frequent in the bottom half of the zone.

Zone B7

Zone B7 includes nine samples between depths of 7.20 and 9.40 m. The zone is characterized by factor 2 loadings in excess of 0.90 and loadings near zero for the other four factors (Plate 2). The dominant microfossils in this zone are Nymphaeaceae leaf hairs, which are several times as abundant as pollen grains (Plate 1). *Brasenia* pollen (not shown) is also important (16.5 ± 18.3 percent), and probably represents the same plants as the leaf hairs. The most important trees are *Pinus* and TCT, but *Quercus* is also present in quantity (Plate 1, Table 2). Other aquatic pollen types include Cyperaceae, *Potamogeton* (1.0 ± 1.1 percent), and "other aquatics" (1.9 ± 1.6 percent).

Zone C7

Zone C7 includes 20 samples at depths ranging between 9.80 and 16.40 m. Loadings for factor 4 are consistently high in this zone, whereas loadings for factor 3 are generally low (Plate 2). Loadings for factor 1 are as high or higher than in zone A7, and show the highest values in the middle of the zone. Factor 5 increases irregularly to values >0.40 near the top of the zone. Only factor 2 behaves consistently throughout the zone, with generally low values. The most common pollen type in zone C7 is Cyperaceae, in marked contrast to all other zones in both cores, which are dominated by *Quercus, Pinus,* or TCT (Plate 1, Table 2). Other aquatic types important in zone C7 are *Typha, Typha/Sparganium,* and Umbelliferae. The major AP types are *Pinus* and TCT. Small amounts of *Abies* (1.0 ± 0.8 percent) are also present, as well as occasional grains of *Picea, Tsuga,* and *Pseudotsuga.*

Zone D7

Zone D7 includes the bottom 28 samples in core CL-73-7, and extends from a depth of 16.60 m to the bottom of the core at 27.10 m. The zone is characterized by abrupt short-term fluctuations in factors 3, 4, and 1, with somewhat smaller fluctuations in factor 5 (Plate 2). Although zone D7 could easily be subdivided into several smaller zones, the general appearance of the zone suggests that this would not be useful, so this zone is characterized here by its variability rather than its uniformity. Important AP types are TCT and *Pinus,* as well as lesser amounts of *Quercus* and *Abies* (1.5 ± 1.2 percent, not shown) (Plate 1, Table 2). *Artemisia* is also rather common, and there are a few minor peaks in the Cheno-Ams curve. Aquatic types are both important and highly variable, and include Nymphaeaceae leaf hairs, Cyperaceae, *Potamogeton* (2.7 ± 4.5 percent), and "other aquatics" (2.4 ± 1.8 percent). *Typha* and *Typha/Sparganium* also appear intermittently, especially at the bottom of the zone.

REFERENCES CITED

Adam, D. P., 1967, Late Pleistocene and Recent palynology in the central Sierra Nevada, California, *in* Cushing, E. J., and Wright, H. E., Jr., eds., Quaternary paleoecology: New Haven, Yale University Press, p. 275–301.

——, 1976, CABFAC/USGS, a FORTRAN program for Q-mode factor analysis of stratigraphically ordered samples: U.S. Geological Survey Open-File Report No. 76-216, 27 p.

——, 1979a, Raw pollen counts from core 4, Clear Lake, Lake County, California: U.S. Geological Survey Open-File Report 79-663, 181 p.

——, 1979b, Raw pollen counts from core 7, Clear Lake, Lake County, California: U.S. Geological Survey Open-File Report No. 79-1085, 92 p.

——, 1987, Palynology of two upper Quaternary cores from Clear Lake, Lake County, California: U.S. Geological Survey Professional Paper (in press).

Adam, D. P., and West, G. J., 1983, Temperature and precipitation estimates through the last glacial cycle from Clear Lake, California, pollen data: Science, v. 219, p. 168–170.

Adam, D. P., Sims, J. D., and Throckmorton, C. K., 1981, 130,000-yr continuous pollen record from Clear Lake, Lake County, California: Geology, v. 9, p. 373–377.

Barrett, S. A., 1908, The ethno-geography of the Pomo and neighboring Indians: University of California Publications in American Archaeology and Ethnology, v. 6, no. 1, 332 p.

Blunt, D. J., Kvenvolden, K. A., and Sims, J. D., 1981, Geochemistry of amino acids in sediments from Clear Lake, California: Geology, v. 9, p. 378–382.

Casteel, R. W., and Beaver, C. K., 1978, Inferred Holocene temperature changes in the North Coast Ranges of California: Northwest Science, v. 52, no. 4, p. 337–342.

Casteel, R. W., Adam, D. P., and Sims, J. D., 1975, Fish remains from Core 7, Clear Lake, Lake County, California: U.S. Geological Survey Open-File Report No. 75-173, 67 p.

——, 1977, Late Pleistocene and Holocene remains of *Hysterocarpus traski* (tule perch) from Clear Lake, California, and inferred Holocene temperature fluctuations: Quaternary Research, v. 7, p. 133–143.

Helley, E. J., Adam, D. P., and Burke, D. B., 1972, Late Quaternary stratigraphic and paleoecological investigations in the San Francisco Bay area, *in* Frizzell, V. A., ed., Unofficial progress report on the U.S.G.S. Quaternary studies in the San Francisco Bay area, Guidebook for Friends of the Pleistocene meeting, October 6-8, 1972, p. 19–30.

Heusser, L. E., 1983, Contemporary pollen distribution in coastal California and Oregon: Palynology, v. 7, p. 19–42.

Heusser, L. E., and Sims, J. D., 1981, Palynology of core CL-80-1, Clear Lake, California: Geological Society of America Abstracts with Programs, v. 13, no. 2, p. 61.

——, 1983, Pollen counts for core CL-80-1, Clear Lake, Lake County, California: U.S. Geological Survey Open-File Report 83-384, 28 p.

Kapp, R. O., 1969, How to know the pollen and spores: Dubuque, Iowa, W. D. Brown, 249 p.

Kloven, J. E., and Imbrie, J., 1971, An algorithm and FORTRAN-IV program for large-scale Q-mode factor analysis and calculation of factor scores: Mathematical Geology, v. 3, p. 61–77.

McLaughlin, R. J., 1981, Tectonic setting of pre-Tertiary rocks and its relation to geothermal resources in the Geysers–Clear Lake area, *in* McLaughlin, R. J., and Donnelly-Nolan, J. M., Research in the Geysers–Clear Lake geothermal area, northern California: U.S. Geological Survey Professional Paper 1141, p. 3–23.

Rymer, M. J., 1981, Stratigraphic revision of the Cache Formation (Pliocene and Pleistocene), Lake County, California: U.S. Geological Survey Bulletin 1502-C, 35 p.

Sandusky, J. C., and Horne, A. J., 1978, A pattern analysis of Clear Lake phytoplankton: Limnology and Oceanography, v. 23, p. 636–648.

Sims, J. D., 1976, Paleolimnology of Clear Lake, California, U.S.A., *in* Horie, S., ed., Paleolimnology of Lake Biwa and the Japanese Pleistocene: Kyoto, Japan, Kyoto University, v. 4, p. 658–702.

——, 1982, Granulometry of core CL-73-4 Clear Lake, California: U.S. Geological Survey Open-File Report No. 82-70, 7 p.

Sims, J. D., and Rymer, M. J., 1975, Preliminary description and interpretation of cores and radiographs from Clear Lake, Lake County, California; Core 4: U.S. Geological Survey Open-File Report 75-666, 19 p.

——, 1976, Map of gaseous springs and associated faults, Clear Lake, California: U.S. Geological Survey Miscellaneous Field Investigations Map MF-721.

Sims, J. D., Rymer, M. J., and Perkins, J. A., 1981, Description and preliminary interpretation of core CL-80-1, Clear Lake, Lake County, California: U.S. Geological Survey Open-File Report 81-751, 175 p.

Manuscript Accepted by the Society September 15, 1986

Printed in U.S.A.

Geological Society of America
Special Paper 214
1988

Correlations of the Clear Lake, California, core CL-73-4 pollen sequence with other long climate records

David P. Adam
U.S. Geological Survey
345 Middlefield Road
Menlo Park, California 94025

ABSTRACT

Clear Lake core CL-73-4 records fluctuating abundances of oak pollen during the last glacial/interglacial cycle that correlate remarkably well with fluctuations in extensive pollen records from Grande Pile in France and Tenaghi Phillipon in Macedonia, as well as with the oxygen-isotope records from deep-sea cores. The record correlates less closely with other extensive records, including those for Lake Biwa, Japan, and Sabana de Bogotá, Colombia. Correlation of the record with the early Weichselian climatic sequence of northwestern Europe is excellent; both sequences show a series of five cryomer/thermomer fluctuations between the end of the last interglaciation (Eemian/Konocti, which is correlated here with the end of marine oxygen-isotope Stage 5e) and the onset of full continental glaciation at the end of Stage 5a. The fluctuations correlate both in their relative durations and in their relative amplitudes.

The Clear Lake record also correlates with various North American sequences. The Sangamon interval of the mid-continent area correlates with the entire Konocti thermomer and early Pomo cryomer interval, and correlations with the glacial sequences of the Sierra Nevada and Rocky Mountains suggest that some Tahoe, Mono Basin, and Bull Lake moraines may be of Sangamon age.

The proposed correlations of the Clear Lake record with other sequences have not been proved. The overall impression, however, is one of remarkable consistency, and it is likely that further work will provide more evidence in support of the sequence of five cryomer/thermomer cycles between the end of the last interglacial period and the onset of full glacial conditions about 70,000 years ago. This sequence is much more complicated than has been generally recognized, although parts of it have been known for many years. The sequence, which has now been found in several widely separated areas, should no longer be ignored.

INTRODUCTION

In another paper in this volume, I described a series of pollen zones from the sediments of Clear Lake core CL-73-4, and defined a series of informal climatic units based upon the pollen zones; the location of Clear Lake and the core is given in that paper. The pollen zones and their correlative climatic units are shown here in Table 1. This paper proposes correlations of the Clear Lake climatic units with other long climatic sequences through all or parts of the last glacial cycle from a number of widely spaced localities. The correlations are based primarily on curve-matching, and should not be regarded as proven. However, the general picture that emerges is one of remarkable consistency among the records, which is particularly interesting because of the great distances involved.

CORRELATIONS

The Clear Lake pollen record is of considerable interest because it covers a long time span, is a continuous record, and comes from a drainage basin that has never been glaciated. The record is especially important because of the detailed sequence of

TABLE 1. CORRELATION BETWEEN INFORMAL CLIMATIC UNITS OF THIS PAPER AND POLLEN ZONES OF ADAM (THIS VOLUME)

Geologic-climate unit	Core CL-73-4 Pollen Zones
Tuleyome thermomer	A
Pomo cryomer	
Late	B, C, D
Cigom 2 cryomer	E
Middle	
Halika thermomers	Fa (in part)
Cigom 1 cryomer	Fb
Early	
Boomli 5 thermomer	G
Tsiwi 5 cryomer	H
Boomli 4 thermomer	I
Tsiwi 4 cryomer	J
(unnamed transitional interval)	K
Boomli 3 thermomer	L
Tsiwi 3 cryomer	M
Boomli 2 thermomer	N
Tsiwi 2 cryomer	O
Boomli 1 thermomer	P
Tsiwi 1 cryomer	Q
(unnamed transitional interval)	R
Konocti thermomer	S
Tsabal cryomer	T, U

climatic changes recorded during the Konocti thermomer and the Early Pomo (=earliest Wisconsinan) cryomer. Pollen records covering various parts of this interval have been available for many years, but correlation of the various overlapping sequences has been difficult, largely because no reliable radiometric age dating methods were available for events prior to 50,000 B.P. The primary emphasis in the correlations given here, however, is on this lower part of the Clear Lake core. Many of the proposed correlations with this part of the section appear to be quite compelling, even without rigorous age control. In the upper part of the Clear Lake record, on the other hand, the character of the minor climatic fluctuations is not as well defined, and the correlations proposed with other parts of the world are more tentative.

In the following text, the Konocti thermomer (Table 1) is first established as equivalent to deep-sea oxygen-isotope Substage 5e and to the Eemian Stage of northern Europe. Next, the detailed correlations of the Clear Lake record with five long, continuous climatic sequences that match the Clear Lake record particularly well are examined (Fig. 1). Finally, correlations of the Clear Lake record with other selected Northern Hemisphere climatic records that include the early part of the last glacial cycle are proposed and presented in Plate 1 (in back cover pocket).

AGE OF THE KONOCTI THERMOMER

The work of Robinson and others (this volume) has extrapolated the clastic sedimentation rate observed at the top of the core (based on radiocarbon ages corrected for reservoir age effects and tree-ring data) to estimate an age of about 131,000 calendar years for the base of core CL-73-4. This indicates that pollen zone S, just above the base of the core and of interglacial character, corresponds to deep-sea oxygen-isotope Stage 5e, which has an estimated age at its lower boundary of 127,000 yr (Shackleton and Opdyke, 1973).

The correlations shown in Figure 1 critically depend on the correctness of correlation of the Konocti thermomer with deep-sea oxygen-isotope Stage 5e and with the Eemian Interglaciation of Europe. The most direct link between the Konocti thermomer and Stage 5e is through correlation of the Clear Lake record with the pollen and oxygen-isotope record described by Heusser and Shackleton (1979) from an oceanic sediment core taken off the Oregon coast. They established that the first pollen assemblage zone below the Holocene that resembles the Holocene is the same unit as oxygen-isotope Stage 5e. Both factor and cluster analyses (Adam, 1987) show that Pollen Zone S (which defines the Konocti thermomer) is the only pollen zone that closely resembles the Holocene deposits of Pollen Zones A and A7; the high-oak pollen zones that correspond to the Boomli thermomers are clearly different from Zones A and A7. The correlations obtained by matching the Konocti thermomer with deep-sea oxygen-isotope Stage 5e are also the same as would be obtained from simple curve-matching.

The correlation of oxygen-isotope Stage 5e with the Eemian was demonstrated by Mangerud and others (1979) at Fjøsanger in Norway. Thus the Konocti thermomer at Clear Lake can be reliably correlated with the Eemian Stage in Europe.

CORRELATION OF THE CLEAR LAKE POLLEN RECORD WITH LONG CONTINENTAL SEQUENCES

The Clear Lake core CL-73-4 oak pollen curve is plotted in Figure 1 along with four other long paleoclimatic curves that span the last 130,000 years. Two curves (Grande Pile, France, and Tenaghi Phillipon, Macedonia) present pollen data; the other two curves present oxygen-isotope data from deep-sea core V28-238 (Shackleton and Opdyke, 1973) and the Greenland Camp Century ice core (Dansgaard and others, 1971). The five curves provide a summary of the climatic behavior of much of the Northern Hemisphere during the last full glacial cycle. Detailed comparison of these curves begins with the correlation of their bases with the Eemian Stage and with deep-sea oxygen-isotope Stage 5e.

The Eemian Stage in the Grande Pile pollen record and Substage 5e in core V28-238 are identified unequivocally (Woillard, 1977; Shackleton and Opdyke, 1973). The correlations of the Grande Pile and V28-238 records with the Macedonian and Camp Century curves shown in Figure 1 are somewhat less certain. The age of the base of the Macedonian curve is dated by extrapolation of estimated sedimentation rates beyond the range of radiocarbon dating; Wijmstra and van der Hāmmen (1974) estimate that the age of the base of the Pangaion interval is about

115,000 yr. Correlation of the Pangaion interval at Tenaghi Phillipon with the Eemian Stage is thus not unreasonable. The dating of the Camp Century record is based on an ice flow model of the ice sheet; the age of the base of the core is estimated as about 120,000 yr (Dansgaard and others, 1971). The correlations of the bases of the Macedonian and Camp Century curves with the other curves shown in Figure 1 are thus based on similarities in estimated ages of the lowest warm intervals in the curves rather than on direct correlations.

The detail shown in the three pollen sections is in part a function of the sedimentation rate. Thus it is not surprising that the Clear Lake record shows more detail than do the other pollen records: the Eemian and Early Weichselian deposits and their correlatives are about 6 m thick at Grande Pile, 14 m thick in the Macedonian section, and nearly 29 m thick at Clear Lake.

Correlations with Grande Pile, France, Tenaghi Phillipon, Macedonia, and the Camp Century ice core

The first few long, continental sequences to record the entire last glacial cycle were from Spain (Florschütz and others, 1971), South America (van der Hammen and Gonzalez, 1960, 1964), and Macedonia (Wijmstra, 1969). Correlations of these sequences with the pollen sequences of the Early Weichselian of northern Europe and with each other were somewhat equivocal because of the great distances and vegetational differences involved. A continuous pollen sequence covering the entire last glacial cycle from a site closer to northern Europe was needed.

Such a site was found at Grande Pile, in northeastern France (Woillard, 1975, 1977, 1978, 1979a, b; Woillard and Mook, 1982). The Grande Pile record provides a continuous climatic sequence since the start of the last interglacial period (Eemian). The site is close enough to northern Europe to permit reliable correlations with the many short pollen records that cover various parts of the Early Weichselian Glaciation, and the entire sequence of climatic fluctuations provides a reasonable basis for correlation with the long Macedonian pollen record (Woillard, 1977).

The Grande Pile pollen record is a critical link in establishing the correlations among the relatively few long, continuous pollen sequences from extraglacial areas and the more abundant short pollen sequences from sites near the continental ice sheets that cover at most only a few stadial/interstadial fluctuations. Many of the correlations of the Grande Pile sequence with other sequences and events have been summarized by Woillard (1977), and the correlations described below draw extensively on her work.

The correspondence between the fluctuations of the Clear Lake oak pollen curve and the fluctuations in the other curves shown in Figure 1 is particularly remarkable in the lower part of the section, corresponding to the Konocti thermomer and the Early Pomo cryomer. The fluctuations in the oak curve are wide and clearly defined and always involve at least two samples.

The pollen curves (Grande Pile, Clear Lake, and Macedonia) show a set of three broad peaks in the frequency of the pollen of warmth-loving plants, starting with the Eemian Stage and its equivalents; the Camp Century record shows three isotopically heavy peaks in a comparable position in the sequence. In the Grande Pile, Macedonian, and Camp Century records these three broad warm peaks appear to be of nearly equal magnitude. This led Woillard (1979) to describe the latter two warm phases (St. Germain I and II) as interglacial periods. Oak pollen frequencies for what are considered equivalent intervals at Clear Lake (the Boomli thermomers) are clearly different from those of both the underlying Konocti thermomer and the Holocene, and it would not be reasonable to give the warmer Boomli intervals (Boomli 2, 3, and 4) interglacial rank at Clear Lake. The correlative St. Germain intervals are considered of interstadial rank in this paper.

The degree of similarity between the Clear Lake, Grande Pile, and Macedonian pollen records during the interval correlated in Figure 1 with oxygen-isotope Stage 5 is remarkable. The broad climatic changes that produced these curves must have affected at least most of the Northern Hemisphere, and perhaps the entire world. If the changes that produced similar climatic curves for France and Macedonia are also found in California, then the rationale for using long-distance climatic correlations is greatly strengthened.

The climatic sequence of the Eemian and Early Weichselian Stages is of particular interest. Many pollen studies in northern Europe cover parts of this interval (for a summary, see Menke and Behre, 1973), but the proper correlation of the various interstadial intervals (such as Amersfoort, Brørup, and Odderade) has not been clear.

The Eemian Stage at Grande Pile and the correlative Konocti thermomer at Clear Lake and Pangaion interval in Macedonia are easily recognizable. Following the Eemian, the correlations are not quite as clear. The Tsiwi I stadial at Clear Lake is equivalent to the Melisey I interval at Grande Pile, and the Boomli I thermomer corresponds to the Doxaton interval in Macedonia and to St. Germain Ia at Grande Pile.

The problem in correlating appears to lie in the presence of the Tsiwi 3 cryomer at Clear Lake. If the Boomli 2 and 3 thermomers are together correlated with oxygen-isotope Stage 5c, one is left with a short but very distinct cold interval in the middle of the Stage 5c sediments in the Clear Lake record. This unit is only 50 cm thick, even though it was deposited at the relatively high sedimentation rate of the Clear Lake core and must represent a brief event.

Two different correlations of the Grande Pile record with the Clear Lake units between the Tsiwi 1 cryomer and the Tsiwi 4 cryomer appear possible; these are shown in Table 2. The Grande Pile climatic periods shown in Figure 1 were named and described according to their apparent importance at that site. The St. Germain I warm interval was split into three parts by the recognition of the Montaigu cold interval that separates St. Germain IA and IC. There is another minor drop in arboreal pollen within the St. Germain IC warm interval at a depth of 1,427.5 cm. This drop affected only a single pollen sample. It was not

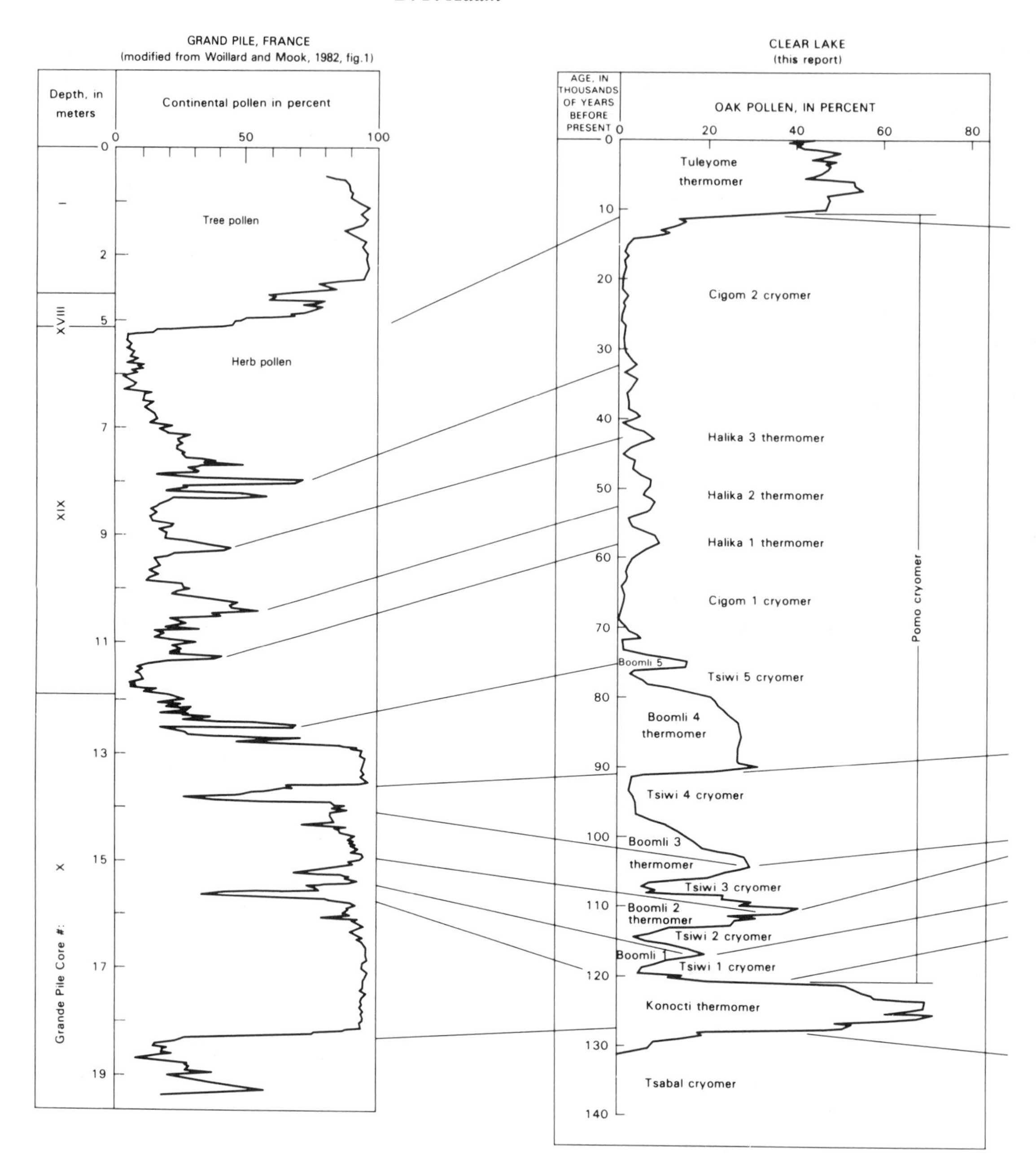

Figure 1 (continued on facing page). Comparison of Clear Lake oak percentage curve with other long climatic sequences. Sources: Grande Pile, Woillard and Mook, 1982, Figure 1; Macedonia, Wijmstra, 1978, Figure 2.4; Camp Century, Dansgaard and others, 1971, Figure 9; core V19-29, from Ninkovich and Shackleton, 1973, with substages of Stage 5 from Adam and others, 1981, Figure 3. Time scale shown for Clear Lake core is that of Robinson and others (this volume).

described as a separate unit in the summary paper that gave the curve shown in Figure 1 (Woillard, 1979b), but was designated as unit St. Germain IC-5b in the more detailed original report (Woillard, 1975).

If St. Germain IC—5b is recognized as a separate cool interval, it is possible to match up the cryomer and thermomer intervals between the Eemian Stage and the Tsiwi 4 cryomer/Melisey II Stade on a one-to-one basis. The correlations that result are shown in the left-hand column of Table 2 (version A).

An alternate correlation (B), shown on the right-hand column of Table 2, attempts to force both the Clear Lake and Grande Pile records into a system in which there are three major warm intervals that correspond to oxygen-isotope Stages 5a, 5c and 5e. If St. Germain I is taken as the equivalent of Boomli

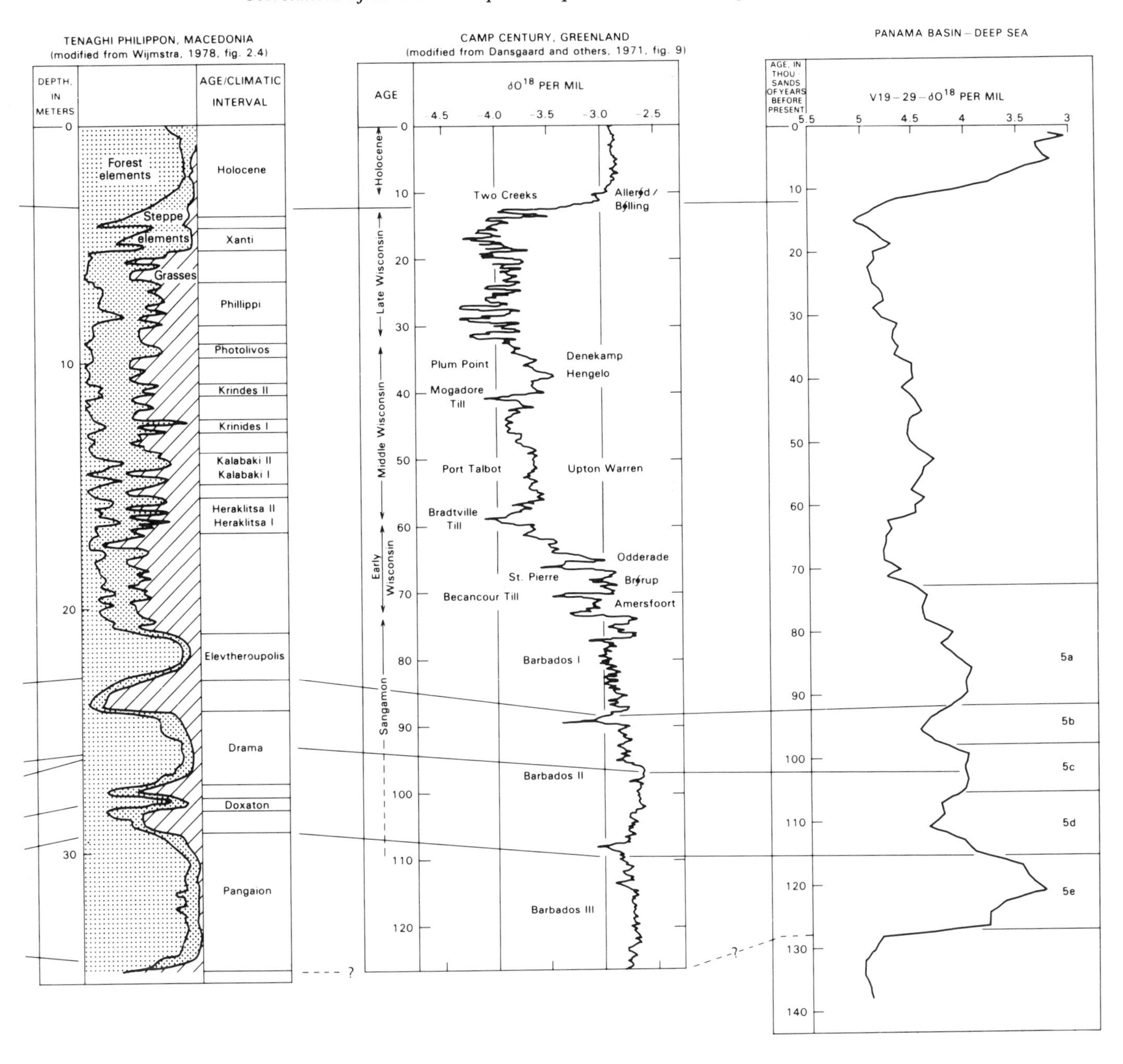

2/Tsiwi 3/Boomli 3 and Stage 5c, then there is no room for an equivalent for the Boomli 1 and Tsiwi 2 intervals in the Grande Pile record.

The version A correlation between the Grande Pile and Clear Lake records is preferred here, because it produces a much better match between the two records. The correlations shown for Grande Pile in Plate 1 include the preferred correlations for the Early Pomo cryomer shown in Table 2. The Ognon I Interstade is correlated with the Boomli 5 thermomer, and Ognon II is correlated with the minor peak in oak pollen that occurs just above the Boomli 5 deposits in core CL-73-4. The correlations shown for the Upper Pomo cryomer are much more tentative than those shown for the older deposits.

The correlation of the Clear Lake and Tenaghi Phillipon records is not as detailed as the correlation with the Grande Pile record. There are four major warm/cold cycles between the Pangaion (=Eemian) and Elevtheroupolis intervals. The correlations preferred here are shown in Plate 1. The Tsiwi 3 cryomer does not have an unequivocal match in the Macedonian record, but there is a sudden decline in the percentage of pollen of forest elements near the middle of the Drama interval that may represent the same event (Fig. 1). No attempt is made here to correlate

TABLE 2. PREFERRED AND ALTERNATE CORRELATIONS BETWEEN GRANDE PILE AND CLEAR LAKE RECORDS

A. Grande Pile (preferred correlation)	Clear Lake	B. Grande Pile (alternate correlation)
Ognon I	Boomli 5	Ognon I
Lanterne I, stadial I	Tsiwi 5	Lanterne I, stadial I
St. Germain II	Boomli 4	St. Germain I I
Melisey II	Tsiwi 4	Melisey II
St. Germain IC-5c, Ic-5d	Boomli 3	St. Germain IC
St. Germain Ic-5b	Tsiwi 3	Montaigu
St. Germain IC-1 through IC-5a	Boomli 2	St. Germain IA
Montaigu (St. Germain IB)	Tsiwi 2	(not recognized)
St. Germain IA	Boomli 1	(not recognized)
Melisey I	Tsiwi I	Melisey I
Eemian	Konocti	Eemian

the Clear Lake record with any of the events between the end of the Elevtheroupolis interval and Holocene in the Macedonian core. The Macedonian site was apparently a much more sensitive recorder of climatic changes during that interval than the Clear Lake site, and the two curves do not match well.

Correlations with the V28-238 and Camp Century oxygen-isotope records

The oxygen-isotope curves shown in Figure 1 vary greatly in the amount of detail shown, but broad features similar to those observed in the three pollen records are present. The correlations of the Camp Century curve with the Barbados I, II, and III sea level maxima of Broecker and others (1968) are those accepted here, but the correlations of the Amersfoort, Brørup, and Odderade Interstadials given by Dansgaard and others (1971) are probably not correct in light of the correlations proposed in this paper. The oxygen-isotope record from core V28-238 (Shackleton and Opdyke, 1973) shows the broad features of the last glacial cycle in a more compressed and blurred form than the other curves shown in Figure 1 because of the much lower sedimentation rate; the record of the last 130,000 yr in core V28-238 is compressed to a 2.1-m-thick sequence.

OTHER CORRELATIONS

Suggested correlations of the Clear Lake climatic sequence with other sequences of various kinds are shown in Plate 1. Each sequence in the plate is correlated with the Clear Lake oak pollen curve independently of the others. These correlations are based mainly on curve-matching, and on the assumption that the Clear Lake record is continuous through the last glacial cycle. The purpose of Plate 1 is to show that the Clear Lake record can be correlated in a plausible way with a variety of climate-related sequences from many localities. The correlations with the Clear Lake record shown in Plate 1 have not been proven independently of climatic inferences, but the general agreement among the many records, as seen in the light of the Clear Lake record, provides a coherent framework for climatic evolution during the last glacial cycle for much of the northern hemisphere.

Generalized Deep-Sea Record

Generalized oxygen-isotope curves for the Caribbean Sea and the Atlantic Ocean (Emiliani, 1958, 1966) were originally interpreted as a record of past fluctuations in surface water temperatures. Emiliani recognized a series of numbered stages, with the odd-numbered stages corresponding to warm stages and the even-numbered stages to cold stages. Stage 1 corresponds to the Holocene, Stage 2 to the full-glacial conditions of the last major continental glaciation of the Northern Hemisphere, Stage 3 to the interstadial conditions that prevailed on the continents before the last major ice advance, Stage 4 to the first major continental ice advance of the last full glacial cycle, and Stage 5 to the last interglacial period. Three warm phases within Stage 5 were clearly recognized (Emiliani, 1958; Fig. 5). Shackleton (1969) formally divided Stage 5 into 5 substages from 5a (youngest) to 5e (oldest); those subdivisions are now in general use and can be readily recognized in many cores (e.g., Emiliani, 1958; Shackleton, 1975, 1977).

The interpretation of the deep-sea oxygen-isotope record as a simple function of sea-surface temperature was challenged by Shackleton (1967), who pointed out that much of the range observed in the isotopic composition of foraminifera through time could be accounted for by the varying isotopic composition of the sea water in which they lived. Evaporation at the sea surface preferentially extracts ^{16}O and leaves the surface water enriched in ^{18}O. During times of growth of continental ice sheets, isotopically light water removed from oceans by evaporation becomes bound in ice sheets, which results in an increase in the proportion of ^{18}O in the oceans. Shackleton (1967) thus interpreted the oceanic oxygen-isotope curves largely as a measure of the amount of ice bound in glaciers, rather than as a paleotemperature curve. Dansgaard and Tauber (1969) estimated that at least 70 percent of the variability of the deep-sea curves can be attributed to changes in oceanic isotopic composition, and at most 30 percent to changes in sea-surface temperature.

The problem is not yet completely resolved. The fluctuations of the Clear Lake oak pollen curve closely resemble fluctuations of the deep-sea oxygen-isotope curves. However, there is no feasible mechanism by which oak pollen frequencies at a continental site could be directly controlled by the isotopic composition of oceanic surface waters. Climate is a more likely controlling mechanism for both curves, and the surface temperature of the ocean may be highly correlated with isotopic composition of the water, at least in some areas (Adam and others, 1981).

My correlation of the Clear Lake climatic record with the deep-sea oxygen-isotope stages is shown in Figure 1 and Plate 1. The particular oxygen-isotope curve shown in Figure 1 is from core V19-29 (Ninkovich and Shackleton, 1975; data are given in Shackleton, 1977); the age estimates for the boundaries between stages shown in Plate 1 are those proposed by Shackleton and Opdyke (1973).

Marine Terraces

Barbados. Three well-dated high stands of sea level that correlate with deep-sea oxygen-isotope substages 5a, 5c, and 5e are recognized from the marine terraces of Barbados (Bender and others, 1979). The high stands were long referred to as Barbados I through Barbados III, and Barbados III was correlated with Stage 5e; Bender and others (1979) have renamed the terraces as shown in Plate 1. The radiometric control for the estimated ages shown in Plate 1 is among the best available for Stage 5.

U.S. Atlantic Coastal Plain. Marine terraces formed along the U.S. Atlantic Coastal Plain between Virginia and South Carolina formed during the high-sea-level periods associated with warm intervals. Cronin and others (1981) used uranium-disequilibrium-series dating of fossil corals to define three terraces formed during the early part of the last glacial cycle. Their 120,000-yr-old terrace is correlated here with the Konocti thermomer; the two younger terraces, with ages of about 94,000 and 72,000 yr, are correlated with the Boomli 2-3 and Boomli 4 thermomers, respectively.

Northwestern Europe

Four versions of the Early Weichselian sequence from northwestern Europe are shown in Plate 1. These are composite sequences for the West Holstein region of Germany (Menke, 1976), Denmark (Andersen, 1980), the Netherlands (van der Hammen and others, 1971), and the sequence from Samerberg in Upper Bavaria described by Grüger (1979).

West Holstein. The most detailed sequence is that from West Holstein (Menke, 1976, p. 54). That composite record matches the Clear Lake sequence in remarkable detail. The Roedebaek Interstadial is described as "very short and unimportant," in agreement with the correlative Boomli 1 thermomer; the final warm phase, the "Keller Interstadial" (also a minor unit) matches the Boomli 5 thermomer. The Brørup Interstadial is divided by a short cold interval into three phases; the latter warm phase is the longer one. This sequence matches the Boomli 2/Tsiwi 3/Boomli 3 series of events at Clear Lake. The other main interstadial, the Odderade, lies above the Brørup, is not interrupted by any cold intervals, and corresponds to the Boomli 4 thermomer. In short, both the major and minor fluctuations in Menke's Early Weichselian climatic history of West Holstein have direct counterparts in the Clear Lake record, both in their durations and their relative magnitudes.

Denmark. The Early Weichselian sequence in Denmark is incomplete; no deposits younger than the Brørup Interstadial have been described. The sequence described by Andersen (1980) is in agreement with the other European sequences shown in Plate 1 for the interval between the Eemian Stage and the Brørup Chronozone, however.

Bavaria. The sequence at Samerberg in Bavaria is of special importance because it established that the palynologic Riss/Würm Interglacial of the Alps is the same as the Eemian Interglacial of northern Europe (Grüger, 1979). The sequence in Bavaria is similar to that from West Holstein, though there are problems in detailed correlations with the Clear Lake record. Two broad interstadial periods follow the Riss/Würm Interglacial (=Eemian), and the older of the two is interrupted by a cold interval. There is also a final small interstadial, incompletely developed at Samerberg, which is found above the two main interstadials. These changes are found in a section about 15 m thick. The pollen diagram is divided into 32 zones; zone 1 is the oldest.

The problems in correlation of Clear Lake with the Samerberg record lie in Grüger's zones 12 through 19. This part of the section appears to correlate with the Tsiwi 1/Boomli 3 part of the Clear Lake record, and is the same part of the section that posed problems in correlating with the Grande Pile record. Grüger (1979, Table 1) correlates his zone 17 with St. Germain IB (Montaigu) at Grande Pile. This correlation then matches Samerberg zones 18 and 19 with the Boomli 2/Tsiwi 3/Boomli 3 sequence at Clear Lake, but there is no zonal equivalent for the Tsiwi 3 cryomer in the Samerberg record.

In particular, the cold intervals in the Samerberg record all show high frequencies of nonarboreal pollen (NAP), and NAP frequencies are close to zero throughout zones 18 and 19. At the base of zone 19, however, there is an oscillation in which the frequency of *Picea* (spruce) drops and the frequency of *Pinus* (pine) rises during a stratigraphic interval of about 15 cm. This change reverses itself, so it is unlikely to be the result of succession or soil depletion. This change in the lower part of zone 19 is tentatively correlated here with the Tsiwi 3 interval at Clear Lake, in order to obtain the same number of cold/warm fluctuations at both sites. This would not be justified if the Samerberg record were the only one available for Europe, as one should not assume *a priori* that the records from two such distant sites should match. However, the remarkable similarity between the other

European records discussed here and the Clear Lake record during the Early Weichselian provides some justification for forcing a match.

The Netherlands. The sequence of climatic stages for the Netherlands is taken from the summary paper of van der Hammen and others (1971); the correlations of the Netherlands and Macedonian sequences are also from that paper. The Amersfoort Interstadial is correlated with Boomli 1, the Brørup with the Boomli 2/Tsiwi 3/Boomli 3 sequence, and the Odderade with Boomli 4. There is no equivalent in the Netherlands record for the minor Boomli 5 thermomer.

Sabana de Bogotá, Colombia

The correlation of the Sabana de Bogotá record with the other sequences is rather poorly controlled; the only finite radiocarbon dates are fairly near the surface. Van der Hammen and Gonzalez (1960) originally correlated the sediments at a depth of 22–24 m with the Eemian Stage; in 1964, however, they considered that the interval from 22 to 31 m probably corresponded to oxygen-isotope Stage 5, and that the Eemian Stage was represented from 29 to 31 m (van der Hammen and Gonzalez, 1964, p. 114). This interpretation was apparently later abandoned, because the summary curve for Bogotá shown by van der Hammen and others (1971, Fig. 12) shows only the portion of the section from 22 to 24 m correlated with the Eemian.

The correlation preferred here is the one given by van der Hammen and Gonzalez in 1964. When that correlation is used, a series of five cold/warm cycles is found above the end of the Eemian Stage equivalent and below the first main cold phase; these stages are correlated here with the Tsiwi/Boomli cycles in the Clear Lake record. Better age control and more rigorous criteria for correlation than simple curve-matching would be desirable, but the correlations shown in Plate 1 appear to provide a good match between the various sequences.

Lake Biwa

The correlation of the Clear Lake pollen record with the record from Lake Biwa, Japan (Fuji, 1976a, b, 1978), is much less clear than the correlations with Europe, Greenland, and the deep-sea oxygen-isotope records. The original interpretations of the Lake Biwa record were given primarily in floristic rather than climatic terms, and the summary diagram curves are for plant groups. The pollen zones shown in Plate 1 are from the initial description of the detailed pollen diagram (Fuji, 1976a, 1978). These zones are not those used by Fuji (1976b).

Because the climatic interpretation of the Lake Biwa record is not clearly stated by Fuji (1976a, b, 1978), I have attempted to correlate the Lake Biwa record with the Clear Lake record using the apparent occurrence of the paleomagnetic Blake Event in the Lake Biwa core at a depth of 50 to 55 m (Fuji, 1976a, b, Kobayashi, 1978). The Blake Event originally described by Smith and Foster (1969) occurs within faunal zone X of Ericson and others (1961). Smith and Foster estimated the time interval included in the Blake event as 108 to 114 ka ±10%. A similar age for the Blake Event at Lake Biwa is supported by fission-track age determinations of 80 ka at a depth of 37 m and 110 ka at a depth of 62 m; Nishimura and Yokoyama (1975) estimated that the fission-track ages from Lake Biwa are accurate to ±20 percent.

More recently, Prell and Hays (1976) estimated that faunal zone X extended from 130 to 84 ka; the upper boundary of Zone X, as shown on their Figure 2, corresponds approximately with the boundary between oxygen-isotope Stages 5a and 5b. If the 50 to 55-m reversal in the Lake Biwa Core is in fact the Blake Event, then the Lake Biwa pollen zones Z-1-a and Z-1-b that correspond to the reversed event should have ages near the middle of the interval from 130 to 84 ka and should correspond to some part of the Clear Lake record between the start of the Konocti thermomer and the end of the Boomli 3 thermomer. Most of the Blake Event records illustrated by Smith and Foster (1969) show the Blake Event about in the middle of the X faunal zone after turbidites are taken into account; thus it seems unlikely that the Blake episode occurred during the Konocti thermomer. It is more likely that Lake Biwa pollen zones Z-1-a and Z-1-b should correlate with some part of the Tsiwi 1-Boomli 3 interval.

If the Blake episode is properly identified in the Lake Biwa core, the sediments deposited during the last interglacial period must lie below a depth of 55 m. Inspection of the summary pollen diagram for that section of the core (Fuji, 1978, Fig. 1) indicates that at only one time during the interval—55 to 110 m—was there a major peak in the frequency of warmth-loving plants; this was during pollen zone Y-2-d, which occurs between depths of about 79 and 82 m. Lesser amounts of warmth-loving plant pollen are also found in pollen zones Y-2-a through Y-2-c at depths between 82 and 97 m.

The interpretation offered here is that Lake Biwa pollen zone Y-2-d, and perhaps zones Y-2-a through Y-2-c as well, should be correlated with the last interglacial interval, and thus with the Konocti thermomer at Clear Lake. This correlation is in conflict with two fission-track ages from the Lake Biwa core. A date at the base of pollen zone Y-2-d has a value of 170,000 yr at a depth of 82 m, and a date at the top of pollen zone Y-1 at a depth of 99 m has a value of 180,000 yr. If the correlation suggested here based on pollen is correct, there is a problem with the Lake Biwa fission-track dates.

Because of the apparent difficulties in correlating the Clear Lake record with that of Lake Biwa, no detailed correlations of the Lake Biwa section above pollen zone Y-2-d with the Clear Lake record are shown in Plate 1. A paleoclimatic interpretation of the important Lake Biwa record is badly needed.

Summary of Intercontinental Correlations

The correlations discussed above show that the Clear Lake pollen record corresponds remarkably well with the available long climatic records from Europe, Greenland, the equatorial Pacific Ocean, and South America. The correlations proposed among the various records appear to be reasonably sound, although independent dating should be done whenever possible.

The Clear Lake pollen record is of major importance because it provides a continuous sequence through the last glacial cycle against which shorter climatic sequences may be evaluated. It suggests that oxygen-isotope curves measured on deep-sea cores record global climatic changes that also had profound effects on terrestrial vegetation, and it establishes that the general climatic sequence observed at Clear Lake and Grande Pile and in Macedonia may be used as a paleoclimatic sequence for at least much of the Northern Hemisphere.

North America

Whereas the Clear Lake record can be correlated with remarkable success with the long, continuous records from distant locations described above, correlation with the various climatic sequences available for North America is much less certain. This is in large part because very few continuous climatic sequences are available for North America; most sequences are defined on the basis of paleosols, tills, and depositional units that cover only relatively short time intervals.

According to the rather ill-defined conventional wisdom, the last glacial cycle in North America consists of the Sangamon interglacial, followed by the Wisconsinan glaciation, followed by the Holocene or postglacial. These terms are widely used in an informal sense throughout North America (Nelson and Locke, 1981), and in a formal sense in the midcontinent area (Frye and others, 1965; Frye and others, 1968).

The Sangamonian Stage. The Sangamonian Stage of Frye and others (1965) is based on the Sangamon Soil, and encompasses the interval from the retreat of the youngest Illinoian glacier in Sangamon County, Illinois, to the onset of the deposition of the Roxana Loess. The status of the Sangamon Soil in its type area has been reviewed by Follmer (1976). The Wisconsinan Stage of Frye and others (1968) includes "all deposits from the contact of Roxana Silt on the A-horizon of the Sangamon soil to the top of the Cochrane till and its contact with the overlying thin discontinuous post-Cochrane deposits in the James Bay Lowland of Ontario, Canada" (Frye and others, 1968, p. E1). The Wisconsinan Stage is divided into the Altonian, Farmdalian, Woodfordian, Twocreekan, and Valderan Substages, as shown in Plate 1.

Given the similarities observed between the Clear Lake record and the other long, continuous climatic sequences described above, the conclusion seems inescapable that the same series of climatic events must have affected the North American midcontinent area in some way. The time-stratigraphic subdivisions shown in Plate 1, however, do not appear to support this conclusion. The transition from the Sangamonian Stage to the Wisconsinan Stage appears to be much too simple in the light of the complex oscillations observed in the Clear Lake record and other long sequences.

The problem lies in the common practices of equating the Sangamonian Stage with the last interglacial interval and regarding the Sangamonian as though it were based on a depositional sequence rather than on a weathering event. The last interglacial interval at Clear Lake is equated here with the Konocti thermomer, which includes only a small part of the time between the end of the penultimate (Illinoian) glaciation and the onset of very cold conditions (correlated here with the development of full continental glaciation) at the end of the Tsiwi/Boomli oscillations. The Sangamonian Stage, on the other hand, includes the entire interval from the last departure of the Illinoian ice sheet to the first arrival of the Altonian ice sheet, as represented by the deposition of wind-blown silt derived from glacial outwash trains. I believe that the Sangamonian Stage must represent more than the last interglacial interval in the strict sense (oxygen-isotope Stage 5e); it must be in at least a general way correlative with all of oxygen-isotope Stage 5, as is shown in Plate 1 (see also Pierce and others, 1976).

Long North American Cores. Long sediment cores have been studied from several localities in North America, including Searles Lake, California (Smith, 1979), Willcox Playa, Arizona (Martin, 1963b), San Augustin Plain, New Mexico (Clisby and Sears, 1956), the Valley of Mexico (Clisby and Sears, 1955; Sears and Clisby, 1955), and south-central Illinois (Grüger, 1970, 1972). All of these cores, with the exception of the Illinois core, are from pluvial lakes and have considerable dating problems.

Illinois. Grüger (1970, 1972) has described a pollen sequence from the Pittsburg Basin in south-central Illinois that he interprets to span the interval from Late Illinoian time through the Holocene. The age of the bottom of the core, inferred on botanic grounds to be below the Sangamon, is based largely on the inferred presence of large amounts of *Taxodium* (bald cypress) pollen in his pollen zones 2a and 2c. The deposits of pollen zone 2 quite probably do correlate with the upper part of the Sangamonian Stage of Frye and others (1965). However, this does not require that they be correlative with the last interglacial period in the strict sense (Stage 5e). It seems more likely that Grüger's zones 2a and 2c postdate Stage 5e, and are correlative with the Clear Lake Boomli 4 and Boomli 2/3 thermomers, as shown in Plate 1. Grüger (1972) specifically acknowledged this possibility, and Frye and Willman (1973) preferred it to Grüger's interpretation.

It is worth noting that if Clear Lake core CL-73-4 had been only 10 m shorter, my interpretation of the Clear Lake record would have been much like that given by Grüger for the Pittsburg Basin. In the absence of the deposits of the Konocti thermomer and their unequivocal interglacial aspect, the most reasonable interpretation of the rest of the record would have been to interpret the Boomli events as the record of the last interglacial period.

Yellowstone National Park. The Yellowstone Park area is chosen here to represent the glacial sequence of the Rocky Mountains because of the series of age determinations available for deposits formed early in the last glacial cycle (Richmond, 1976). The Yellowstone sequence is of particular importance for this study because it reports at least two major glacial advances during oxygen-isotope Stage 5. Richmond has correlated these advances with the Bull Lake Stage, which has long been assumed to be the equivalent of the Tahoe Stage in the Sierra Nevada.

Richmond's Till 5 is correlated here with the Tsabal cryomer, Till 6 with the Tsiwi 1/Tsiwi 2 interval, Till 7 with the Tsiwi 4 cryomer, Till 8 with the Tsiwi 5 cryomer(?), and Till 9 with the Cigom 1 cryomer (Plate 1). Caldera lake silt D is correlated with the Konocti thermomer, lake silt E with the Boomli 2-Boomli 3 interval, and lake silt F with the Boomli 4 thermomer. The age estimates available for the lake silt units are in good agreement with the other correlations proposed for the Clear Lake record.

The summary of the Yellowstone record shown in Plate 1 also shows Richmond's correlations of the Yellowstone tills with the glacial deposits of the Wind River Mountains (Sacagawea Ridge, Bull Lake, and Pinedale Glaciations). The correlations given here do not necessarily apply to all Rocky Mountain glacial deposits that have been correlated with the Bull Lake and Pinedale deposits, but only to the Yellowstone sequence as described by Richmond (1976). Some deposits that have been correlated with the Bull Lake Glaciation may well predate the Konocti thermomer (Pierce and others, 1976), and the correlations with the Clear Lake record proposed here should not be used to refute such a possibility.

Puget Sound Area. Several long pollen sequences estimated to cover the last 80,000 years have been described from the Puget Sound and Olympic Peninsula region of Washington (Heusser, 1972, 1977). Recently, the Salmon Springs Drift, thought to be of mid-Wisconsinan age, was discovered to be of middle Pleistocene age at its type locality, based upon fission-track and zircon ages and magnetic polarity data (Easterbrook and others, 1981). Some of Heusser's pollen zones were attributed to the Salmon Springs Glaciation. Because of the present uncertainty concerning the local correlation of stratigraphic units in the western Washington area, no attempt is made to correlate Heusser's important sections with the Clear Lake record.

Pluvial Lakes. In many areas, pluvial lakes are the only sites where it is possible to recover a continuous record of sedimentation covering a long time interval. For purposes of pollen analysis, however, pluvial lakes present many difficult problems that confound the interpretation of pollen diagrams from their sediments. Such lakes are found in areas in which present-day precipitation is not great enough to keep lake basins overflowing regularly. When the ratio between precipitation (P) and evaporation (E) drops below a critical value, the lake ceases to overflow, and the lake level then drops below the level of the outlet until a balance is reached between the amount of evaporation from the (reduced) lake surface and the amount of water reaching the lake through precipitation and inflow from the drainage basin. If the P/E ratio is sufficiently low, the lake will dry up completely.

The size of a lake influences the way in which the lake incorporates pollen from various components of the surrounding vegetation into its sediments (for example, see Faegri and Iversen, 1964, p. 104; Moore and Webb, 1978, p. 108; Jacobson and Bradshaw, 1981). The climatic changes that cause pluvial lakes to change size and the vegetation around them to shift also change the way in which the pollen record is related to the vegetation. The combined effect of these processes is to make pollen records from fluctuating pluvial lakes rather difficult to interpret.

The other major difficulty faced in dealing with pluvial lake pollen records is that when a lake dries up, the pollen in the surficial sediments becomes exposed to oxidation, and the sediments themselves may be subjected to wind erosion. Hiatuses in such pollen records are not uncommon.

Willcox Playa. Martin (1963b) described a long pollen record from a 42-m-long core from the Willcox Playa in southeastern Arizona. The top 2 m of the core did not contain pollen because of surface oxidation; below the oxidized surface zone was a thick sequence of reduced, pollen-bearing sediments that ranged in depth from 2 to 21 m. Between the depths of 21 and 29.3 m, there were four oxidized zones separated by three reduced zones; the middle reduced zone, at a depth of 24.4 m, consisted of only a single sample, but all of the other zones were at least 1 m thick.

Martin (1963b, Fig. 3) correlated the interval between 23.5 and 29.3 m, containing the lowest three of the four oxidized zones just mentioned, with the Sangamon interval on the assumption that the oxidized zones represent times when the lake was dry and interglacial conditions prevailed. He included the oxidized zone at 21–22 m within the Wisconsin glacial interval.

In the light of the correlations of the Clear Lake record mentioned above, it seems reasonable that the oxidized zones in the Willcox Playa core correspond to the Konocti thermomer and the Boomli 2, 3, and 4 thermomers at Clear Lake. If so, the minor Boomli 1 and 5 thermomers were not warm or dry enough to allow Willcox Playa to dry up. The suggested correlations are shown in Plate 1, along with Martin's division of his sequence into the Illinoian, Sangamon, and Wisconsin.

San Augustin Plains. A long (150 m) pollen profile (not shown in Plate 1) from the San Augustin Plains in western New Mexico was described by Clisby and Sears (1956) in very general terms. High frequencies of spruce (*Picea*) pollen were equated with relatively cold intervals, and high frequencies of nonarboreal pollen types (NAP) were taken as a measure of aridity; no other curves were shown. The section was not well dated, but the top 15 m of the core were shown to correspond to the full glacial conditions of the Upper Wisconsinan and the Holocene. It appears likely that the top 90 m of the San Augustin Plains core covers about the same interval as the Clear Lake record, but the original diagram for the San Augustin Plains record is not detailed enough to permit precise correlations.

Searles Lake. The Quaternary sequence of lacustrine and playa sediments from Searles Lake, California, has been summarized by Smith (1979). He presented a summary curve of lake fluctuations during the last full glacial cycle; however, the summary curve is not shown in relation to the stratigraphy of the core, and the various lake phases are not named. His correlations of the Searles Lake record with other long records are shown using a series of six low stands labeled A through F. Low Stand A is correlated here with the early part of the Tuleyome thermomer at Clear Lake, Low Stands B and C are correlated with the Halika thermomers, Low Stands D and E are correlated with the Boomli 4 and 3 thermomers, respectively, and Low Stand F is correlated with the Konocti thermomer.

Lakes Bonneville and Lahontan. The Lake Bonneville and Lake Lahontan sequences correlated here with the Clear Lake record are taken from the summary paper by Morrison and Frye (1965, Fig. 2). The lower half of the two sequences in their Figure 2 show a lower soil, correlated with the last interglaciation, overlain by a series of five lacustrine units, and then by an alluvial/colluvial/aeolian unit on which is developed another soil.

The lower soil units (Dimple Dell and Cocoon Soils) are correlated with the Sangamonian Stage of the midcontinent by Morrison and Frye, and the succeeding five lake phases are correlated with the Altonian Substage. A different interpretation is offered here; the lower soils are correlated with the Konocti thermomer at Clear Lake, and thus with isotopic Substage 5e, rather than with the entire Sangamonian Stage, and at least some of the overlying lake deposits are correlated with the upper part of the Sangamonian Stage and with the Tsiwi cryomers at Clear Lake.

The correlation of the series of climatic fluctuations above the Dimple Dell and Cocoon Soils with the Clear Lake record depends on how the mid-Wisconsinan Promontory and Churchill Soils are correlated. Two reasonable possibilities exist, but both create some problems. The first possibility is to consider the Promontory and Cocoon Soils as correlative with the soil-forming interval that occurred between the Tahoe and Tioga Glaciations in the Sierra Nevada, when those glaciations are considered in the broad sense used by Burke and Birkeland (1979). In this case, the middle soils are considered correlative with the Boomli 4 thermomer at Clear Lake, and the pluvial periods recorded by the Alpine and Eetza Formations are correlative as a group with the Tsiwi cryomers 1 through 4. The two prominent lake phases above the middle soils would then be correlative with the Cigom 1 and Cigom 2 cryomers. This correlation seems unlikely for several reasons. Radiocarbon ages for the Churchill Soil indicate an age of about 25,000 yr for the end of the soil-forming interval (Morrison and Frye, 1965), which seems much too young to allow interpretation of the Churchill as equivalent to the Tahoe/Tioga soil-forming interval. (This correlation would also require that the five lake advances of the Alpine and Eetza Formations be correlated with only three apparent cool intervals at Clear Lake.)

The other interpretation, which is preferred here, is that the Promontory and Churchill Soils are correlative with the Halika thermomers. The lacustrine advances of the Alpine and Eetza Formations would then correlate with the Tsiwi cryomers and possibly the Cigom 1 cryomer, and the lake deposits of the Bonneville and Sehoo Formations would correlate with the Cigom 2 cryomer. This interpretation is consistent with that proposed by Davis (1978, Fig. 3). It is not clear in this interpretation whether the Wyemaha Formation, which lies below the Churchill Soil and above the lacustrine units of the Eetza Formation, should be considered the equivalent of the Cigom 1 cryomer or not. If it is, that implies that climatic conditions during Cigom 1 were markedly less pluvial than during the other stadial phases, either because of low precipitation or high evaporation (or both). If it is not, not all of the Tsiwi cryomers were matched by high lake levels in the Bonneville and Lahontan Basins, or not all of the Boomli thermomers produced lake recessions. The latter seems more likely; in particular, the Boomli 5 thermomer might not have been long enough to produce a significant lake recession.

Sierra Nevada. The climatic sequence of the last glacial cycle in the Sierra Nevada is still far from well understood. Two largely independent glacial histories have been developed for the western and eastern slopes of the range. The eastern side has been more intensively studied (Russell, 1887; Blackwelder, 1931; Putnam, 1949; Sharp and Birman, 1963; Birkeland, 1964; Birman, 1964; Dalrymple, 1964; Sharp, 1969, 1972; Curry, 1971; Birkeland and others, 1976; Burke and Birkeland, 1979). Vegetation is sparser because of the climatic rain shadow, and active tectonism and volcanism have helped to preserve a more detailed record of glacial advances on the east side than has been observed on the west side.

Unfortunately, the added detail preserved on the eastern side of the range has not produced a definitive glacial sequence. Burke and Birkeland (1979) concluded that the criteria used to separate many of the previously described glacial advances along the eastern side of the Sierra Nevada are not adequate to permit reliable definition of those advances as separate events. They suggested "mapping only multiple Tioga and Tahoe deposits until better criteria for further subdivision are developed" (1979, p. 49). Those authors were concerned primarily with relative dating techniques and the problems involved in correlating glacial sequences from one drainage basin to another along the eastern side of the Sierra. Thus they placed only minor emphasis on geomorphic criteria that could only be used to provide relative ages for different deposits within a single basin.

Given the warning by Burke and Birkeland that many of the glacial advances previously recognized as separate events in the Sierra Nevada cannot be distinguished from each other on the basis of weathering criteria, any attempt to correlate the Sierra Nevada glacial record with the continuous climatic sequence at Clear Lake must be regarded as tentative.

The degree of weathering of Tahoe moraines in the Sierra Nevada is much greater than for Tioga Stage moraines. Potassium-argon dates from Sawmill Canyon indicate that the Tahoe Stage occurred before 53,000 ± 44,000 yr B.P. (Dalrymple and others, 1982), and Bailey and others (1976) reported dates of 62,000 ± 13,000 and 126,000 ± 25,000 yr B.P. for two basalt flows that overlie and underlie the Casa Diablo till near Mammoth. Burke and Birkeland (1979) believed that the Casa Diablo till is of Tahoe age.

The available radiometric dates are thus consistent with either a pre-Konocti or a post-Konocti age for the Tahoe Glaciation. A post-Konocti age is preferred here, primarily in order to provide enough glacial events to match the cryomers in the Clear Lake record. Not all workers would agree with a post-Konocti age for the Tahoe, however; for example, Colman and Pierce (1981) assumed that the Tahoe is pre-Konocti, with an age of about 140,000 yr. From the point of view of interpreting the Clear Lake record, the difficulty with a pre-Konocti age for the Tahoe Glaciation is that one is left with a complicated series of events that had a major impact on the vegetation of the Coast Ranges and yet left little or no record in the glacial history of the Sierra Nevada. The present climate is marginally able to support tiny glaciers in the Sierra Nevada, and it seems probable that climatic changes such as those recorded during the Early Pomo cryomer at Clear Lake would have produced sizable glaciers in the Sierra.

One of the most detailed morainal sequences exposed along the eastern side of the Sierra Nevada is in the Bloody Canyon–Sawmill Canyon area near the south end of Mono Lake (Sharp and Birman, 1963; Wahrhaftig and Sharp, 1965; Burke and Birkeland, 1979). At that locality, both the Tahoe Stage and Tiogo Stage deposits as defined by Burke and Birkeland (1979) are represented by multiple moraines. The oldest moraine is the Mono Basin moraine of Sharp and Birman (1963); it forms the sides of Sawmill Canyon and was originally deposited by a glacier that flowed out of Bloody Canyon. The other moraines were deposited after a drainage shift that diverted the ice flowing out of Bloody Canyon away from Sawmill Canyon and northeast into the present course of Walker Creek. The moraines along Walker Creek are the Tahoe, Tenaya, and Tioga moraines of Sharp and Birman (1963); Burke and Birkeland (1979) considered the Tenaya moraine to be of Tioga age.

The Mono Basin moraine cannot be distinguished from the Tahoe moraine that cuts it using the relative age dating techniques of Burke and Birkeland (1979). This makes it unlikely that a major weathering interval occurred between the deposition of the two moraines. On the other hand, there was enough time between the two glacial advances for a significant change in the local drainage to occur. There has also been a significant weathering interval that affected both the Tahoe and the Mono Basin moraines, but did not affect the Tenaya or Tioga moraines.

The correlations proposed here between the glacial advances of the Bloody Canyon–Sawmill Canyon area and the Clear Lake climatic sequence are shown in Plate 1. The Mono Basin advance most probably correlates with the Tsiwi 1, 2, or 3 cryomers. These are all relatively brief events at Clear Lake, and the oak pollen frequencies are all somewhat higher than during the Tsiwi 4 cryomer, which is correlated here with the Tahoe advance. The short duration of the cold intervals and the oscillations of climate during the Tsiwi 1/Tsiwi 3 interval probably account for the fact that the Mono Basin advance was not as large as the subsequent Tahoe advance, and was thus overridden in most localities. The Tahoe advance was larger because conditions were somewhat cooler and the cold interval was also of somewhat greater duration, assuming that the thickness of the intervals in the Clear Lake section provides a reasonable estimate of their duration.

The Boomli 4 thermomer was the longest of the warm intervals during the Early Pomo cryomer, and therefore is likely to have been a time during which significant weathering occurred. Climatic conditions at Clear Lake appear to have been similar to those that prevailed during the Boomli 2 and 3 thermomers, but apparently those intervals did not persist long enough to produce a significant difference between the weathering characteristics of the Mono Basin and Tahoe moraines in the Sierra Nevada.

The Tenaya and Tioga advances are correlated here with the Cigom 1 and 2 cryomers, respectively. The Tioga advance is generally recognized as the last major ice advance in the Sierra Nevada, so correlation of that advance with the Cigom 2 cryomer seems straightforward. The Tenaya advance is still not well understood in the Sierra Nevada, but where Tenaya deposits occur they are clearly older than the Tioga Stage and younger than the Tahoe Stage. The Tenaya advance is correlated here with the Cigom 1 cryomer because that seems to be the most reasonable solution, rather than because of any compelling data or arguments.

If the Tahoe and Mono Basin glacial advances are both older than the Konocti thermomer, the glacial advances of the Tenaya and Tioga stages must have been large enough to obliterate any deposits left by the glacial advances of the first half of the last glacial cycle. If that is so, then the morainal deposits that have been used to infer the glacial history of the Sierra Nevada are in fact very poorly suited to that task. The glacial history of the Sierra must instead be developed using well-dated sedimentary sequences (preferably continuous) that include outwash deposits of upstream glaciations. Such deposits exist on both sides of the Sierra, although they will be difficult to sample.

San Joaquin Valley. The alluvial stratigraphy of the San Joaquin Valley was described by Marchand and Allwardt (1981). They described a series of upward-coarsening alluvial deposits of Wisconsinan and earlier age that they designated as the Modesto and Riverbank Formations. Periods of coarse alluviation probably correspond to deglacial phases in the Sierra Nevada; interglacial periods are characterized by erosional intervals and soil formation on the alluvial deposits.

The Modesto Formation appears to correspond with the Middle and Late Pomo cryomer at Clear Lake. The Modesto is divided into an Upper Member and a Lower Member, which are separated by a soil-forming interval. These probably correlate with the Cigom 2 and Cigom 1 cryomers. As noted above, if the maximum alluvial deposition occurred during the melting of the glaciers, the Modesto units probably correspond with the latter half of the Cigom units.

The Modesto Formation is underlain by the Riverbank Formation, which Marchand and Allwardt (1981) subdivided into three members. The age of the Riverbank Formation is not well understood, but Marchand and Allwardt believed that it probably ranges between about 130,000 and 450,000 years. Soils developed on the Upper Unit of the Riverbank suggested to Marchand and Allwardt that the Upper Unit was deposited before the last interglaciation, which they correlated with marine isotope Stage 5.

Hansen and Begg (1970) reported open-system uranium-series ages averaging 103,000 ± 6,000 yr on vertebrate remains from the Teichert gravel pits east of Sacramento. The date was regarded as too young by Marchand and Allwardt (1981), in part because the bones were attributed to the Middle Unit of the Riverbank Formation. However, the site was attributed to the upper Riverbank Formation by Shlemon (1972). Study of pollen and plant macrofossils from the site indicated a somewhat cooler climate (Ritter and Hatoff, 1977) which would be consistent with an interstadial environment of the Early Pomo cryomer.

The interpretation preferred here is that the uranium-series dates of Hansen and Begg (1970) are valid, and the upper Riverbank Formation is considered here to include deposits that post-date the last interglacial *sensu stricto.* The upper Riverbank thus includes sediments produced by the melting of the Tahoe and Mono Basin glaciers in the Sierra Nevada, and the soil-forming interval recorded at the top of the upper Riverbank corresponds to the weathering interval that separates Tahoe-age from younger deposits in the Sierra (see below). The correlation of the upper Riverbank with the Sierran glacial sequence is the same as that given by Ritter and Hatoff (1977).

CONCLUSION

The correlations of the Clear Lake record with other sequences proposed above are unproven. The overall impression, however, is one of remarkable consistency, and it seems likely that further work will provide more evidence in support of the observed sequence of five cryomer/thermomer or stadial/interstadial cycles between the end of the last interglacial period and the onset of full glacial conditions about 70,000 yr ago. This sequence is much more complicated than has been generally recognized, although parts of it have been well known for many years. The sequence has now been found in several widely separated continental areas, and should be explicitly recognized as a widespread climatic phenomenon and studied in greater detail.

REFERENCES CITED

Adam, D. P., 1987, Palynology of two Upper Quaternary cores from Clear Lake, Lake County, California, with a chapter on dating by Stephen W. Robinson: U.S. Geological Survey Professional Paper 1363.

Adam, D. P., Sims, J. D., and Throckmorton, C. K., 1981, 130,000-yr continuous pollen record from Clear Lake, Lake County, California: Geology, v. 9, p. 373–377.

Anderson, S. T., 1980, Early and Late Weichselian chronology and birch assemblages in Denmark: Boreas, v. 9, p. 53–69.

Bailey, R. A., Dalrymple, G. B., and Lanphere, M. A., 1976, Volcanism, structure, and geochronology of Long Valley caldera, Mono County, California: Journal of Geophysical Research, v. 81, p. 725–744.

Bender, M. L., Fairbanks, R. G., Taylor, F. W., Matthews, R. K., Goddard, J. G., and Broecker, W. L., 1979, Uranium-series dating of the Pleistocene reef tracts of Barbados, West Indies: Geological Society of America Bulletin, pt. I, v. 90, p. 577–594.

Birkeland, P. W., 1964, Pleistocene glaciation of the northern Sierra Nevada, north of Lake Tahoe, California: Journal of Geology, v. 72, p. 810–825.

Birkeland, P. W., Burke, R. M., and Yount, J. C., 1976, Preliminary comments on Late Cenozoic glaciations in the Sierra Nevada, *in* Mahaney, W. C., ed., Quaternary stratigraphy of North America: Stroudsburg, Pennsylvania, Dowden, Hutchinson & Ross, p. 283–295.

Birman, J. H., 1964, Glacial geology across the crest of the Sierra Nevada, California: Geological Society of America Special Paper 75, 80 p.

Blackwelder, E., 1931, Pleistocene glaciation in the Sierra Nevada and Basin Ranges: Geological Society of America Bulletin, v. 42, p. 865–922.

Bloom, A. L., Broecker, W. S., Chappell, J.M.A., Matthews, R. K., and Mesolella, K. J., 1974, Quaternary sea level fluctuations on a tectonic coast: New $^{230}Th/^{234}U$ dates from the Huon Peninsula, New Guinea: Quaternary Research, v. 4, p. 185–205.

Broecker, W. S., Thurber, D. L., Goddard, J., Ku, T. L., Matthews, R. K., and Mesolella, K. J., 1968, Milankovitch hypothesis supported by precise dating of coral reefs and deep-sea sediments: Science, v. 159, p. 297–300.

Burke, R. M., and Birkeland, P. W., 1979, Reevaluation of multiparameter relative dating techniques and their application to the glacial sequence along the eastern escarpment of the Sierra Nevada, California: Quaternary Research, v. 11, p. 21–51.

Clisby, K. H., and Sears, P. B., 1955, Palynology of southern North America, Pt. III, Microfossil profiles under Mexico City correlated with the sedimentary profiles: Geological Society of America Bulletin, v. 66, p. 511–520.

Clisby, K. H., and Sears, P. B., 1956, San Augustin Plains; Pleistocene climatic changes: Science, v. 124, p. 537–539.

Colman, S. M., and Pierce, K. L., 1981, Weathering rinds on andesitic and basaltic stones as a Quaternary age indicator, western United States: U.S. Geological Survey Professional Paper 1210, 56 p.

Cronin, T. M., Szabo, B. J., Ager, T. A., Hazel, J. E., and Owens, J. P., 1981, Quaternary climates and sea levels of the U.S. Atlantic Coastal Plain: Science, v. 211, p. 233–240.

Curry, R. R., 1971, Glacial and Pleistocene history of the Mammoth Lakes Sierra, California; A geologic Guidebook: Missoula, Montana Department of Geology, Geological Serial Publication 11, 49 p.

Dalrymple, G. B., 1964, Potassium-argon dates of three Pleistocene interglacial basalt flows from the Sierra Nevada, California: Geological Society of America Bulletin, v. 75, p. 753–758.

Dalrymple, G. B., Burke, R. M., and Birkeland, P. W., 1982, Note concerning K-Ar dating of a basalt flow from the Tahoe–Tioga Interglaciation, Sawmill Canyon, southeastern Sierra Nevada, California: Quaternary Research, v. 17, p. 120–122.

Dansgaard, W., and Tauber, H., 1969, Glacier oxygen-18 content and Pleistocene ocean temperatures: Science, v. 166, p. 499–502.

Dansgaard, W., Johnsen, S. J., Clausen, H. B., and Langway, C. C., Jr., 1971, Climatic record revealed by the Camp Century ice core, *in* Turekian, K. K., ed., The Late Cenozoic glacial ages: New Haven, Yale University Press, p. 37–56.

Davis, J. O., 1978, Quaternary tephrochronology of the Lake Lahontan area, Nevada and California: Nevada Archeological Survey, Research Paper no. 7, 137 p.

Easterbrook, D. J., Briggs, N. D., Westgate, J. A., and Gorton, M. P., 1981, Age of the Salmon Springs glaciation in Washington: Geology, v. 9, p. 87–93.

Emiliani, Cesare, 1958, Paleotemperature analysis of core 280 and Pleistocene correlations: Journal of Geology, v. 66, no. 3, p. 264–275.

——, 1966, Paleotemperature analysis of Caribbean cores P6304-8 and P6304-9 and a generalized temperature curve for the past 425,000 years: Journal of Geology, v. 74, no. 2, p. 109–126.

Ericson, D. B., Ewing, M., Wollin, G., and Heezen, B. C., 1961, Atlantic deep-sea sediment cores: Geological Society of America Bulletin, v. 72, p. 193–286.

Faegri, K., and Iverson, J., 1964, Textbook of pollen analysis, 2nd ed., rev.: New York, Hafner, 237 p.

Florschütz, F., Menendez Amor, J., and Wijmstra, T. A., 1971, Palynology of a thick Quaternary succession in southern Spain: Palaeogeography, Palaeoclimatology, and Palaeoecology, v. 10, p. 233–264.

Follmer, L. R., 1976, The Sangamon Soil in its type area; A review, *in* Mahaney, W. C., ed., Quaternary soils: Norwich, England, Geo Abstracts Ltd., p. 125–165.

Frye, J. C., and Willman, H. B., 1973, Wisconsin climatic history interpreted from Lake Michigan Lobe deposits and soils: Geological Society of America Memoir 136, p. 135–152.

Frye, J. C., Willman, H. B., and Black, R. F., 1965, Outline of glacial geology of Illinois and Wisconsin, *in* Wright, H. E., Jr., and Frey, D. G., eds., The Quaternary of the United States: Princeton, Princeton University Press, p. 43–61.

Frye, J. C., Willman, H. B., Rubin, M., and Black, R. F., 1968, Definition of Wisconsinan Stage: U.S. Geological Survey Bulletin 1274-E, 22 p.

Fuji, N., 1976a, Palaeoclimatic and palaeovegetational changes around Lake Biwa, central Japan, during the past 100,000 years, *in* Horie, S., ed., Paleolimnology of Lake Biwa and the Japanese Pleistocene: Kyoto, Japan, Kyoto University, v. 4, p. 316–356.

——, 1976b, Palynological investigation on a 200-meter core samples from Lake Biwa in central Japan, *in* Horie, S., ed., Paleolimnology of Lake Biwa and the Japanese Pleistocene: Kyoto, Japan, Kyoto University, v. 4, p. 357–421.

——, 1978, Paleovegetational and paleoclimatic changes around Lake Biwa, central Japan, during the past 100,000-270,000 years, *in* Horie, S., ed., Paleolimnology of Lake Biwa and the Japanese Pleistocene: Kyoto, Japan, Kyoto University, v. 6, p. 235–262.

Grüger, E., 1970, Development of the vegetation of southern Illinois since late Illinoian time (preliminary report): Revue de Geographie physique et de geologie dynamique, v. 12, fasc. 2, p. 143–148.

——, 1972, Late Quaternary vegetation development in south-central Illinois: Quaternary Research, v. 2, p. 217–231.

——, 1979, Die Seeablagerungen vom Samerberg/Obb. und ihre Stellung im Jungpleistozän: Eiszeitalter und Gegenwart, v. 29, p. 23–34.

Hansen, R. O., and Begg, E. L., 1970, Age of Quaternary sediments and soils in the Sacramento area, California, by uranium and actinium series dating of vertebrate fossils: Earth and Planetary Science Letters, v. 8, p. 411–419.

Heusser, C. J., 1972, Palynology and phytogeographical significance of a Late Pleistocene refugium near Kalaloch, Washington: Quaternary Research, v. 2, p. 189–201.

——, 1977, Quaternary palynology of the Pacific Slope of Washington: Quaternary Research, v. 8, p. 282–306.

Heusser, L. E., and Shackleton, N. J., 1979, Direct Marine–Continental correlation: 150,000-year oxygen isotope-pollen record from the North Pacific: Science, v. 204, p. 837–839.

Jacobson, G. L., Jr., and Bradshaw, R.H.W., 1981, The selection of sites for paleovegetational studies: Quaternary Research, v. 16, p. 80–96.

Kobayashi, K., 1978, Correlation of short geomagnetic polarity episodes between Lake Biwa and deep-sea sediments, *in* Horie, S., ed., Paleolimnology of Lake Biwa and the Japanese Pleistocene: Kyoto, Japan, Kyoto University, v. 6, p. 88–100.

Mangerud, J., Sønstegaard, E., and Sejrup, H.-P., 1979, Correlation of the Eemian (interglacial) Stage and the deep-sea oxygen-isotope stratigraphy: Nature, v. 277, p. 189–192.

Marchand, D. E.,and Allwardt, A., 1981, Late Cenozoic stratigraphic units, northeastern San Joaquin Valley, California: U.S. Geological Survey Bulletin 1470, 70 p.

Martin, P. S., 1963, Geochronology of pluvial Lake Cochise, southern Arizona, II; Pollen analysis of a 42-meter core: Ecology, v. 44, p. 436–444.

Menke, B., 1976, Neue Ergebnisse zur Stratigraphie und Landschaftsentwicklung im Jungpleistozän Westholsteins: Eiszeitalter und Gegenwart, v. 27, p. 53–68.

Menke, B., and Behre, K., 1973, History of vegetation and biostratigraphy: Eiszeitalter und Gegenwart, v. 23/24, p. 251–267.

Moore, P. D., and Webb, J. A., 1978, An illustrated guide to pollen analysis: New York, Halstead, 133 p.

Morrison, R. B., and Frye, J. C., 1965, Correlation of the Middle and Late Quaternary successions of the Lake Lahontan, Lake Bonneville, Rocky Mountain (Wasatch Range), southern Great Plains, and eastern Midwest areas: Nevada Bureau of Mines, Report 9, 45 p.

Nelson, A. R., and Locke, W. W., III, 1981, Quaternary stratigraphic usage in North America; A brief survey: Geology, v. 9, p. 134–137.

Ninkovitch, D., and Shackleton, N. J., 1975, Distribution, stratigraphic position, and age of ash layer "L" in the Panama Basin region: Earth and Planetary Science Letters, v. 27, p. 20–34.

Nishimura, S., and Yokoyama, T., 1975, Fission-track ages of volcanic ashes of core samples of Lake Biwa and the Kobiwako Group (2), *in* Horie, S., ed., Paleolimnology of Lake Biwa and the Japanese Pleistocene: Kyoto, Japan, Kyoto University, v. 6, p. 138–142.

Pierce, K. L., Obradovich, J. D., and Friedman, I., 1976, Obsidian hydration dating and correlation of Bull Lake and Pinedale Glaciations near West Yellowstone, Montana: Geological Society of America Bulletin, v. 87, p. 703–710.

Prell, W. L., and Hays, J. D., 1976, Late Pleistocene faunal and temperature patterns of the Colombia Basin, Caribbean Sea: Geological Society of America Memoir 145, p. 201–220.

Putnam, W. C., 1949, Quaternary geology of the June Lake district, California: Geological Society of America Bulletin, v. 60, p. 1281–1302.

Richmond, G. M., 1976, Pleistocene stratigraphy and chronology in the mountains of western Wyoming, *in* Mahaney, W. C., ed., Quaternary stratigraphy of North America: Stroudsburg, Pennsylvania, Dowden, Hutchinson & Ross, p. 353–379.

Ritter, E. W., and Hatoff, B. W., 1977, Late Pleistocene pollen and sediments; An analysis of a central California locality: Texas Journal of Science, v. 29, p. 195–207.

Russell, I. C., 1887, Quaternary history of Mono Valley, California: U.S. Geological Survey 8th Annual Report, Pt. I, p. 261–394.

Sears, P., and Clisby, K. H., 1955, Palynology in southern North America, Pt. IV: Pleistocene climates in Mexico: Geological Society of America Bulletin, v. 66, p. 521–530.

Shackleton, N. J., 1967, Oxygen isotope analyses and Pleistocene temperatures re-assessed: Nature, v. 215, p. 15–17.

——, 1969, The last interglacial in the marine and terrestrial records: Proceedings of the Royal Society of London, series B, v. 174, no. 1034, p. 135–154.

——, 1975, The stratigraphic record of deep-sea cores and its implications for the assessment of glacials, interglacials, stadials, and interstadials in the Mid-Pleistocene, *in* Butzer, K. W., and Isaac, G. L., eds., After the Australopithecines; Stratigraphy, ecology, and culture change in the Middle Pleistocene: The Hague, Mouton, p. 1–24.

——, 1977, The oxygen isotope stratigraphic record of the Late Pleistocene: Philosophical Transactions of the Royal Society of London, Series B, v. 280, p. 169–182.

Shackleton, N. J., and Opdyke, N. D., 1973, Oxygen isotope and paleomagnetic stratigraphy of equatorial Pacific core V28-238; Oxygen isotope temperatures and ice volumes on a 10^5 year and 10^6 year scale: Quaternary Research, v. 3, no. 1, p. 39–55.

Sharp, R. P., 1969, Semiquantitative differentiation of glacial moraines near Convict Lake, Sierra Nevada, California: Journal of Geology, v. 77, p. 68–91.

——, 1972, Pleistocene glaciation, Bridgeport Basin, California: Geological Society of America Bulletin, v. 83, p. 2233–2260.

Sharp, R. P., and Birman, J. H., 1963, Additions to classical sequence of Pleistocene glaciations, Sierra Nevada, California: Geological Society of America Bulletin, v. 74, p. 1079–1086.

Shlemon, R. J., 1972, The lower American River area, California; A model of Pleistocene landscape evolution: Association Pacific Coast Geographers Yearbook, v. 34, p. 62–86.

Smith, G. I., 1979, Subsurface stratigraphy and geochemistry of Late Quaternary evaporites, Searles Lake, California: U.S. Geological Survey Professional Paper 1043, 130 p.

Smith, J. D., and Foster, J. H., 1969, Geomagnetic reversal in Brunhes Normal Polarity Epoch: Science, v. 163, p. 565–567.

Van der Hammen, T., and Gonzalez, E., 1960, Upper Pleistocene and Holocene climate and vegetation of the "Sabana de Bogota" (Colombia, South America): Leidse Geologische Mededelingen, v. 25, p. 261–315.

——, 1964, A pollen diagram from the Quaternary of the Sabana de Bogota (Colombia) and its significance for the geology of the Northern Andes: Geologie en Mijnbouw, v. 43, p. 113–117.

Van der Hammen, T., Wijimstra, T. A., and Zagwijn, W. H., 1971, The floral record of the Late Cenozoic of Europe, *in* Turekian, K. K., ed., The Late Cenozoic glacial ages: New Haven, Yale University Press, p. 391–424.

Wahrhaftig, C., and Sharp, R. P., 1965, Sonora Pass Junction to Bloody Canyon, *in* Wahrhaftig, C., Morrison, R. B., and Birkeland, P. W., eds., Guidebook for Field Conference I, Northern Great Basin and California, VII INQUA Congress: Lincoln, Nebraska Academy of Sciences, p. 71–88.

Wijmstra, T. A., 1969, Palynology of the first 30 metres of a 120 m deep section in northern Greece: Acta Botanica Neerlandica, v. 18, p. 511–527.

Wijmstra, T. A., and van der Hammen, T., 1974, The last interglacial-glacial cycle; State of affairs of correlation between data obtained from the land and from the ocean: Geologie en Mijnbouw, v. 53, n. 6, p. 386–392.

Willman, H. B., and Frye, J. C., 1970, Pleistocene stratigraphy of Illinois: Illinois State Geological Survey Bulletin 94, 204 p.

Woillard, G. M., 1975, Recherches palynologiques sur le Pleistocene dans l'est de la Belgique et dans les Vosges Lorraines: Acta Geographica Lovaniensia, v. 14, 118 p.

——, 1977, Comparison between the chronology from the beginning of the classical Eemian to the beginning of the classical Würm in Grande Pile peat-bog, and other chronologies in the world: International Geological Correlation Program, Project 73/1/24, Quaternary glaciations in the Northern Hemisphere, Report 4, Session in Stuttgart–Hohenheim, September 5-13, 1976, 12 p.

——, 1978, The last interglacial-glacial cycle at Grande Pile in northeastern France: Travaux du Laboratoire de Palynologie et Phytosociologie, Universite de Louvain (Belgium), 21 p.

——, 1979a, Abrupt end of the last interglacial s.s. in north-east France: Nature, v. 281, p. 558–562.

——, 1979b, Grande Pile peat bog; A continuous pollen record for the last 140,000 years: Quaternary Research, v. 9, p. 1–21.

Woillard, G. M., and Mook, W. G., 1982, Carbon-14 dates at Grande Pile; Correlation of land and sea chronologies: Science, v. 215, p. 159–161.

Manuscript Accepted by the Society September 15, 1986

Printed in U.S.A.

Geological Society of America
Special Paper 214
1988

Diatom biostratigraphy and the paleolimnology of Clear Lake, Lake County, California

J. Platt Bradbury
U.S. Geological Survey
Box 25046, MS 919
Denver Federal Center
Denver, Colorado 80225

ABSTRACT

Fossil diatoms from a 177-m core (CL-80-1) taken near the center of the main basin (Upper Arm) of Clear Lake, California, provide evidence about the stratigraphic relationships, age, and environmental history of these lacustrine deposits. In general, diatom assemblages from the core are dominated by planktonic genera such as *Stephanodiscus, Cyclotella,* and *Melosira.* Shallow-water species of *Fragilaria* and *Amphora* are common and sometimes abundant. Several planktonic diatoms from the core are also found in the Kelseyville Formation, which is exposed on the southern margin of Clear Lake and is inferred to underlie the modern lacustrine deposits. The presence of these taxa in the same stratigraphic order in both the Clear Lake core and the Kelseyville Formation suggests partial correlation between the two and implies a relationship between the Kelseyville Formation and the lacustrine sediments beneath Clear Lake.

In the upper 50 m of core CL-80-1, diatom assemblages apparently reflect late Pleistocene and Holocene paleoenvironmental changes, although their environmental significance may be obscured by reworking of diatoms from older sediments, by tectonically caused changes in patterns and rates of sedimentation, and by the impact of volcanism. Nevertheless, the diatoms indicate that lacustrine environments have been characterized by fresh, moderately deep, nutrient-rich water throughout much of their sedimentary history. Cooler climatic and lacustrine environments of the late Pleistocene were characterized by a codominance of *Stephanodiscus* and *Melosira* species, implying a mesotrophic to eutrophic, stratified lake. After the change from Pleistocene to Holocene climates, Clear Lake became yet more eutrophic and turbid. Stratification was short-term and irregular, and warm-water conditions extended throughout a greater portion of the growing season although there is evidence for a middle Holocene return to cooler and moister conditions.

The modern limnology of Clear Lake, which is characterized by massive blooms of blue-green algae and by the abundance of *Melosira granulata,* apparently began about 15,000 years ago.

INTRODUCTION

A biostratigraphic and paleolimnologic study of diatoms from a 177-m core of the lacustrine deposits underlying Clear Lake, Lake County, California, was designed to provide a paleolimnologic background for the description and interpretation of Quaternary paleoenvironments and paleoclimates in the region. Analysis of pollen (Adam and others, 1981; Sims and others, 1981; and Adam and West, 1983) and fish remains (Casteel and others, 1977), as well as studies of volcanic ash, sedimentology, and geochronology (Sims, 1976; Blunt and others, 1981) have provided ancillary paleoenvironmental information.

Fossil diatoms are often common or abundant in the lacustrine facies of the Kelseyville Formation that outcrops in the drainage basin of Clear Lake (Rymer, 1981). Their study documents the earlier paleolimnologic history of the basin and its relationship to volcanism and tectonism associated with the development of the Clear Lake Volcanics, and helps evaluate the possibility of resedimentation of older diatom frustules with the younger deposits of Clear Lake.

The focus of this paper is to provide an integrated paleoenvironmental and geologic history of the Clear Lake basin that incorporates data from earlier investigations and relies significantly on modern limnologic and biologic observations of this very eutrophic lake. Consequently, the modern limnologic environment of Clear Lake is summarized first, especially in relation to diatom productivity. This is followed by a description of the diatom assemblages in the Kelseyville Formation south of Clear Lake and in the sediments cored beneath the modern lake and by a discussion of the implications of possible correlations between the assemblages. Plates 1 through 4 illustrate characteristic centric diatoms from Clear Lake that are important for correlations and paleolimnology. Finally, a paleolimnologic history of Clear Lake based on diatoms and related studies is presented in the context of Quaternary climate change and tectonic and volcanic impact on the lake basin and its drainage area.

LIMNOLOGY, ALGAL PRODUCTIVITY, AND SEDIMENTATION IN CLEAR LAKE

Clear Lake is a large (17,670 ha), shallow (mean depth, 9.6 m), eutrophic lake with a comparatively intricate shoreline that partially restricts its three basins (Fig. 1) and allows them to function to some extent as individual limnologic systems. Throughout the lake, water temperatures are comparatively high (annual range, 8° to 25°C), and large inputs of nutrients from agricultural fertilizers and municipal wastes foster massive blooms of algae. The sedimentation and decomposition of the algae make the sediments and immediately overlying water anoxic from July to September, the period of weak and sporadic thermal stratification of the water (Horne and Goldman, 1972). During the winter rainy season, suspended matter from inflowing streams causes high inorganic turbidity in the lake.

The water in Clear Lake is ionically typical of lakes of open river systems. The hydrochemistry is dominated by Ca^{+2} and HCO_3^{-2} + CO_3^{-2} and has lesser amounts of Mg^{+2}, Na^{+1}, and K^{+1}, and of Cl^{-1} and SO_4^{-2}. The water is classified as hard and has a conductivity between 200 and 300 micromhos/cm. Variations in water chemistry result from productivity, rainfall, inflowing streams, and input from mineral springs. Mineral springs can provide algal nutrients, specifically N and SiO_2 (Thompson and others, 1981), and may influence algal productivity in restricted areas of the lake.

The phytoplankton dynamics and ecology of Clear Lake are variable and complicated. The large nutrient inflow, coupled with a rapid cycling of nutrients—particularly phosphorus—after short periods of stratification, suggest that phytoplankton productivity is seldom nutrient-limited (Goldman and Wetzel, 1963). However, because nutrients arrive mostly during the rainy season via stream flow, excessive inorganic and organic turbidity may cause limitation of light for some species. Indeed, the highest productivity levels appear to be related to the first storms of the season (early fall), when adequate illumination is still present and turbidity has not reached maximum levels (Goldman and Wetzel, 1963).

During late spring, summer, and early fall, the productivity of Clear Lake is dominated by blue-green algae, principally *Anabaena* and *Aphanizomenon,* whose ability to fix and utilize atmospheric nitrogen and to float in the upper, well-illuminated part of the water column produces nuisance blooms of extraordinarly high productivity (Horne and Goldman, 1972).

The blue-green algal productivity of Clear Lake varies from year to year and in the different basins of the lake, presumably because of the differential input of essential nutrients and the variable depth and morphometry of the subbasins. The principal inflows are by Scotts Creek (Rodman Slough) and Kelsey Creek; the creeks enter the shallow, extensive main basin of Clear Lake and probably account for the generally high productivity in this part of the lake (Horne and Goldman, 1972). The Oaks Arm, however, also has very high productivity on occasion, perhaps reflecting the greater transparency of the water in this deep region of the lake (Goldman and Wetzel, 1963).

The massive blooms of blue-green algae in Clear Lake overshadow the primary productivity of planktonic diatoms, partly because blue-green algae can fix needed atmospheric nitrogen when phosphate is plentiful and because they do not require silica and hence are not limited by low supplies of this requisite diatom nutrient. Additionally, many blue-green algae can regulate their buoyancy and thereby float in the photic zone. Planktonic diatoms sink to the dark and unproductive bottom waters between periods of turbulent mixing.

Sediment traps, installed by R. Y. Anderson in both the main basin and Highlands Arm of Clear Lake (Fig. 1), contain abundant cells with chloroplasts of the diatoms *Melosira granulata* and *Stephanodiscus niagarae,* and many additional species, without chloroplasts usually in small numbers (Table 1). Primary productivity studies (Horne and others, 1972) show significant positive relationships between carbon fixation (productivity) and the diatom *Stephanodiscus niagarae* (originally identified as *Coscinodiscus* sp.) and between phytoplankton biomass and *Melosira* sp.

Although modern diatom phytoplankton dynamics in Clear Lake have not been studied, the following observations can be gleaned from the limited data available. The productivity studies by Horne and Goldman (1972) and Horne and others (1972) concentrated on nitrogen fixation by blue-green algae, but their analyses indicated that on August 26, 1970, *Melosira* (presumably *M. granulata*) was common at the same time the blue-green alga *Anabaena* was actively fixing atmospheric nitrogen and total algal biomass was high. This observation correlates with the known distribution of *M. granulata* in other lakes where it tends to be a summer-to-late-summer–dominant in shallow, turbid

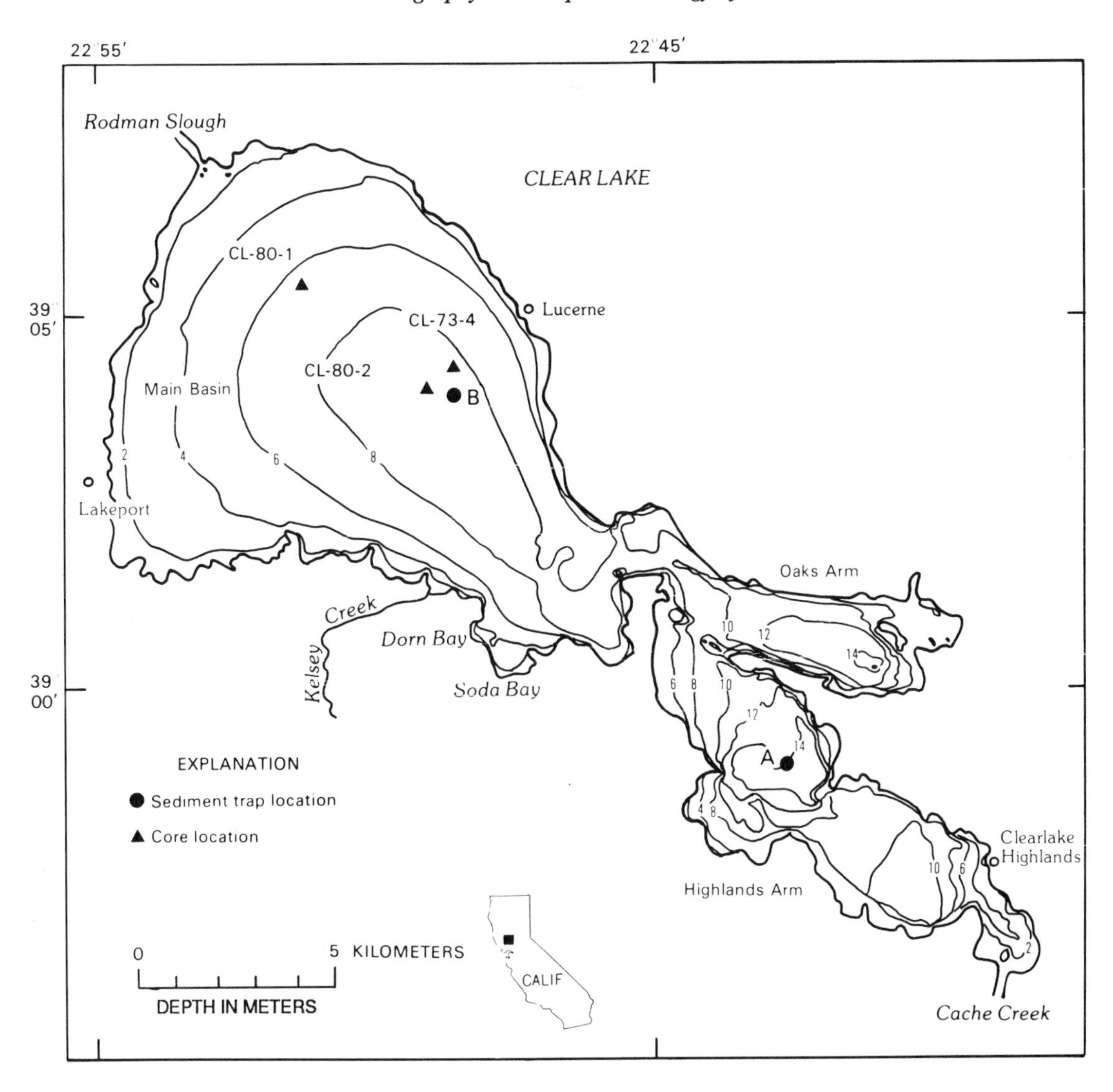

Figure 1. Bathymetric map of Clear Lake, Lake County, California. (After Goldman and Wetzel, 1963.)

lakes with abundant nutrients, especially silica (Kilham and Kilham, 1975; Lund, 1962). Aside from high requirements for silica, the nutrient needs for *M. granulata* are not well known, although it probably prefers rather high levels of nitrogen and phosphorus as well, because it is commonly present in eutrophic lakes. For example, in a shallow eutrophic Scottish lake (Loch Leven), *M. granulata* became a dominant member of the summer phytoplankton when nitrate levels increased (Bailey-Watts, 1982). Active turbulence is equally important for *M. granulata* because this heavily silicified diatom will sink below the photic zone unless periodic turbulence suspends it. Turbulence is also required to recycle nutrients needed by this diatom.

Stephanodiscus niagarae is the second most common diatom in the Clear Lake phytoplankton. Horne and others (1972) identified this diatom as *Coscinodiscus* (a marine genus with superficial similarities), and noted that its presence was positively correlated with carbon fixation (a measure of productivity) on September 21, 1970, when high algal productivity occurred in the Highlands Arm region of the lake.

The limnologic and algal productivity studies that focused on nitrogen fixation of blue-green algae are insufficient to allow much speculation about the ecology and population dynamics of *S. niagarae* in Clear Lake. However, its occurrence in the fall is consistent with observations of this and closely related species in other lakes (Bradbury, 1978; Bailey-Watts, 1973). *Stephanodiscus niagarae,* like several other species of the genus, competes successfully for silica when phosphorus is abundant (S. Kilham, oral communication, 1983; Kilham and Kilham, 1978). When phosphorus levels in the water are high, *Stephanodiscus* effectively utilizes low levels of silica and becomes the predominant diatom. As phosphorus levels decline, other diatoms become more efficient competitors, and bloom successively according to

TABLE 1. MODERN DIATOM ASSEMBLAGES FROM CLEAR LAKE, CALIFORNIA

Diatoms	Dorn Bay, Littoral Sample, 1-m depth (%)	Highlands Arm, Sediment Trap A* (1976-1977) (%)	Main Basin Sediment Trap B* (1976-1977) (%)
Melosira varians	24	+	
Fragilaria vaucheriae	22	+	+
Cocconeis palcentula	14	+	+
Fragilaria construens var. *venter*	6	+	
Navicula menisculus	6		
Navicula confervacea	4		
Achnathes lanceolata	3		
Melosira granulata	3	88	68
Navicula graciloides	2		
Synedra ulna	2	+	+
Amphora perpusilla	1	+	+
Cocconeis pediculus	1		
Eunotia pectinalis	1		
Nitzschia amphibia	1		
Stauroneis phoenicentron	1		
Rhoicosphenia curvata	1		+
Gomphonema parvulum	1		
Eunotia monodon	1		
Navicula huefleri	1	+	+
Cyclotella glomerata			1
Amphora veneta	+		+
Amphora ovalis	+	+	
Bacillaria paradoxa?	+		
Biddulphia laevis	+		
Caloneis bacillum	+		
Cyclotella meneghiniana	+	+	+
Cymbella muelleri	+	+	
Cymbella mexicana	+	+	
Diatoma vulgare	+	+	
Gomphonema gracile	+	+	6
Melosira distans	+	+	6
Melosira sp.	+		+
Navicula pupula var. *rectangularis*	+		+
Navicula exigua	+		+
Navicula sp. cf. *N. trypunctata*	+		
Navicula cuspidata	+		
Nitzschia palea	+		
Nitzschia frustulum	+		
Rhopalodia gibberula	+		
Stephanodiscus rotula var. *intermedia*	+	+	3
Stephanodiscus rotula var. *minutula*	+	+	5
Melosira ambigua	+	2	1
Stephanodiscus			
subtilis		+	6
hantzschii		+	+
alpinus?	+	+	
sp.		+	1
Fragilaria crotonensis		+	+

*Mean of ten sediment samples. Plus sign (+) indicates less than 1 percent.

nutrient availability and their growth strategy. *Stephanodiscus* often blooms in large numbers during or just after periods of lake overturn (in the spring and fall in dimictic lakes). However, in shallow lakes that are not ice covered during the winter and that mix to the bottom sporadically throughout the year, it is unlikely that blooms of *Stephanodiscus* have such regular periodicity. In such systems as Clear Lake, *Stephanodiscus* appears when sufficient phosphorus is available and blue-green algae are unable to thrive because of some other limitation. This situation occurs more or less regularly in the fall when decreasing temperatures and lower illumination limit the blue-green algae, and the first storms of the winter rainy season bring additional nutrients to the lake (Goldman and Wetzel, 1963). However, because of the irregularity of turbulent mixing, storms, and related nutrient influxes in shallow eutrophic lakes, the population dynamics of diatom blooms are sporadic and difficult to characterize seasonally (e.g., Bailey-Watts, 1978). Nevertheless, *Stephanodiscus niagarae* was common in the nearshore waters of Clear Lake in the fall of 1978 (Tomas Adams, oral communication, 1983).

SEDIMENT-TRAP STUDIES

In order to better understand the depositional processes in Clear Lake and to assess aspects of the diatom phytoplankton succession, self-timing sediment traps were installed at two localities in the lake as part of a program to study lake sedimentation processes (Anderson, 1977). The traps were set to automatically mark increments of sedimentation at 10-day intervals. One trap was placed about 3 km southwest of the town of Lucerne in the main basin of Clear Lake at a depth of about 5 m in water about 8 m deep (locality B, Fig. 1). The second trap was placed near the middle of the Highlands Arm at a depth of about 8 m in water 14 m deep (locality A, Fig. 1). The traps were installed on July 31 and August 1, 1976. They were recovered by Anderson 1 yr later.

Because of turbulent resuspension of bottom sediments and high productivity in Clear Lake, sediment overflowed the traps after about 50 days at locality B (main basin) and after 90 to 100 days at locality A (Highlands Arm). Consequently, the trap record of sedimentation and phytoplankton succession at Clear Lake spans only the beginning of the rainy season. Nevertheless, valuable information about plankton diatom succession and productivity was obtained for this season that amplifies the isolated information presented in earlier limnologic studies of the lake (Goldman and Wetzel, 1963; Horne and Goldman, 1972; Horne and others, 1972).

During the fall, the diatom phytoplankton of Clear Lake is dominated by *Melosira granulata*; this species comprises about 70 percent of the diatom cell numbers at locality B and 80 to nearly 100 percent at locality A (Fig. 2). *Stephanodiscus niagarae* is always subdominant, rarely attaining more than 15 percent of the total diatom count. At locality B, small-sized species of *Stephanodiscus* (*S. rotula* var. *minutula, S. subtilis, S. hantzschii*) often are present as well, but these forms, in contrast to *M. granulata* and *S. niagarae,* were not observed to contain chloroplasts; possibly they were reworked from earlier periods of productivity.

The sediment trap studies (Fig. 2) also indicate that diatoms are less abundant at locality B than at locality A. Diatoms were quantified by mounting one drop of water-saturated sediment on a coverslip and counting all identifiable diatom frustules along a transect of known length. At locality B, an enormous amount of diatomaceous hash and silt collected in the trap during the fall of 1976. Presumably this material primarily represents sediment resuspended by turbulent mixing in this open-water, shallow part of the lake. The trap samples in the deep part of the Highlands Arm, however, generally contain abundant, well preserved diatoms. This probably reflects the deeper water and less active turbulence at this site, and may also relate to lower turbidity and greater productivity at this season in this part of the lake (Goldman and Wetzel, 1963).

The record of diatom blooms in the fall of 1976 shows that the main basin and the Highlands Arm of the lake have different limnologic environments, although numbers of *Stephanodiscus niagarae* increase irregularly in both basins during the fall (mid- to late September). This seems to support the observation of Horne and others (1972) about the correlation of this discoid diatom with productivity on September 21, 1970.

LITTORAL DIATOM ASSOCIATIONS

Like the planktonic diatoms, littoral diatom communities in Clear Lake have received almost no attention. Diatoms in this habitat live attached to substrates such as rocks, wood, or stems of macrophyte aquatic vegetation, or simply on the mud surface, where they can move about in search of favorable places for growth. Because of the high productivity and dominance of blue-green algae during much of the summer growing season in Clear Lake, production of littoral algae is limited by lack of light. The low-light transmissivity of Clear Lake is due to the capability of blue-green algae to float to the surface and thereby shade the water and lake bottom beneath, and to the high turbudity that results from frequent periods of mixing and seasonal inflow of rivers (Goldman and Wetzel, 1963).

Despite apparently unfavorable conditions, a sample from 1 m depth at Dorn Bay (Fig. 1) contains a comparatively diverse (although sparse) littoral diatom assemblage (Table 1). The dominant diatom of this assemblage, *Melosira varians,* characterizes eutrophic and especially mildly polluted shallow water (Lowe, 1974). This diatom is also commonly attached to various substrates in slow-moving eutrophic rivers, such as the lower reaches of the Sacramento River, where it is often the dominant diatom (Britton, 1977). The subdominant diatom, *Fragilaria vaucheriae,* also known as a pollution-tolerant diatom, achieves high reproduction rates in mildly polluted water (Lange-Bertalot, 1979). The same trophic conditions apply to many of the less common species, such as *Navicula menisculus* (Hustedt, 1957) and *Navicula confervacea* (Cholnoky, 1968; Lowe, 1974).

A single littoral sample from Clear Lake cannot characterize the littoral diatom flora of this complex environment, especially in areas in which marginal mineralized springs enter the lake (Thompson and others, 1981). Yet the sample nevertheless provides important baseline information about the modern trophic status of Clear Lake that can be useful in paleolimnologic comparisons.

DIATOM ASSEMBLAGES IN LACUSTRINE DEPOSITS AROUND CLEAR LAKE

Lacustrine and fluvial deposits of Pliocene and Pleistocene age are exposed around Clear Lake basin, particularly east and southeast of the Oaks and Highlands Arms, and south of the main basin in the drainage area of Kelsey Creek (Rymer, 1981). These deposits have been assigned by Rymer (1981) to the Cache Formation (generally east of the Oaks Arm and within the upper

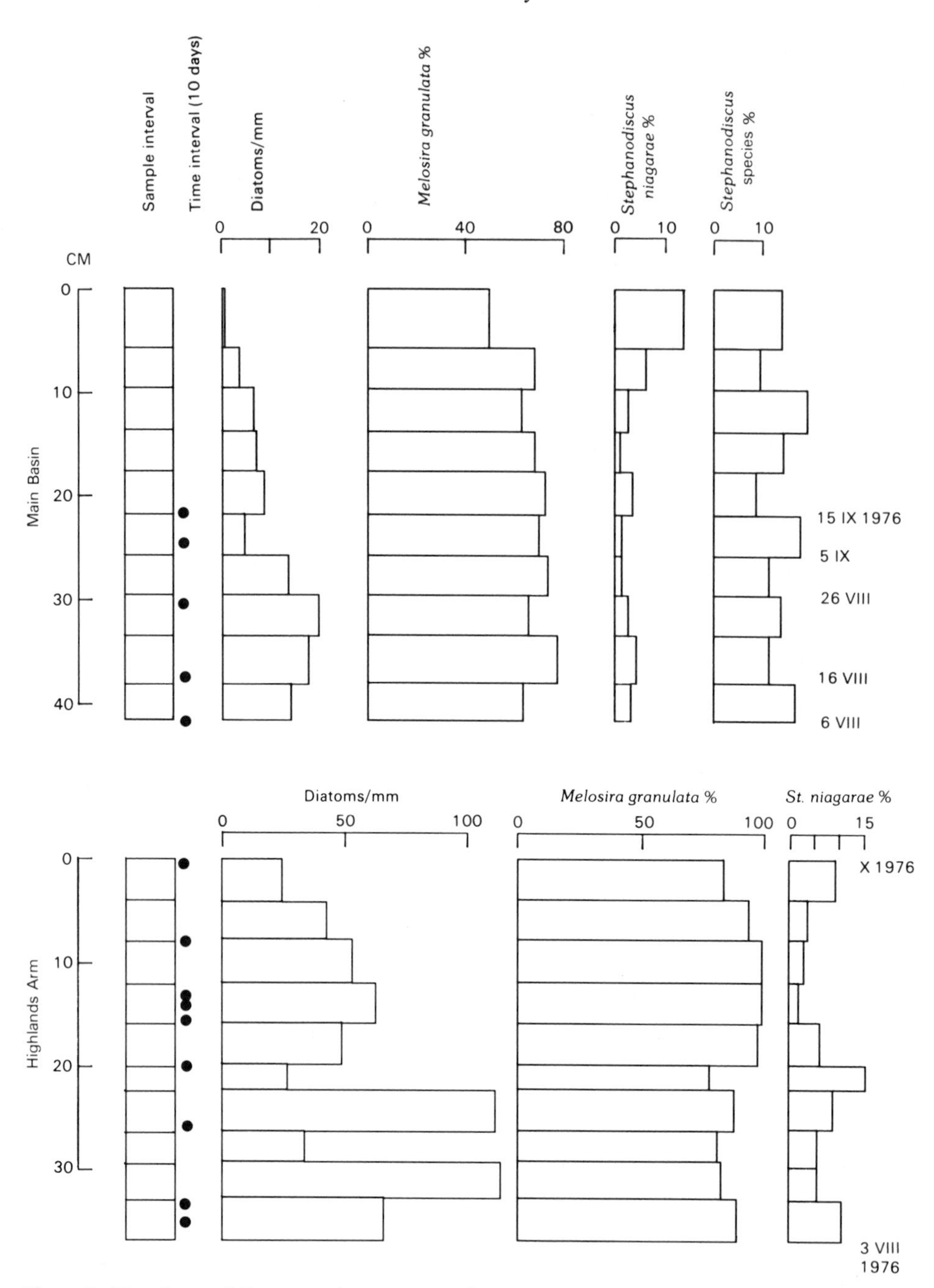

Figure 2. Abundance of diatoms and percentages of dominant species in sediment-trap samples representing fall 1976 from two localities in Clear Lake (Fig. 1). Ten-day collection intervals marked by layers of Teflon granules (black dots).

drainage of the North Fork of Cache Creek), the Lower Lake Formation (a restricted band of deposits east, southeast, and south of the Highlands Arm), and the Kelseyville Formation (in the drainage of Kelsey Creek south of the main basin). The Cache Formation is of Pliocene and early Pleistocene age, and the Lower Lake and Kelseyville Formations, although not well dated, are presumably middle Pleistocene in age.

The deposits contain a number of different lithologies, including beds of volcanic tuff, conglomerates, fluvial silts and sands, marls, and diatomites. Because the Kelseyville Formation outcrops in and underlies a major drainage basin of Clear Lake in the area of Big Valley and Kelsey Creek and its tributaries, analysis of the diatoms in this formation is essential to assess the likelihood of reworking of fossil diatoms into the deposits of the

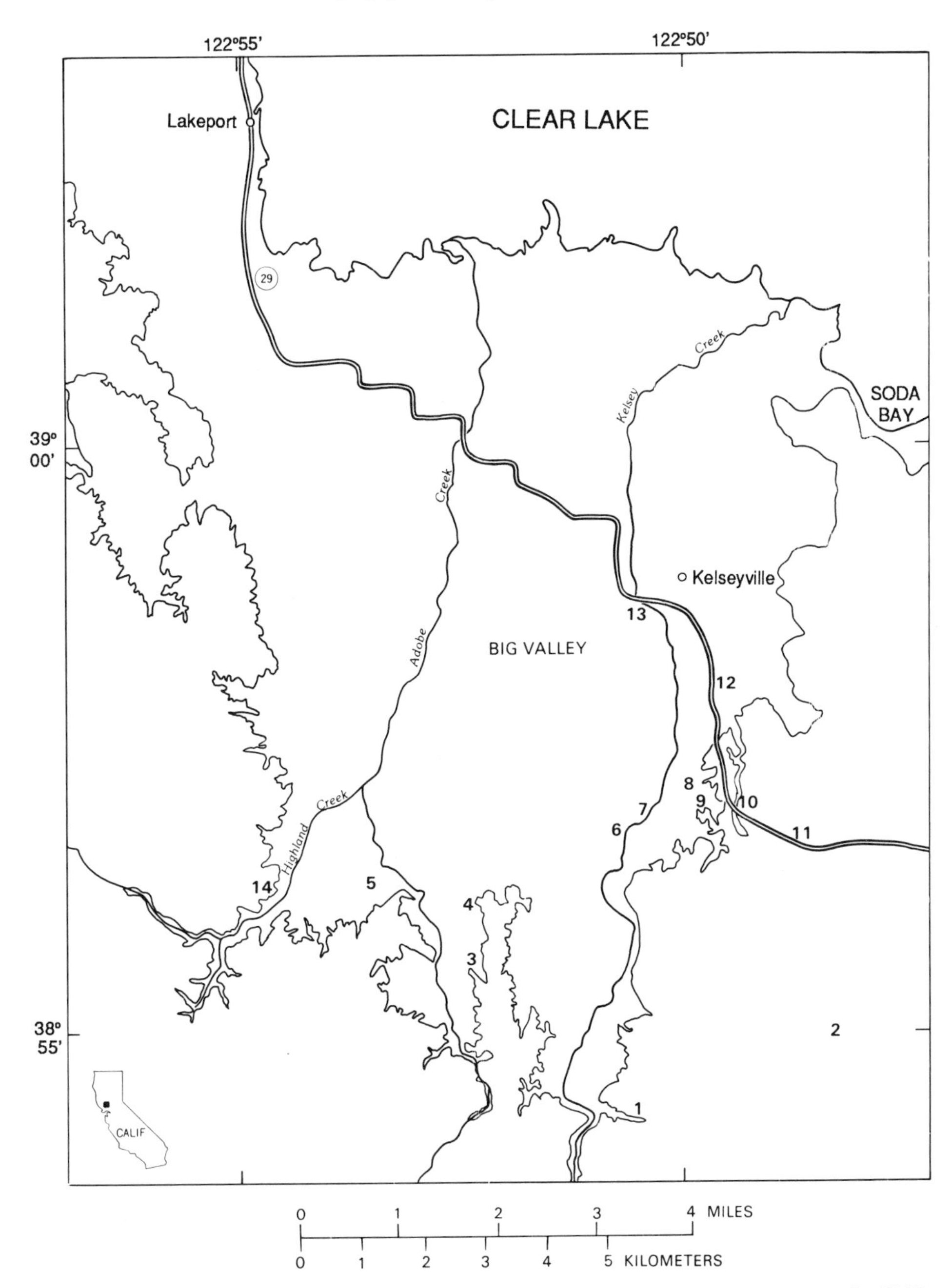

Figure 3. Index map of sample collection localities in Kelseyville Formation, Big Valley area. See Table 2 for specific locations.

present Clear Lake basin. In addition, stratigraphic studies (Rymer, 1981) suggest that the Kelseyville Formation might intergrade with the Clear Lake lacustrine deposits at depth; diatom analyses could help examine this possibility.

The Kelseyville Formation is exposed south of the town of Kelseyville, where the drainages of Kelsey Creek, Adobe Creek, and Cole Creek enter the Big Valley area (Fig. 3). Most of the outcrop area is lowland that represents an upthrown block surrounded by normal faults. The distribution of a marker unit, the Kelsey Tuff Member, within the Kelseyville Formation indicates that the Kelseyville dips to the north at a gentle angle of about 15° (Rymer, 1981). To the north the deposits are truncated by the Big Valley fault, which has displaced the Kelseyville Formation downward about 67 m. Rymer (1981) suggested that the formation was about 500 m thick, thinned considerably to the south, and consisted largely of fine- to medium-grained clastic

TABLE 2. LOCATIONS, LITHOLOGY, AND STRATIGRAPHIC POSITION OF OUTCROP SAMPLES FROM THE KELSEYVILLE FORMATION, BIG VALLEY AREA

Sample Location Number*	Sample Number	Lithology	Location	Stratigraphic position
1	Ost-5	Gray medium sand, very diatomaceous	SW1/4NW1/4SEC.10, T12N,R9W	Lowest part of formation 380 m below unit Qkk
2	4RO36A	White diatomite	NW1/4SE1/4SEC.26, T12N,R9W	
3	7RO30A	Diatomaceous fine gray silty clay	SW1/4NE1/4SEC.26, T13N,R9W	East margin of basin, 210 m below unit Qkk
4	6ROO8A	Diatomaceous granulitic silt with concretionary limonite	SW1/4NW1/4SEC.33, T13N,R9W	240 m below unit Qkk
	6ROO9A	Diatomaceous fine, gray silty clay	SW1/4NW1/4SEC.33, T13N,R9W	240 m below unit Qkk
5	6RO11A	Fine orange sand	SW1/4NW1/4SEC.32, T13N,R9W	West margin of basin, 210 m below unit Qkk
6	7ROO1A	Gray silty diatomite	NW1/4SE1/4SEC.27, T13N,R9W	200 m below unit Qkk
7	Ost-7	Fine gray silty sand with rare diatoms	SE1/4NE1/4SEC.27, T13N,R9W	92 m below unit Qkk
	Ost-6	Gray lithic sand with fish scales	SE1/4NE1/4SEC.27, T13N,R9W	85 m below unit Qkk
8	25 IX 78-6	Fine orange silty sand	NE1/4NW1/4SEC.26, T13N,R9W	3 m above unit Qkk
9	4RO30A	White diatomaceous silt	SW1/4NE1/4SEC.26, T13N,R9W	5 cm above unit Qkk
10	4ROO3E	Gray silt with rare diatoms	SE1/4NE1/4SEC.26, T13N,R9W	10 cm below unit Qkk
11	5RO21A	Coarse sand with diatoms	NE1/4SW1/4SEC.25, T13N,R9W	10 m above unit Qkk
	25 IX 78-5	Diatomite beds	NE1/4SW1/4SEC.25, T13N,R9W	10-20 m above unit Qkk. Beds about 2 m apart; A is lowest
12	4RO57A	Coarse gray sand with fine sand matrix	NE1/4SW1/4SEC.23, T13N,R9W	5 m below unit Qkk
13	5RO20B	Orange sand with rare diatoms	SE1/4NE1/4SEC.15, T13N,R9W	67 m above unit Qkk
14	8 IX 83-1	Silty diatomite	350 m E of SW corner SEC.30, T13N,R9W	Lower third of Kelseyville Formation?

*See Figure 3.

rocks. Outcrops of diatomite and fine-grained lacustrine sediments occur locally, and well preserved diatoms often occur sparsely in the coarser material as well.

Plant macrofossils and pollen in the lower part of the Kelseyville Formation indicate deposition during a warm climate, but the upper part contains fossils suggestive of a cold climate (Rymer, 1981).

Because much of the outcrop area of the Kelseyville Formation is drained by streams that flow into Clear Lake, there is a possibility of reworking of both pollen and diatoms from these sediments into the younger sediments of Clear Lake, and samples from the Kelseyville Formation were examined for diatoms (Fig. 3) to evaluate this possibility. The specific diatom locations (Table 2) can be approximately arranged in stratigraphic order with respect to the Kelsey Tuff Member (Qkk) (Rymer, 1981). Faulting and poor exposures make this stratigraphic arrangement tentative, however, and the placement of individual samples may err by as much as 100 m in the lower half of the exposed section. Nevertheless, the relative stratigraphic position of samples is reasonable (Fig. 4).

Fish fossils from the Kelseyville Formation (Casteel and Rymer, 1981) suggest a fluvial-lacustrine environment, but the comparative rarity of autochthonous lacustrine sediment components (diatoms, carbonates, gyttja, etc.) in the deposits and the dominance of fine- to medium-textured clastic sediment indicates that fluvial sediment sources prevailed in the part of the ancient basin represented by outcrops. Additionally, the discovery of tree stumps in growth position (Rymer, 1981) demonstrates subaerial environments during some intervals. Probably, hiatuses and erosion of previously deposited sediments happened at such times; consequently the depositional environments represented in the Kelseyville Formation are likely to be complex and discontinuous. In particular, the bedded diatomites above the Kelsey Tuff Member (Fig. 4) are interpreted to be significantly younger than the clastic rocks of the Kelseyville Formation at equivalent elevations because their lithology contrasts so greatly with the bulk of the Kelseyville Formation.

The percentage distribution of diatoms in the Kelseyville Formation (Fig. 4) indicates that several taxa are restricted stratigraphically; others occur more or less commonly throughout the

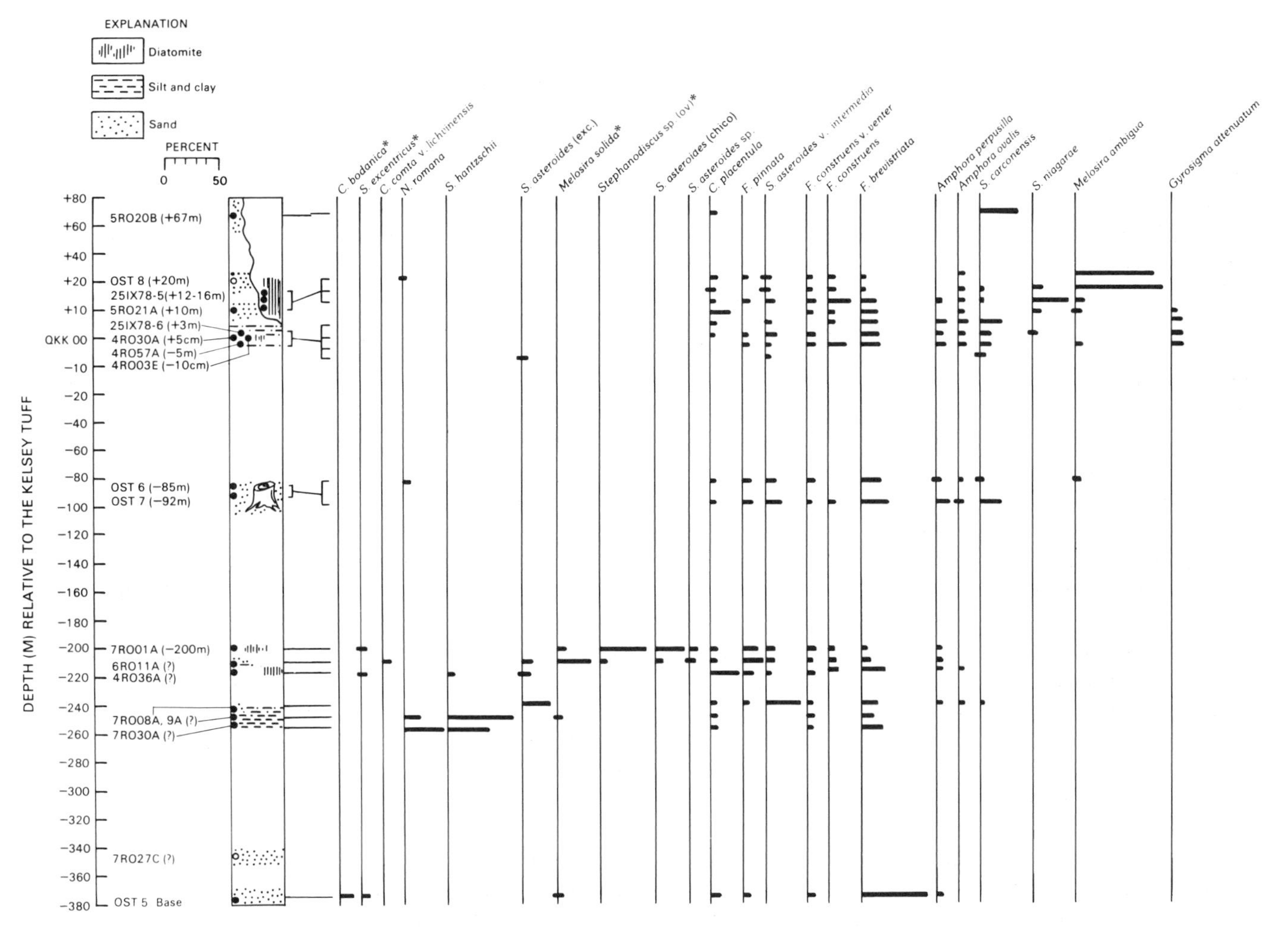

Figure 4. Dominant and(or) diagnostic diatoms in Kelseyville Formation arranged on partial lithologic section showing approximate stratigraphic location of samples relative to Kelsey Tuff Member (Qkk) of Kelseyville Formation. Open circles represent barren sample; scales for taxa marked with asterisk are expanded two times. (Data from Rymer, 1981, and M. J. Rymer, written communication (1982.)

unit. Some of the diatoms that are widely distributed throughout the formation, such as *Fragilaria brevistriata, F. construens, Cocconeis placentula,* and *Amphora ovalis,* are also common in the Cache and Lower Lake Formations (Rymer and others, this volume). In fact, these species are commonly distributed today in a wide range of shallow aquatic habitats. Other species, particularly of the genera *Stephanodiscus, Melosira,* and *Cyclotella* (Plates 1 through 4), which are generally planktonic, imply the presence of deeper-water environments. Their presence suggests that the Kelseyville Formation represented by outcrops was the marginal facies of a larger lacustrine system that has either been removed by erosion or been buried by younger deposits of Clear Lake. Because diatoms are usually rare, and broken or corroded in the coarser parts of the formation, it is possible that they have been reworked from earlier deposits of lacustrine sediments as a consequence of the variable depositional environments associated with fluvial-lacustrine sedimentation.

With the exception of *Stephanodiscus niagarae, S. hantzschii,* and *Melosira ambigua,* the other species of these genera are unknown or only rarely found living in modern environments. Specifically, the *Stephanodiscus asteroides* varieties and forms are thought to be extinct, both having disappeared from the Gadeb region of Ethiopia before 2 Ma (Gasse, 1980). Similarly, *Cyclotella comta* var. *lichvinensis* is presumed to have been extinct since the early Pleistocene (Loginova, 1982), and *Melosira solida*—though common in many Neogene lacustrine deposits throughout the world—apparently did not survive after the early Pleistocene in North America and Europe.

Stephanodiscus sp. cf. *S. carconensis* (Plate 3, Figs. 1–8) is also common in the Kelseyville Formation, especially in the upper part of the unit (Fig. 4). *Stephanodiscus carconensis* is often regarded as a Pliocene or Pleistocene form because it is widely reported in deposits of that age in the western United States and other parts of the world (Bradbury and Krebs, 1982).

There are occasional reports of this species being extant, such as at Lake Biwa, Japan, where it is dominant in the winter plankton (Negoro, 1960). Nevertheless, reports of this species living in North American lakes, such as at Klamath Lake, Oregon (Gasse, 1980; Van Landingham, 1967), are difficult to substantiate and apparently result from a misreading of K. E. Lohman's account in Moore (1937). Reports of the species in Devil's Lake, North Dakota (Elmore, 1921; Young, 1924), also require verification. *Stephanodiscus carconensis* was not recorded in a study of Devil's Lake algae (Verch and Blinn, 1972), nor was the species encountered in a paleolimnologic analysis of a 7-m core of lake sediments from this site (Stoermer and others, 1971). However, a somewhat similar species, *S. niagarae* var. *magnifica,* was found very rarely in the surface sediments and in greater percentages at depth in the core. A restudy of Elmore's original material might resolve the possibility of a mididentification of *S. carconensis.*

Stephanodiscus is an extraordinarily variable genus, and many species are superficially similar. Illustrations of *Stephanodiscus* species (cf. Gasse, 1980; Plates 2, 3) show that the characters used to separate different species are to some degree variable and transitional. Identifications are thus often difficult and commonly must be qualified. This problem becomes an important consideration when identifying a species as extinct or extant. For example, although *Stephanodiscus asteroides* is an extinct species (Gasse, 1980), it has a variety, *S. asteroides* var. *intermedia,* that is morphologically very close to an extant species, *S. rotula* var. *intermedia,* and is linked to it by transitional forms (Gasse, 1980). Consequently, it can become a matter of opinion whether a species is extinct or living. This problem is not new to paleontology, especially where polymorphic organisms such as diatoms are concerned, and for this reason statements about the stratigraphic and geographic distribution of fresh-water diatoms must be made with considerable caution.

These distinctive species, varieties, and forms of *Stephanodiscus* often occur in lacustrine deposits associated with volcanic terranes and(or) contemporaneous volcanism. Crater and caldera lakes, basins formed from damming by lava flows, and graben basins with associated volcanism are all geologic environments in which lacustrine deposits of Pliocene to Pleistocene age may contain diatoms resembling *Stephanodiscus carconensis, S. asteroides,* and eccentric, elliptical, or oval forms and varieties of these and other species (Plates 2, 3). The relationship between these distinctive diatom morphologies and the volcanic environment is so consistent that it is tempting to speculate that these diatoms may be ecophenotypes that develop in response to characteristic chemical and nutrient environments associated with volcanic lakes. If this is so, this group of diatoms may be of biostratigraphic value only inasmuch as they indicate a distinctive volcanic environment. When volcanism is widespread and suitable environments exist for the development of these distinctive *Stephanodiscus* species, they become abundant and characterize lacustrine deposits of similar age throughout the region. When volcanism ends and the distinctive lacustrine habitats disappear, these diatoms die off locally, although some may survive as isolated relict populations in suitable habitats. An example may be the occurrence of a *S. carconensis*-like diatom in Lake Biwa, Japan, where deposition of volcanic ashes has occurred throughout its approximately 5-m.y. history (Ueno, 1975; Ikebe and Yokoyama, 1976).

The *Stephanodiscus* species present in the Kelseyville Formation—particularly *S. carconensis* and its varieties, *S. asteroides,* and the undescribed oval or elliptical forms of *Stephanodiscus*—are consistent with a presumed middle Pleistocene age for the formation (Rymer, 1981). Yet, as the example from Lake Biwa indicates, this assemblage minimally implies any fresh-water, lacustrine habitat, associated with volcanism that included the deposition of volcanic ash and possibly influenced mineral and(or) thermal springs. This type of habitat is consistent with the stratigraphic and geologic evidence of the Kelseyville Formation.

Less information is available for the other distinctive diatoms of the Kelseyville Formation. The *Melosira* species (Plate 1, Figs. 1–13) that occurs most abundantly in sample 6R011A (Fig. 4) is characterized by wide, lump- or costae-like thickenings of the valve collar (Plate 1, Figs. 11-13), a feature common in Miocene and Pliocene species of this genus. This thick-walled, heavily structured diatom closely resembles *Melosira solida* Eulenstein (Van Heurck, 1885). Similar forms are found in several Pliocene localities in the western United States, such as the Glenns Ferry Formation in the western Snake River Basin, Idaho, in Alturas, California, and in the Carson Sink, Nevada. *Melosira solida* is one of the planktonic diatoms that now lives in Lake Biwa, (Mori, 1974; Negoro, 1960), but, like *Stephanodiscus carconensis,* it apparently has not been recorded in other regions. In Lake Biwa, *M. solida* dominates the plankton in the spring and fall (Negoro, 1960), presumably during periods of lake overturn. It is also found in the Uji River, which flows from Lake Biwa near Kyoto, Japan. Population studies of *M. solida* in this river (Zoriki, 1976) indicate that it is intolerant of high organic nutrient loading and that it is more common in the late fall, the same time it is blooming in Lake Biwa.

The presence of *Cyclotella comta* v. *lichvinensis* in sample 6R011A provides a unique stratigraphic marker for the Kelseyville Formation (Fig. 4) but little biochronologic or paleolimnologic information. The same is true for *Cyclotella bodanica* at the base of the formation (sample Ost-5; Plate 4, (Figs. 1–9). *C. comta* var. *lichvinensis* has not been found in modern diatom assemblages, although the variety is morphologically close to the species; *C. comta* is a widespread planktonic diatom in fresh-water oligotrophic to mesotrophic lakes (Lowe, 1974). The co-occurrence of *C. comta* var. *lichvinensis* with *M. solida* and *Stephanodiscus* species is not unusual in light of what is known about the ecology of *C. comta,* which is most common in cool water of low productivity. *C. comta* is recorded throughout much of the Lake Biwa core (Mori, 1975) but because the diatom has not been figured, it cannot be compared with *C. comta* v. *lichvinensis* of the Kelseyville Formation.

Cyclotella bodanica is an extant planktonic diatom, characteristic of oligotrophic to mesotrophic lakes in north temperate regions (Molder and Tynni, 1968). However, its maximum geographic distribution is not well established because this species has been inconsistently identified. As with many diatoms, the morphology of *C. bodanica* varies considerably, particularly in the relative length of the striae and the size of the central area. The specimens from the Kelseyville Formation have unusually long striae and small central areas (Plate 4, Figs. 1–9), and are very similar to populations found in oligotrophic lakes and ponds of the Canadian Yukon (identified as *C. comta* in Bradbury and Whiteside, 1980) and to specimens from Scandinavia (Molder and Tynni, 1968; Cleve-Euler, 1951). Some specimens from the Kelseyville Formation have such long striae that they nearly eliminate the central area (Plate 4, Figs. 3a, b). These forms closely approach the morphology of an extinct species, *Cyclotella temperei,* from Pliocene deposits of France (Ehrlich, 1969) and Russia (Loseeva, 1982). The large variability of these species and the fact that similar species are extant limits their biochronologic value. Probably they, like *Stephanodiscus carconensis,* reflect distinctive limnologic environments that are rare or inadequately studied today.

The presence of *Cyclotella bodanica* in the lowest part of the Kelseyville Formation implies cool to cold oligotrophic lacustrine environments, according to present information on modern analogs. Cool conditions in the lowest Kelseyville Formation preceded the warm climates indicated by plant megafossils found stratigraphically higher than sample Ost-5 (Rymer, 1981).

In contrast to the majority of the diatom assemblages studied from the Kelseyville Formation, two suites of samples contain diatoms that are readily comparable to Holocene diatom floras. Samples 6R009A and 7R030A are dominated by *Stephanodiscus hantzchii* and *Nitzschia romana,* and samples 25 IX 78-5A, -5B, and -5C have abundant *Melosira ambigua* and/or *S. niagarae.* Both assemblages are widely known from temperate lakes. *Stephanodiscus hantzschii* is a common planktonic diatom in eutrophic lakes (e.g., Bradbury and Winter, 1976). and *N. romana* is a comparatively tolerant diatom that characterizes nutrient-enriched portions of streams, rivers, and marginal lacustrine habitats (Lange-Bertalot, 1979). The *Melosira ambigua, Stephanodiscus niagarae* assemblage suggests mesotrophic to eutrophic lacustrine open-water environments. The dominance of these planktonic species and reduced numbers of shallow-water benthic or epiphytic diatoms implies a lake of moderate to large size (⩾10 ha) and depth (tens of meters) with a littoral zone at some distance away from the site of deposition.

This diatom assemblage contrasts so completely with the typical Kelseyville Formation diatoms (*Stephanodiscus carconensis, S. asteroides,* and distinctive *Cyclotella* and *Melosira* species) that it is difficult to envision these samples as belonging to the same lacustrine depositional system; this is especially true for samples 25 IX 78-5A, -5B, and -5C. These samples consist of fine-grained diatomaceous silts and diatomite that differ from the coarse-textured clastics of most of the Kelseyville Formation. Because of these differences, the *Stephanodiscus niagarae* and *Melosira ambigua* assemblages are interpreted to rest unconformably on the Kelseyville Formation, and are probably significantly younger in age (Fig. 4). Similar conclusions were reached by M. J. Rymer (oral communication, 1983) on the basis of stratigraphic correlations.

Paleolimnologic interpretation of the Kelseyville Formation is complicated by inadequate knowledge of the ecology of the abundant species of *Stephanodiscus* and the distinctive *Melosira solida* and *Cyclotella* species. Lake Biwa, Japan, seems to be the only modern site where some of these species now exist; Lake Biwa is a comparatively oligotrophic lacustrine system in a cool temperate climate, and the Kelseyville Formation was deposited in a similar environment. However, both *Melosira solida* and *Stephanodiscus carconensis* have persisted in Lake Biwa for perhaps hundreds of thousands or even millions of years. In that time span the lake is known to have experienced both warmer and cooler climatic periods (Fuji, 1974); consequently, it is inappropriate to apply a specific climatic interpretation to the presence of these species alone. The proportional distribution of *M. solida* and *S. carconensis* in the Lake Biwa core (Mori and Horie, 1975) suggests that *S. carconensis* was more dominant during the last cold period of the Pleistocene (22,000 to about 12,000 yr ago) and that *M. solida* characterized more temperate periods (Mori, 1975). A tentative application of the model to the Kelseyville Formation suggests that somewhat cooler conditions may be reflected by the abundance of *S. carconensis* in the upper half of the formation, and that *M. solida* (sample 6R011A) could imply more temperate environments. This interpretation is consistent with the paleoclimatic evidence provided by fossil plant studies (Rymer, 1981), but considerably more work must be done on the Kelseyville Formation before a satisfactory paleolimnologic and paleoclimatic history of the formation can follow. The fluvial, marginal lacustrine depositional system inferred for the Kelseyville Formation on the basis of sedimentary structures and lithology (Rymer, 1981) would be difficult to interpret in conventional paleolimnologic terms, because deposition in such environments is highly variable and often controlled by tectonic rather than by limnologic or climatic factors.

DIATOM ASSEMBLAGES IN LACUSTRINE DEPOSITS BENEATH CLEAR LAKE

The principal focus of this study is to evaluate and interpret the diatom assemblages and the biostratigraphy of the lacustrine sediments underlying Clear Lake in order to: (1) supplement earlier paleolimnologic studies of the lake (Sims, 1976); and (2) provide insights into the paleoclimatic history of the area, based on pollen studies of cores of the lake sediments (Adam and others, 1981; Adam and West, 1983).

A representative suite of 70 samples from core CL-80-1, which extends from the modern sediment-water interface to a depth of 177 m, was studied. Core CL-80-1 was taken from near the center of the Upper Arm of Clear Lake in a water depth of 7.5 m (Fig. 1; Sims and others, this volume). Although the

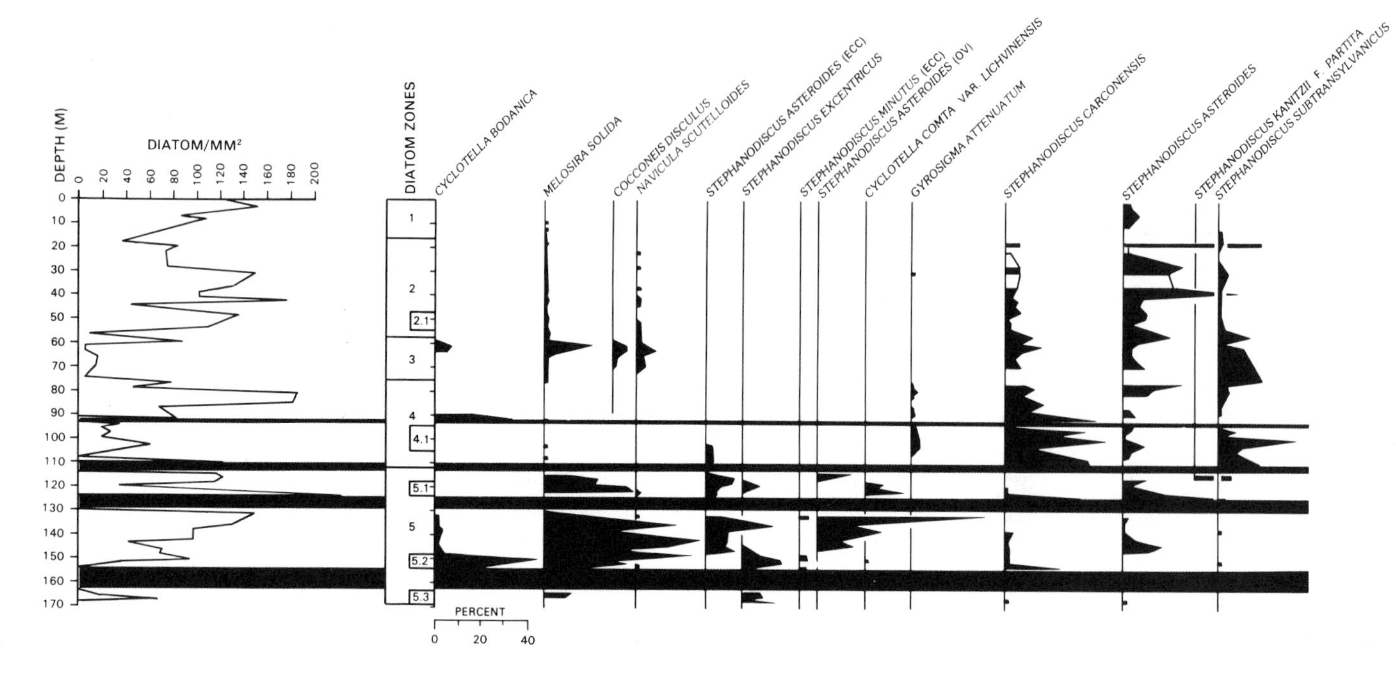

Figure 5. Diatom stratigraphy of core CL-80-1, Clear Lake, California. Shaded zones represent samples essentially barren of diatoms. Continued on facing page.

sediment was continuously cored, total recovery was only 66.5 percent, leaving some parts of the sequence sparsely represented (Sims and others, this volume). Nevertheless, the samples are, in general, adequately spaced and provide a reasonable picture of the diatom biostratigraphy of the core (Fig. 5).

Most of the sediments of core CL-80-1 are highly diatomaceous although diatom preservation is variable. The dry samples were broken down in distilled water, and an aliquot of water-saturated mud was dried on a coverslip and mounted in Hyrax mounting medium with a refractive index of 1.65. Generally, 200 to 500 diatoms were identified and enumerated from each sample; however, this was not possible for samples that had few or no diatoms. Diatom concentration was estimated by calculating the number of diatoms in a unit area of mounted sediment aliquots. Selected samples from core CL-73-4, the principal core used for pollen analysis (Adam and others, 1981; Adam, this volume), were analyzed to investigate the stratigraphic relation between it and core CL-80-1.

Diatom taxa, including species, varieties, and informally designated forms, which make up at least 5 percent of one sample and occur in at least three adjacent samples, have been used to zone the sediment section represented by core CL-80-1. These taxa have been plotted as percentages of the total diatom assemblage (Fig. 5). Exceptions to this general practice and minor taxa of importance are noted in the text.

In general, centric diatoms, specifically those belonging to *Cyclotella, Stephanodiscus,* and *Melosira,* dominate the sediments beneath Clear Lake. Annotated illustrations of the forms characteristic of both the deposits beneath Clear Lake and the Kelseyville Formation document this interesting assemblage (Plates 1 through 4). The only pennate diatoms of consistent importance in the core are taxa of the genera *Fragilaria* and *Amphora.* The percentage distribution of diatoms in core CL-80-1 is highly variable; the abundances of individual taxa change from very high to low percentages or to absence over comparatively short stratigraphic distances. This variable stratigraphic distribution may be caused by selection of available sample intervals or by rapid variations in sediment character and diatom preservation. Rapid paleolimnologic changes and related abrupt changes in diatom percentages cannot be ruled out in some cases, but this possibility is difficult to evaluate when the sample intervals are widely spaced.

The basal zone, zone 5, is characterized by large percentages of *Melosira solida* and variable (but often large) percentages of eccentric or ovate forms of *Stephanodiscus,* particularly *S. asteroides* and *S. astraea* v. *minutula.* Such forms are indicated in Figure 5 by the abbreviations ECC (eccentric) and OV (ovate) after the species name.

Three subzones (5.1, 5.2, and 5.3) are based on the restricted occurrence of distinctive diatoms or on abnormally large percentages of more widely distributed diatoms. The basal subzone (5.3) is dominated by *Fragilaria brevistriata* and *Fragilaria construens.* Subzone 5.2 contains large percentages of *Cyclotella bodanica;* subzone 5.1 is characterized by the restricted occurrence of *Cyclotella comta* var. *lichvinensis.* Zone 5 has two intervals that are barren of diatoms and is terminated at the top by a third. Highly corroded fragments of diatoms are present in some of the assemblages from these intervals. The barren sediment probably results

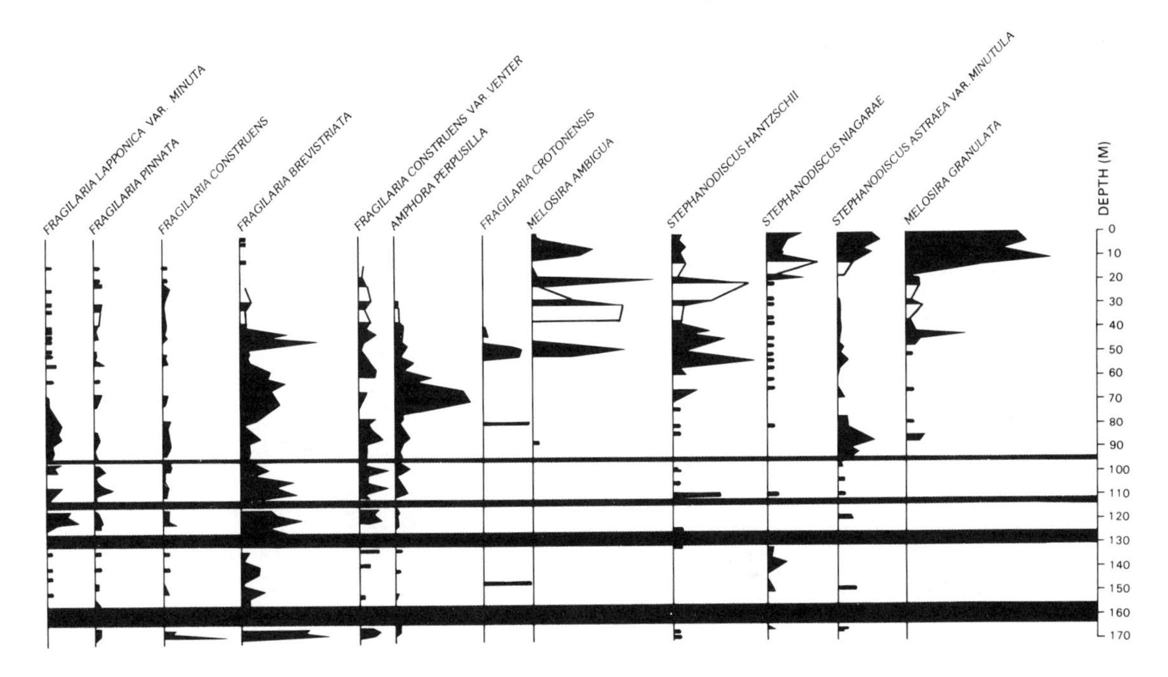

from a combination of postdepositional destruction of diatoms and an influx of fluvial sediments. The lowest barren zone occurs in the part of the core that contains several coarse clastic units (Sims and others, this volume); these sediments suggest fluvial deposition. The barren zones between 124 and 130 and between 110 and 113 m contain samples with abundant ostracodes; the ostracodes apparently lived in or originated from shallow-water environments (Forester, this volume) where diatom preservation is usually poor.

Diatom zone 4 is characterized by large percentages of *Stephanodiscus carconensis* and *S. subtransylvanicus.* Subzone (4.1) has low but consistent percentages *Gyrosigma attenuatum.* The top of this subzone (92.7 m) is marked by an interval barren of diatoms, and, like some of the previous levels barren of diatoms; this subzone contains abundant ostracodes. *Cyclotella bodanica* reappears just above this level and may have been reworked from older lacustrine sediment, although the frustules are generally well preserved.

Diatom zone 3 is characterized by high percentages of *Amphora perpusilla* and *Cocconeis disculus* and by the appearance, near the top of the zone (61.3 m), of degraded *Melosira solida* cells. Low diatom concentration and poor preservation at this level suggest that these cells might have been reworked into the sediments from previously deposited material. In general, diatom concentrations are very low throughout zone 3, and the lowest level (75 m) contains abundant ostracodes.

Diatom zone 2 is characterized by large but fluctuating percentages of *Stephanodiscus hantzschii* and *Melosira ambigua.* A subzone (2.1) near the base of this zone (45–51 m) is characterized by large percentages of *Fragilaria crotonensis,* a species abundant only in isolated samples below this depth. Large intervals between samples as a consequence of poor core recovery in zone 2 makes zonation of this part of the core difficult.

Diatom zone 1, the uppermost zone in core CL-80-1, contains many of the same diatom species that presently live in Clear Lake. The characteristic diatoms of this zone are *Stephanodiscus niagarae* and *Melosira granulata. Stephanodiscus astraea* var. *minutula, S. hantzschii,* and *Melosira ambigua* are also present. In both zones 1 and 2, diatom concentration is high and preservation generally good.

Before discussing the diatom stratigraphy of the lacustrine deposits beneath Clear Lake, it is appropriate to emphasize the important relationships between diatom concentration, diatom preservation, and percentage of individual taxa (Fig. 5). The percentage diagram graphically equates samples of greatly different diatom concentrations; consequently, interpretation of a high percentage of a given species must be tempered by consideration about its actual concentration. Samples that contain less than 50 diatom frustules per square millimeter should probably be considered suspect and interpreted with caution.

Diatom preservation often, but not necessarily, parallels diatom concentration. Samples with few diatoms commonly have only poorly preserved, but still identifiable, fragments, usually of the more heavily silicified or more abundant species. Although differences in preservation are difficult to quantify, under some circumstances it is possible to recognize badly corroded fragments, probably reworked from older deposits, mixed with well-preserved diatoms of syngenetic origin. Unfortunately, recognition of reworking is not always—and perhaps even rarely—possible. Diatomaceous rocks are usually lightweight and easily eroded; therefore, comparatively large pieces can move quickly downstream, become redeposited in lakes, and subsequently be broken down by burrowing organisms and other lacustrine processes. Diatoms on the surface of such fragments will likely be broken or destroyed, but those inside make the journey unscathed and will appear in a sample in surprisingly good condition. There is no question that this is happening in Clear Lake today. Kelsey Creek sediments entering the lake commonly contain pieces of water-rounded diatomite or diatomaceous rocks that originate from outcrops of the Kelseyville Formation in the drainage area of Kelsey Creek. Regardless of this, the modern sediments beneath Clear Lake contain comparatively few reworked diatoms, principally because diatom productivity in the lake is high and overshadows the influx of reworked frustules. However, during periods of low diatom productivity or high influx of sediment from rivers, reworked diatoms could constitute the bulk of the fossil flora in the sediments.

Because of these factors, the diatom assemblages in the basal half of zone 4 and in zone 3 where the low concentrations of diatoms occurs should be interpreted cautiously. This is particularly true for zone 3, where the large fluctuations in percentage of *Stephanodiscus carconensis, S. asteroides,* and *S. subtransylvanicus* are based on very few specimens that are often broken or corroded.

Despite the inevitability of some reworking of previously deposited diatoms from the Kelseyville Formation into the sediments of core CL-80-1, it is unlikely that reworking completely explains or invalidates the Clear lake diatom biostratigraphy. The distinctive succession of diatom assemblages in core CL-80-1, although probably modified locally by redeposition, mainly represents real biostratigraphic changes reflecting paleolimnologic changes in Clear Lake and the ancient lake basins that preceded it. These changes are discussed below in the context of apparent correlations between the Clear Lake diatom biostratigraphy of cores CL-80-1 (Fig. 5) and CL-73-4, and the diatomaceous deposits surrounding Clear Lake, principally from the Kelseyville Formation.

CORRELATIONS BETWEEN DIATOM BIOSTRATIGRAPHY OF CORES CL-80-1, CL-73-4, AND CL-80-2

The pollen from two long cores from Clear Lake, CL-73-4 and CL-80-1, were studied to establish a long paleoenvironmental record of late Quaternary time. The percentages of *Quercus* pollen in the upper 50 m of core CL-80-1 (Heusser and Sims, 1983) correlate closely to the percentages of these pollen types at the same depths in core CL-73-4 (Adam and others, 1981) (Fig. 6). Below this depth, however, the peaks of *Quercus* do not. This is puzzling because the cores are only about 3 km apart; both cores are from the central region of the main body of Clear Lake and were taken in about the same water depth (CL-73-4 in 7.5 m and CL-80-1 in 8.4 m).

Diatom assemblages from 11 samples from core CL-73-4 were used to evaluate the correlation between this core and CL-80-1. The similar assemblages (Table 3, Fig. 6) confirm a close

TABLE 3. DIATOM ASSEMBLAGES AT SELECTED DEPTHS IN CORES CL-80-1 AND CL-73-4

CL-80-1 Depth (m)	Diatom Taxa	Cl-73-4 Depth (m)
10.5	Melosira granulata D S. astraea var. minutula A Stephanociscus niagarae C	10
19.4	Melosira ambigua A Stephanodiscus niagarae A M. granulata var. angustissima C Stephanodiscus sp. C Cyclotella meneghiniana F	20
40+	S. asteroides var. intermedia A Stephanodiscus carconensis C S. asteroides C S. subtransylvanicus F	30
48	Fragilaria crotonensis A Melosira ambigua A S. asteroides A Stephanodiscus subtransylvanicus F S. carconensis F	60
51	S. hantzschii A Melosira ambigua C Stephanodiscus asteroides C S.subtransylvanicus F S. carconensis F	64.7
62	Stephanodiscus asteroides A S. sybtransylvanicus C S.carconensis F Melosira solida F Cocconeis disculus F Cyclotella bodanica R	71.9
65	S. asteroides A Stephanodiscus sybtransylvanicus C S. carconensis C Cocconeis disculus F Navicula scutelloides C	79.1
70	S. carconensis var. pusilla A Stephanodiscus asteroides C Cocconeis disculus F Stephanodiscus hantzschii? F Navicula scutelloides F Melosira solida R	86.25
75	Amphora perpusilla C S. subtransylvanicus C S. carconensis Stephanodiscus asteroides F Melosira solida F Cocconeis disculus R Gyrosigma attenuatum R	90
82	Stephanodiscus asteroides D S. subtransylvanicus A Diploneis ovalis C Amphora proteus C Epithemia turgida var. C Navicula walkeri R	106.8
98	Stephanodiscus asteroides D S. subtransylvanicus A S. carconensis F Amphora proteus F Epithemia turgida var. F Navicula walkeri F Gyrosigma attenuatum F Diploneis ovalis F Melosira solida R	115.2

*Explanation of symbols:
D = Dominant; A = Abundant; C = Common; F = Few; R = Rare.

+Poor depth control.

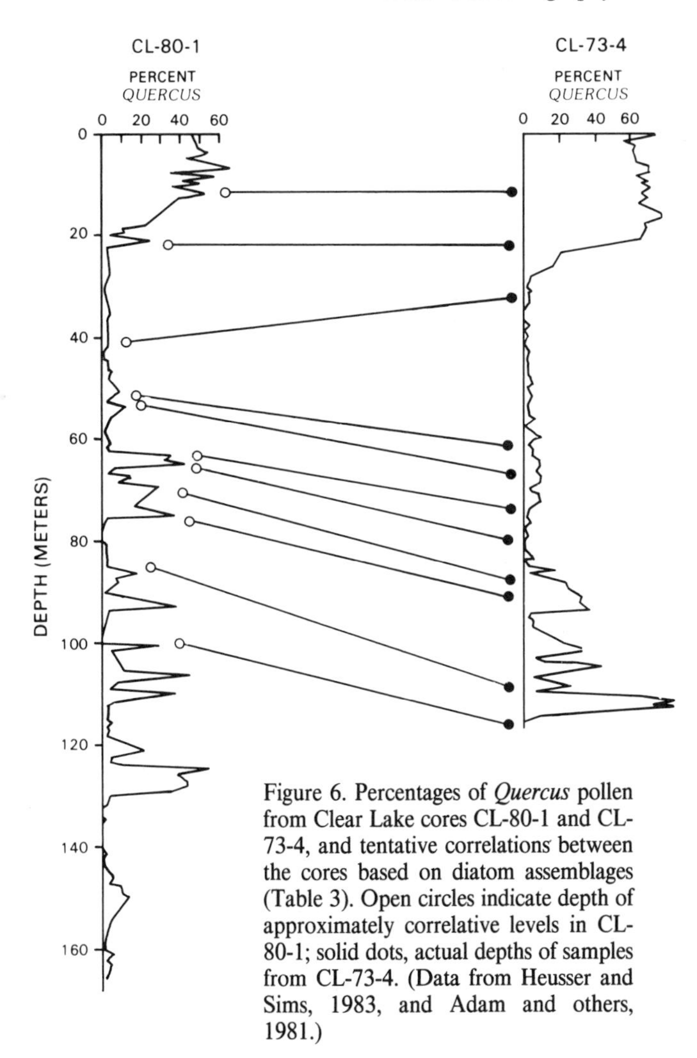

Figure 6. Percentages of *Quercus* pollen from Clear Lake cores CL-80-1 and CL-73-4, and tentative correlations between the cores based on diatom assemblages (Table 3). Open circles indicate depth of approximately correlative levels in CL-80-1; solid dots, actual depths of samples from CL-73-4. (Data from Heusser and Sims, 1983, and Adam and others, 1981.)

correlation between both cores on their upper parts, especially above 30 m, and a general depth agreement (to within 15 m) between the two cores below 30 m depth. Although some of these correlations are imprecise, several of them are constrained by a restricted stratigraphic distribution of a distinctive species, and therefore appear probable.

The tentative correlation between the two cores based on diatoms is confirmed by the presence of a distinctive volcanic ash at 54 m in CL-80-1 and 68 m in CL-73-4 (Sarna-Wojcicki, this volume). The correlation indicates that the rates of sediment accumulation at the two localities have not been the same through time and that stratigraphically equivalent levels are 12–25 m deeper in CL-73-4 than they are in CL-80-1. The questionable correlation between CL-73-4 at 30 m and CL-80-1 at 40 m may result from poor core recovery and inadequate sample interval control.

Percentage profiles of a dominant pollen type, *Quercus* (Fig. 6), can be approximately correlated in each core by following along the tie lines provided by the distinctive ash and diatom assemblages. This practice appears to be more or less successful down to a depth of 80 m in CL-80-1, but a detailed relative match of the peaks is impossible at any depth, especially below 80 m. The poor pollen correlation between these two cores indicates that their depositional environments must have been very different periodically throughout their lacustrine history.

Distinctive diatom assemblages were used to correlate both CL-80-1 and CL-73-4 to a third core, CL-80-2, that was taken less than 1 km from CL-73-4 (Fig. 1). The abundance of *Fragilaria crotonensis* in all three cores indicates that a depth of about 50 m in CL-80-1 correlates with a depth of 60 m and about 66 m in cores CL-73-4 and CL-80-2, respectively (Fig. 7). The 15-m deeper correlation to CL-80-2 from CL-80-1 may represent the natural slope of the ancient lake bottom at that time, but subsequent correlations indicate that faulting and(or) tectonic warping has affected the relative stratigraphic placement of the cores. For example, *Cocconeis disculus* and *Navicula scutelloides* occur in a zone of low diatom concentration between 60 and 70 m in core CL-80-1 (Fig. 5). The same assemblage is found at a depth of 79-86 m in CL-73-4 (Table 3) and at a depth of 104 m in CL-80-2. This correlation indicates a downward displacement of the CL-73-4 section of about 21 m relative to CL-80-1, and a downward displacement of the CL-80-2 section of about 20 m relative to CL-73-4 or a total of 41 m between CL-80-2 and CL-80-1 (Fig. 7). Greater displacements result from the correlations based on the distinctive species of *Cyclotella, C. ocellata* (Plate 3, Figs. 9–16) and *C. kutzingiana* v. *radiosa* (Plate 4, Figs. 10–15). In core CL-80-1 they occur at 90 and 94 m, respectively, separated by a barren zone at 93 m. The same sequence appears in CL-80-2 at depths of 156 (*C. ocellata*), 158, (barren zone) and 161 m (*C. kutzingiana* v. *radiosa*). The base of core CL-73-4 correlates below these distinctive assemblages, and the next highest sample studied (106.8 m, Table 3) presumably lies above them. These correlations indicate a downward displacement of 19 m of CL-73-4 relative to CL-80-1 and of 48 m between CL-73-4 and CL-80-2, making a total displacement of 67 m between CL-80-1 and CL-80-2 (Fig. 7).

The different amounts of downward displacement suggested by these correlations indicates that several periods of faulting or warping occurred, particularly between CL-73-4 and CL-80-2. These cores are only about 800 m apart, and whatever was responsible for the displacement of one relative to the other quite probably resulted in a change of depositional environments and very likely disturbance of sediments that included reworking, slumping, and perhaps turbidity flow.

If tectonic movement within the Clear Lake basin is the reason for the poor correlation of *Quercus* peaks between CL-80-1 and CL-73-4, this movement would affect the patterns of pollen distribution and might cause reworking; both factors are difficult to assess with the present data. It is clear that within core CL-80-1, high percentages of *Quercus* pollen correspond to zones of low diatom concentration (Fig. 8). This relationship is consistent enough to suggest a common controlling factor. The most obvious factor, fluvial input, could supply large quantities of *Quercus* pollen derived from oak-dominated communities that grow

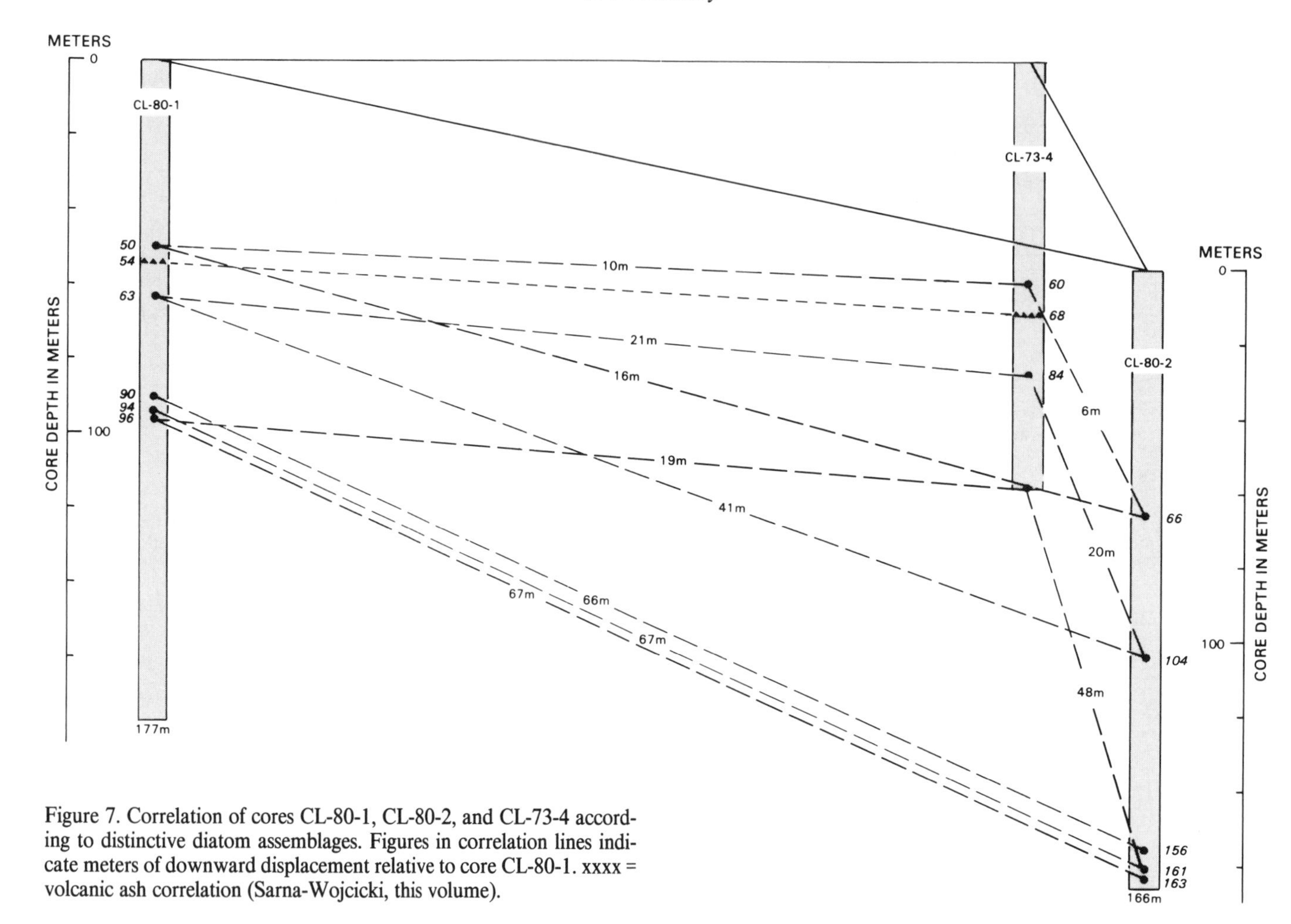

Figure 7. Correlation of cores CL-80-1, CL-80-2, and CL-73-4 according to distinctive diatom assemblages. Figures in correlation lines indicate meters of downward displacement relative to core CL-80-1. xxxx = volcanic ash correlation (Sarna-Wojcicki, this volume).

in abundance along streams and valleys, and at the same time, streams could supply clastic sediments low in diatoms. In this hypothetical relationship, climatic change and(or) tectonics by controlling stream flow, could control diatom concentration and *Quercus* pollen input. If there are two potential sources of *Quercus* pollen, one from stream input and another from air-fall pollen rain, and if their actual contribution of pollen to the record is a function of climate, tectonics, and basin morphometry (depth), which is also a function of tectonism, it is small wonder that a correlation between records from the same basin would be difficult.

The distinctive diatoms growing in the lake or reworked from unique sources provide useful stratigraphic markers like volcanic ashes. Distinctive diatoms have apparently been reworked into core CL-80-1 (Fig. 5). Specifically, these are the zones in which *Cyclotella bodanica* briefly reappears at 92 and 61 m, where it co-occurs with *Melosira solida, Navicula scutelloides,* and *Cocconeis disculus.* The low diatom concentration in these zones—in addition to the stratigraphically sudden increase and decrease of the percentage record of these species—strongly suggest reworking. These species are heavily silicified, which increases their probability of successful reworking and may account for their reappearance without the former members of their assemblage, the more weakly silicified eccentric *Stephanodiscus* species. The presence of *Cyclotella bodanica* and *Melosira solida* in core CL-73-4 aids correlation between the cores, and it is likely that a close interval sampling and study of the diatoms in CL-73-4 would provide fixed correlations based on these reworked diatoms.

CORRELATION BETWEEN DIATOM BIOSTRATIGRAPHY OF CL-80-1 AND SURROUNDING DEPOSITS

Potential sources for these reworked diatoms are deposits within the drainage basin of Clear Lake, and logical candidates for such deposits are the Cache, Lower Lake, and Kelseyville Formations described separately (Rymer and others, this volume) and unnamed lacustrine deposits, possibly correlative with the Kelseyville Formation, north of the main basin of Clear Lake (M. J. Rymer, written communication, 1983). Of these, only the

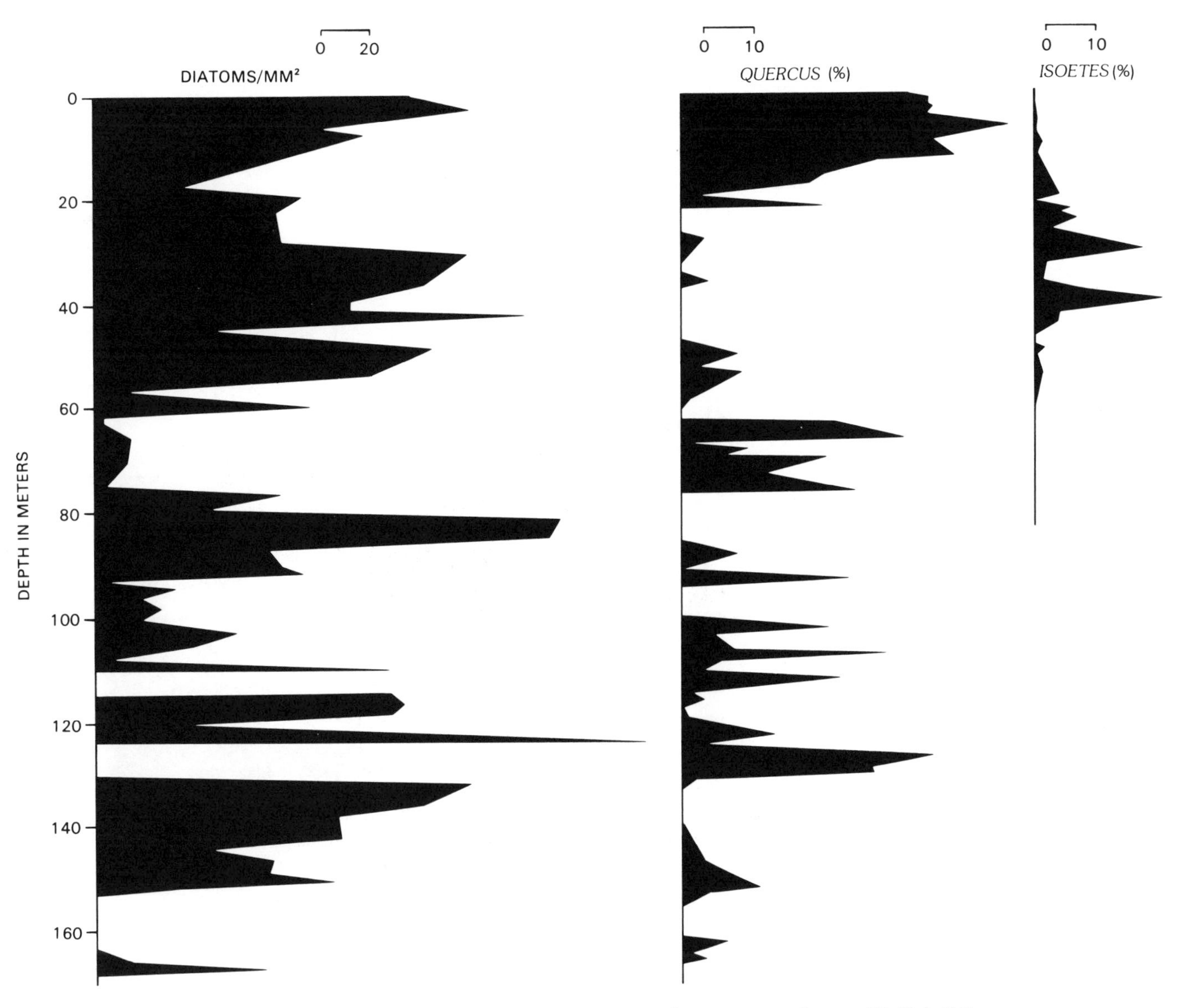

Figure 8. Diatom concentration and *Quercus* and *Isoetes* pollen percentages for core CL-80-1. Pollen profiles from Heusser and Sims (1983).

Lower Lake and Kelseyville Formations are truly suitable, because most of the Cache Formation is not diatomaceous and lies outside the drainage basin of Clear Lake. The diatomaceous outcrops of the Cache Formation that are within the Clear Lake drainage basin are eroded only by small, ephemeral streams (M. J. Rymer, written communication, 1982). In any case, reworking of diatoms from the Cache Formation, although technically possible today, is unlikely to be detected because the diatom flora of this formation contains only species that are not readily distinguished from modern species or from diatoms that occur abundantly in the Lower Lake and Kelseyville Formations and in the modern sediments of Clear Lake.

The upper parts of the Lower Lake Formation are highly diatomaceous, and their close proximity to Clear Lake indicates that some reworking is possible. However, like the Cache Formation, this formation does not contain many distinctive diatoms.

The Kelseyville Formation is the most likely source of diatoms that could potentially be reworked into the sediments accumulating in Clear Lake. This formation crops out extensively within the drainage of Kelsey Creek, and it flows directly into the main basin of Clear Lake. Although much of the Kelseyville Formation consists of clastic sediment that is barren or has low diatom concentrations, some units are highly diatomaceous, and fragments of these units occur in the sediment load of Kelsey Creek today. In addition, the diatom assemblages of the Kelseyville Formation contain distinctive species such as *Cyclotella bodanica, C. comta* var. *lichvinensis, Melosira solida,* and eccentric *Stephanodiscus* species; these species are also present in the lacus-

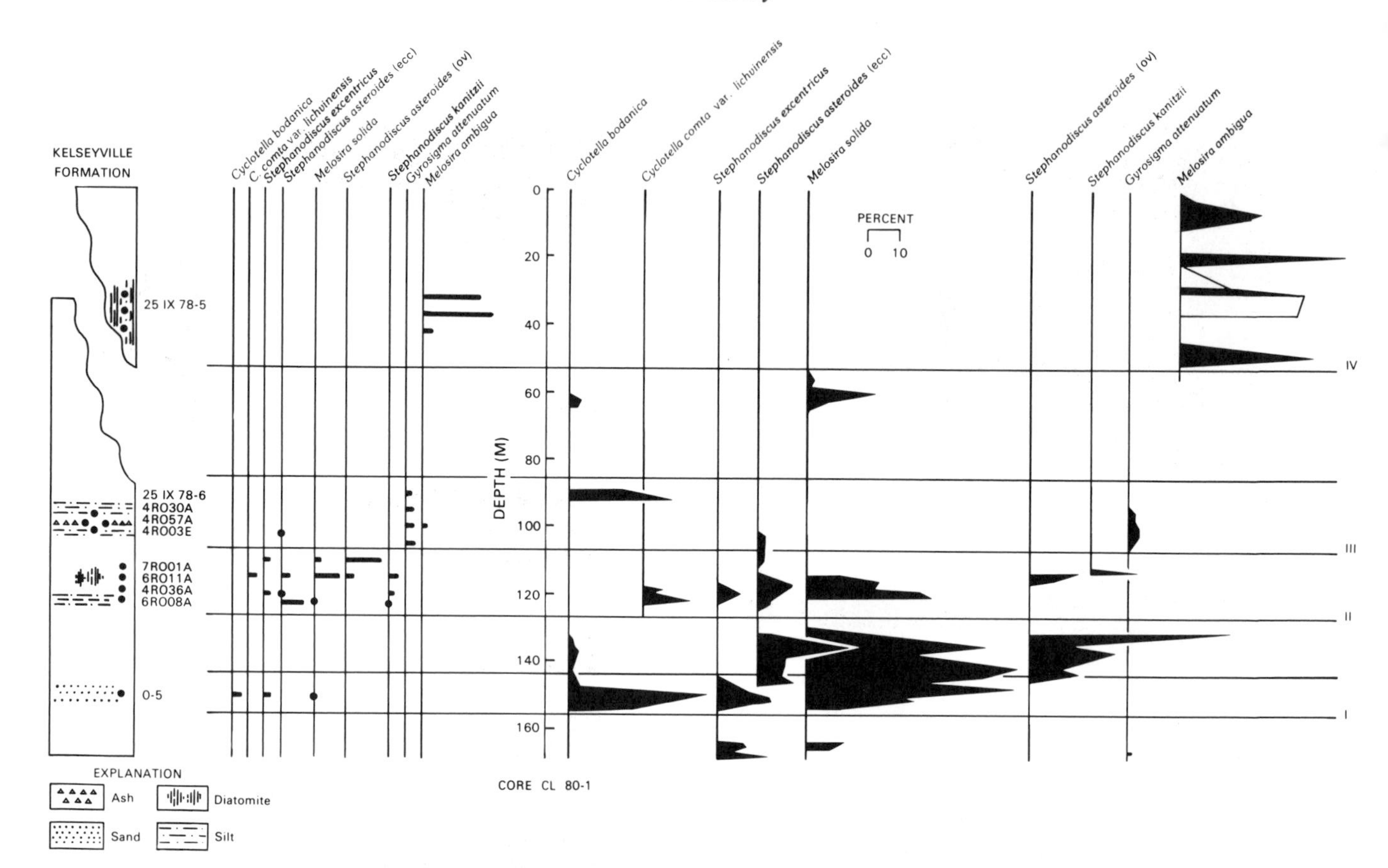

Figure 9. Percentages of selected diatoms in Kelseyville Formation and core CL-80-1. Black dots indicate present (<1%); open triangles, Kelsey Tuff Member (Qkk) of Kelseyville Formation. Samples from Kelseyville Formation have been maintained in their appropriate stratigraphic order but placed opposite their assumed correlations with same assemblages in core CL-80-1. Horizontal boundaries (I-IV) indicate zones of assumed correlations.

trine sediments beneath Clear Lake (Figs. 4, 5). Small percentages of stratigraphically restricted occurrences of these distinctive diatoms could reasonably be ascribed to a reworked source in the Kelseyville Formation. However, the larger percentages of these species that occur throughout much of the lower part of core CL-80-1 require additional consideration.

The abundance of *Melosira solida, Cyclotella bodanica, C. comta* var. *lichvinensis,* and eccentric *Stephanodiscus* species in both the lower part of core CL-80-1 and of the Kelseyville Formation suggests that they may be correlated. These species are especially abundant in the lower third of the Kelseyville Formation and the basal 70 m of core CL-80-1 (diatom zone 5), where they occur in the same general stratigraphic succession (Fig. 9). Four major, probably correlative, zones can be recognized. The first (I) is a basal zone characterized by *Cyclotella bodanica* (Kelseyville Formation, sample Ost-5 and Core CL-80-1, 150 m, diatom subzone 5.2). A second (II), intermediate zone is characterized by *Cyclotella comta* var. *lichvinensis,* eccentric *Stephanodiscus* species, and *Stephanodiscus kanitzii* (Kelseyville Formation samples 7R001A, 6R011A, 4R036A, and 6R008A, and core CL-80-1, 114 to 125 m in diatom subzone 5.1). This zone is overlain by a third zone (III) that contains small but significant percentages of *Gyrosigma attenuatum* (Kelseyville Formation samples 25 IX 78-6, 4R030A, 4R075A, and 4R003E, and core CL-80-1, 94 to 112 m in diatom subzone 4.1). Although the mutual presence of some species such as *Cyclotella comta* var. *lichvinensis,* which has a restricted stratigraphic distribution in both sections, implies that close correlation may be possible between the Kelseyville Formation and core CL-80-1; in general the overlapping ranges of the diatoms in core CL-80-1 and the lack of sufficient stratigraphically controlled samples from the Kelseyville Formation make it impossible to correlate units exactly. Even if such control were available from the Kelseyville Formation, the marginal lacustrine depositional environment represented by this deposit could be difficult to correlate with the deeper water record contained in CL-80-1.

Another potentially correlative diatom zone (IV) exists in the upper part of the Kelseyville Formation and core CL-80-1. The sample series 25 IX 78-5 A, B, C in the uppermost part of the Kelseyville Formation, which is characterized by large percentages of *Melosira ambigua,* could relate stratigraphically to the upper 50 m of core CL-80-1 where this species is also locally abundant (Fig. 9). Inadequate sample control in core CL-80-1 makes it impossible to make more accurate correlation, although the rarity of *Melosira granulata* in the Kelseyville samples suggests that the most likely correlation to core CL-80-1 is between 20 and 40 m in depth, where *M. granulata* is present only in low percentages (Fig. 5).

Although the present analyses are insufficient to demonstrate the correlation in all desired detail, some sort of correlation must exist to explain the identical stratigraphic sequences of distinctive diatoms. Correlations of volcanic ashes in core CL-80-1 with approximately dated tephras in the western United States (Sarna-Wojcicki, this volume) indicate that the basal 30 m (below 148 m) of the core is about 450,000 yr old, and therefore in the proper time range to correlate with the estimated age of the Kelseyville Formation, which is less than 600,000 yr old (Rymer, 1981). In addition, one ash, tephra II (sample 6R008) has a probable correlative in core CL-80-1 at a depth of 123 m (M. J. Rymer, oral communication, 1984). The ash correlations and etching of mafic minerals in core CL-80-1 suggest a hiatus at a depth of about 115 m (Sarna-Wojcicki, this volume); they substantiate correlations based on diatoms between CL-80-1 and the Kelseyville Formation. Several important considerations result from this correlation.

First, the Kelsey Tuff Member of the Kelseyville Formation (Rymer, 1981) lies within sediments containing *Gyrosigma attenuatum* (Fig. 9). This unit could reasonably occur in or just below diatom subzone 4.1 (94–105 m) of core CL-80-1 (Fig. 5); the core also has significant percentages of this restricted species. The Kelsey Tuff Member has not been recognized at this depth in core CL-80-1, although it is unlikely that it would necessarily have the same thickness and character as it does in outcrop. In the Kelseyville Formation, the Kelsey Tuff Member consists of a probable fine-grained airfall unit of about 30 cm in thickness, which is overlain by 1.4 m of coarse, unsorted pumiceous and basaltic andesite lapilli tuff that may have been redeposited from the slopes of Mt. Konocti, near the presumed source of the tuff (Rymer, 1981). The fluvial and marginal lacustrine Kelseyville Formation would logically contain thicker equivalents of this unit than might the deeper water lacustrine sediments of core CL-80-1. Alternatively, the apparent absence of the tuff in core CL-80-1 may be related to its removal by erosion during the time of the suspected hiatus at 115 m (Sarna-Wojcicki, this volume).

Second, if the correlation of the upper, highly diatomaceous deposits of the Kelseyville Formation (samples 25 IX 78-5 A, B, C) with the upper 50 m of core CL-80-1 on the basis of occurrence of *Melosira ambigua* is correct, it suggests the presence of another hiatus in the Kelseyville Formation that is at least equivalent to about 45 m of lacustrine deposition in Clear Lake (50-95 m, core CL-80-1). Possibly this hiatus was caused by tectonic adjustments in the Clear Lake–Kelseyville basin because the Kelseyville Formation samples come from tilted lake beds that are presently 100 m above the surface of Clear Lake and about 130–150 m above the presumably correlative units in CL-80-1. Tectonic events on this scale certainly have affected the limnology and depositional patterns in Clear Lake and should be taken into account in paleoenvironmental interpretations of lacustrine records from Clear Lake. This hypothetical event would have postdated the deposition of *M. ambigua*–rich sediments and presumably predated the deposition of Holocene diatom assemblages rich in *M. granulata.* It may have occurred as part of the tectonic activity that changed the Highlands Arm of Clear Lake from a shallow-water marsh to the deepest (16 m) sub-basin of the system (Casteel and others, 1977).

Third, because the Kelseyville Formation has units containing fossil leaves that imply climatic change (Rymer, 1981), a correlation between this unit and core CL-80-1 can provide paleoclimatic insights into the deposits of Clear Lake. The fossil leaves, especially of *Quercus* (oak) and *Platanus* (sycamore) in the lower part of the formation (about 200 to 240 m below the Kelsey Tuff Member) suggest a warm, dry climate; the fossils from samples near the Kelsey Tuff Member—*Pinus* (pine), *Picea* (spruce), and *Abies* (fir)— suggest cooler and moister environments. The warm and cool units of the Kelseyville Formation, respectively, correlate with parts of core CL-80-1 that contain either large percentages of *Melosira solida* or of *Stephanodiscus carconensis* (Figs. 4, 9). Insofar as the ecology of these diatoms is known from the occurrence of similar forms in Lake Biwa, Japan, their climatic interpretation compares favorably with that based on fossil leaves and cones in the Kelseyville. In Lake Biwa, *Melosira solida* is dominant in the Holocene and in parts of the core that are considered to indicate warmer climates, whereas *Stephanodiscus carconensis* achieved maximal percentages in the cooler, late Pleistocene parts of the core (Mori, 1974, 1975).

PALEOLIMNOLOGY OF CLEAR LAKE

Although the foregoing discussion suggests caution in interpreting the lacustrine record beneath Clear Lake in paleoclimatic and paleolimnologic terms, a paleolimnologic interpretation remains a potentially interesting and useful exercise as long as the record is reasonably continuous and the instances of diatom redeposition can be identified. In general, the latter is possible and the former at least can be evaluated in terms of diatom concentration (Fig. 5). Incomplete core recovery, especially between 15 and 40 m in core CL-80-1, will partially frustrate interpretations of the Pleistocene to Holocene transition.

The paleolimnologic interpretation will begin at the base of core CL-80-1, go upward, and in general follow the diatom zones of Figure 5; emphasis is placed on samples with high concentrations of environmentally significant diatoms.

The basal diatom subzone (5.3) of core CL-80-1 is dominated by *Fragilaria brevistriata* and *F. construens.* Smaller percentages of *Melosira solida* and *Stephanodiscus excentricus* also occur. The high percentages of *Fragilaria construens* that occur only in this zone, and the abundance of *F. brevistriata,* are similar to the occurrences of these species in the basal units of the Lower Lake Formation (Rymer and others, this volume). However, *Mel-*

osira solida and eccentric forms of *Stephanodiscus* are absent in the Lower Lake Formation; therefore, it is unlikely that this basal subzone is correlative with the Lower Lake Formation.

The dominance of *F. brevistriata* and *F. construens* implies shallow water and the presence of aquatic macrophytes. The water, which was essentially fresh, well oxygenated, and clear, suggests a marginal lacustrine environment. Fluvial environments were probably nearby, and the lake level and low salinities were maintained by one or more through-flowing streams. The presence of low percentages of *Melosira solida* and *Stephanodiscus excentricus* suggests the presence of deeper, open-water environments with moderate to high nutrient concentrations at least seasonally near this predominantly shallow-water environment.

About 10 m (154–164 m) of sediment essentially barren of diatoms overlies diatom subzone 5.3. This unit probably represents a fluvial deposit that accumulated rapidly and may have eroded previously deposited lacustrine sediments in the process.

Diatom subzone 5.2 is characterized by large percentages of *Cyclotella bodanica* and *Melosira solida. Stephanodiscus excentricus* is of less importance, and the initially large percentage of *S. carconensis* reflect only a few fragmentary remains in an interval of very low diatom concentration.

The diatom assemblage of this zone is dominated by planktonic species; their abundance implies the presence of open water deep enough to exceed the photic limit. In this case the lake depth was probably more than 10 m. *Cyclotella bodanica,* as suggested in the discussion of its presence in the Kelseyville Formation, probably indicates generally fresh, cool, oligotrophic water. *Melosira solida,* according to its presence in Lake Biwa and in general agreement with the distribution of this genus in temperate lakes, indicates warm, eutrophic conditions. The co-occurrence of these two diatoms is not inconsistent because they probably bloom under different limnologic conditions that recur on a seasonal basis. Unfortunately, not enough is known about the specific ecology of either species to clarify this relationship. *Cyclotella bodanica* probably blooms after *Melosira solida* when nutrient levels have been reduced by *M. solida* and summer algal production. Recorded blooms of *C. bodanica* occur often in the late summer or fall in subalpine lakes (Huber-Pestalozzi, 1942).

The ecologic requirements of *Stephanodiscus excentricus* are unknown because the species has been recorded only as a fossil. However, its requirements are probably not substantially different than those of other *Stephanodiscus* species; they tend to bloom in the spring or late fall when phosphorus/silica ratios are high (Kilham and Kilham, 1978). Related species of *Stephanodiscus, Cyclotella,* and *Melosira* occur today, blooming in different seasons and under different limnologic and nutrient conditions in temperate and north-temperate lakes. The co-occurrence of distinctive forms of these species in fossil assemblages is therefore not surprising, but these forms are difficult to interpret paleolimnologically. The most probable analog for diatom subzone 5.2 would be a cool to temperate, possibly dimictic, lake. Because many occurrences of this form of *C. bodanica* are from northern latitudes or from high-elevation lakes, it is reasonable to propose cooler climates for this interval of core CL-80-1 than for subsequent levels.

Above diatom subzone 5.2 lies 16 m of highly diatomaceous sediment dominated by *M. solida, Stephanodiscus asteroides,* and the eccentric and oval forms of *S. asteroides* (ECC and OV). These species of *Stephanodiscus* are apparently extinct, although they are related very closely to *S. carconensis,* which currently lives in Lake Biwa; further studies may show these taxa to be conspecific. The predominance of *M. solida,* according to the Lake Biwa analog, and the modern distribution of similar extant *Melosira* species, such as *M. granulata,* suggest an open-water, turbulent environment of a moderately eutrophic, seasonally warm, fresh-water lake. The very reduced representation of *Cyclotella bodanica* presumably indicates the reduced occurrence of cool, oligotrophic lacustrine environments, although the *Stephanodiscus* species and forms imply continued climatic seasonality and somewhat higher trophic levels. The appearance of *S. niagarae,* an extant form presently occurring in Clear Lake, is consistent with this interpretation. Warmer climates may ultimately have been responsible for the dominance of *M. solida,* as Mori (1974, 1975) proposed for its distribution in Lake Biwa.

Core CL-80-1 is barren of diatoms between 124 and 130 m, but above this 6-m-thick layer, in diatom subzone 5.1, highly diatomaceous sediments occur. As with the underlying deposits, subzone 5.1 is characterized by large numbers of *M. solida* and *Stephanodiscus* species. An extinct diatom, *Cyclotella comta* var. *lichvinensis,* is restricted to this subzone and serves as a distinctive stratigraphic marker. The ecologic significance of this diatom can only be inferred from its fossil distribution in north-temperate European lacustrine deposits (Loseeva, 1982), and from the modern distribution of closely related *C. comta.* This planktonic species apparently prefers cool fresh water of low to moderate trophic status (Lowe, 1974), especially natural forest lakes of the north-central United States (Stoermer and Yang, 1970). Although the presence of *C. comta* var. *lichvinensis* could imply somewhat fresher, cooler, and less eutrophic conditions in Clear Lake, its occurrence with species of the earlier diatom assemblages suggests that a substantial limnologic change was unlikely. The increased representation of *Fragilaria brevistriata* and *F. construens* var. *venter* indicates increased input from the littoral zone of the lake.

Stephanodiscus sp. cf. *S. asteroides* (cf. *S. kanitzii* f. *partita* Gasse, 1980) is present in a single sample (114.5 m) in the interval between diatom subzone 5.1 and the top of diatom zone 5. Except for this sample, the interval (110–113 m) is barren of diatoms. As with *Cyclotella comta* var. *lichvinensis, Stephanodiscus* sp. cf. *S. asteroides* provides a distinctive stratigraphic marker, but little paleolimnologic information. The stratigraphically lower assemblage of *M. solida* and the eccentric and oval *Stephanodiscus* species that occurs with *Stephanodiscus* sp. cf. *S. asteroides* suggest a continuation of similar limnologic characteristics. It is possible that large amounts of silica and assorted micronutrients or other elements in the water related to volcanogenic mineral springs played an important role in the development of distinctive ecophenotypes of *Stephanodiscus.* Physiologic studies of modern diatoms in different nutrient and biochemical environments (e.g., Schmid, 1979) show that the morphology of diatom valves is highly affected by such environmental changes. The rapid change in abundance of distinctive *Stephanodiscus* species in diatom zone 5 may have been caused by the change in distinctive chemical environments rather than by large-scale limnologic and climatic factors.

The zones of sediment barren of diatoms and the rapidly fluctuating diatom concentrations in diatom zone 5 probably relate to fluvial input and tectonic disturbance of the basin. Some penecontemporaneous redeposition of diatoms is very probable in this zone, and the more these factors determine the character of the diatom biostratigraphy of zone 5, the more regional paleolimnologic and paleoclimatic interpretations are compromised.

Diatom zone 4 is dominated by *Stephanodiscus carconensis. Stephanodiscus astraea* var. *minutula, S. subtransylvanicus,* and *Fragilaria brevistriata* are common. *Gyrosigma attenuatum* is present in diatom subzone 4.1 (94–105 m) and *Cyclotella bodanica,* possibly reworked from earlier deposited sediment, appears abundantly in two samples (90 and 92 m) above a barren zone at a depth of 93 m. At present it is not possible to determine whether *C. bodanica* is reworked. Although the frustules are comparatively well preserved, the abrupt appearances and disappearances of the specimens in this 2-m interval arouses suspicion of reworking. If reworking did occur, it happened without the addition of significant numbers of *M. solida,* and that is perplexing because *M. solida* is a heavily silicified diatom that should survive even vigorous transport. It may be that not only this interval of *C. bodanica* is reworked, but also the lower occurrence in diatom subzone 5.2, and that somewhere, probably within the drainage of Kelsey Creek, a diatomaceous deposit dominated only by *C. bodanica* is periodically reworked into the Kelseyville Formation and(or) the lacustrine sediments of Clear Lake. This intriguing possibility will have to be investigated before the paleolimnologic interpretation of this species in Clear Lake can be confidently accepted.

It is reasonable, but still speculative, to interpret the dominance of *Stephanodiscus carconensis*(?) in terms of its presence in Lake Biwa, Japan. There it predominates in parts of the 200-m Lake Biwa core that represent cool Pleistocene environments; Mori (1975) suggested that its dominance may be caused by active circulation of the lake in winter. If *S. carconensis,* like some other species of *Stephanodiscus,* requires high ratios of phosphorus/silica (Kilham and Kilham, 1978), then limnologic circumstances that promote such nutrient ratios will favor *S. carconensis.* High P/Si ratios can be maintained by extending periods of lake turnover, which resupply the photic zone with phosphorus generated in the anoxic bottom sediments of a lake's profundal region. Long, cool spring and fall seasons will likely be accompanied by extended periods of turnover, which should provide sufficient phosphorus for large blooms of *Stephanodiscus* if adequate illumination and turbulence are also available. The large representation of *S. subtransylvanicus* is also consistent with this interpretation because the closely related extant species *S. transylvanicus* prospers under similar conditions. In the Great Lakes, *S. transylvanicus*—which apparently prefers a water temperature about 6°C—is most abundant during the cold months of the year (Stoermer and Ladewski, 1976).

The small but significant representation of *Gyrosigma attenuatum* in diatom subzone 4.1 probably does not reflect substantial limnologic changes because the dominance of *Stephanodiscus* species throughout the subzone does not imply substantial shallowing of Clear Lake; therefore, it seems probable that *G. attenuatum* either entered Clear Lake from rivers or was transported to deeper water from the lake margins. *Gyrosigma attenuatum* may represent an interval of increased river discharge into Clear Lake. This species is indifferent to moderate current velocities; consequently, it is often found in slow-moving rivers and streams, as well as in ponds and the shallow areas of lakes.

The assemblages in diatom zone 3 contain very low concentrations of diatoms but are relatively dominated by *Amphora perpusilla* and *Fragilaria brevistriata.* These shallow-water benthic diatoms (Bradbury and Winter, 1976) imply (1) a significantly reduced volume of the Clear Lake basin, and (2) probably an increased influence of river inflow. The planktonic diatom assemblage, although of secondary importance, contains the species of *Stephanodiscus* that abundantly characterized diatom zone 4. Consequently, significantly warmer conditions do not seem likely during diatom zone 3, although the very low representation of all diatoms makes interpretation of this zone speculative. For example, it is possible that much of the diatom assemblage of zone 3 represents input from rivers and ponds mixed with reworked planktonic diatoms of zone 4. The few *Cocconeis disculus* in the upper part of zone 3 coincide with probably reworked frustules of *Melosira solida* and *Cyclotella bodanica. Cocconeis disculus* attached to various substrates (Johnson and others, 1975) inhabits shallow-water regions of lakes or streams, and its presence in zone 3 may result from the input from rivers suggested for this zone. Increased river input coincident with shallower water suggests tectonism played some role in determining the volumetric capacity of Clear Lake.

Diatom zone 2 (17–60 m) includes sediments that can be correlated on the basis of pollen and diatom stratigraphy with the radiocarbon-dated core CL-73-4 (Sims and others, 1981); accordingly, this zone contains sediments that are at least 15,000 yr old at the top and may be as much as 60,000 yr old at the base if constant sedimentation rates are assumed.

Large but rapidly fluctuating percentages of *Stephanodiscus hantzschii* and *Melosira ambigua* characterize zone 2. In subzone 2.1 (depth, 46–51 m) *Fragilaria crotonensis* is abundant, and at a depth of about 42 m *Melosira granulata* becomes abundant locally. All these diatoms are planktonic species common in mesotrophic to eutrophic temperate lakes, where they often bloom seasonally in response to nutrient changes. The *Melosira* species and *F. crotonensis* tend to predominate in the summer and fall, presumably when phosphorus levels are comparatively low relative to silica, and *S. hantzschii* tends to predominate in early spring when turnover substantially increases phosphorus loading (Kilham and Kilham, 1978; Kilham and Van Donk, 1982). Some evidence suggests that *F. crotonensis* prefers somewhat higher phosphorus levels than *M. ambigua* because the former is known to replace the latter as some lakes become artificially enriched (Bradbury, 1978; Brugam, 1978). Diatom concentration in zone 2 is very high, and the dominance of planktonic species implies deeper water and(or) less sediment input from rivers relative to zone 3. Large numbers of *S. hantzschii* suggest cool conditions during this period, but the summer-blooming diatoms indicate seasonality, and it is not unreasonable to suspect that at times during this interval Clear Lake became dimictic or at least stratified for periods long enough to allow nutrient depletion and seasonal diatom succession. These lacustrine changes correlate with pollen studies in which large percentages of coniferous pollen are believed to imply cooler and moister climates of the last glacial cycle (Adam and others, 1981; Adam, this volume). The large percentages of *Isoetes* pollen that occur in diatom zone 2 probably indicate clear, cool, nutrient-poor water, at least during some seasons. The *Isoetes* curve for core CL-80-1 (Fig. 8) probably represents the least ambiguous lacustrine evidence of the last glacial cycle because most California species of this plant are presently located in ponds, streams, and shallow parts of lakes at high elevations (>2,000 m) or live in the northern part of the state (Jepson, 1923). Peak percentages of *Isoetes* correlate with peak percentages and concentrations of *Stephanodiscus asteroides*; the coincidence of these peaks imply that *S. asteroides* may indicate cooler and moister environments associated with the late Wisonsinan.

Diatom zone 1 (17–0 m) encompasses the Holocene part of the Clear Lake record. The zone is dominated by *Melosira granulata, M. ambigua,* and *Stephanodiscus* species, particularly *S. astraea* v. *minutula* and *S. niagarae.* All these species are present in Clear Lake today (Table 1, Fig. 2), and the Holocene dominant, *M. granulata,* also currently dominates the Clear Lake diatom plankton.

The similarity between the Holocene diatom assemblages of Clear Lake and the modern diatom community suggests that, in general, the modern ecologic conditions of Clear Lake are a reasonable analog for Holocene environments. An exception to this generalization is the strong middle Holocene increase and subsequent decrease in *Melosira ambigua* at a depth of 7 m. Because the distribution of this species is more characteristic of the late Pleistocene parts of core CL-80-1, its reappearance in the middle Holocene may imply a return to cooler and moister conditions for a brief period. Unfortunately, little is known about the specific ecology of *M. ambigua* relative to *M. granulata.* Because this environmental change does not appear in the pollen record (Adam and others, 1981), the climatic change either was not large enough or did not last long enough to alter terrestrial vegetation significantly. Nevertheless, a middle Holocene climatic shift to moister conditions has some precedent in western North America. Byrne (1982) found such a change in cores from the Gulf of California, and Adam (1967) found that a middle Holocene increase in riparian and aquatic vegetation occurs in Osgood Swamp, in the central Sierra Nevada about 240 km east of Clear Lake. Pollen studies from a small lake in the Puget Sound area (Barnowsky, 1981) suggest the presence of cooler, wetter, and more maritime climates in that area during the last 5,500 yr. Close interval sampling and an adequate radiocarbon chronology are required to further investigate this change indicated by the diatom biostratigraphy in Clear Lake. The large area of Clear Lake and its drainage basin (1,370 km^2; Sims and others, 1981), which consists of a wide range of elevations and vegetation types, may make the Clear Lake cores palynologically insensitive to small-scale climatic changes.

CONCLUSIONS

The diatom biostratigraphy of the 177-m core, CL-80-1, from Clear Lake, contains a paleolimnologic record that has been affected by climatic changes, tectonic changes, and probably by volcanic-hydrologic changes associated with magmatism in this seismically active area. Limnologic environments, as suggested by varying percentages and concentrations of diatom species, have differed considerably in Clear Lake throughout the period of time represented by the core. These differing environments imply variable sedimentation rates, but probably far greater variability in processes and rates of sedimentation was caused by volcano-tectonic disturbances. The consequence of this variability is that a chronology for core CL-80-1, based on downward extrapolation of sedimentation rates derived from correlations with radiocarbon-dated horizons in core CL-73-4, is unlikely to be accurate to the degree necessary for paleoclimatic correlations to other records.

The lower 70 to 80 m of core CL-80-1 seem to correlate with the Kelseyville Formation, based on diatom analyses, and if that formation can be satisfactorily dated paleontologically or radiometrically, it may provide some information for the age of the basal part of the core. Above 80 m, and particularly above 50 m, core CL-80-1 appears to have a valid record of Pleistocene paleoenvironmental changes. The zone of low diatom concentration and dominance of shallow-water diatoms (at 60–75 m depth) suggests a higher than usual influx of clastic materials; the influx may also be recorded in the magnetic remanence of the core, which increases 10-fold below 50 m depth (Liddicoat, 1981).

The paleoenvironmental changes in Clear Lake core CL-80-1 for the late Wisconsinan and Holocene suggest a transition from cooler and probably wetter climates in the late Pleistocene to warmer and drier climates of the Holocene, although fluctuations in the diatom stratigraphy indicate variability and climatic or limnologic instability throughout this period. Poor core recovery complicates the paleoenvironmental interpretation in this part of the core, and the likelihood of tectonism impacting on the paleoenvironment of Clear Lake is difficult to evaluate. It is critical that such nonclimatic environmental effects be considered before this record can be unambiguously interpreted in paleoclimatic terms.

PLATE SECTION

Light micrographs of stratigraphically characteristic diatoms from Pleistocene sediments of Clear Lake, Lake County, California. Scale bar = 10 μm.

PLATE 1

Figures 1–13. *Melosira (Aulacosira) solida* Eulest. Clear Lake Core CL-80-1, 132.0 m. a, Focus on mantle exterior; b, internal focus for optical section. Figures 11-13, view of "ring-leiste" showing internal, costae-like struts.

Figure 14. *Stephanodiscus* sp. cf. *S. subtransylvanicus* Gasse. Clear Lake Core CL-80-2, 161.43 m.

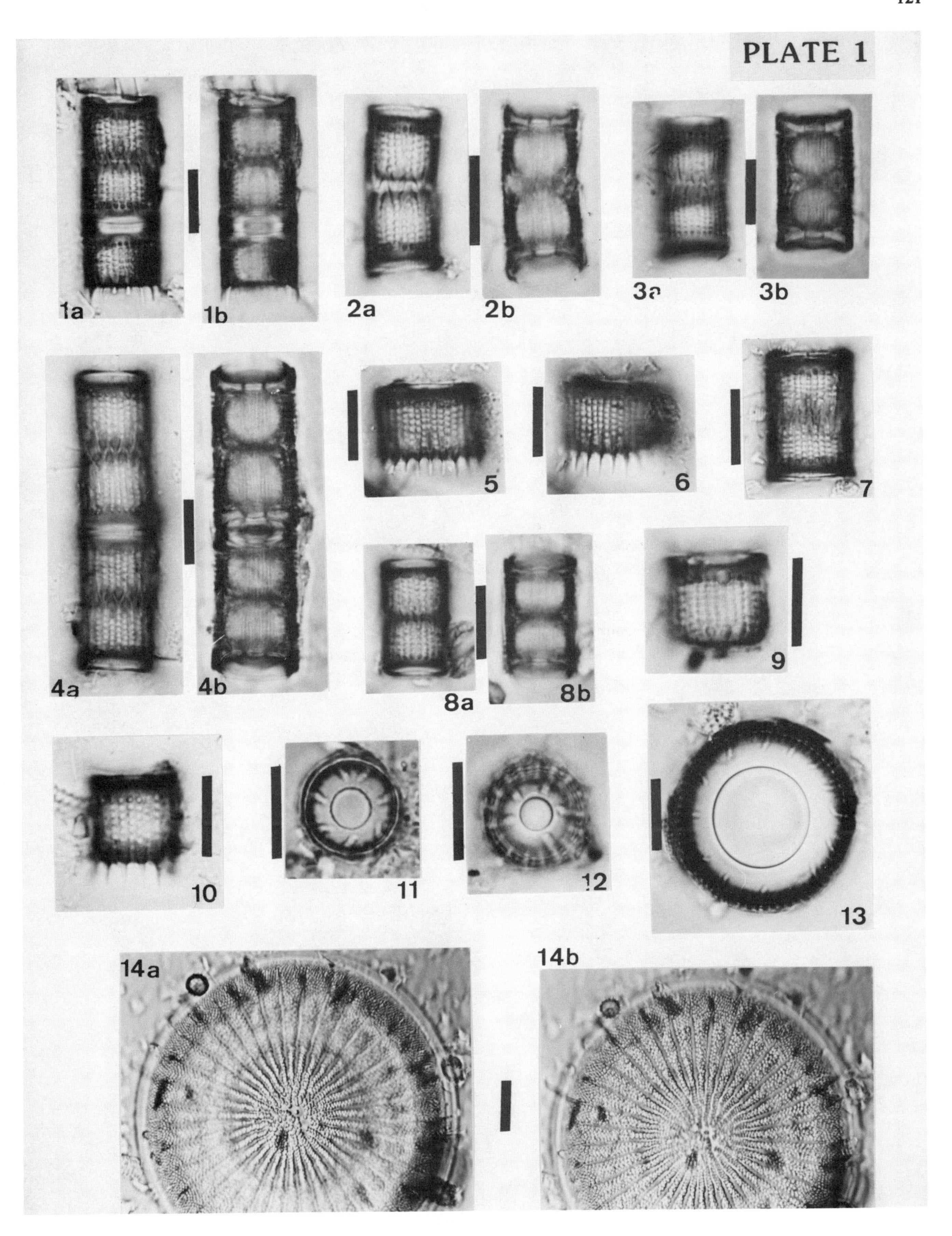
PLATE 1
1a
1b
2a
2b
3a
3b
4a
4b
5
6
7
8a
8b
9
10
11
12
13
14a
14b

PLATE 2

Figures 1-15. *Stephanodiscus* sp. cf. *S. asteroides* Gasse. Clear Lake Core CL-80-1, 132.0 m. These figures illustrate an ovate form of the species.

Figures 16-20. *Stephanodiscus* sp. cf. *S. asteroides* Gasse. Figures 16, 17, Clear Lake Core CL-80-1, 132.0 m, Figures 18-20, Clear Lake Core CL-80-1, 150.5 m. a- b- c- = Low, intermediate, and high focus, respectively. These figures illustrate an eccentric form of the species.

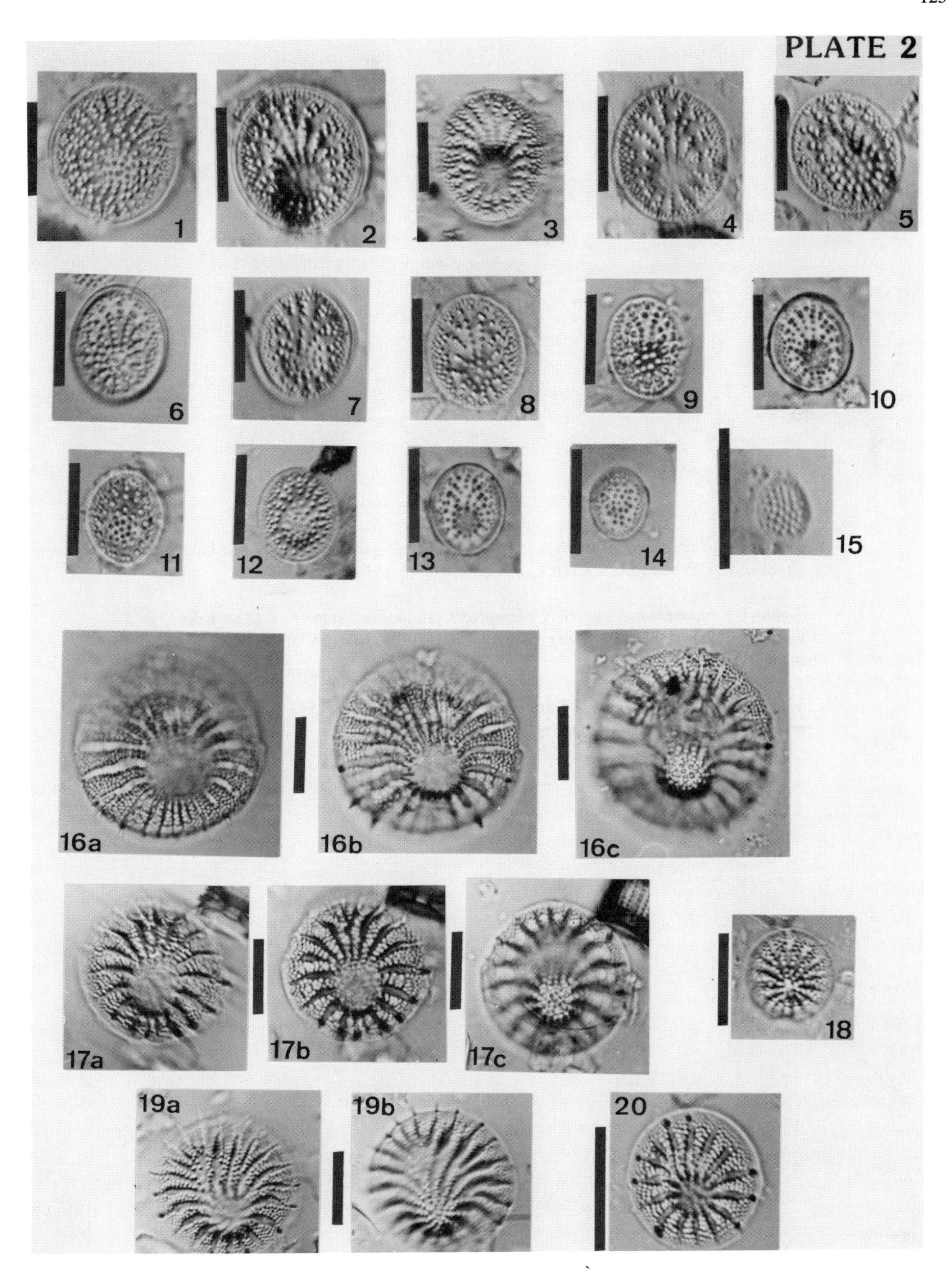
PLATE 2
1
2
3
4
5
6
7
8
9
10
11
12
13
14
15
16a
16b
16c
17a
17b
17c
18
19a
19b
20

PLATE 3

Figures 1–5. *Stephanodiscus* sp. cf. *S. carconensis* Grun. Figures 1, 3, 4, 5, Clear Lake Core CL-80-2, 156.03 m. Figure 2, Clear Lake Core CL-80-2, 161.43 m.

Figures 6-8. *Stephanodiscus* sp. cf. *S. carconensis* var. *pusilla.* Figures 6, 7, Clear Lake Core CL-80-1, 132.0 m. Figure 8, Clear Lake Core CL-80-2, 156.03 m. a, Intermediate focus; b, high focus.

Figures 9-16. *Cyclotella ocellata* Pant. Clear Lake Core CL-80-2, 156.03 m. For Figure 11, a indicates low focus; b, high focus; and c, oblique illumination.

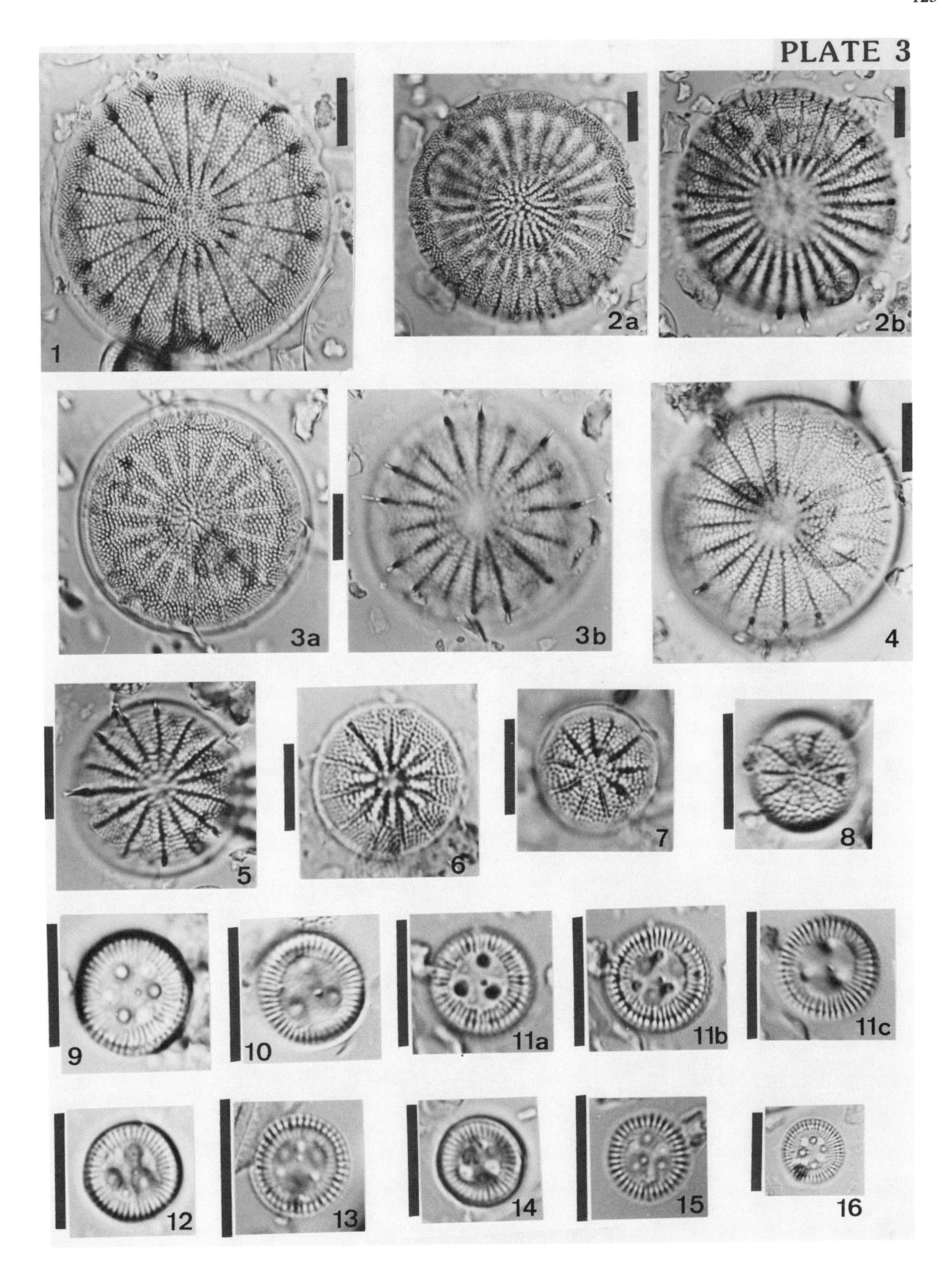
PLATE 3
1
2a
2b
3a
3b
4
5
6
7
8
9
10
11a
11b
11c
12
13
14
15
16

PLATE 4

Figures 1-9. *Cyclotella* sp. cf. *C. bodanica* Eulest. Figures 1-3 and 6-9, Clear Lake Core CL-80-1, 150.5 m. Figures 4, 5, outcrop, Highland Springs Reservoir, Lake County, California. a, High focus; b, low focus.

Figures 10-15. *Cyclotella* sp. cf. *C. kutzingiana* var. *radiosa* Fricke. Clear Lake Core CL-80-2, 161.43 m.

Figures 16-18. *Stephanodiscus* sp. cf. *S. asteroides* Gasse. Clear Lake Core CL-80-1, 150.5 m. a, Low focus; b, high focus.

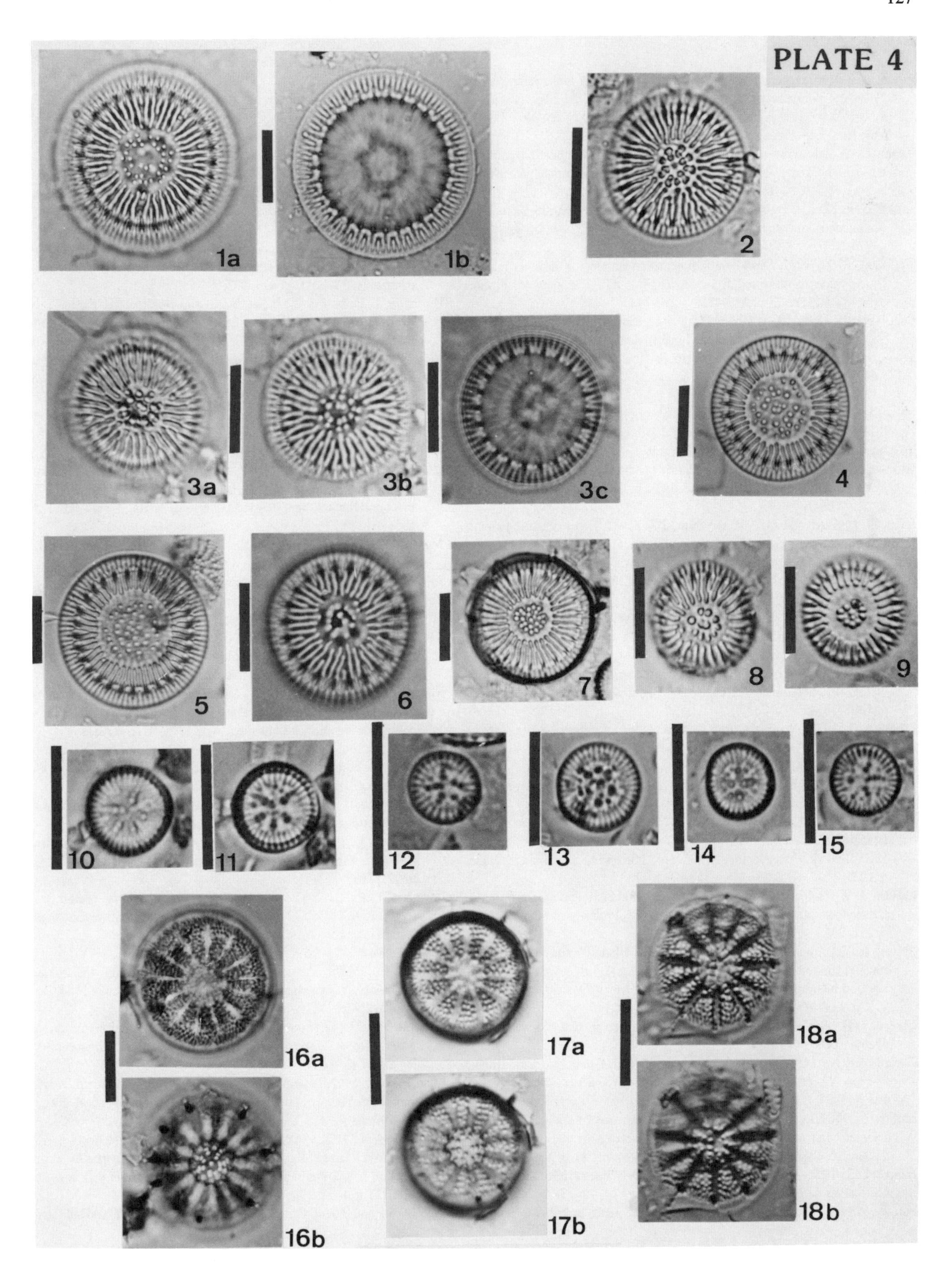
PLATE 4
1a
1b
2
3a
3b
3c
4
5
6
7
8
9
10
11
12
13
14
15
16a
16b
17a
17b
18a
18b

REFERENCES CITED

Adam, D. P., 1967, Late Pleistocene and Recent palynology in the central Sierra Nevada, California, *in* Cushing, E. J., and Wright, H. E., eds., Quaternary Paleoecology, Proceedings of the VII Congress of INQUA: New Haven, Yale University Press, v. 7, p. 275–301.

Adam, D. P., and West, G. J., 1983, Temperature and precipitation estimates through the last glacial cycle from Clear Lake, California, pollen data: Science, v. 219, p. 168–170.

Adam, D. P., Sims, J. D., and Throckmorton, C. K., 1981, 130,000-Yr continuous pollen record from Clear Lake, Lake County, California: Geology, v. 9, p. 373–377.

Anderson, R. Y., 1977, Short-term sedimentation response in lakes in western United States as measured by automated sampling: Limnology and Oceanography, v. 22, no. 3, p. 423–433.

Bailey-Watts, A. E., 1973, Observations on a diatom bloom in Loch Leven, Scotland: Biological Journal of the Linnean Society, v. 5, no. 3, p. 235–253.

—— , 1978, A nine-year study of phytoplankton of the eutrophic and non-stratifying Lock Leven (Kinross, Scotland): Journal of Ecology, v. 66, p. 741–771.

—— , 1982, The composition and abundance of phytoplankton in Lock Leven (Scotland) 1977-1979 and a comparison with the succession in earlier years: International Revue der Gesamten Hydrobiologie v. 67, no. 1, p. 1–25.

Barnowsky, C. W., 1981, A record of Late Quaternary vegetation from Davis Lake, Southern Puget Lowland, Washington: Quaternary Research, v. 16, p. 221–239.

Blunt, D. J., Kvenvolden, K. A., and Sims, J. D., 1981, Geochemistry of animo acids in sediments from Clear Lake, California: Geology, v. 9, p. 378–382.

Bradbury, J. P., 1978, A paleolimnological comparison of Burntside and Shagawa Lakes, northeastern Minnesota: Environmental Protection Agency, Ecological Research Series, EPA-600/3-78-004, 50 p.

Bradbury, J. P., and Whiteside, M. C., 1980, The paleolimnology of two lakes in the Klutlan Glacier region, Yukon Territory, Canada: Quaternary Research, v. 14, p. 149–168.

Bradbury, J. P., and Winter, T. C., 1976, Areal distribution and stratigraphy of diatoms in the sediments of Lake Sallie, Minnesota: Ecology, v. 57, no. 5, p. 1005–1014.

Britton, L. J., 1977, Periphyton phytoplankton in the Sacramento River, California, May 1972–April 1973: U.S. Geological Survey, Journal of Research, v. 5, no. 5, p. 547–559.

Brugam, R. B., 1978, Human disturbance and the historical development of Linsely Pond: Ecology, v. 59, no. 1, p. 19–36.

Byrne, R., 1982, Preliminary pollen analysis of Deep Sea Drilling Project Leg 64, Hole 480, Cores 1-11, Initial reports of the Deep Sea Drilling Project, vol. LXIV, pt. 2: Washington, D.C., U.S. Government Printing Office, p. 1225–1237.

Casteel, R. W., and Rymer, M. J., 1981, Pliocene and Pleistocene fishes from the Clear Lake area: U.S. Geological Survey Professional Paper 1141, p. 231–235.

Casteel, R. W., Adam, D. P., and Sims, J. D., 1977, Late-Pleistocene and Holocene remains of *Hysterocarpus traski* (Tule perch) from Clear Lake, California, and inferred Holocene temperature fluctuations: Quaternary Research, v. 7, p. 133–143.

Cholnoky, B. J., 1968, Die Okologie der Diatomeen in Binnengewassern: J. Cramer, Lehre, 699 p.

Cleve-Euler, A., 1951, Die Diatomeen von Schweden und Finnland: Kungl. Svenska Vetenskapsakademiens Handlingar, Fjarde Serien, Band 2, No. 1, 163 p.

Ehrlich, A., 1969, Revision de l'espece *Cyclotella temperei* Peragallo et Heribaud; Examen compare aux microscopes: Photonique, electronique et electronique a balayage: Cahiers de Micropaleontologie, ser. 1, no. 11, p. 1–11.

Elmore, C. J., 1921, The diatoms (Bacillarioidae) of Nebraska: Nebraska University Studies, v. 21, nos. 1-4, p. 22–214.

Fuji, N., 1974, Palynological investigations on 12-meter and 200-meter core samples of Lake Biwa in central Japan; *in* Horie, S., ed., Paleolimnology of Lake Biwa and the Japanese Pleistocene: Kyoto, Japan, Kyoto University, v. 2, p. 227–235.

Gasse, F., 1980, Les Diatomees Lacustres Plio-Pleistocenes du Gadeb (Ethiopie) systematique, paleoecologie, biostratigraphie: Revue Algolgique, Memoire hors-serie no. 3, 249 p.

Goldman, C. R., and Wetzel, R. G., 1963, A study of the primary productivity of Clear Lake, Lake County, California: Ecology, v. 44, no. 2, p. 283–294.

Heusser, L. E., and Sims, J. D., 1983, Pollen counts for core CL-80-1 Clear Lake, Lake County, California: U.S. Geological Survey Open-File Report 83-384, 30 p.

Horne, A. J., and Goldman, C. R., 1972, Nitrogen fixation in Clear Lake, California; I. Seasonal variation and the role of heterocysts: Limnology and Oceanography, v. 17, no. 5, p. 678–692.

Horne, A. J., Dillard, J. E., Fujita, D. K., and Goldman, C. R., 1972, Nitrogen fixation in Clear Lake, California; II. Synoptic studies on the autumn *Anabaena* bloom: Limnology and Oceanography, v. 17, no. 5, p. 693–703.

Huber-Pestalozzi, G., 1942, Das Phytoplankton des Susswassers; Systematik und Biologie, *in* Thienemann, A., Die Binnengewasser: Stuttgart, E. Schweizerbartische Verlagsbuchhandlung, v. xvi, Teil 2, Halfte 2, p. 367–549.

Hustedt, F., 1957, Die Diatomeeflora des Flussystems der Weser im Gebiet der Hansestadt Bremen: Abhandlungen des naturwissenschaftlichen Vereins Bremen, Band 34, No. Heft 3, p. 181–440.

Ikebe N., and Yokoyama, T., 1976, General explanation of the Kobiwako Group; Ancient Lake deposits of Lake Biwa, *in* Horie, S., Paleolimnology of Lake Biwa and the Japanese Pleistocene: Kyoto, Japan, Kyoto University, v. 4, p. 31–51.

Jepson, W. L., 1923, Annual of the flowering plants of California: Berkeley, Associated Students Bookstore, University of California, 1238 p.

Johnson, R., Richards, T., and Blinn, D. W., 1975, Investigation of diatom populations in rhithron and potamon communities in Oak Creek, Arizona: Southwestern Naturalist, v. 20, no. 2, p. 197–204.

Kilham, P., and Kilham, S. S., 1978, Natural community bioassays: Predictions of results based on nutrient physiology and competition: Verhandlungen des Internationalen Vereins für Limnologie, vol. 20, p. 68–74.

Kilham, S. S., and Kilham, P., 1975, *Melosira granulata* (Ehr.) Ralfs: morphology and ecology of a cosmopolitan freshwater diatom: Verhandlungen des Internationalen Vereins für Limnologie, vol. 19, p. 2716–2721.

Kilham, S. S., and Van Donk, E., 1982, Succession in Lake Maarsseveen I; A laboratory approach [abs.]: American Society of Limnology and Oceanography, Annual Meeting, Program and Abstracts, p. 32.

Lange-Bertalot, H., 1979, Pollution tolerance of diatoms as a criterion of water quality estimation: Nova Hedwigia, Beiheft 64, p. 285–304.

Liddicoat, J. C. , 1981, Paleomagnetism of a 177-m core from Clear Lake, California: Geological Society of America Abstracts with Programs, vol. 13, no. 2, p. 67.

Loginova, L. P., 1982, The Likhvin diatom flora from the central part of the east-European Plain, its paleogeographical and stratigraphic significance: Acta Geologica Academiae Scientiarum Hungaricae, v. 25, nos. 1-2, p. 149–160.

Loseeva, E. I., 1982, Atlas of Late Pliocene diatoms from the Prikamya: XI International Quaternary Association Congress, Moscow, 1982, Leningrad, Nauka, 204 p.

Lowe, R. L., 1974, Environmental requirements and pollution tolerance of freshwater diatoms: Environmental Protection Agency, Environmental Monitoring Series, EPA-670/4-74-005, 334 p.

Lund, J.W.G., 1962, Phytoplankton from some lakes in northern Saskatchewan and from Great Slave Lake: Canadian Journal of Botany, v. 40, p. 1499–1514.

Moiseeva, A. I., 1971, Atlas of the Neogene diatomaceous deposits of the Primorskova region: Ministerstvo Geologii SSSR Vsesoyuzni Oroena Lemina Naucho-Issledovatelskii Geologicheskii Institut (VSEGEI) new series vol. 171, 151 p.

Molder, K., and Tynni, R., 1968, Uber Finnlands rezente und subfossil Diatomeen II: Bulletin of the Geological Society of Finland, v. 40, p. 151–170.

Moore, B. N., 1937, Non-metallic mineral resources of eastern Oregon: U.S. Geological Survey Bulletin 875, 180 p.

Mori, S., 1974, Diatom succession in a core from Lake Biwa, *in* Horie, S., ed., Paleolimnology of Lake Biwa and the Japanese Pleistocene: Kyoto, Japan, Kyoto University, v. 2, 288 p.

——, 1975, Vertical distribution of diatoms in core samples from Lake Biwa-Ko: Proceedings of the Japan Academy, v. LI, no. 8, p. 675–679.

Mori, S., and Hori, S., 1975, Diatoms in a 197.2-meter core sample from Lake Biwa-ko: Proceedings of the Japan Academy, v. 51, no. 8, p. 675–679.

Negoro, K., 1960, Studies of the diatom vegetation of Lake Biwa-Ko (First Report): Japanese Journal of Limnology, v. 21, p. 200–220.

Rymer, M. J., 1981, Stratigraphic revision of the Cache Formation (Pliocene and Pleistocene), Lake County, California: U.S. Geological Survey Bulletin 1502-C, 35 p.

Schmid, A. M., 1979, Influence of environmental factors on the development of the valve in diatoms: Protoplasma, v. 99, p. 99–115.

Sims, J. D., 1976, Paleolimnology of Clear Lake, California, U.S.A., *in* Horie, S. H., ed., Paleolimnology of Lake Biwa and the Japanese Pleistocene: Kyoto, Japan, Kyoto University, v. 4, p. 658–702.

Sims, J. D., Adam, D. P., and Rymer, M. J., 1981, Late Pleistocene stratigraphy and palynology of Clear Lake: U.S. Geological Survey Professional Paper 1141, p. 219–230.

Sims, J. D., Rymer, M. J., and Perkins, J. A., 1981, Description and preliminary interpretation of core CL-80-1, Clear Lake, Lake County, California: U.S. Geological Survey Open-File Report No. 81-751, 175 p.

Stoermer, E. F., and Ladewski, T. B., 1976, Apparent optimal temperatures for the occurrence of some common phytoplankton species in southern Lake Michigan: Ann Arbor, University of Michigan, Great Lakes Research Division, Publication no. 18, 49 p.

Stoermer, E. F., and Yang, J. J., 1970, Distribution and relative abundance of dominant planktonic diatoms in Lake Michigan: Ann Arbor, University of Michigan, Great Lakes Research Division, Publication no. 16, 64 p.

Stoermer, E. F., Taylor, S. M., and Callender, E., 1971, Paleoecological interpretation of the Holocene diatom succession in Devils Lake, North Dakota: Transactions of the American Microscopical Society, v. 90, no. 2, p. 195–206.

Thompson, J. M., Sims, J. D., Yadav, S., and Rymer, M. J., 1981, Chemical composition of water and gas from five nearshore subaqueous springs in Clear Lake: U.S. Geological Survey Professional Paper 1141, p. 215–218.

Ueno, M., 1975, Evolution of life in Lake Biwa; A biogeographical observation *in* Horie, S., ed., Paleolimnology of Lake Biwa and the Japanese Pleistocene: Kyoto, Japan, Kyoto University, v. 3, 577 p.

Van Heurck, H., 1885, Synopsis des diatomees de Beligique: Reprint, 1981, Amsterdam, Holland, Linneaus Press, Anvers. 235, p. 132 plates.

Van Landingham, S. L., 1967, Paleocology and microfloristics of Miocene diatomites from the Otis Basin–Juntura region of Harney and Malheur counties, Oregon: Beihefte zur Nova Hedwigia, Heft 26, 77 p.

Verch, R., and Blinn, D. W., 1972, Seasonal investigations of algae from Devil's Lake, North Dakota: Prairie Naturalist, v. 3, nos. 3 and 4, p. 67–79.

Young, R. T., 1924, The life of Devil's Lake, North Dakota: Publication of the North Dakota Biological Station, 116 p.

Zoriki, T., 1976, On the relation of the diatom flora and water quality in the Uji River, Kyoto: Japanese Journal of Limnology, v. 37, no. 1, p. 29–36.

Manuscript Accepted by the Society September 15, 1986

Printed in U.S.A.

Geological Society of America
Special Paper 214
1988

The Clear Lake, California, ostracode record

Richard M. Forester
U.S. Geological Survey
Denver Federal Center, MS 919
Denver, Colorado 80225

ABSTRACT

Modern-day Clear Lake is a turbulent, turbid, permanent, polymictic lake. It is also a highly productive fresh-water lake whose dominant solutes are $Mg^{2+} + Ca^{2+} - HCO_3^-$. The lake has a diverse and abundant limnetic community, yet has a depauperate benthic community. The benthic community structure appears to be ecologically simple as the result of turbulence-induced substrate instability coupled with unpredictable periods of anoxia induced by the oxygen-consuming organic matter. Ostracodes, which are ubiquitous, largely benthic, environmentally sensitive, and diverse organisms, are represented in Clear Lake only by the nektic species *Cypria ophtalmica.* The substrate conditions provide adequate reason for the absence of most ostracodes from the modern lake, but their absence in the fossil record suggests that the modern lacustrine environment existed in the past despite known climate changes. This seeming paradox can be explained by considering the influence of various types of climatic change on the lacustrine environment; certain types of climate-environmental changes would maintain a lacustrine environment unsuited to ostracodes. These hypothesized climate-lacustrine environmental changes would favor the modern (Holocene) and other oak-dominated periods to be the warmest and driest in Clear Lake history, whereas the pine-TCT (Taxodiaceae, Cupressaceae, and Taxaceae) periods would be cooler and wetter than today. The largely barren ostracode record, coupled with rare ostracode occurrences, would support a glacial-interglacial Clear Lake climatic history characterized primarily by changes in the annual precipitation-evaporation budget.

INTRODUCTION

Clear Lake (Fig. 1) is a large, warm-polymictic, turbulent, turbid, permanent, shallow, lacustrine system (Goldman and Wetzel, 1963; Horne and Goldman, 1972). The large surface area (114 km^2) and shallow depth (average 6 m) provide for frequent wind mixing of surface and bottom waters, but the bottom-most waters and sediments can become oxygen-deficient on calm days and may remain anoxic from July through November (Horne and Goldman, 1972). Moreover, sublittoral sediments are rich in oxygen-depleting organic material and are often fluid (M. Rymer, personal communication, 1982). The nearly continual turbulence produces an oxic, turbid littoral zone that has an unstable mobile substrate.

The solute composition of Clear Lake is characterized by $Mg^{2+} = Ca^{2+} \geqslant Na^+ >> K^+$ and $HCO_3^- >> SO_4^{2-} \geqslant Cl^-$; the salinity is low, usually less than 200 ppm (Thompson and others, 1981a). By contrast, many of the local springs that discharge into Clear Lake have solute compositions dominated by $Mg^{2+} + Na^+ - HCO_3^-$ or $Na^+ - Cl^-$, and have much higher salinities than Clear Lake (Thompson and others, 1981b). Clear Lake is relatively dilute, especially considering its location in an area in which precipitation and evaporation are roughly equal; the lake has a large surface area, which enhances evaporation, and several springs contribute concentrated solutes to the lake. The $Mg^{2+} + Ca^{2+} - HCO_3^-$ solutes and dilute water of Clear Lake indicate that dissolved ions are flushed from the lake, implying the lake is not in evaporative phase with the atmosphere. In this respect, Clear Lake behaves chemically more like a fluvial than a lacustrine system.

The nutrient-rich, turbulent waters of Clear Lake support a diverse and dense phytoplankton community that contributes significantly to the lake's turbidity. High primary productivity supports a large nektic-consumer population (Copepoda, Clad-

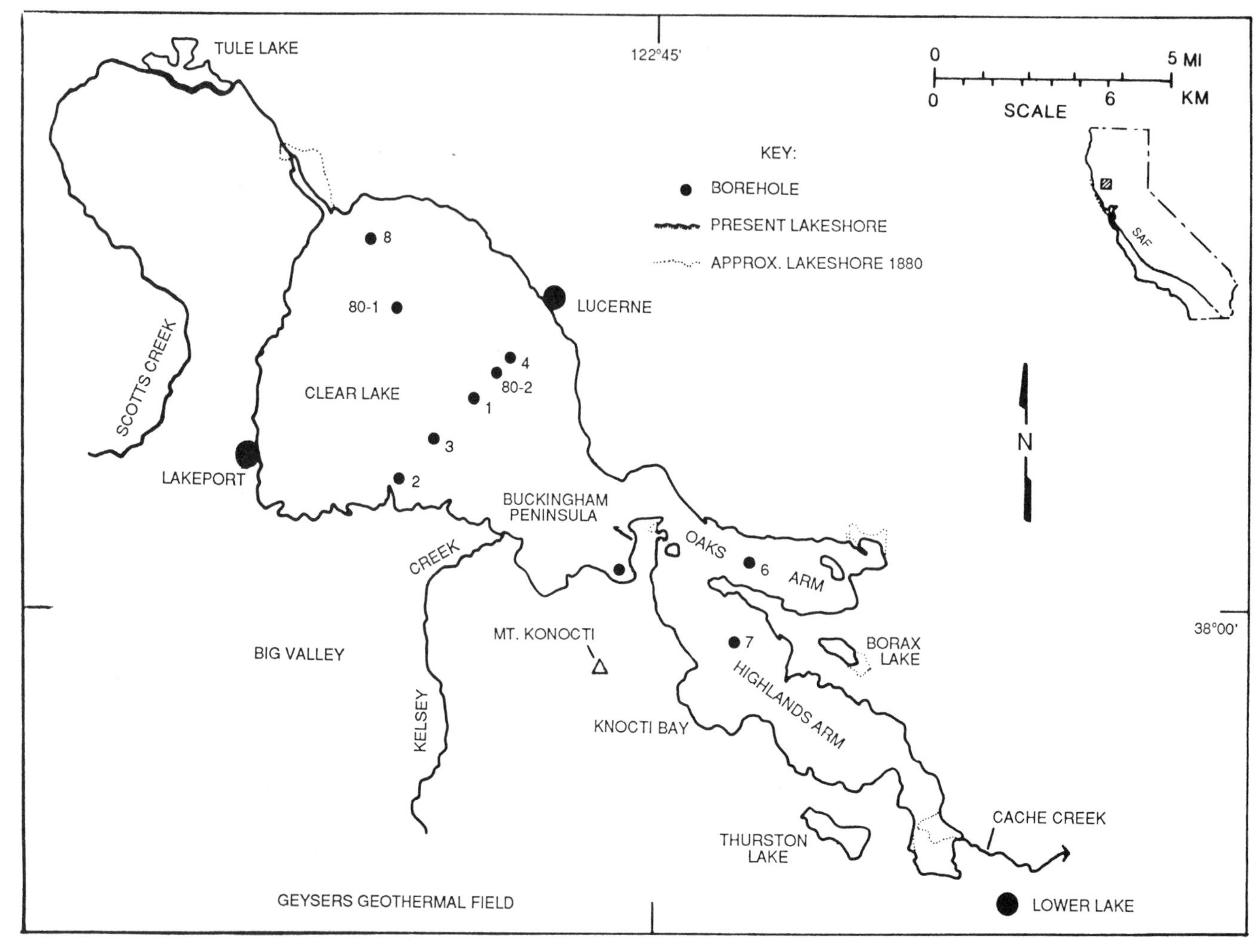

Figure 1. Map showing location of the Clear Lake area in California with respect to San Andreas Fault zone (SAF on inset). Location of cores taken from Clear Lake for this study shown with two designations: single numbers 1 through 8 represent location of series of cores taken in 1973, which bear prefix CL-73- in text; locations 80-1 and 80-2 represent cores taken in 1980, which bear the prefix-CL- in the text. Approximate 1880 shoreline of Clear Lake is shown for reference (after Becker, 1888).

ocera), which in turn supports a large fish population. By contrast, the benthic community in Clear Lake appears to be relatively simple, composed of low-oxygen–tolerant organisms such as midge fly larvae. The littoral zone lacks subaquatic macrophytes due to the turbidity-induced opaqueness of the water column, but it does contain a number of emergent macrophyte taxa (Goldman and Wetzel, 1963). As far as is known, no ostracodes other than the nektic *Cypria ophtalmica* are living in Clear Lake proper. Today the rarity of benthic plants and animals in Clear Lake is readily understood in terms of the low-light and low-oxygen conditions. Examination of 132 samples from five cores reveals that, with few exceptions, the fossil record in Clear Lake is also barren of benthic organisms. The absence of benthic organisms (especially ostracodes) from a lake over a long period of time is very unusual, because one assumes that with climatic (Adam and others, 1981) or other environmental changes, the present limnologic conditions of the lake would change sufficiently to permit the survival of a benthic community.

The work herein examines the rare ostracode occurrences and the other available information from the Clear Lake cores in an attempt to define relative environmental boundary conditions that might prohibit uniquitous organisms such as ostracodes from living in the lake throughout its known existence. Even though the resulting paleolimnologic interpretations are based on negative evidence and are thus speculative, they have some credence because, whereas the possible reasons that ostracodes might be absent from a short core interval are large, the possibilities decrease with an increasing core sample interval and an increasing number of cores sampled. The paleolimnologic conclusions presented here are supported in part by other studies (see, for example, Bradbury, this volume).

MATERIALS AND METHODS

Appendix 1 lists all sample intervals examined from each core (Fig. 1). The longest core, CL-80-1, was the most extensively sampled. An average of one sample was examined from every recovered meter of sediment, for a total of 72 samples. This coarse sample interval was intended to provide a gross distribution of ostracodes that would aid in more refined sampling strategies. Appendix 2 lists the ostracodes recovered from the Clear Lake core samples; it shows that only 3 of the 72 samples from core CL-80-1 contain abundant ostracodes. Based on these meager results, further sampling of core CL-80-1 seemed unwarranted.

I examined samples from other cores, particularly those cores taken from more shoreward localities than CL-80-1, to determine if the absence of ostracodes in CL-80-1 was due to a local phenomenon such as central lake profundal anoxia. Cores CL-73-2, -4, -7, and -8 (Sims and others, 1981) were sampled at regular bisected intervals, producing a minimum of 4 samples in core CL-73-2 and a maximum of 17 samples in core CL-73-8 (see Appendix 1). All of these samples, except for one sample from 20.25 m in core CL-73-7, were barren of calcareous microfossils. Thus, only 4 of a total of 132 samples taken from five different cores contained abundant ostracodes, and an additional 15 samples had sparse numbers of ostracodes present.

The sediment from each sample was disaggregated in warm water containing about a tablespoon of baking soda and was then washed over a standard 100-mesh sieve (150 μm openings). The >150 μm residue was air-dried in paper towels and poured through a nested set of sieves; the resulting size fractions were examined under a binocular microscope, and ostracodes and other organisms were hand-picked from the residue into standard micropaleontologic slides for identification. Sediment sample volumes ranged from 50 cc to more than 100 cc, where a 50-cc sample is, based on my experience, about 2.5 to 10 times the sediment volume needed to obtain tens to thousands of adult ostracode valves.

MODERN LACUSTRINE OSTRACODES—OCCURRENCE AND DISTRIBUTION

Nonmarine ostracodes are ubiquitous, occurring commonly in nearly all oxic aquatic environments. Recent studies (Delorme and others, 1977; Delorme and Zoltai, 1984; De Deckker, 1981, 1982; Forester, 1983, 1985, 1986; Forester and Brouwers, 1985) show that nonmarine ostracodes are environmentally sensitive organisms. A brief discussion about nonmarine ostracodes is provided to accentuate their commonness and by so doing to emphasize their rarity in the Clear Lake fossil record.

Ostracodes are microscopic crustaceans that have a bivalved carapace composed of low-magnesium calcite. Some taxa are capable of limited swimming, although four or five species can become nektic. Most ostracodes crawl on or in the substrate and are therefore most profoundly influenced by environmental parameters at the sediment-water interface. Most ostracodes have a long and complex life cycle, ranging from a month to more than a year, depending on the species. In contrast to other small, short-lived crustaceans, the life cycle of ostracodes appears to be synchronized with the annual rather than with seasonal lacustrine cycles.

Ostracodes live in virtually all aquatic environments, but are most abundant and commonly preserved in lakes or environments dominated by ground-water discharge, including ponds, marshes, and springs. The highest species diversity is often found in the low-energy littoral zones of fresh-water lakes; a 20-cc sample from a core taken in these zones may contain as many as 20 to 40 species and hundreds (or rarely thousands) of adult ostracode valves. A 20-cc sample from saline lakes, by contrast, may contain only one or two species, but thousands of adult valves.

Lacustrine ostracodes inhabit fresh water (salinity $<3,000$ ppm total dissolved solids [tds]) and saline water (tds $>3,000$ ppm). The geographic distribution of fresh-water ostracodes is often limited to narrow latitudinal ranges, suggesting that various aspects of water temperature (seasonal maxima or minima, annual profile) are important to the life cycle of those ostracodes (Delorme and Zoltai, 1984; Forester, 1985). Thus fresh-water ostracodes are especially valuable for determining relative paleowater temperature, which for lakes in phase with climate, provides relative air temperature.

Hydrochemistry (solute composition, salinity) plays a vital role in the occurrence and productivity of lacustrine ostracodes (Delorme, 1969; Forester, 1983, 1986; Forester and Brouwers, 1985). Fresh-water lacustrine ostracodes are often restricted to Ca^{2+} + (Mg^{2+}) – HCO_3^- solute types as well as to low salinities. Fresh-water environments not dominated by Ca^{2+} + (Mg^{2+}) – HCO_3^- solutes often lack "fresh-water" ostracodes, but do contain "saline" ostracode taxa. Moreover, particular saline ostracode taxa are restricted to particular solute types or groups of solute types independent from salinity (Forester, 1983), but a species' upper-salinity tolerance may be anion specific (Forester, 1986). Ostracodes thereby provide a means of interpreting paleohydrochemistry, which, for paleolakes that were in phase with local climate, provides information about the annual atmospheric precipitation-evaporation budget.

Hydrochemical parameters and water temperature are among the most important factors that determine whether an ostracode can survive in a particular lake. The amount of oxygen at the sediment-water interface normally defines the areas within a lake where ostracodes can survive. Small productive lakes that undergo thermal stratification usually have near-anoxic or anoxic conditions below the thermocline, and thus ostracodes are often absent from sublittoral or profundal sediments. Small oligotrophic lakes or large lakes that have oxic conditions at the sediment water interface often have abundant and diverse ostracode communities in the sublittoral to profundal zones. The littoral zone of most lakes and marginal lacustrine environments is normally oxygenated during the entire year or most of the year and thereby supports an abundant, diverse ostracode community whose species composition is determined by hydrochemical and thermal parameters.

Biologic factors such as food type, predators, and parasites may also determine ostracode occurrence, but more likely play a role in ostracode productivity. Most ostracodes are believed to be nonselective detritus feeders, implying that all lakes should provide satisfactory food supplies.

OSTRACODE OCCURRENCES IN CLEAR LAKE CORES

Core CL-80-1

Appendix 1 lists the core depths of all samples processed for ostracodes. Appendix 2 lists the taxa recovered from the entire sample suite. The original sampling strategy for core CL-80-1 is set forth in the section above on materials and methods. When I determined that core CL-80-1 was essentially barren of ostracodes, the sampling strategy was redesigned to examine samples from as many cores as possible to determine if the general absence of ostracodes might be due to local anoxia at the sediment-water interface. Anoxia, either within the sediment or at the sediment-water interface, is common in sublittoral or profundal areas of productive lakes and serves as a barrier for ostracodes or other benthic organisms. To act as a barrier, the period(s) of anoxia need not be continuous in the case of long-lived organisms like ostracodes. As is evident from Appendices 1 and 2, most of the Clear Lake record is barren of ostracodes, and thus local regions of anoxia within the lake do not explain the absence of ostracodes.

The samples from core CL-80-1 in which ostracodes are abundant, namely at 92.70, 124.00 and 130.00 m, provide insights into the probable paleoenvironment of Clear Lake. The thousands of ostracodes present in these samples largely belong to *Candona* n.sp. 1 and are represented by a disproportionate number of juvenile valves. Ostracodes usually have eight instars (juveniles whose size increases geometrically during molting) and an adult stage. Thus, an ideal life assemblage has an adult to juvenile ratio of 1:8. The smaller, younger instars are generally removed during the laboratory washing process, so that the ratio of adults to juveniles of a life assemblage becomes 1:4 or 1:3. The *Candona* n.sp. 1 in the above three samples have juveniles that far outnumber the adults; due to numerous broken valves, an accurate count cannot be made. Several counts of different groups of 100 valves shows that the ratio of adults to juveniles ranges from as small as 1:8 to as large as 1:16; 1:12 is a common value. Studies of marine ostracodes by Van Harten (1986) show that assemblages with a disproportionate number of juveniles are often produced by differential valve transportation. When juvenile and adult ostracode valves are suspended, the smaller juveniles are preferentially kept in suspension and are often transported farther than the adult valves, producing a transported ostracode assemblage that is enriched in juveniles. Similar differential transportation is known from lacustrine systems (R. M. Forester, unpublished data), and I believe that transport is the best explanation for the observed ratio of adults to juveniles in these three Clear Lake CL-80-1 samples.

A few other samples from CL-80-1 contained ostracodes, but the abundance was always low, consisting of fewer than 20 valves per 50 to 100 cc of sediment. All remaining samples were barren of calcareous microfossils.

Cores CL-73-2, -4, -7, and -8

Samples from these four cores were examined for ostracodes; with the exception of sample 20.25 m from core CL-73-7, all other samples were barren or contained only rare ostracodes. Core CL-73-7, sample 20.25 m, contains a diverse assemblage of ostracodes including all of the taxa found in the samples from core CL-80-1. Core CL-73-7 was taken from the Highlands Arm of Clear Lake, in an open lacustrine environment (Sims and others, 1981). The uppermost 6.7 m of the core is composed of what Sims and others (1981) described as a sapropelic mud, presumably deposited under environmental conditions similar to those presently existing at the core site. The remainder of the core was described by those authors as being an "interbedded fibrous brown (5YR5/3) peat, mud, clay, and volcanic ash," which was presumably deposited under marginal lacustrine conditions. Moreover, Sims and others (1981) recovered seeds of numerous subaquatic macrophytes and *Nuphor* (water lily) from the lower part of the core, supporting the interpretation of shallow-water pond or marsh environments. This rapid transition from marginal lacustrine environments to open lacustrine environments is believed to have been the result of tectonic activity (Sims and others, 1983).

The ostracode assemblage from core CL-73-7, sample 20.25 m, appears to be a life assemblage, based on the following observations: (1) there are numerous adults of all of the species; (2) the adult to juvenile ratio is on the order of 1:4; and (3) the taxa are environmentally compatible with a fresh-water marginal-lacustrine setting. The modern biogeography of *Candona* n.sp. 1, *C.* n.sp. 2, and *Limnocythere* n.sp. appears to include only central and southern California; therefore these taxa are not living in regions colder than those in the modern-day Clear Lake area, which suggests these taxa are thermophilic or eurythermic. *Cypridopsis vidua* is a eurythermic ostracode, whereas *Cypria ophtalmica* is probably a cryophilic species. This sort of species mixture suggests seasonal variability in air and water temperature. The absence of numerous other taxa that only live in colder areas today suggests that the water temperature at this site was not significantly colder than the modern values.

The environmental tolerances of these ostracodes further suggest that the physical environment was a low-turbulence, relatively clear, shallow pond-marsh environment. The hydrochemical environment was dominated by Ca^{2+} + (Mg^{2+}) – HCO_3^- solutes, and salinity was low, probably less than 500 ppm. The presence of *Limnocythere* n.sp. suggests seasonal hydrochemical variability, whereas *Cypria ophtalmica* implies the opposite. These sorts of species mixtures often occur in water bodies having relatively constant hydrochemical conditions during one or more seasons (often winter-spring) and variable conditions during one or more seasons (often summer-fall). Perhaps *C. ophtalmica* lived during the winter-spring season, which implies above-freezing temperatures and relatively constant hydrochemistry, whereas *L.* n.sp. lived during the summer, implying variable hydrochemistry. Seasonal hydrochemical variability of this type would suggest cool-wet winters and warm-dry summers.

Other samples from core CL-73-7 taken from the marginal-lacustrine sediments contain common insect parts, rare fish material, cladocera ephippia, seeds, and plant material, but not ostracodes. The presence of aquatic and terrestrial organisms in a peat-dominated sediment containing no ostracodes suggests either that a dense stand of emergent vegetation existed without any open pools of water, or that an ephemeral marsh was inundated with water for short periods of time.

MODERN CLEAR LAKE OSTRACODE OCCURRENCES

Modern littoral and sublittoral Clear Lake sediments were not extensively sampled for ostracodes because a preliminary suite of four samples collected by Michael Rymer in 1981 proved to be barren of ostracodes. The absence of ostracodes in these samples, coupled with the known limnology of Clear Lake, indicates that the lake is presently inhospitable for most ostracodes.

In addition to the littoral sediment samples, the sediment collected in two sediment traps set to collect suspended sediments (R. Y. Anderson, personal communication) was also processed to determine if any ostracodes currently are part of the suspended sediment load. Interestingly, the sediment trap samples did not contain dead ostracode valves, although they did contain the complete remains of *Cypria ophtalmica. Cypria ophtalmica* is a strong swimmer, one of the few ostracode taxa that can lead a nektic or pseudonektic mode of life. It has a weakly calcified carapace that is not commonly preserved in sediments, which explains its absence in the modern Clear Lake sediments and provides a reason for its absence in the fossil sediments. This ostracode is very abundant in the two sediment traps, implying that it is abundant throughout the lake.

DISCUSSION

Physical Barriers to Ostracode Habitation in the Modern Lake

The combination of several physical parameters forms an inhospitable environment in Clear Lake for many benthic organisms. Continual turbulence, relatively shallow water, and mild temperatures combine to produce a polymictic lake that readily and often cycles nutrients from the sediment and bottom waters into the surface waters, resulting in high organic productivity. The high organic production of the surface water produces an oxygen-consuming organic rain that leaves the sediments and bottom-most waters in a near-anoxic and/or intermittent anoxic state throughout much of the year. The short and often unpredictable oxic periods make the sediment surface unsuitable for all but short-lived and/or low-oxygen–tolerant benthic organisms. Moreover, the continual turbulence produces a highly mobile littoral substrate that, coupled with the opaqueness of the water, makes the more predictably oxic littoral zone unsuitable for benthic plants and animals.

Ostracodes are ubiquitous aquatic organisms, although any lake may have environmental conditions that exclude ostracodes or minimizes their productivity for variable periods of time. The most common barrier to ostracodes is anoxic conditions, which often occur in productive lakes below a thermocline. However, although ostracodes may not live in certain areas of a lake, it is very unusual to find a circumneutral or alkaline lake with no ostracodes. Even regions of a lake in which ostracodes do not live will receive transported ostracodes from other areas in the lake. Therefore, the general absence of all ostracode taxa, with the single exception of *Cypria ophtalmica* from modern-day Clear Lake, is unusual. Environmental parameters, however, can combine in numerous ways, as they do in modern-day Clear Lake, to render an environment uninhabitable. The absence of ostracodes from a lacustrine system at any one point in time is unusual, but not extremely rare.

Ostracode Absences in the Paleolake

The absence of ostracodes from the long Clear Lake record (core CL-80-1) and from numerous other areas of Clear Lake (cores CL-73-2, -4, -7, and -8) for the time represented by those cores is very rare; to the best of my knowledge this has not been observed in other lakes regardless of size. The most obvious explanation for this generally barren ostracode record is that the modern-day turbulent, shallow, high-productivity conditions extend throughout the history of the lake recorded by the core. These conditions would maintain the present intermittent bottom-water–sediment anoxia, together with the unstable littoral substrates that prove to be the most effective modern barriers to ostracode habitation.

Postburial solution of the ostracode valves is an alternate explanation for the general absence of ostracodes from the Clear Lake cores. Solution may well explain the absence of the thin, poorly calcified *Cypria ophtalmica* valves from Clear Lake sediments. *Cypria ophtalmica* is abundant in the modern lake and probably was abundant in the paleolake, but is not commonly preserved in modern or fossil sediments. Other ostracode taxa have a much more calcified carapace than *C. ophtalmica,* so their valves are much less likely to be dissolved in the sediments.

Moreover, numerous adult and juvenile ostracodes are preserved in a few places in core CL-80-1 in organic-rich sediments. Similarly, an abundant, well-preserved ostracode assemblage is preserved in core CL-73-7 in muddy peats. If postburial solution is erasing the ostracode record in Clear Lake sediments, it is doing so in a selective manner within sediments that are texturally and compositionally similar. Therefore, I regard carbonate preservational problems to be an unlikely solution for the largely barren Clear Lake ostracode record. The thin-shelled and easily dissolved *Cypria ophtalmica* is the exception; the absence of this taxon is believed to be due to solution on deposition.

Paleoclimate and the Paleolacustrine Environment-Ostracode Habitability

Physical and chemical lacustrine parameters are controlled by or correlative with regional climate to varying degrees. A large change in the regional Clear Lake climate would produce corresponding changes in the physio-chemical lacustrine parameters, producing different lacustrine environments than exist today. Palynologic studies (Adam and others, 1981; Adam and West, 1983) clearly show that the paleoclimate of Clear Lake was different than today, and diatom studies (Bradbury, this volume) show that limnologic changes partly correlative with the palynologic changes have also occurred. Conceivably, a change from the modern lacustrine environment could produce a paleoenvironment that was suited to ostracodes. This suggests that either climate-limnologic change was not sufficient to alter the anoxia problem, or, if anoxia was eliminated, the resulting environment was still unsuited to ostracode survival. In order to evaluate these possibilities, we need to consider what limnologic changes relative to the modern lacustrine environment could be produced by particular climatic changes.

Turbulence and Nutrient Cycling

If wind-induced turbulence were less than today, nutrients would be cycled less frequently, presumably reducing organic productivity and the oxygen-consuming organic matter sedimentation, which might provide a seasonal window when ostracodes could live in some areas of the lake. A similar reduction in nutrient cycling could be attained by making the lake deeper, thereby isolating the nutrient-rich bottom waters from surface waters, except for periods of very high turbulence.

Air Temperature and Seasonality

If seasonal air temperatures were different from those of today with cooler (warmer) winters (summers), the water temperatures should change accordingly, producing substantial thermal and chemical changes in the lacustrine environment. Warmer seasonal or annual air temperatures, combined with the modern annual precipitation, should produce a higher rate of evaporation, which should reduce inflow and outflow and increase the salinity of the lake. If the warmer conditions were significant enough to reduce water depth to the point at which the lake was no longer stratifying—ceasing to be polymictic and consequently becoming more saline but oxygenated throughout—the lacustrine environment should support abundant ostracodes. Goose Lake, Oregon, is a partial modern analog for this hypothetical situation. The Clear Lake diatoms (Bradbury, this volume) indicate that present-day environments roughly coincide with the Holocene, which in turn coincides with oak dominance (Sims and others, 1981), suggesting the Holocene period is one of the warmest and/or driest periods in Clear Lake history. If previous climates were both warmer and wetter, the inflow-outflow rate might not reduce: the salinity would not increase, and the polymictic conditions would remain. These conditions would still prove to be a barrier to ostracodes, and thus the absence of ostracodes could not be used to distinguish a warmer and wetter climate from a warmer than modern climate. This possibility, however, does not appear probable, because even though the oak-pollen–dominated periods (Holocene and earlier) appear to be the warmest periods in Clear Lake history (Adam and West, 1983), they do not appear to be wetter, because a summer-moisture–intolerant fungus would kill the oak tree roots if that were the case (Sims and others, 1981).

Cooler seasonal or annual air temperatures, especially on the scale suggested by Adam and West (1983), should change the lake from its polymictic state to a warm monomictic or dimictic state. Cooler air temperature might also result in lower evaporation and thus a higher inflow/outflow rate or a deeper lake. Any lake that stratifies may have low oxygen levels at the sediment-water interface below the thermocline. An absence of ostracodes cannot be used to suggest that this type of change has not occurred; however, any ostracodes that did occur should reflect the overall colder nature of the water-column. Adam and West (1983) suggested that the coldest periods in Clear Lake history are dominated by pine-TCT pollen. The only ostracodes recovered from a pine-TCT period occurred in a sample from 20.25 m, core CL-73-7. Five ostracode species were present, three of which are not known from areas colder than Clear Lake today. Numerous other ostracode species are restricted to regions colder than the Clear Lake area today and are known to live or even thrive in the kinds of hydrochemistries common to Clear Lake. The absence of ostracode taxa that are restricted to colder water (from core CL-73-7, sample 20.25 m) and the presence of taxa that do not live in colder water suggest that in this one instance the ostracodes show that the pine-TCT period is not substantially colder than today. The 6° to 8°C cooling suggested by Adam and West (1983) would make the paleoclimate in the Clear Lake area similar to regions in Washington, east and west of the Cascades. Ostracodes collected from Washington lakes show that numerous cryophilic taxa live in those lakes that are not part of the Clear Lake fossil record.

Precipitation-Evaporation budgets

The balance between annual (seasonal) precipitation (P) and evaporation (E) is another important climatic parameter whose effects are closely tied to and often parallel with temperature, making the distinction between temperature and P/E difficult. This problem is evident from the discussion above regarding the probable impact of warmer temperatures on modern Clear Lake, involving either greater or equal levels of precipitation. That argument can be reversed, and we can query about the effects of modern temperatures but a lower P/E ratio upon Clear Lake. As with higher temperatures and existing precipitation, the lake would become shallower and more saline and perhaps not stratify. The absence of ostracodes and the presence of freshwater diatoms (Bradbury, this volume) suggests that P//E ratios have not been lower than they are now.

If the annual precipitation were greater than today, but air temperature were similar to or cooler than modern values, the lake level should rise due to less evaporation, or the lake should have a higher rate of outflow. The absence of open lacustrine sediments in core CL-73-7 prior to about 10,000 years ago suggests that the lake level of Clear Lake was not higher than today during the late Pleistocene other than accounting for Holocene sediment filling; this implies that the dam height has not changed much during this period. Therefore, an increase in precipitation would have to result in a higher outflow, which would produce a salinity that is lower than modern salinity. The modern lake has a low salinity and behaves chemically like a fluvial system; that is, it does not appear to have a hydrochemistry in evaporative phase with the atmosphere, nor does it retain the ions from the more concentrated springs surrounding the lake. Ostracode diversity and abundance are always low in dilute Ca-HCO_3-dominated waters. The combination of low salinity due to higher outflow and the modern turbulent, turbid, oxygen-consuming organic productivity would be a formidable barrier to ostracodes. Thus the paleoclimate in the Clear Lake region could experience major changes in precipitation, ranging from present conditions to much wetter conditions, without making the lake habitable for ostracodes. Adam and West (1983) suggested that precipitation could be as much as 2 m higher during the pine-TCT period than today. Even though they do not put much faith in their precipitation estimates, their values show that the palynologic profile could be interpreted as indicating P/E changes as well as or rather than temperature changes.

Probable Paleoclimate History

This discussion has provided potential paleolimnologic responses to changes, in particular climatic parameters, in order to understand what types of changes might be possible without Clear Lake becoming habitable to ostracodes. The absence of ostracodes from much of the Clear Lake fossil record is readily understood if changes in P/E control the lake's limnology. The largely barren ostracode record, combined with the palynologic, diatom, and sedimentologic records, supports this story. The pollen and diatom records suggest that the modern climate may be among the warmest and driest in the history of Clear Lake, whereas the limited ostracode and sediment records suggest that the paleoclimate was not substantially colder than today. The palynologic record does support a generally wetter climate in the past. Thus, the Wisconsinan climate and perhaps the other pine-TCT periods appear to have been largely wetter periods than today. Adam and West (1983, p. 170) pointed out that the absence of spruce from this record suggests periods of summer drought, which suggests that the wetter climate means long, rainy (or snowy), cloud-covered winters, rather than year-round moisture. The seasonal variability of the hydrochemical environment interpreted from the ostracode assemblage in a sample from core CL-73-7, 20.25 m, would support this possibility (see the discussion in the section on cores CL-73-2, -4, -7, and -8). The cloud cover might have moderated the winter temperatures, preventing both extreme highs and lows, and probably resulted in somewhat cooler conditions.

SUMMARY

Samples from several Clear Lake cores are largely barren of ostracodes, which is unusual because ostracodes are nearly ubiquitous aquatic organisms. The primary environmental parameter that prevents ostracodes from living in any aquatic environment is permanent or unpredictable anoxia. The sediment-water interface in Clear Lake is rarely oxic because of the high organic productivity, and thus modern-day Clear Lake is unsuited to ostracode habitation. Given probable limnologic responses to various changes in particular climatic parameters, the lake would remain unsuited to ostracode habitation if the climate did not change or if climatic change was largely in the form of increases or decreases in P/E values. Existing pollen, diatom, and sedimentologic data support, but do not exclusively demonstrate, the validity of this interpretation.

APPENDIX 1. SAMPLES PROCESSED FOR OSTRACODES, LISTED IN STRATIGRAPHIC ORDER FOR EACH CORE EXAMINED

Sample	Depth (m)	Sample	Depth (m)
Core CL-80-1			
0004	1.00	0209	65.75
0007	3.25	0203	68.55
0014	6.50	0231	69.95
0022	8.50	0232	74.60
0029	10.50	0241	76.85
0036	12.50	0243	76.14
0041	18.60	0249	78.85
0043	19.41	0255	81.00
0050	22.00	0263	83.00
0055	28.09	0271	85.00
0059	30.54	0277	87.25
0073	36.75	0282	88.50
0085	38.75	0295	90.00
0099	41.35	0303	92.00
0104	42.50	0307	92.70
0117	44.60	0312	94.00
0128	47.30	0320	96.10
0135	48.90	0328	98.10
0152	51.00	0349	100.00
0169	53.50	0372	102.50
0176	56.50	0397	105.00
0182	59.60	0405	108.00
0189	61.35	0412	109.75
0202	63.35	0431	112.00
0441	114.00	0569	144.70
0450	116.75	0584	146.75
0485	118.75	0592	148.50
0463	120.00	0600	150.50
0466	122.00	0604	151.50
0481	124.00	0609	153.90
0489	126.00	0612	155.75
0501	128.00	0618	158.75
0508	130.00	0626	161.75
0516	132.00	0634	163.75
0538	136.00	0647	165.75
0546	138.00	0657	167.75
Core CL-73-8			
1	0.00	10	11.25
2	1.25	11	12.50
3	2.50	12	13.75
4	3.75	13	15.00
5	5.00	14	16.25
6	6.25	15	17.50
7	7.50	16	18.75
8	8.75	17	20.00
9	10.00		
Core Cl-73-2			
1	6.13	3	7.88
2	7.00	4	13.13
Core CL-73-4			
1	7.19	9	64.71
2	14.38	10	71.88
3	21.57	11	79.09
4	28.75	12	86.25
5	35.95	13	93.47
6	43.13	14	100.63
7	50.33	15	107.85
8	57.50	16	115.00
Core CL-73-7			
1	0.00	8	18.57
2	3.38	9	19.41
3	6.75	10	20.25
4	10.13	11	21.08
5	13.50	12	21.93
6	16.88	13	22.77
7	17.74	14	23.63

APPENDIX 2. OSTRACODE FAUNAL LIST FROM CLEAR LAKE CORE SAMPLES

Owing to low species diversity, all taxa recovered are listed and given a species number, which appears with the sample listing. Abundance data are based on total counts of whole adult valves and follow the scheme: rare ≤3; present = 4 to 8; common = 9 to 20; very common = 21-55; abundant ≥56.

Ostracodes from Clear Lake Cores	Species No.
Candona n.sp. 1	I
Candona n.sp. 2	II
Limnocythere n.sp.	III
Cypria ophtalmica (Jurine, 1820)	IV
Physocypria globula Furtos, 1933	V
Cypridopsis vidua (Mueller, 1776)	VI

Sample	Depth (m)	Ostracode Taxa (abundance)
Core CL-80-1		
0232	74.60	I (rare)
0243	76.14	I (common, III (present)
0282	88.50	I (common)
0303	92.00	I (rare)
0307	92.70	I (abundant), III (rare), IV (common), s, f*
0312	94.00	I (rare), III (rare)
0320	96.10	I (rare)
0328	98.10	I (rare)
0372	102.50	I (rare)
0397	105.00	I (rare)
0405	108.00	I (common), III (present)
0431	112.00	I (common), f
0481	124.00	I (abundant), III (rare), s
0489	126.00	I (rare)
0508	130.00	I (abundant), V (rare)
Core CL-73-4		
14	100.63	I (rare)
15	107.85	I (rare)
Core CL-73-7		
10	20.25	I (abundant), II (present), III (abundant), IV (very common), VI (abundant)

*The presence of snails and/or otoliths is indicated by the letters s and f, respectively, following the ostracode species listing. In addition to the organisms listed, the majority of samples contained plant stem fragments (commonly carbonized) and fish remains (usually bones), as well as insect remains (typically chironomid head capsules).

REFERENCES CITED

Adam, D. P., and West, G. J., 1983, Temperature and precipitation estimates through the last glacial cycle from Clear Lake, California, pollen data: Science, v. 219, p. 168–170.

Adam, D. P., Sims, J. D., and Throckmorton, C. K., 1981, 130,000-yr continuous pollen record from Clear Lake, Lake County, California: Geology, v. 9, p. 373–377.

De Deckker, P., 1981, Taxonomy and ecological notes of some ostracods from Australian inland waters: Tranactions of the Royal Society of South Australia, v. 105, pt. 3, p. 91–138.

——, 1982, Australian aquatic habitats and biota; Their suitability for paleolimnological investigations: Transactions of the Royal Society of South Australia, v. 106, pt. 3, p. 145–153.

Delorme, L. D., 1969, Ostracodes as Quaternary paleoecological indicators: Canadian Journal of Earth Sciences, v. 6, p. 1471–1476.

Delorme, L. D., and Zoltai, S. C., 1984, Distribution of an Arctic ostracod fauna in space and time: Quaternary Research, v. 21, p. 65–73.

Delorme, L. D., Zoltai, S. C., and Kalas, L. L., 1977, Freshwater shelled invertebrate indicators of paleoclimate in northwestern Canada during late glacial times: Canadian Journal of Earth Sciences, v. 14, no. 9, p. 2029–2046.

Forester, R. M., 1983, Relationship of two lacustrine ostracode species to solute composition and salinity; Implications for paleohydrochemistry: Geology, v. 11, p. 435–438.

——, 1985, *Limnocythere bradburyi* n.sp.; A modern ostracode from central Mexico and a possible Quaternary paleoclimate indicator: Journal of Paleontology, v. 59, p. 8–20.

——, 1986, Determination of the dissolved anion composition of ancient lakes from fossil ostracodes: Geology, v. 14, p. 796–798.

Forester, R. M., and Brouwers, E. M., 1985, Hydrochemical parameters governing the occurrence of estuarine and marginal estuarine ostracodes; An example from south-central Alaska: Journal of Paleontology, v. 59, p. 344–369.

Goldman, C. R., and Wetzel, R. G., 1963, A study of the primary productivity of Clear Lake County, California: Ecology, v. 44, no. 2, p. 283–294.

Horne, A. J., and Goldman, C. R., 1972, Nitrogen fixation in Clear Lake California. I. Seasonal variation and the role of heterocysts: Limnology and Oceanography, v. 17, no. 5, p. 678–692.

Sims, J. D., Adam, D. P., and Rymer, M. J., 1981, Late Pleistocene stratigraphy and palynology of Clear Lake: U.S. Geological Survey Professional Paper 1141, p. 219–229.

Sims, J. D., Rymer, M. J., and Perkins, J. A., 1983, Late Quaternary stratigraphy and paleolimnology, Clear Lake, California: Geological Society of America Abstracts with Programs, v. 15, no. 5, p. 278.

Thompson, J. M., Goff, F. E., and Donnelly-Nolan, J. M., 1981a, Chemical analyses of waters from springs and wells in the Clear Lake volcanic area: U.S. Geological Survey Professional Paper 1141, p. 183–191.

Thompson, J. M., Sims, J. D., Yadav, S., and Rymer, M. J., 1981b, Chemical composition of water and gas from five nearshore subaqueous springs in Clear Lake: U.S. Geological Survey Professional Paper, 1141, p. 215–218.

Van Harten, D., 1986, Use of ostracodes to recognize downslope contamination in paleobathymetry and a preliminary reappraisal of paleodepth of the Prasas Marls (Pliocene) Crete, Greece: Geology, v. 14, p. 856–859.

Manuscript Accepted by the Society September 15, 1986

Printed in U.S.A.

Geological Society of America
Special Paper 214
1988

Correlations and age estimates of ash beds in late Quaternary sediments of Clear Lake, California

Andrei M. Sarna-Wojcicki
Charles E. Meyer
David P. Adam
John D. Sims
U.S. Geological Survey
345 Middlefield Road
Menlo Park, California 94025

ABSTRACT

We have identified ash beds in sediment cores of Clear Lake, California, by the chemistry of their volcanic glasses and petrography. These identifications enable us to correlate between cores, and to correlate three ash beds to several localities outside the Clear Lake basin where they have been isotopically dated or their ages estimated by stratigraphically bracketing dates. The three dated ash beds are ash bed 1 (Olema ash bed), estimated to be between 55 and 75 ka, in two deep cores CL-80-1 and CL-73-4, and two ash beds in core CL-80-1, ash bed 6 (Loleta ash bed), estimated to be between 0.30 and 0.39 Ma, and ash bed 7, estimated to be about 0.4 Ma.

Available age control from extrapolation of radiocarbon ages downward in the two cores, age constraints from correlations of ash beds, and etching of mafic minerals in ash beds at depths below about 118 m in core CL-80-1 suggest the following depositional histories for the two cores: in core CL-73-4, sedimentation appears to have been rapid (about 1 mm/yr) and continuous from about 120 ka to the present, corresponding to a depth interval from about 115 m to the present lake bottom. In the deeper core CL-80-1, sedimentation took place at a relatively moderate rate (0.4 mm/yr) from about 460 ka until sometime between about 300 and 140 ka, corresponding to a depth interval from about 168 to 118 m. Slow deposition or erosion took place sometime during the interval from about 300 to 140 ka, corresponding to an inferred hiatus at a depth of about 118 m. From 140 ka to the present, rapid sedimentation took place at about the same rate (about 0.8 mm/yr) as in core CL-73-4, corresponding to a depth interval from about 118 m to the present lake bottom.

The age of sediments in Clear Lake is not well constrained within the depth interval of about 70 to 130 m in the two deep cores, and the duration of the putative hiatus at about 118 m in the core CL-80-1 may be shorter than we propose. The presence of a hiatus at about 118 m depth in this core, however, is suggested by etching of mafic minerals in tephra layers below this level but not above, indicating that a period of subaerial exposure, or exposure above the groundwater table, had occurred for sediments below this level.

INTRODUCTION

We present results of analyses bearing on the correlation of ash beds in upper Quaternary sediments of Clear Lake, California. Robinson and others (this volume) present evidence for a time scale for Clear Lake sediments in core CL-73-4, based on radiocarbon ages corrected for contamination and on the downward extrapolation of sedimentation rates beyond radiocarbon age control. On the basis of this time scale, these authors have correlated variations in oak pollen abundances in core CL-73-4 with climatically controlled oxygen isotopic variations in the Pacific Ocean (Ninkovich and Shackleton, 1975). This correlation, if valid, has important implications for on-land studies of late Quaternary climatic variations in the western United States because it establishes a direct link between worldwide secular sea-level fluctuations inferred from oxygen-isotopic data and climatically controlled secular variations in a continental pollen record. Our study provides independent evidence for the ages of the sediments in Clear Lake as a check on the time scale proposed by Robinson and others.

We find that the corrected radiocarbon ages in core CL-73-4 at Clear Lake down to a depth of about 40 m (Robinson and others, this volume) and the downward extrapolation of sedimentation rates to the bottom of the hole are consistent with the calculated sedimentation rates and depositional history that we have derived from the correlation of ash beds in the Clear Lake cores, and with the correlations of these beds to other sites in the western conterminous United States and adjoining regions of the northeastern Pacific Ocean. Sedimentation in core CL-73-4 appears to have been continuous; sediments at the bottom of the core are estimated to be about 120,000 yr old at a depth of 115 m. The sedimentation rate in Clear Lake core CL-80-1 (Sims and others, this volume) appears similar to that of core CL-73-4 down to about 118 m, corresponding to an estimated age of about 140,000 yr old. A hiatus probably exists at this depth, with older sediments below ranging in age from about 300 ka at 118 m, to about 460 ka at a depth of 168 m.

Using the Loleta ash bed (Wagner, 1980) identified in cores and surface outcrops, we documented a direct correlation between Clear Lake core CL-80-1 and deep-ocean marine sediments in DSDP hole 36, located about 500 km west-northwest of Cape Mendocino, California, in the eastern Pacific Ocean. This ash bed, situated at about 130 m in core CL-80-1, may enable a direct comparison of climatic character between the marine and continental environments, should the data from sediments in the northeastern Pacific Ocean become available. Its age is estimated to be about 300 to 390 ka on the basis of dated age horizons in DSDP hole 36 and an assumption of a uniform sedimentation rate, and isotopic ages of near-source tephra (Sarna-Wojcicki and others, 1987; A. M. Sarna-Wojcicki and J. K. Nakata, unpublished data, 1987).

METHODS

Samples of tephra layers were obtained from Clear Lake cores CL-73-3, –4, –7, and CL-80-1 (Fig. 1). Volcanic glass shards and lithic grains with glassy groundmass were separated from ash samples using methods described by Sarna-Wojcicki (1976). Glass shards were analyzed by electron microprobe for nine major and minor oxides (SiO_2, Al_2O_3, Fe_2O_3, MgO, MnO, CaO, TiO_2, Na_2O, and K_2O), using methods described by Sarna-Wojcicki and others (1984). Glass shards of other analyzed tephra layers from approximately coeval deposits in the western conterminous United States and adjacent areas of the northeastern Pacific Ocean were analyzed previously or together with the Clear Lake samples. Chemical compositions of glass shards of these tephra layers were compared using a similarity coefficient to determine correlations (Borchardt and others, 1972; Sarna-Wojcicki, 1976; Sarna-Wojcicki and others, 1979, 1984). The similarity coefficient is given by:

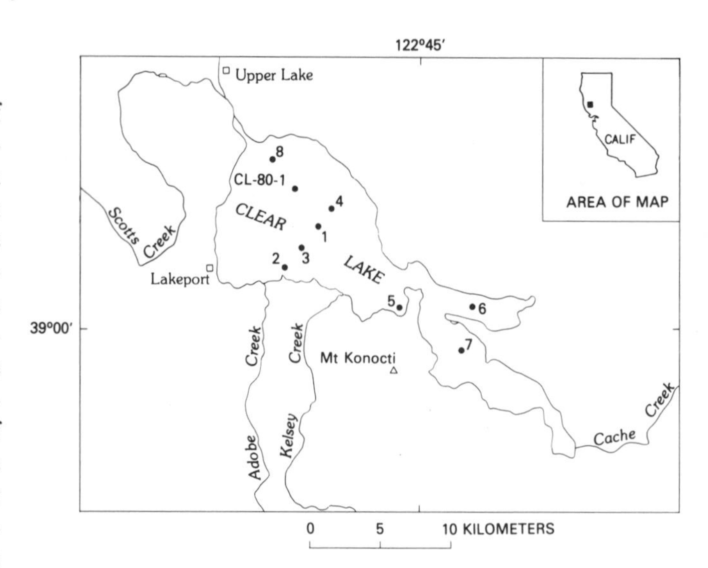

Figure 1. Locations of drilling sites in Clear Lake, California. Sites with single-digit numbers are from Clear Lake coring program (Sims, 1976) and have prefixes CL-73-. (After Sims and others, 1981a; modified from Sims, 1976.)

$$d(\mathrm{A, B}) = \frac{\sum_{i=1}^{n} R_i}{n}$$

where $d_{(A,B)} = d_{(B,A)}$ is the similarity coefficient for comparison between sample A and sample B, i is the element number, n is the number of elements, R_i is the $X_i\mathrm{A}/X_i\mathrm{B}$ if $X_i\mathrm{B} \geqslant X_i\mathrm{A}$ (otherwise $X_i\mathrm{B}/X_i\mathrm{A}$), $X_i\mathrm{A}$ is the concentration of element i in sample A, and $X_i\mathrm{B}$ is the concentration of element i in sample B. MgO, MnO, and TiO_2 were not used in matching glass chemistry of tephra samples because these oxides are present in low concentrations, near the limit of sensitivity of the electron probe, and would introduce scatter into calculations of the similarity coefficient.

In addition to the chemical composition of glass shards and lithic grains, we used other criteria such as the morphology of glass shards, the mineralogy, and particularly the stratigraphic position and sequence in determining correlations among ash beds. We estimated ages of correlated tephra beds at several localities from sedimentation rate curves and interpolation between dated horizons.

We analyzed nine ash beds from core CL-80-1, and one each from cores CL-73-3, -4, and -7 (Table 1). The correlations and age estimates presented here are considered preliminary, pending further analyses by more sensitive techniques. This is because the electron microprobe technique lacks the resolution of some other chemical methods of tephra analysis, such as x-ray fluorescence and instrumental neutron activation for minor and trace elements (Sarna-Wojcicki and others, 1984; Sarna-Wojcicki and others, 1985), and because the sedimentation rate curves we used to estimate ages of ash beds involve the assumption of constant sedimentation rate between dated horizons.

RESULTS OF ANALYSES

Cores CL-80-1 and CL-73-4

Samples of ash beds obtained from core CL-80-1 are of two types: (1) vitric, silicic tephra (samples 1, 4, 6, and 7), mostly composed of glass shards from ash beds produced by plinian eruptions; and (2) lithic, generally less silicic, tephra composed of microporphyritic or microlitic grains with a glassy or devitrified groundmass, produced either by phreatic or phreatomagmatic eruptions, maars or dome blowouts (ash beds 2, 3, 5, 8, and 9) (Fig. 2). Tephra of the latter type is generally much more heterogeneous than the vitric tephra and was much harder to analyze by electron microprobe. Some samples (ash beds 5, 8, and 9) contain both types of tephra.

Correlatives of ash beds 1, 6, and 7 have been identified from other localities in the western conterminous United States outside the Clear Lake basin, and from the northeastern Pacific Ocean. Ash beds 2, 3, 5, 8, and 9 were probably produced by small, local eruptions, most likely from the Clear Lake volcanic field. Mafic maar deposits are common in the vicinity of the southeastern shores of Clear Lake. The vitric components of ash beds 5, 8, and 9 are products of magmatic eruptions. These eruptions were also probably small and local, as indicated by the presence of coarse, pyroclastic material such as pumice lapilli, and no known correlative ash beds at localities outside of the Clear Lake basin.

None of the nine ash beds in Clear Lake core CL-80-1 has been dated directly. Three of these ash layers (1, 6, and 7), however, are overlain and underlain by dated ash beds or other kinds of age datums at sites outside the Clear Lake basin. We estimated ages of these ash beds using this age control and sedimentation rate curves. Ash beds of core CL-80-1 are described below in sequence, from the shallowest to the deepest.

Ash bed 1, core CL-80-1

Ash bed 1 (sample CL-0172), at a depth of 53.70 m in Clear Lake, is 7 cm thick, and correlates on the basis of its shard compositions with an ash (sample AH-4) 13 cm thick at a depth of 68.63 m in core CL-73-4 (Table 1; Fig. 2). This ash bed is composed primarily of clear, clean, angular bubble-wall glass shards, with subordinate tubular pumice shards containing spindle-shaped veiscles, some partly filled with liquid. Some glass shards have a light brown tint in transmitted light. The ash bed also contains shredded fragments of biotite and small clinopyroxene grains.

Identification of ash bed 1 in cores CL-80-1 and CL-73-4 provides a direct age tie between sediments of the two cores. The similarity coefficient (SC) between samples of this ash bed in the two cores, based on results of analysis by electron microprobe, is 0.996 (where the maximum value, 1.00, represents a perfect match). The eruptive source of this ash bed is not known.

Correlation and age estimates of ash bed 1

An estimate of the minimum age of ash bed 1. A minimum-age estimate of ash bed 1 can be obtained from ages above the ash bed in the Clear Lake cores, or from other sites where the presence of this ash bed can be documented. Ash bed 1 (sample AH-4) in core CL-73-4 lies 28 m below the oldest reliable radiocarbon date obtained in this core, 25 ka (Robinson and others, this volume). We have identified ash bed 1 in a recently drilled core at Tulelake, in northeastern Siskiyou County, just south of the California-Oregon border (sample T-199 in core T-2, Figs. 2, 4; SC = 0.99). At Tulelake, ash bed 1 (sample T-199) is situated at a depth of 13 m in the core, 6 m below the Trego Hot Springs bed (samples T-175, -176), dated at 23,400 ^{14}C yr b.p. (Davis, 1978; Verosub and others, 1980), confirming the very rough minimum age constraint on ash bed 1.

Ash bed 1, however, has also been identified in a section of upper Pleistocene lake beds exposed at Summer Lake, in south-central Oregon (sample DR-24; Davis, 1985; Figs. 2, 4; SC = 0.99). At Summer Lake, this ash lies about 8 m below the top of the section, and 4.4 m below the Pumice Castle ash bed (sample GS-57; Davis, 1985; Figs. 4, 5) erupted from the Crater Lake, Oregon, area (Bacon, 1983). A potassium-argon age on the latter ash bed at its type locality at Crater Lake, Oregon, was initially reported as about 52 ka (Bacon, 1983), but has been revised to 74 ka, in light of new data (C. A. Bacon, written communication, 1984). These data place a further minimum-age constraint on ash bed 1.

An estimate of the age of ash bed 1 from sedimentation rates. We can estimate the age of ash bed 1 by interpolating its age between age control points from a sedimentation rate curve derived for the Tulelake core, assuming that the sedimentation rate was constant. We derive the slope for this curve from the age and depth of the overlying Trego Hot Springs ash bed (samples T-175, -176, 8 m; 23,400 ^{14}C yr b.p.), and the age and depth of an underlying ash bed and associated magnetostratigraphic excursion that is most likely the Blake event (about 110 ka; H. Rieck, USGS, Flagstaff, Arizona, unpublished data, 1987). From these data, we estimate ash bed 1 (sample T-199, 13.01 m) to be

TABLE 1. ELECTRON MICROPROBE ANALYSES OF GLASS SHARDS OR MICROLITIC-MICROPORPHYRITIC LITHIC GRAINS OF ASH BEDS FROM CORES IN CLEAR LAKE, CALIFORNIA, AND COMPARATIVE COMPOSITIONS OF SOME OTHER TEPHRA LAYERS

Core	Sample	Depth 1/ (m)	SiO_2	Al_2O_3	Fe_2O_3 2/	MgO	MnO	CaO	TiO_2	Na_2O	K_2O	T_o 3/	T_r 4/	Size fraction (mesh) 5/
Ash Beds in Clear Lake Cores 73-3 and 73-7														
CL-73-3	CL-3-12	11.37	74.37*	13.79	1.93	0.46	0.05	1.91*	0.32	4.56	2.61*	94.97	100.00	(-100+200)
CL-73-7	CL-40	12.33	74.35*	13.83	1.92	0.44	0.04	1.88*	0.34	4.48	2.72*	94.56	100.00	"
Ash Beds in Clear Lake Cores CL-80-1, CL-73-4, and Correlative Ash Beds														
Ash Bed 1 and Correlative Ash Beds														
CL-80-1	CL-0172	53.7	75.29	13.36	1.72	0.09	0.04	0.48	0.14	4.72	4.16	94.65	100.00	(-100+200)
CL-73-4	AH-4(1)	68.5	75.25	13.37	1.75	0.11	0.06	0.48	0.16	4.68	4.14	94.42	100.00	"
"	AH-4(2)	"	75.06	13.56	1.77	0.08	0.05	0.51	0.18	4.56	4.03	93.30	100.01	"
"	AH-4(3)	"	75.01	13.63	1.76	0.11	0.06	0.50	0.17	4.48	4.05	92.63	100.01	"
Tulelake 2	T-199	13.0	75.22	13.26	1.74	0.10	0.07	0.49	0.21	4.83	4.08	94.80	100.00	"
---	DR-24	----	74.87	13.53	1.75	0.12	0.05	0.49	0.17	4.91	4.11	95.93	100.00	"
Ash Bed 2														
CL-80-1	CL-0315	93.0	73.62*	14.20*	2.10	0.50	0.04	2.22*	0.35	4.13*	2.84*	92.17	100.00	"
"	"	"	73.96	13.98	2.00	0.45	0.02	2.17	0.40	4.29	2.73	94.48	100.00	(-200+325)
Ash Bed 3														
CL-80-1	CL-0353	99.6	73.13	13.21	1.22	0.14	0.03	0.99*	0.17	3.18	4.93	94.64	100.00	"
Ash Bed 4														
CL-80-1	CL-0437	110.6	75.13	14.01	1.66*	0.30	0.05	1.37*	0.23	4.90	2.34*	93.94	99.99	(-100+200)
"	CL-0437	"	75.18	13.89	1.66*	0.29	0.05	1.39*	0.22	4.87	2.28*	94.01	100.00	(-200+325)
Ash Bed 5														
CL-80-1	CL-0491A	123.7	76.06	13.27	1.19*	0.10	0.04	0.94*	0.19	3.17	5.05*	96.20	100.01	(-100+200)
"	"	"	76.88	12.79	1.05*	0.08	0.02	0.77*	0.15	3.07	5.20*	95.77	100.01	(-200+325)
"	CL-0491B	"	77.60*	12.16	0.98	0.06*	0.01	0.72*	0.20	2.98	5.28*	95.66	99.99	"
"	"	"	69.01*	16.03	2.92	0.97*	0.06	3.31*	0.51	4.01	3.19*	93.13	100.01	(-100+200)
Ash Bed 6 and Correlative Ash Beds														
CL-80-1	CL-0519	130.2	74.69	14.06	1.96	0.12	0.06	0.75	0.14	5.05	3.18	95.33	100.01	(-100+200)
"	CL-0519	"	75.17	13.49	1.89	0.11	0.07	0.75	0.14	5.11	3.28	94.13	100.01	(-200+325)
DSDP-36	1-5(88)	8.0	74.11	14.01	1.98	0.12	0.05	0.74	0.14	5.42	3.14	94.44	100.00	"
"	1-5(57)	"	73.97	14.02	2.06	0.12	0.06	0.73	0.17	5.16	3.20	94.41	100.00	"
---	SM-ASH-4	----	75.45	13.88	2.00	0.11	0.06	0.73	0.17	5.16	3.20	94.14	100.00	(-100+200)
---	SM-ASH-5	----	74.49	13.81	1.99	0.10	0.06	0.74	0.17	5.26	3.13	94.37	100.00	"
---	CRANNELL	----	74.32	13.81	2.02	0.12	0.03	0.78	0.14	5.33	3.17	96.62	99.99	"
Ash Bed 7 and Correlative Ash Bed														
CL-80-1	CL-0590	148.0	76.82*	12.83	1.07	0.07	0.05	0.45	0.06	3.82	4.83	95.10	100.00	"
Tulelake 2	T-296	59.8	76.54*	13.06	1.06	0.06	0.05	0.48*	0.09	3.93	4.74	95.14	100.01	(-200+325)
"	T-296	"	76.97	12.63	1.10	0.07	0.04	0.48*	0.11	3.94	4.67	95.10	100.01	(-100+200)
Ash Bed 8														
CL-80-1	CL-0616	150.91	70.50*	16.26*	2.09*	0.22*	0.04	3.38*	0.42	3.59*	3.50*	97.09	100.00	"
Ash Beds and Pumice 9														
CL-80-1	CL-0661	168.07	74.35*	14.22	1.53*	0.28*	0.05	1.36*	0.21	3.46	4.55*	95.60	100.01	"
"	CL-0661	"	75.36*	13.51	1.31*	0.17*	0.03	1.18*	0.20	3.35	4.87*	95.06	99.98	(-200+325)
"	CL-0662	168.12	75.65	13.59	1.26	0.15	0.04	0.95*	0.16	3.50	4.69*	96.14	99.99	(-100+200)
"	CL-0663	168.25	75.28	13.57	1.31	0.15	0.03	0.94	0.19	3.54	4.82	94.83	100.00	"
Kelsey Tuff Member of the Kelseyville Formation, Clear Lake Basin														
---	KELSEY-1	---	57.36	18.29	6.63	4.39	0.13	8.44	0.76	3.08	1.03	100.11	100.00	"
Tuff below Kelsey Tuff Member in Kelseyville Formation														
---	6R008A	---	69.98*	15.75	2.74	0.77*	0.05	2.82*	0.52	3.82	3.54*	97.43	99.99	"
Tulelake T-64 Ash Bed and Two Thin Ash Zones in DSDP Hole 173														
Tulelake 1	T-64	17.0	70.57	14.69	3.16	0.72	0.08	2.34	0.65	4.87	2.92	96.00	100.00	"
DSDP-173	1-4(2)	4.3	70.35	14.83	3.21	0.70	0.04	2.28	0.62	4.84	2.91	98.39	99.99	"
DSDP-173	1-3(2)	3.7	70.17	14.97	3.26	0.68	0.08	2.24	0.66	4.78	2.92	98.71	100.01	"

Note: Concentrations in oxide weight percent, recalculated to 100 percent on a fluid-free basis. Values with asterisk indicate glass is not homogenous with respect to the oxide indicated. C. E. Meyer, U.S. Geological Survey, Menlo Park, California, analyst.

1 Depth to base of tephra layer.

2 Total iron calculated as Fe_2O_3.

3 Original total on analysis.

4 Total recalculated to 100 percent on fluid-free basis.

5 Nylon screen openings are 140 μm for the 100-mesh and 80 μm for 200-mesh screens.

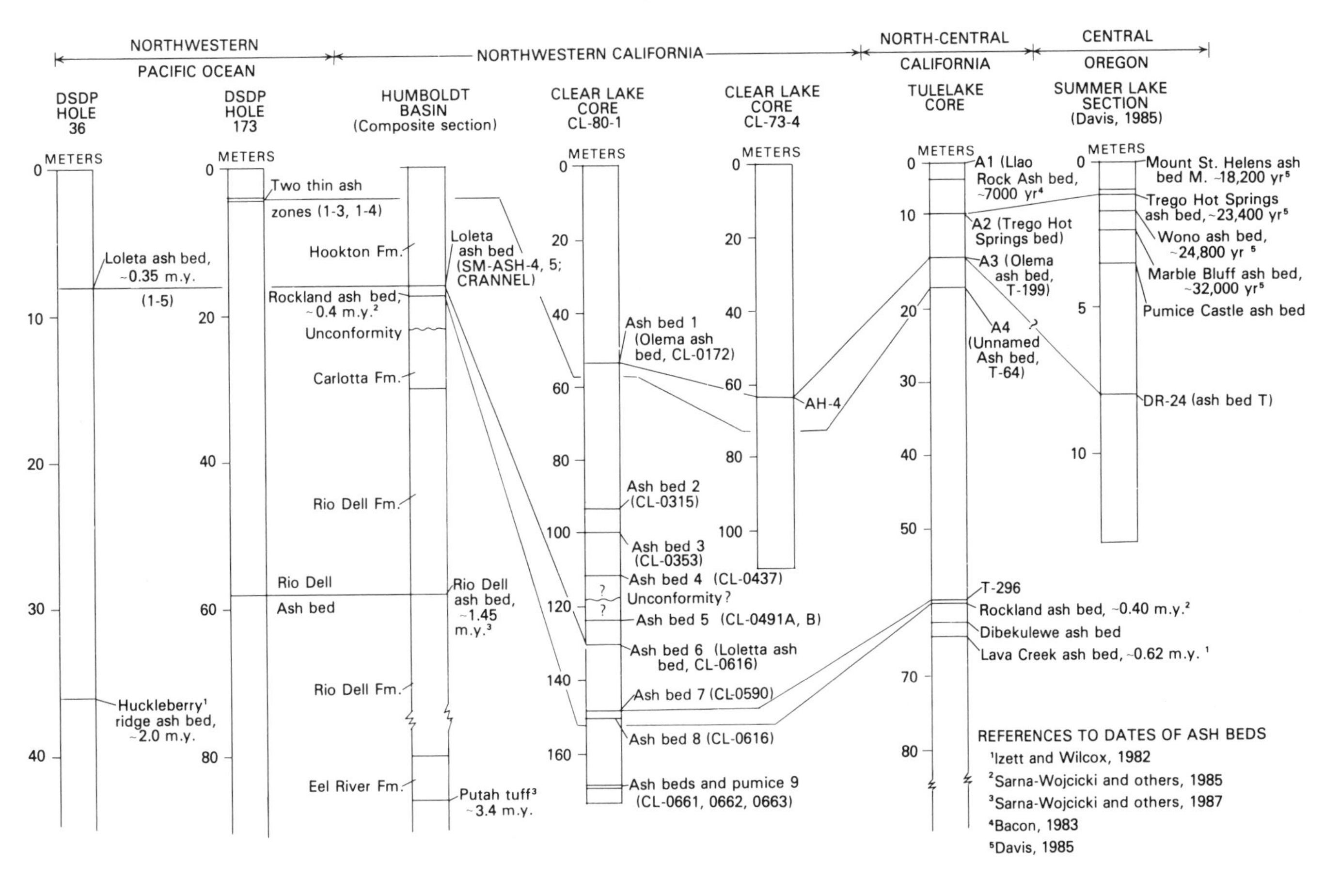

Figure 2. Correlation of ash beds in Clear Lake cores CL-80-1 and CL-73-4 with ash beds at other locations. Solid lines indicate correlation; broken lines indicate that a particular ash is absent or has not been found at a specific locality.

between about 55 and 75 ka. This estimate entails some uncertainty because of our constant sedimentation rate assumption between the age control points. The chronology of Robinson and others (this volume) for core CL-73-4 based on corrected radiocarbon ages, extrapolation of sedimentation rates downward in the core beyond the range of radiocarbon dating, and correlation of oak pollen frequencies with the oxygen-isotopic variations in the deep ocean, suggest an age of about 55 ka for this ash bed. This age is reasonably close, when statistical errors are taken into account, to an amino acid age estimate of 55 ± 13 ka on organic material obtained at a depth of 79.55 m, about 11 m below ash bed 1 (sample AH-4 in core CL-73-4; Blunt and others, 1981). The latter age is at the younger end of our estimate based on sedimentation rates.

An estimate of the maximum age of ash bed 1. An estimate of the maximum age of ash bed 1 can be obtained from tephra layers that underlie this bed. At Tulelake, ash bed 1 (sample T-199 at 13 m) overlies an unnamed ash bed (sample T-64, 17 m; Fig. 2), for which we have no direct age. We correlated the latter ash bed, however, with two thin, disseminated ash zones situated about 85 cm apart in DSDP hole 173 west of Cape Mendocino (ash zones at about 3.7 and 4.6 m; Figs. 2, 3; Table 1). Volcanic glass shards of the two ash zones are chemically indistinguishable on the basis of probe analysis (SC = 0.98), and both correlate equally well with the ash bed (sample T-64, at 17 m) at Tulelake. The two ash zones in DSDP hole 173 may represent a single ash bed that has been repeated in the core by drilling disturbance. We have estimated the ages of these two ash zones in DSDP core 173 by sedimentation rate curves, assuming that sedimentation rates between age control points in this core were constant. The ages of the two zones are estimated to be between about 140 and 180 ka (Fig. 4). This estimate entails considerable uncertainty, again due to an assumption of a constant sedimentation rate in DSDP hole 173, and does not provide a firm maximum age for ash bed 1. The best estimate of the age of ash bed 1 in our opinion is between 55 and 75 ka, or 65 ± 10 ka, based on data from the Tulelake core.

Ash bed 2, core CL-80-1

Ash bed 2 (sample CL-0315), 1 cm thick, is a crystal-lithic ash with a minor vitric component, situated at a depth of 92.97 m

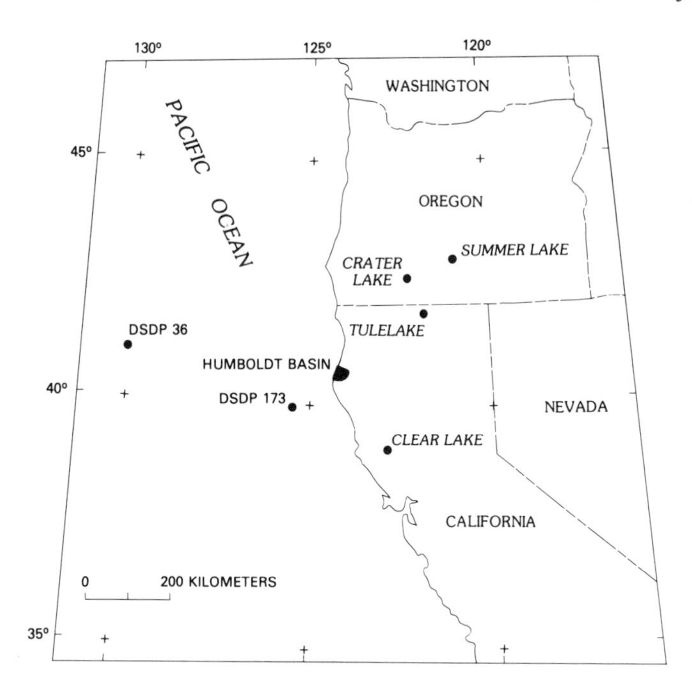

Figure 3. Locations of cores and stratigraphic sections where tephra layers are found that provide direct or indirect age control for ash beds in sediments of Clear Lake cores CL-73-4 and CL-80-1.

in core CL-80-1. It is composed of mostly microlitic grains with a glassy groundmass and large, euhedral crystals of hornblende, hypersthene, and feldspar. In addition, some vesiculated pumice shards are present. The source and the age of this ash bed are unknown. Ash bed 2 is also chemically similar to ash bed CL-40, situated in core 7 at a depth of 12.33 m (SC = 0.96). From the sedimentation rate curve for core CL-80-1 (Fig. 4), we have estimated this ash to be 110,000 yr old.

Ash bed 3, core CL-80-1

Ash bed 3 (sample CL-0353), 8 cm thick, situated at a depth of 99.68 m, is a lithic crystal ash. The lithic grains are mottled brownish-green under the petrographic microscope, and are composed of isotropic to low-birefringent material, possibly devitrified glass. The ash bed is chemically most similar to ash bed 5 lower down in the core (CL-0491A, Table 1; SC = 0.98), and to ash beds and pumice near the bottom of the core (ash beds 9, samples CL-0661 through CL-0663, similarity coefficients of 0.95 to 0.96), all probably erupted within the Clear Lake volcanic field. Thus, there is a recurrence of very similar grain and shard types vertically within core CL-80-1. With the exception of the chemically distinct ash bed 8, this recurrence is restricted to those ash beds (2, 3, 5, and 9) that we consider were produced by local eruptions. This recurrence may result, at least in part, from reworking of tephra layers within the Clear Lake basin. Using the sedimentation rate curve for core CL-80-1 (Fig. 4), we have estimated the age of this ash bed to be 120 ka.

Ash bed 4, core CL-80-1

Ash bed 4 (sample CL-0437), 5 cm thick, situated at a depth of 110.68 m in core CL-80-1, is a vitric ash containing both well-vesiculated pumice shards and solid, clear, angular bubble-wall glass shards. Glass of this ash bed, as determined by electron probe analysis, is chemically very similar to a very much younger ash bed derived from Mount St. Helens (ash bed We; Crandell and Mullineaux, 1978; SC = 0.98) and may have a Cascade source. A pumice-lapilli ash erupted from Mt. Jefferson in central Oregon is also similar to ash bed 4 (SC = 0.95). This ash is undated but is believed to be pre-Wisconsinan in age (W. E. Scott, oral communication, 1983). New data (J. O. Davis, Desert Research Institute, unpublished data, 1987) indicate that ash bed 4 is also chemically similar to an ash bed in Summer Lake that is undated, but which may be as old as 200 to 300 ka. We have estimated ash bed 4 to be 130,000 yr old on the basis of the sedimentation rate curve for core CL-80-1 (Fig. 4).

Ash bed 5, core CL-80-1

Ash bed 5 (sample CL-0491), 1.5 cm thick, is situated at a depth of 123.72 m, and is composed of microlitic and microporphyritric lithic grains having a vitric or devitrified groundmass. This bed also contains angular pumice lapilli about 7.5 to 10 mm in their longest dimension. Judging by size, these lapilli were probably derived from a nearby source. They could also be reworked from older tephra layers within the Clear Lake basin. Ash bed 5 contains augite and strongly pleochroic hypersthene, both strongly etched. Chemically, this ash bed consists of two types. The lapilli (Cl-0491A) are composed of silicic glass with high K_2O and low Fe_2O_3 and CaO concentrations. These lapilli are similar in composition to the presumably local ash beds 3 and 9 (SC = 0.98 to 0.96). The lithic grains in this ash bed (CL-0491B) are less silicic; have high Al_2O_3, Fe_2O_3, and CaO concentrations; and are low in K_2O. These grains are somewhat similar in composition to the tuff near the base of the Kelseyville Formation east of Clear Lake, within the Clear Lake drainage basin (sample 6R008A; Rymer, 1981; similarity coefficient, 0.94), but not sufficiently so to be considered correlative on the basis of chemical similarity alone. The composition of these more mafic grains in ash bed 5, however, is very heterogeneous (Table 1), and a correlation between these units cannot be precluded, as we may not have analyzed a sufficient number of grains to define the entire range of compositions or possible multiple compositional modes in both units. From the sedimentation rate curve for core CL-80-1 (Fig. 4), we estimate ash bed 5 to be 325,000 yr old (see below).

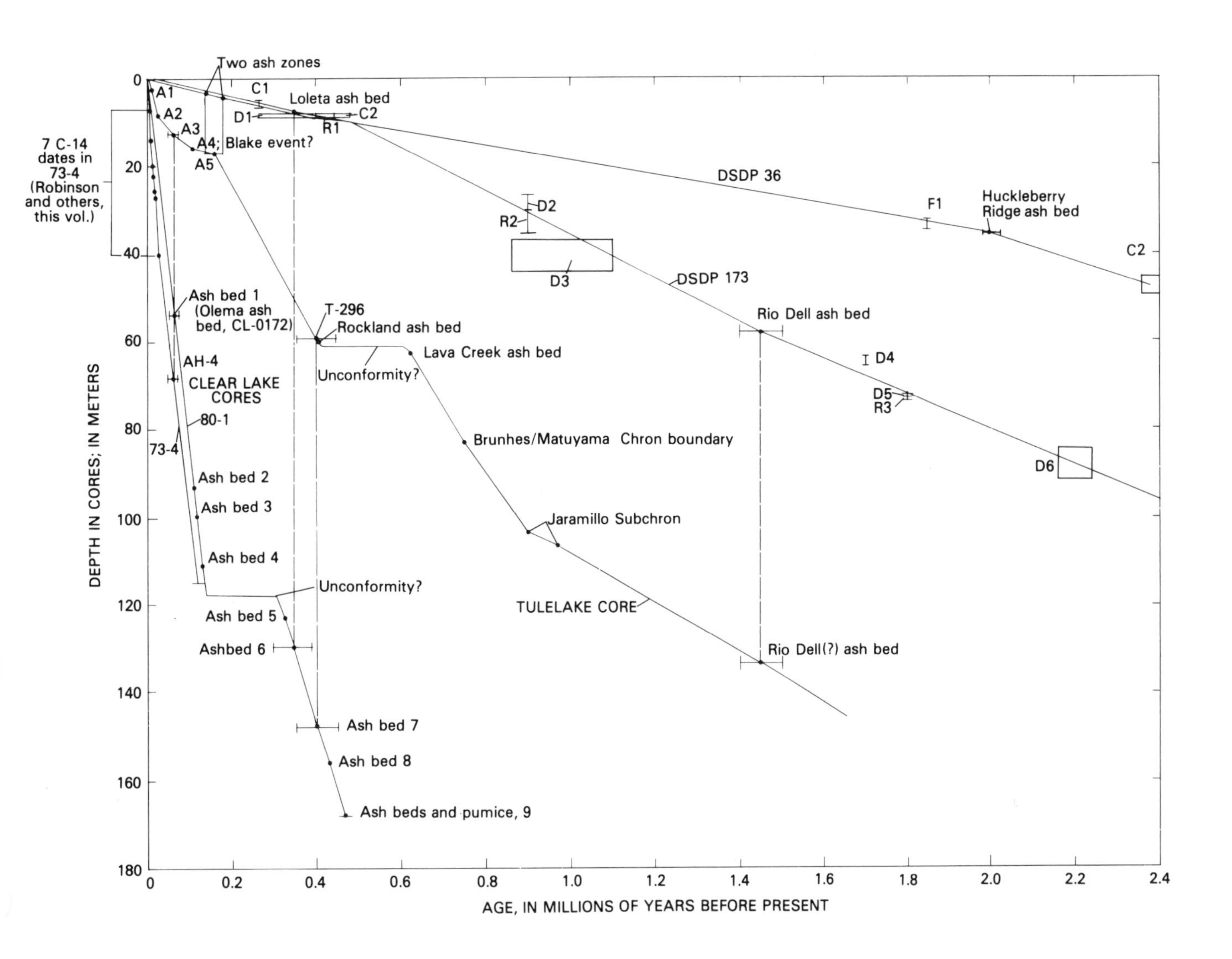

Figure 4. Sedimentation rate curves for five cores based on available age control from correlated, dated ash beds and biostratigraphy. A1–A5, ash beds in Tulelake core. A1—Llao Rock ash bed, ca. 7 ka (Bacon, 1983); A2—Trego Hot Springs ash bed, ca. 23 ka (Davis, 1985); A3—Olema ash bed, equivalent to ash bed 1 in Clear Lake cores CL-80-1 and ash bed AH-4 in CL-73-4, estimated to be 55,000 to 75,000 yr old; A4—unnamed ash bed associated with magnetostratigraphic excursion assigned to the Blake event, ca. 110 ka (H. Rieck, USGS, Flagstaff, Arizona, written communication, 1987); A5—unnamed ash bed, correlated with two ash zones in DSDP hole 173, estimated to be between 140 and 180 ka. C, D, F, and R, biostratigraphic age datums: C—calcareous nannofossil; D—diatom; F—foraminifer; R—radiolarian. Bar and box symbols indicate range of age and stratigraphic uncertainty of datums. For specific references to biostratigraphic datums, see Sarna-Wojcicki and others, 1987, Figures 5 and 6. Other datums (dots and bars) as labelled. Positions of the Brunhes/Matuyama Chron boundary and Jaramillo Subchron boundaries from H. Rieck, USGS, Flagstaff, Arizona, written communication, 1987.

Ash bed 6, core CL-80-1; the Loleta ash bed of Wagner (1980)

Ash bed 6 (sample CL-0519), situated at a depth of 130 m, occurs as thin partings, 0.25 to 0.5 mm thick. These partings are composed of clear, angular, solid bubble-wall glass shards, bubble-wall junction shards, and some tubular pumice shards with spindle-shaped vesicles. The sample contains abundant octohedra of authigenic pyrite, some of which coat the shards. Ash bed 6 correlates very well with the Loleta ash bed (Wagner, 1980), exposed at several localities in the Humboldt basin of Humboldt County, northwestern California (Fig. 3; SC = 0.99 to 0.98). We have recently identified this ash bed in samples of deep-ocean sediments of DSDP hole 36, about 500 km west-southwest of Cape Mendocino, California (Sarna-Wojcicki and others, 1987).

Age of ash bed 6 from stratigraphic relations and sedimentation-rate estimates

In the Humboldt basin, the Loleta ash bed closely overlies the Rockland ash bed, dated by Meyer and others (1980) at about 0.45 ± 0.08 Ma using the fission track method on zircons. Recent work (A. M. Sarna-Wojcicki and C. E. Meyer, unpub. data, 1984), indicates that the Rockland ash bed is younger, about 0.40 Ma. Thus, the age of the Loleta ash bed in the Humboldt basin is somewhat younger than 0.40 Ma (Fig. 2). We estimate the age of the Loleta ash bed in DSDP hole 36 (depth, 8 m), using the age of an identified ash bed lower in the hole (depth, 36 m; Figs. 2, 4); this is the Huckleberry Ridge ash bed, erupted from the Yellowstone area of northwestern Wyoming and eastern Idaho, and dated in the Midwest of the United States at about 2.0 Ma by the K-Ar and fission-track methods (Christiansen and Blank, 1972; Izett and Wilcox, 1982). We assume that sedimentation rates at DSDP site 36 were relatively constant during approximately the last 2 m.y. Site DSDP 36 lies in the deep ocean basin of the northeastern Pacific Ocean, 500 km west of northern California and about 200 km west of the Gorda Ridge. Using the age of the Huckleberry Ridge ash bed, we obtained an estimate of 0.39 ma for the Loleta ash bed (Sarna-Wojcicki and others, 1987). This estimate does not conflict with our recently determined age for the Rockland ash bed or with observed stratigraphic relations between the Rockland and Loleta ash beds in the Humboldt basin (Fig. 2). Recent K-Ar ages of the Bend Pumice, the proximal correlative of the Loleta ash bed near Bend, Oregon, average about 0.3 Ma (A. M. Sarna-Wojcicki and J. K. Nakata, unpublished data, 1987). Thus, the age of the Loleta ash bed is probably between about 0.3 and 0.39 Ma.

Ash bed 7, core CL-80-1

Ash bed 7 (sample CL-0590), situated at a depth of about 148.02 m, is composed of blocky, clear, bubble-wall junction shards with straight ribs, and minor, larger shards of tubular pumice with spindle-shaped vesicles. The vesicles are commonly 80 to 100 percent filled with fluid. Conical and cylindrical vesicles are also present in the pumiceous shards. The ash sample contains hypersthene and biotite, probably not cogenetic to the ash layer (no glass coats were observed on the grains and some grains were rounded). Ash bed 7 correlates well with an ash bed (sample T-296) situated at a depth of 59.76 m in Tulelake core 2 (SC = 0.98 to 0.97). This ash bed lies 11 cm above the Rockland ash bed (sample T-291) in this core. By this correlation, ash bed 7 is very close in age to the Rockland ash bed, or about 0.40 Ma (Fig. 2). These stratigraphic relations are in agreement with those we find in Clear Lake core CL-80-1 and elsewhere for ash bed 6.

Ash bed 8, core CL-80-1

Ash bed 8 (sample CL-0616), 1 cm thick at a depth of 156.22 m, is composed of microlitic and microporphyritic lithic grains having a glassy or devitrified groundmass similar to those of ash bed 5. In addition, crystals of etched augite are fairly abundant. This ash does not match chemically with any other samples we have analyzed. From the sedimentation rate curve for core CL-80-1, we estimate the age of this ash bed to be about 425 ka.

Ash beds and pumice 9, core CL-80-1

These ash beds and pumice (samples CL-0661 through -0663) occur in close stratigraphic proximity within a 20-cm interval (sample CL-0661 at 168.05 to 168.07 m, CL-0662 at 168.12 m, and CL-0663 at 168.18–168.25 m). They were probably formed by an eruptive episode of short duration within the Clear Lake volcanic field, judging by the coarseness of the pyroclasts within this interval.

Sample CL-0661 consists of both microlitic and microporphyritic grains having a glassy matrix, as well as some angular, clear, bubble-wall, and bubble-wall junction shards. Some tubular pumice shards are present, with spindle-shaped vesicles 10 to 15 percent filled with liquid. Rounded, blocky glass shards (perlitic obsidian?) with abundant microlites are also present. Abundant clinopyroxene and hornblende are also present; some hornblende (pleochroic brown to greenish-brown) is cogenetic to the lithic grains.

Sample CL-0662 is composed of coarse pumice lapilli, which are as much as 3 or 4 cm in their longest dimension. Fragments scraped from them are bubble-wall junction and pumiceous shards, both tubular with cylindrical- to spindle-shaped vesicles, and irregular pumice shards with spherical to ovoid vesicles. The sample contains cogenetic, pleochroic dark-green hornblende and a poorly pleochroic, nearly straight-extinguishing or straight-extinguishing pyroxene or amphibole. The crystals of the latter mineral are tabular to acicular or prismatic, with yellow body color, little or no pleochroism, and moderate birefringence. Some clinopyroxene, etched slightly at the terminations, is also present.

Sample CL-0663 is composed of tubular pumiceous glass shards and smaller, clear, bubble-wall junction shards with straight ribs. Spindle-shaped vesicles in the tubular pumice shards are partly filled with liquid; the smaller ones are more than 90 percent filled. This sample also contains hypersthene, brown pleochroic hornblende, and an elongate clinopyroxene. Samples CL-0662 and 0663 are similar to each other (SC = 0.99) and to other ash layers in core CL-80-1 of presumed local provenance (ash beds 3 and 5; SC = 0.96). From our sedimentation rate curve for core CL-80-1 (Fig. 4), we estimate the age of this tephra zone, and the bottom of the core, to be about 460 ka.

Correlation of cores CL-73-3 and CL-73-7

A 2.5-cm-thick ash bed (sample CL-3-12), situated at a depth of 11.37 m in core CL-73-3, is composed of frothy, highly vesiculated pumice shards, both the tubular variety with spindle-shaped, cylindrical, and conical vesicles, as well as irregular to equant pumice shards with spherical and ovoid vesicles. Small, brown, pleochroic hornblende and plagioclase feldspar grains are contained within the pumice shards and thus are probably cogenitic with the tephra layer. This ash bed correlates very well on the basis of glass shard chemistry with an ash bed (sample CL-40), situated at a depth of 12.33 m in core CL-73-7 (SC = 0.99). We have not found this ash bed in Clear Lake cores CL-73-4 or CL-80-1. Glass shard chemistry of this ash suggests a source in the Cascade Range. Pumice shards in sample CL-40 have the same characteristics as those in sample CL-3-12, except that the latter appear to be somewhat more frothy than the former. CL-40 contains the same type of small brown hornblende and feldspar grains as CL-3-12. An age of 17,660 ± 340 ^{14}C yr b.p. (I-7756; Sims and others, 1981a) was obtained on organic material stratigraphically just above this ash bed. The correlation documented here supersedes that suggested by Sims and others (1981a), which was based on stratigraphic position and physical appearance of the ash beds.

Correlation of core CL-73-4

The ash bed (sample AH-4) at 68.63 m in this core correlates with ash bed 1 of core CL-80-1, and is discussed above under the latter heading.

DISCUSSION

Age estimates of tephra layers in Clear Lake cores based on tephra correlations

The sedimentation rate curves and depositional history derived for upper Quaternary sediments of Clear Lake based on tephra correlations are consistent with corrected radiocarbon ages in core CL-73-4 and downward extrapolation of sedimentation rates calculated from the radiocarbon ages (Robinson and others, this volume). The age of ash bed 1 in Clear Lake core CL-80-1 (and consequently that of correlative samples of this ash bed, AH-4 in Clear Lake core CL-73-4, T-199 in Tulelake core 2, and DR-24 at Summer Lake, Oregon) is not well constrained by available age control and tephra correlations. Tephra correlations, stratigraphic relations, and age data indicate that ash bed 1 is definitely older than 25 ka. Estimates from a sedimentation rate curve at Tulelake suggest that the age of this ash bed is between 55 and 75 ka, which is in reasonable agreement with oak pollen data and correlation to the deep-sea record (Adam, this volume *b*; Robinson and others, this volume).

By our correlations and estimates, ash bed 6 in Clear Lake core CL-80-1 is about 0.30 to 0.39 Ma, and ash bed 7 is about 0.40 Ma (Figs. 2, 4).

Age and stratigraphic relations in the Clear Lake cores, the Tulelake cores, the two deep-ocean cores, and the composite stratigraphic section exposed in Humboldt basin support the correlations presented here (Fig. 2). Correlations of the three identified ash beds of Clear Lake core CL-80-1 also agree with biostratigraphic data and with age estimates derived for these ash beds using sedimentation rate curves and interpolation between dated horizons.

The age ranges of sediments in Clear Lake core CL-80-1 in the depth interval 54 to 130 m, and in core CL-73-4 in the interval 78 m to the bottom of the core, are not well constrained. The slope of the sedimentation rate curves down to the inferred unconformity in core CL-80-1, and down to the bottom of the core in core CL-73-4, are extrapolated from the estimated range and position of ash bed 1, the Olema ash bed (Fig. 4). If ash bed 4 in core CL-80-1 is indeed as old as 200,000 to 300,000 yr, then the duration of the inferred unconformity at a depth of about 118 m may be much shorter than we have shown, or it may not exist at all. Another line of evidence, however, which suggests that a hiatus exists in core CL-80-1 at about this depth, is obtained from the relative weathering of minerals in the tephra layers of core CL-80-1.

Hiatus in core CL-80-1

Mafic mineral phenocrysts (hypersthene, augite, and, to a lesser extent, hornblende) in Clear Lake ash bed 5 (CL-0491; depth, 123.71 m) and below are etched in some samples, but those of Clear Lake ash bed 4 (CL-0437; depth, 110.6 m; Fig. 2) and above in core CL-80-1 are not. Etching in mafic minerals occurs in the weathering zone above the water table; such etching is thus an indication of subaerial exposure. These observations suggest that a hiatus may exist between ash beds 4 and 5 in core CL-80-1 (Figs. 2, 4); this hiatus would be situated somewhere between 110.6 and 124 m in the core. We have arbitrarily drawn this hiatus at about 118 m in Figure 4.

Late Quaternary depositional history in the Clear Lake basin

The recent depositional history at the sites of cores CL-80-1 and CL-73-4 in Clear Lake, as inferred from the presence of the proposed hiatus and the correlations and ages of ash beds in core CL-80-1 and CL-73-4, indicate deposition of sediments in the basin for some undetermined period of time prior to about 460 ka, until sometime between about 300 and 140 ka. According to Hearn and others (this volume), the depositional basin of Clear Lake probably formed about 0.6 Ma, possibly as a consequence of subsidence following volcanic eruptions that occurred at about this time from the Clear Lake volcanic field. Our data on the correlation of tephra layers and the presence of etching in minerals below 118 m in hole CL-80-1 suggest that some erosion, or a period of no deposition, occurred sometime during the interval between about 300 and 140 ka. Rapid deposition began again about 140 ka and continued to the present (sedimentation rate curve for core CL-80-1 in Fig. 4). This chronology is still tentative, pending further analyses of tephra layers in sediments of Clear Lake and other geochronologic work.

Presently available data suggest that no hiatus is present in core CL-73-4, and that the sediments at the bottom of this core are about 120,000 yr old (Fig. 4). This is in reasonable agreement with estimates based on the extrapolation of sedimentation rates calculated from corrected radiocarbon ages obtained in the upper part of this core (Robinson and others, this volume).

Sedimentation rates in both cores CL-80-1 and CL-73-4 above a depth of about 115 m appear to be similar (0.8 and 1.0 mm/yr, respectively) but more rapid in the upper part of the latter, as indicated by the lower position of ash bed 1 (AH-4) in core CL-73-4 relative to its position in core CL-80-1 (Figs. 2, 4). Sedimentation rates in Clear Lake were more rapid than in any of the other sites at which the same ash beds were found, probably because rainfall and topographic relief have been higher in the Clear Lake drainage basin than at the other land sites. Age control obtained for the Clear Lake cores, combined with other observations presented here, suggest the following history for this basin: sudden blockage or subsidence of the Clear Lake basin, accompanied and followed by rapid filling; this was followed by non-deposition or erosion. A second episode of blockage or subsidence and rapid filling ensued, continuing to the present.

ACKNOWLEDGMENTS

We thank Julie Donnelly-Nolan and J. Alan Bartow for their contributions to this manuscript.

REFERENCES CITED

Backman, J., and Shackleton, J. J., 1983, Quantitative biochronology of Pliocene and early Pleistocene calcareous nannofossils from the Altantic, Indian, and Pacific Oceans: Marine Micropaleontology, v. 8, p. 141–170.

Bacon, C. R., 1983, Eruptive history of Mount Mazama and Crater Lake Caldera, Cascade Range, U.S.A.: Journal of Volcanology and Geothermal Research, special issue on arc volcanism, v. 18, p. 57–115.

Blunt, D. J., Kvenvolden, K. A., and Sims, J. D., 1981, Geochemistry of amino acids in sediments from Clear Lake, California: Geology, v. 9, p. 378–382.

Borchardt, G. A., Aruscavage, P. J., and Millard, H. T., Jr., 1972, Correlation of the Bishop ash, a Pleistocene marker bed, using instrumental neutron activation analysis: Journal of Sedimentary Petrology, v. 42, no. 2, p. 301–306.

Christiansen, R. L., and Blank, H. R., Jr., 1972, Volcanic stratigraphy of the Quaternary rhyolite plateau in Yellowstone National Park: U.S. Geological Survey Professional Paper 729-B, 18 p.

Crandell, D. R., and Mullineaux, D. R., 1978, Potential hazards from future eruptions of Mount St. Helens Volcano, Washington: U.S. Geological Survey Bulletin 1383-C, 26 p.

Davis, J. O., 1978, Tephrochronology of the Lake Lahontan area, Nevada and California: Nevada Archeological Research Paper 7, 137 p.

—— , 1985, Correlation of late Quaternary tephra layers in a long pluvial sequence near Summer Lake, Oregon: Quaternary Research, v. 23, p. 38–53.

Izett, G. A., and Wilcox, R. E., 1982, Map showing localities and inferred distributions of the Huckleberry Ridge, Mesa Falls, and Lava Creek ash beds (Pearlette Family ash beds) of Pliocene and Pleistocene age in the western United States and Southern Canada: U.S. Geological Survey Miscellaneous Investigations Series Map I-1325, 1 sheet, scale 1:400,000.

Keller, G., 1978, Late Neogene planktonic foraminiferal biostratigraphy and paleoceanography of the northeastern Pacific; Evidence from DSDP sites 173 and 310 at the north Pacific front: Journal of Foraminiferal Research, v. 8, no. 4, p. 332–349, pl. 1–5.

Meyer, C. E., Woodward, M. J., Sarna-Wojcicki, A. M., and Naeser, C. W., 1980, Zircon fission-track age of 0.45 million years on ash in the type section of the Merced Formation, west-central California: U.S. Geological Survey Open-File Report 80-1071, 6 p.

Ninkovich, D., and Shackleton, N. J., 1975, Distribution, stratigraphic position, and age of ash layer "L" in the Panama Basin region: Earth and Planetary Sciences Letters, v. 27, p. 20–34.

Rymer, M. J., 1981, Stratigraphic revision of the Cache Formation (Pliocene and Pleistocene), Lake County, California: U.S. Geological Survey Bulletin 1502-C, 35 p.

Sarna-Wojcicki, A. M., 1976, Correlation of late Cenozoic tuffs in the central Coast Ranges of California by means of trace- and minor-element chemistry: U.S. Geological Survey Professional Paper 972, 30 p.

Sarna-Wojcicki, A. M., Bowman, H. R., Meyer, C. E., Russell, P. C., Woodward, M. J., McCoy, G., Rowe, J. J., Jr., Baedecker, P. A., Asaro, F., and Michael, H., 1984, Chemical analyses, correlations, and ages of upper Pliocene and Pleistocene ash layers of east-central and southern California: U.S. Geological Survey Professional Paper 1293, 40 p.

Sarna-Wojcicki, A. M., Bowman, H. R., and Russell, P. C., 1979, Chemical correlation of some late Cenozoic tuffs of northern and central California by neutron activation analysis of glass and comparison with x-ray fluorescence analysis: U.S. Geological Survey Professional Paper 1147, 15 p.

Sarna-Wojcicki, A. M., Meyer, C. E., Bowman, H. R., Hall, N. T., Russell, P. C., Woodward, M. J., and Slate, J. L., 1985, Correlation of the Rockland ash bed, a 400,000-year-old stratigraphic marker in northern California and western Nevada, and implications for middle Pleistocene paleogeography of Central California: Quaternary Research, v. 23, p. 236–257.

Sarna-Wojcicki, A. M., Meyer, C. E., and Slate, J. L., 1983, The Lava Creek, Bishop, and Huckleberry Ridge ash beds in Pacific coast Quaternary marine deposits—on land and in deep-ocean cores: Geological Society of America Abstracts with Programs, v. 15, no. 5, p. 389.

Sarna-Wojcicki, A. M., Morrison, S. D., Meyer, C. E., and Hillhouse, J. W., 1987, Correlation of upper Cenozoic tephra layers between sediments of the western United States and eastern Pacific Ocean and comparison with biostratigraphic and magnetostratigraphic data: Geological Society of America Bulletin, v. 98, p. 207–223.

Sims, J. D., 1976, Paleolimnology of Clear Lake, California, U.S.A., *in* Horie, Shoji, ed., Paleolimnology of Lake Biwa and the Japanese Pleistocene: Kyoto, Japan, Kyoto University, v. 4, p. 658–702.

Sims, J. D., Adam, D. P., Rymer, M. J., 1981a, Late Pleistocene stratigraphy and palynology of Clear Lake, *in* McLaughlin, R. J., and Donnelly-Nolan, J. M., eds., Research in the Geysers–Clear Lake geothermal area, northern California: U.S. Geological Survey Professional Paper 1141, p. 219–230.

Sims, J. D., Rymer, M. J., and Perkins, J. A., 1981b, Description and preliminary interpretation of core CL-80-1, Clear Lake, Lake County, California: U.S. Geological Survey, Open-File Report 81-751, 175 p.

Verosub, K. L., Davis, J. O., and Valastro, S., Jr., 1980, A paleomagnetic record from Pyramid Lake, Nevada, and its implications for proposed geomagnetic excursions: Earth and Planetary Science Letters, v. 49, p. 141–148.

Wagner, J. R., 1980, Summary of regional stratigraphy and geologic structure, *in* Evaluation of the potential for resolving the geologic and seismic issues at the Humboldt Bay Power Plant Unit No. 3, prepared for the Pacific Gas and Electric Company, Appendix A: San Francisco, Woodward-Clyde Consultants, p. A1–A73.

Manuscript Accepted by the Society September 15, 1986

Printed in U.S.A.

Geological Society of America
Special Paper 214
1988

Radiocarbon content, sedimentation rates, and a time scale for core CL-73-4 from Clear Lake, California

Stephen W. Robinson
David P. Adam
John D. Sims
U.S. Geological Survey
345 Middlefield Road
Menlo Park, California 94025

ABSTRACT

Radiocarbon dating of disseminated organic matter from 10 horizons in Clear Lake core CL-73-4 produced apparent ages ranging from 4,230 to 32,650 B.P. Old carbon from lake sediments and springs beneath the lake adds about 4,200 years to the apparent age of each Holocene sample. A significant component of younger carbon—which cannot be completely removed by cleaning in sodium hydroxide solution—makes the dates older than 20,000 yr unacceptable.

The younger dates are corrected for the old carbon effect, calibrated to the dendrochronologic time scale, and then used to derive a sedimentation rate for the Holocene part of the core. Sediment accumulation is expressed as the mass in kilograms per square centimeter of noncombustible overburden above a given level in the core in order to compensate for variations in degree of sediment compaction and organic content.

The Holocene sedimentation rate, when applied to the entire core, yields an estimated core-bottom age of 133 ka. This independent evidence is consistent with the correlation of the high oak-pollen zone just above the base of the core with the last interglaciation. When the oak pollen maxima at the top and bottom of the core are equated with the Holocene and the last interglacial, the larger intervening fluctuations in the oak curve show a marked similarity to the climatic record preserved in deep ocean sediments. We correlate the major fluctuations of the oak pollen curve with their counterparts in the deep-sea record, and further refine the Clear Lake time scale by adjusting the age of the apparent Stage 5/4 boundary to 73 ka. The revised time scale indicates that sedimentation rates during the last glacial and interglacial were slightly higher and lower, respectively, than during the Holocene.

According to the revised time scale, interstadial events in the Clear Lake pollen record appear synchronous with prominent radiocarbon-dated interstadials in other areas, as well as with high sea stands dated by uranium-series disequilibrium methods.

INTRODUCTION

The pollen record from Clear Lake is one of the very few that spans the entire last glacial cycle. Previous dating of the last glacial and last interglacial deposits beneath Clear Lake (Sims, 1976; Adam and others, 1981; Sims and others, 1981) has depended primarily upon curve-matching between the oak pollen vs. depth curve (Fig. 1a) and the fluctuations in $\delta^{18}O$ recorded in deep-sea cores. This paper presents the results of radiocarbon measurements on CL-73-4 sediments that provide independent confirmation of the dating of the core described above.

Because woody material is very rare in CL-73-4, radiocarbon dating has been performed on disseminated organic matter, which amounted to approximately 9 to 25 percent of the dry

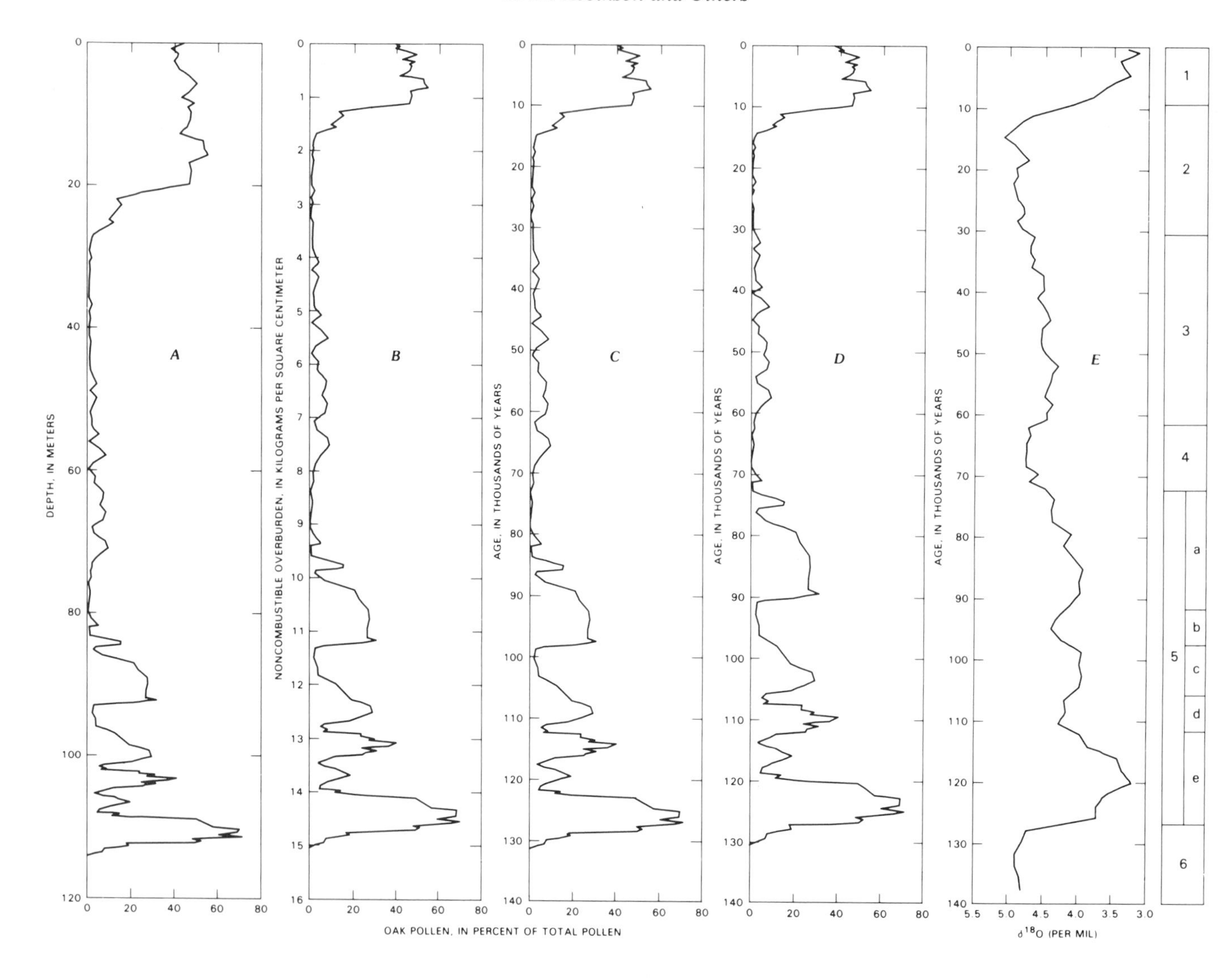

Figure 1. Clear Lake core CL-73-4 oak pollen expressed as percent of total pollen, plotted against: (a) depth in meters, (b) weight of noncombustible overburden, (c) age in calendar years as calculated by extrapolating sedimentation rates to core base, and (d) age in calendar years after adjusting age of boundary correlated with Stage 4/5 boundary in deep-sea record to 73,000 yr. Curve (e) shows ^{18}O record from deep-sea core V19-29 in the Panama Basin, after Ninkovitch and Shackleton (1975).

weight. Detailed data on the genesis of this organic matter is not available, but we assume that it originated primarily from biological productivity in the lake and secondarily from productivity in the drainage basin. We further assume that the biological production of the organic matter was essentially contemporaneous with the deposition of the sediment.

The standard or "DeVries" method of chemical decontamination of samples for radiocarbon analysis consists of leaching the sample sequentially in acidic, basic, and finally acidic solutions at elevated temperatures for periods of 8 to 16 hours each. The initial acidification removes any carbonate detritus present; the basic leach removes organic decomposition products known as humic and fulvic acids which, by their mobility, may introduce carbon from other horizons; the final acid leach neutralizes any residual hydroxide and prevents subsequent absorption of carbon dioxide from the air.

All samples received acid treatment to remove carbonates, but for some samples the basic (sodium hydroxide) leach was omitted. Samples from three depths (0.7, 58, and 71 m) were split and analyzed both with and without the basic leach. For the 0.7-m sample the sodium hydroxide-solution fraction was also analyzed. The carbon in the pretreated sediments was combusted to carbon dioxide in a quartz tube under oxygen flow at a temperature of about 900°C. The purified carbon dioxide was analyzed for radiocarbon by beta decay counting in an underground laboratory described elsewhere (Robinson 1979). The

TABLE 1. RESULTS OF RADIOCARBON ANALYSES OF ORGANIC CARBON IN CLEAR LAKE CORE CL-73-4 SEDIMENT*

USGS	A	B	C	D	E	F
-614A	0.68	0.029	Acid only	4,560 ± 90	—	
-614B	0.68	0.029	Base soluble	4,230 ± 110	—	—
-614C	0.68	0.029	Acid and base	4,475 ± 50	260	310-430†
-446	8.1	0.340	Acid and base	6,930 ± 90	2,715	2,750-3,000§
-445	15.0	0.775	Acid and base	10,400 ± 80	6,185	6,910-7,230§
-319	19.8	1.133	Acid only	14,120 ± 160	9,905	10,600**
-320	22.1	1.309	Acid only	14,460 ± 120	10,245	11,000**
-321	26.9	1.702	Acid only	16,600 ± 140	—	—
-444	41.0	3.245	Acid and base	26,350 ± 440	—	—
-193A	58.0	5.527	Acid only	23,500 ± 300	—	—
-193B	58.0	5.527	Acid and base	29,600 ± 580	—	—
-322A	71.0	7.532	Acid only	29,300 ± 390	—	—
-322B	71.0	7.532	Acid and base	32,650 ± 670	—	—
-447	80.1	9.217	Acid and base	32,200 ± 750	—	—

*Key to table columns identified by letter:
A = Core depth (m)
B = Noncombustible overburden (kg/cm^2)
C = Sample pretreatments
D = Conventional ^{14}C age
E = Analysis adjusted by subtraction of core-top age (geothermal effect)
F = Analysis calibrated to dendrochronologic (absolute) time scale

†From Stuiver (1982).

§From Klein and others (1982).

**From Stuiver (1971).

results of Core CL-73-4 radiocarbon dating are presented in Table 1.

The relationshp between depth and age in a lacustrine deposit is controlled by: (1) the rate of delivery of sediment to the core site, (2) diagenetic sediment compaction, and (3) variations in production of organic matter in the lake and its preservation in the sediment. To eliminate the effects of variables (2) and (3), one can replace the depth parameter by the mass of noncombustible sediment per unit area above a given depth (hereafter termed "noncombustible overburden"; Fig. 1b). The new scale will deviate from linearity against age only if the inorganic sediment influx changes with time. This technique has been applied by Stuiver (1970, 1971) in the investigation of atmospheric radiocarbon variation. He found that rates of accumulation of noncombustible sediment are remarkably uniform in lakes in equatorial Africa and Taiwan far from the geomorphic and climatic influence of ice sheets. However, Stuiver did observe a significant change in sedimentation rate in the one lake record that includes the major glacial/nonglacial transition at 14 to 15 ka.

The bulk dry density of the core CL-73-4 deposits varies fairly smoothly from about 0.5 g/cm^3 at the top to about 2.0 g/cm^2 at the core bottom (Fig. 2a). Of the original bulk density samples, 38 were selected for measurement of weight loss on combustion in order to construct a noncombustible overburden scale for the core. The sediment pellets (2–4 g) were powdered, and then combusted at 550°C for 35 min. The weight-loss results (Fig. 2b) appear to show a rapid diagenetic loss of organic matter in the upper part of the core, and below a depth of 30 m, a uniform organic content of 9 to 10 percent. The relationship between noncombustible overburden (X) and core depth (z) is given by the equation:

$$X(z) = \int_0^z \rho(z0)\,(1-F(z))\,dz,$$

where ρ is the bulk density of dry sediment and F is the fractional loss on ignition. For X we adopt units of kilograms of noncombustible sediment per square centimeter. It should be noted that this noncombustible sediment also contains minor contributions of

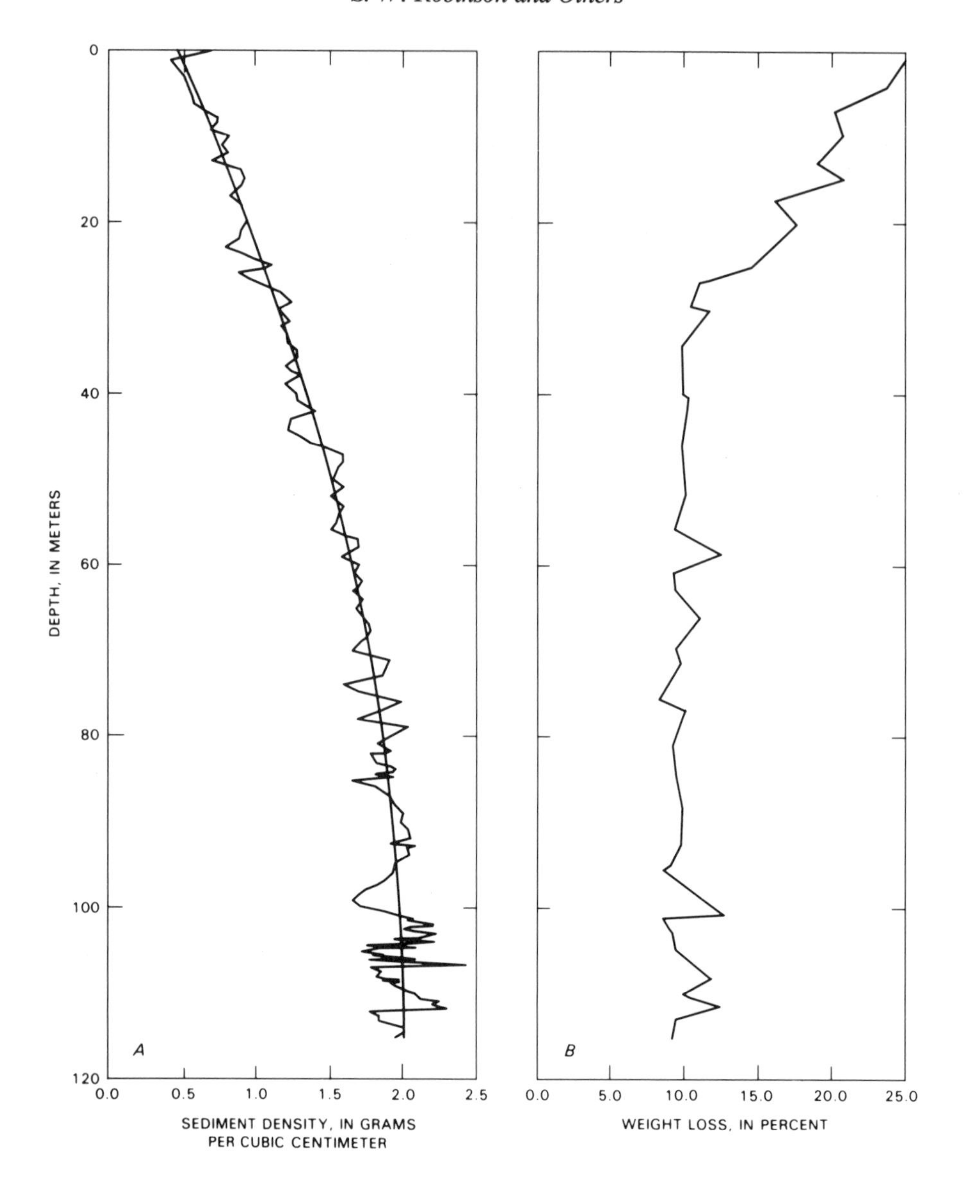

Figure 2. Plots of (a) sediment dry bulk density; and (b) percent weight loss on ignition, versus depth in meters for Clear Lake core CL-73-4.

biogenic silica and carbonates, which we assume are negligibly small.

Conventional radiocarbon ages are plotted versus the non-combustible overburden depth scale in Figure 3. Prominent features in the figure are the linear relationship between X and radiocarbon age for the upper four or five dates, the core-top age of about 4,200 radiocarbon yr, the deviation of the deeper ages from the straight line observed for the top four or five samples, and the older ages obtained for NaOH–treated sediment samples.

THE OLD CARBON EFFECT

Clear Lake has an alkalinity of about 2.5 Moles of carbon per m^3 and an average depth of about 8 m, so the mean residence time of dissolved inorganic carbon in the lake due to gas exchange (5–30 M/m^2-yr; Peng and Broecker, 1980) with atmospheric carbon dioxide is of the order of a few years. With such a short residence time, the radiocarbon content of the lake's dissolved inorganic carbon, and hence of the phytoplankton, would be expected to yield a conventional radiocarbon age very close to zero years. The unusually old core-top age is due to two factors: old inorganic carbon from the sediments, and even older carbon entering the lake from gaseous springs beneath the lake. The flux of inorganic carbon from the mineralization of organic matter in the sediments may be estimated from the variation of organic matter (weight loss) with depth, and its radiocarbon content is approximately 25 percent of that of modern carbon. The effect of this carbon source is significant in Clear Lake because of the ex-

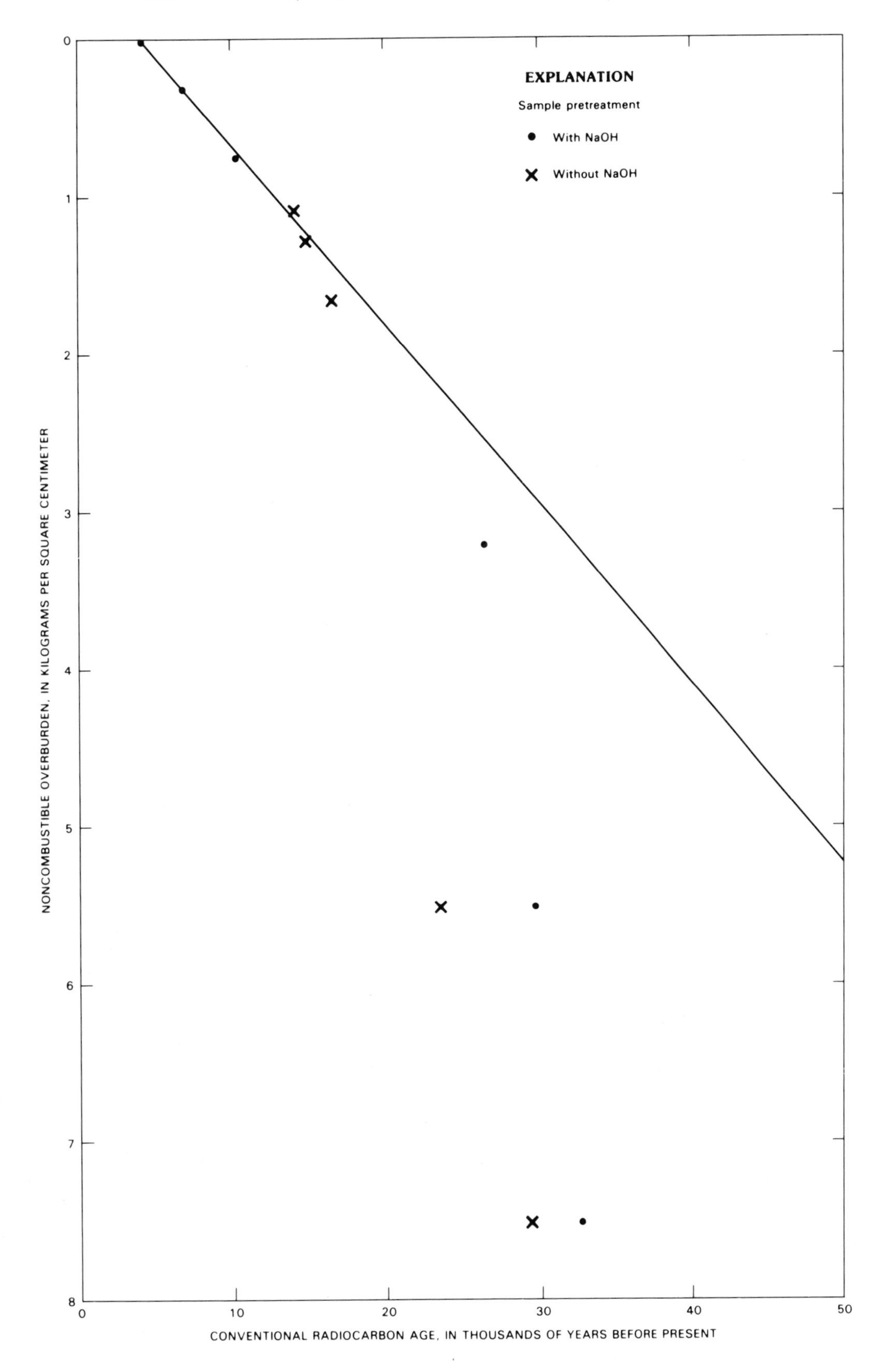

Figure 3. Conventional radiocarbon age plotted against weight of noncombustible overburden for Clear Lake core CL-73-4.

ceptionally high primary productivity in its waters. Clear Lake overlies a large number of gaseous springs (Sims and Rymer, 1976) that emit carbon dioxide gas and biocarbonate-rich waters (Thompson and others, 1981). The inorganic carbon contained in geothermal fluids is known to contain very little or no radiocarbon due to its apparent isolation from the atmosphere for many half-lives. For example, six measurements performed in the U.S. Geological Survey Menlo Park Radiocarbon Laboratory on geothermal fluids from Steamboat Springs, Nevada, Yellowstone Park, Wyoming, and Cerro Prieto, Mexico, have yielded radiocarbon contents ranging from 0.09 to 1.64 percent of modern activity, corresponding to apparent ages of from 56,000 to 33,000 yr BP.

A further test of the existence of an old carbon effect on radiocarbon in dissolved inorganic carbon in Clear Lake was performed on the shell of a modern pelecypod collected in 1922 at an unknown location in the lake. The apparent age of the shell is 6,300 ± 50 radiocarbon yr (USGS-644) (this specimen was made available by B. Roth, California Academy of Science, San Francisco).

There are two possible interpretations of the discrepancy between this result and the core-top age of 4,200 yr. One is that the mollusk lived closer to sources of geothermal carbon than the phytoplankton whose debris reached the lake bottom at the core site. The other is that the lake may be well mixed with respect to the geothermal input and the discrepancy is due to terrigenous organic detritus. This organic component would have zero-effective-age carbon and would reduce the apparent core-top age. If this is the case, it implies that about 24 percent of the core organic carbon is terrigenous. To use the radiocarbon measurements on the core as age indicators it is necessary to subtract this net 4,200-yr old carbon effect from each date, assuming a constant old carbon effect over the period for which the radiocarbon dates are used.

The upper five radiocarbon dates from core CL-73-4 (Fig. 4) indicate a constant sediment accumulation rate of 0.1224 ± 0.0014 kg/cm^2-1,000 yr (1,000 *conventional* radiocarbon yr), and a core-top age of 4215 ± 60 yr. This accumulation rate would give an age of 123 ± 1.4 ka (conventional radiocarbon years) for the core bottom.

Below a level of X = 1.309 kg/cm^2 the radiocarbon dates suggest a markedly greater rate of sediment accumulation (Fig. 3). In addition, the samples pretreated with NaOH lie closer to the line of constant accumulation rate than untreated samples. This younger age of the samples not subjected to NaOH leaching is an indication of contamination with younger carbon, since in an uncontaminated sample all the chemical species of organic carbon would have the same radiocarbon age. Unfortunately, there is no assurance that the NaOH treatment removed all the contaminant carbon, and the dates below the X = 1.309 kg/cm^2 level are discarded as unreliable. In the sample pairs USGS-193A and B and USGS-322 A and B, the excess activities of the splits not treated with NaOH are 0.9 and 2.8 percent, respectively, of modern activity level. The most likely cause of this contamination is the incorporation of modern carbon by bacterial activity during the 4 to 6 yr the sediment was stored in a wet condition before the radiocarbon analyses were performed. A similar case of contamination of a marine core has been documented by Geyh and others (1974). They believed that atmospheric carbon dioxide is absorbed by the core pore water and subsequently assimilated into the cells of sulfate-reducing bacteria. The ages of the younger samples will be less affected by the incorporation of the same amount of younger contamination because of their higher radiocarbon and carbon contents. Due to the excess radiocarbon in the atmosphere from nuclear weapons testing, the atmospheric CO_2 contamination could have contained up to 50 percent more radiocarbon than the 1950 modern standard.

Since the deeper part of the core appears to extend well beyond the range of the radiocarbon dating method, it is desirable to convert the chronologic data thus far obtained from conventional radiocarbon years (Stuiver and Polach, 1977) into absolute years so that they can be compared to time scales based on other dating methods. This is done using calibration data for radiocarbon in three-ring–dated wood over the last 8,000 yr (Stuiver 1982; Klein and others, 1982). For the two older samples an approximate extrapolation is used from the calibration of radiocarbon in Lake of the Clouds (Minnesota) sediments against a varve chronology that extends back to 9,200 conventional radiocarbon years (Stuiver, 1971). The conversion formula used here for dates in the range 8,000 to 13,000 conventional radiocarbon yr is $T_a = 1.0291\ T_c + 700$, where T_a is the estimated absolute age and T_c is the conventional radiocarbon age. The upper five dates converted to absolute years in this manner give a noncombustible sediment accumulation rate of 0.1144 ± 0.0019 kg/cm^2-ka, which projects back to a core bottom (15 kg/cm^2) age of 132.9 ± 2.2 ka.

The above calculations assume an approximately uniform sediment accumulation rate over the period of deposition of Clear Lake core CL-73-4. The reasonableness of this assumption can be tested by comparing the climatic signal of the oak pollen curve (Fig. 1) with other climatic records for which a time scale has been more firmly established.

First we compare (Fig. 1c versus 1e) with the ^{18}O record of benthic foraminifera in deep-sea cores dated by uranium-series disequilibrium, correlation with high sea levels, magnetostratigraphy, and comparison with the periodicities of the earth's orbital mechanics. If the oak pollen curve is compared with ^{18}O in core V19-29 from the eastern equatorial Pacific Ocean (Ninkovitch and Shackleton, 1975), we find good general agreement, with high oak pollen percentage correlating with low $^{18}O/^{16}O$ ratio (Interglacials). However, the last interglacial (isotope stage 5) seems too short in the Clear Lake core, while the Wisconsinan (isotope stages 2 through 4) seems too long. This is probably a consequence of our projecting the accumulation rate derived for the present interglacial stage through the last glacial. We can now refine our estimates of Clear Lake sediment accumulation rates by correlation with isotope stage boundaries 4/5 and 5/6, which are well dated (Hays and others, 1976; Kominz and others, 1979)

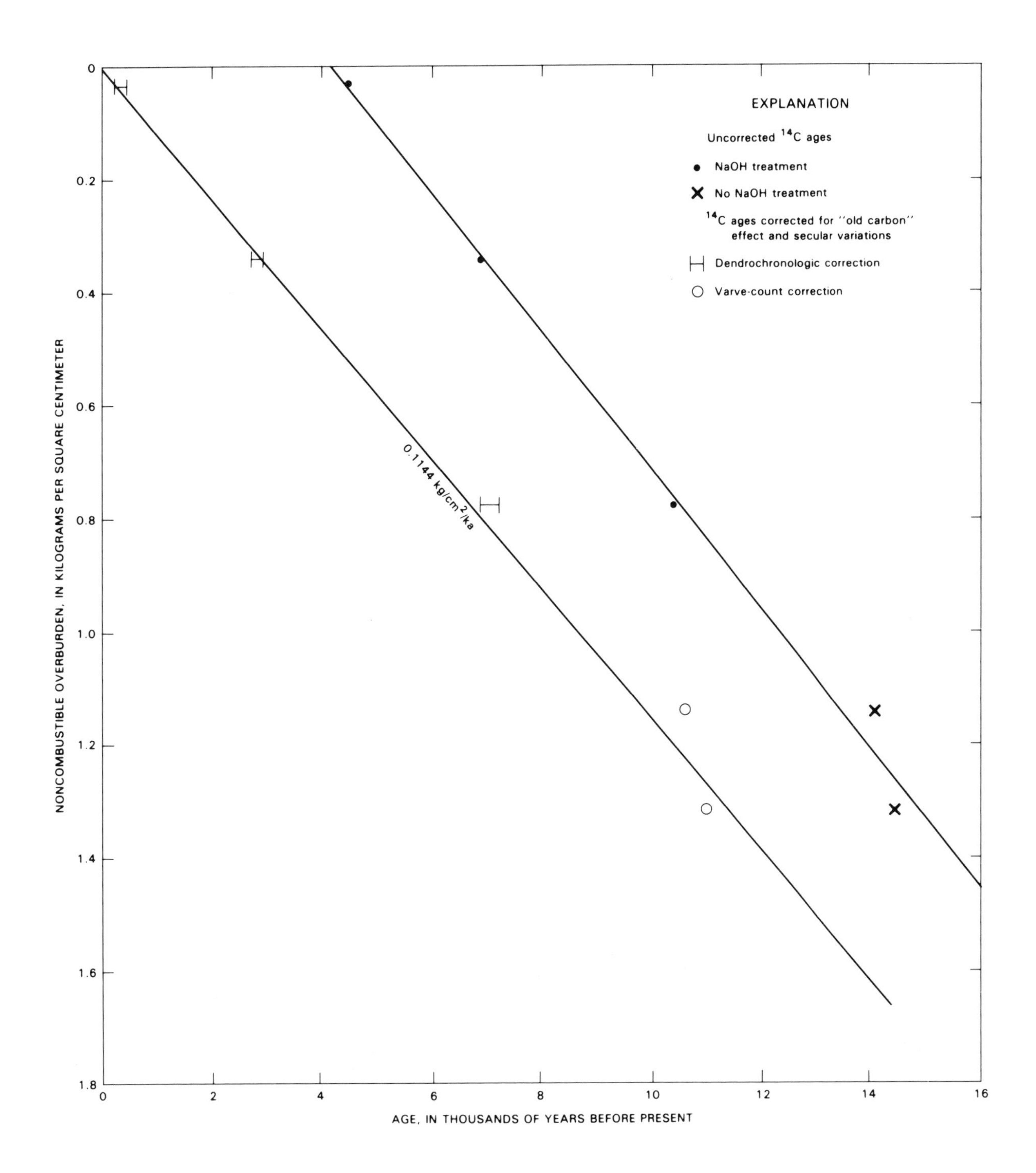

Figure 4. Conventional radiocarbon ages (upper curve) and the same data after correction for apparent reservoir age and secular variations in atmospheric radiocarbon activity (lower curve). Sedimentation rate shown by lower curve is used to estimate age of core base.

TABLE 2. COMPARISON OF BØLLING-DRYAS-ALLERØD COMPLEX AGES IN EUROPE AND IN CLEAR LAKE CORE CL-73-4*

	A	B	C
--------------	10.0 ka	11.0 ka	---------------------
Younger Dryas (cold)			11.4 ka (1.31 kg/cm²)
--------------	11.0 ka	12.0 ka	---------------------
Allerød (warm)			11.9 ka (1.37 kg/cm²)
--------------	11.8 ka	12.8 ka	---------------------
Older Dryas (cold)			13.2 ka (1.54 kg/cm²)
--------------	12.0 ka	13.0 ka	---------------------
Bølling (warm)			13.4 ka (1.58 kg/cm²)
--------------	12.5 ka	13.6 ka	---------------------

*Key to table columns identified by letter:
A = Conventional ^{14}C age of boundary in Europe (Andersen, 1981).
B = Estimated absolute ages of the boundaries shown in column A, using the formula ^{T}absolute = 1.029 ^{T}radiocarbon - 700 (see text).
C = Estimated absolute ages of pollen samples that appear to correlate with the European intervals, calculated from the rate of noncombustible sediment accumulation.

at 73 ka and 127 ka, respectively. Choosing the Clear Lake stage 4/5 boundary at X = 9.7225 kg/cm^2 and the stage 5/6 boundary at X = 14.7125 kg/cm^2 from the oak pollen curve (Fig. 1b), we obtain an accumulation rate of 0.1367 kg/cm^2-ka for the last glacial period (isotope stages 2 through 4) and a rate of 0.0924 kg/cm^2-ka for the last interglacial (isotope stage 5). A higher accumulation rate during the last glacial interval is not unreasonable in a regime of higher precipitation and greater geomorphic instability. It is not clear why the rate during the last interglacial should have been so much less than during the present one.

Other possibilities for comparison are the Bølling-Dryas-Allerød complex of climatic oscillations and the mid-Wisconsinan interstadials, which are well dated at from 10.0 to 12.5 ka (conventional radiocarbon years) in northwestern Europe (Andersen, 1981). Correlative events may be present in the Clear Lake oak pollen record between X = 1.3 and 1.65 kg/cm^2, just before the onset of the present interglacial. For purposes of comparison, the European radiocarbon ages are approximately converted to absolute ages by the same scheme used for USGS-319 and 320. The comparisons (Table 2, Fig. 5) show general agreement, although the gap in the core CL-73-4 pollen record between 1.37 and 1.54 kg/cm^2 probably included much of Allerød time.

In Table 3 the ages of events in the Clear Lake oak pollen record are compared with correlative interstadials radiocarbon dated at Grande Pile in Northeastern France (Woillard and Mook, 1982), and relative high sea-level stands dated by uranium-series disequilibrium in corals from raised terraces in the Huon Peninsula, New Guinea (Chappel and Veeh 1978; Bloom and others, 1974; Chappell, 1974). The conventional radiocarbon ages from Grande Pile have been converted to a half-life of 5,730 yr, but no further data are available to enable conversion to absolute years. Again, if the correlations are accepted, the agreement is excellent. It may be noted that, whereas at Grande Pile the Denekamp or "30 ka event" is the most prominent intersta-

TABLE 3. COMPARISON OF SOME DATES FOR CLIMATIC EVENTS IN ISOTOPE STAGE 3

Clear Lake, California, Minor Oak Peaks*	Grand Pile, France Interstadials† (^{14}C t1/2 = 5730)	High Sea Stands, Huon Peninsula, New Guinea, (uranium-series)§
32.3-34.3 ka (4.1-4.4 kg/cm^2)	"Denekamp" 29.8-31.9 ka	31.0 ± 2.5
39.6-42.8 ka (5.1-5.5 kg/cm^2)	"Hengelo" 41.2 ka	40.0 ± 3.0
48.6-57.6 ka (6.3-7.5 kg/cm^2)	"Moershoofd" 48-62 ka	ca. 60 ka
59 ka (Stage 4/3 boundary)	61 ka in marine record**	

*Based on Clear Lake time scale T = 11.36 + (X-1.3)/0.1347, T (ka), X (kg/cm^2).
†From Woillard and Mook (1982).
§From Chappell and Veeh (1978), Chappell (1974), and Bloom and others (1974).
**From Kominz and others (1979).

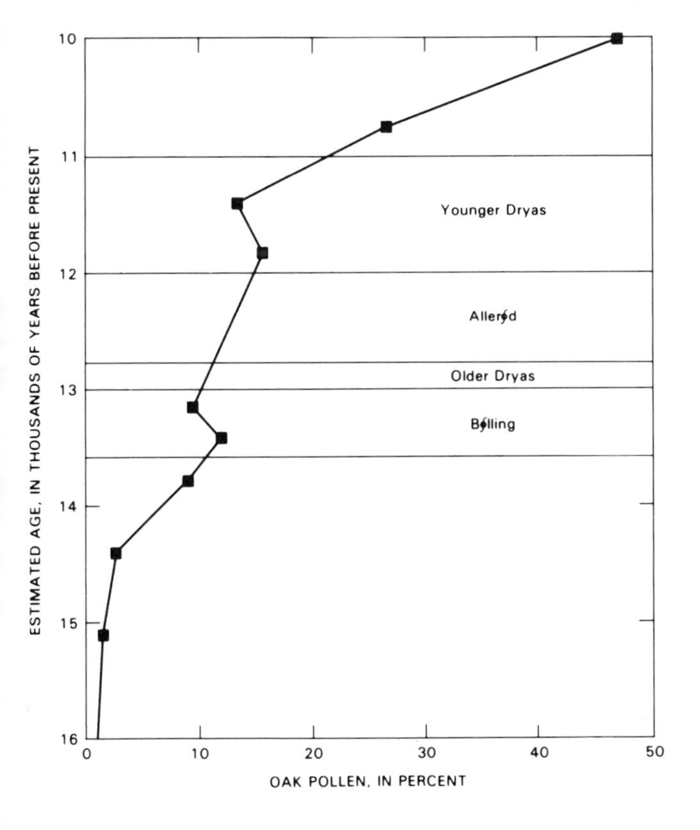

Figure 5. Oak pollen frequency vs. estimated age in calendar years for Clear Lake core CL-73-4 for the period from 10,000 to 16,000 BP. Ages of Younger Dryas, Allerød, Older Dryas, and Bølling intervals are shown on right; those ages have been converted from radiocarbon to calendar years using same method as was used for Clear Lake ages.

dial, it is the least pronounced in the Clear Lake oak curve. Another possible comparison is with the uranium-series dated high sea-level stands at 82, 105, and 125 ka (Bender and others, 1979), which correspond to values of X = 10.4, 12.5, and 14.4 kg/cm^2, in the Clear Lake core. Each corresponds to a time of high oak pollen percentage, respectively, but one major and one minor oak peak in stage 5 are left without an associated high sea stand.

The radiocarbon data presented here require extrapolation of sedimentation rates far beyond the range of our valid data in order to estimate the age of the base of core CL—73-4. The estimated age of 133,000 yr for the base of the core by extrapolation is nevertheless remarkably consistent with the age obtained by curve-matching of the oak pollen record with well-dated deep-sea oxygen-isotope records. We interpret the consistency of the results obtained by the two independent methods as confirmation of a last-interglacial age for the high-oak pollen zone just above the base of the core.

REFERENCES CITED

Adam, D. P., Sims, J. D., and Throckmorton, C. K., 1981, 130,000-Yr continuous pollen record from Clear Lake, Lake County, California: Geology, v. 9, no. 8, p. 373–377.

Andersen, B. J., 1981, Late Weichselian ice sheets in Eurasia and Greenland, *in* Denton, G. H. and Hughes, T. J., eds., The last great ice sheets: New York, Wiley-Interscience, p. 3–65.

Bender, M. L., Fairbanks, R. G., Taylor, F. W., Matthews, R. K., Goddard, J. G., and Broecker, W. S., 1979, Uranium-series dating of the Pleistocene reef tracts of Barbados, West Indies: Geological Society of America Bulletin, v. 90, p. 577–594.

Bloom, A. L., Broecker, W. S., Chappell, J.M.A., Matthews, R. K., and Mesolella, K. J., 1974, Quaternary sea level fluctuations on a tectonic coast: New $^{230}Th/^{234}U$ dates from the Huon Peninsula, New Guinea: Quaternary Research, v. 4, p. 185–205.

California Department of Water Resources, 1966, Clear Lake water quality investigation: Department of Water Resources Bulletin No. 143-2, 202 p.

Chappell, J., 1974, Geology of coral terraces, Huon Peninsula, New Guinea; A study of Quaternary tectonic movements and sea-level changes: Geological Society of America Bulletin, v. 85, p. 553–570.

Chappell, J., and Veeh, H. H., 1978, $^{230}Th/^{234}U$ age support of an interstadial sea level of -40m at 30,000 yr BP: Nature, v. 276, p. 602–604.

Geyh, M. A., Krumbein, W. E., and Kudrass, H.-R., 1974, Unreliable C-14 dating of long-stored deep-sea sediments due to bacterial activity: Marine Geology 14, p. M45–M50.

Hays, J. D., Imbrie, J., and Shackleton, N.J., 1976, Variations in the earth's orbit; Pacemaker of the ice ages: Science, v. 194, p. 1121–1132.

Klein, J., Lerman, J. C., Damon, P. E., and Ralph, E. K., 1982, Calibration of radiocarbon dates: Radiocarbon, v. 24, p. 103–150.

Kominz, M. A., Heath, G. R., Ku, T.-L., and Pisias, N. G., 1979, Brunhes time scales and the interpretation of climatic change: Earth and Planetary Science Letters, v. 45, p. 394–410.

Ninkovitch, D., and Shackleton, N. J., 1975, Distribution, stratigraphic position, and age of ash layer "L" in the Panama Basin region: Earth and Planetary Science Letters, v. 27, p. 20–34.

Peng, T.-H., and Broecker, W. S., 1980, Gas exchange rates for three closed-basin lakes: Limnology and Oceanography, v. 25, p. 789–796.

Robinson, S. W., 1979, Radiocarbon dating at the U.S. Geological Survey, Menlo Park, California, *in* Berger, R., and Suess, H. E., eds., Radiocarbon dating: Los Angeles, University of California Press, p. 268–273.

Ruddiman, W. F., and McIntyre, A., 1981, Oceanic mechanisms for amplification of the 23,000-year ice-volume cycle: Science, v. 212, p. 617–627.

Shackleton, N. J., 1977, The oxygen isotope stratigraphic record of the Late Pleistocene: Philosophical Transactions of the Royal Society of London, series B, v. 280, p. 169–182.

Sims, J. D., and Rymer, M. J., 1976, Map of gaseous springs and associated faults in Clear Lake, California: U.S. Geological Survey Miscellaneous Field Studies Map MF-721, scale 1:48,000.

Sims, J. D., Adam, D. P., and Rymer, M. J., 1981, Late Pleistocene stratigraphy and palynology of Clear lake, *in* McLaughlin, R. J., ed., Research in The Geysers–Clear Lake geothermal area, northern California: U.S. Geological Survey Professional Paper 1141, p. 219–230.

Stuiver, M., 1970, Long-term C-14 variations, *in* Olsson, I. U., ed., Radiocarbon variations and absolute chronology, Nobel Symposium 12: Stockholm, Almqvist and Wiksell, p. 197–213.

——, 1971, Evidence for the variation of atmospheric C^{14} content in the Late Quaternary, *in* Turekian, K. K., ed., The Late Cenozoic glacial ages: New Haven, Yale University Press, p. 57–70.

——, 1982, A high-precision calibration of the AD radiocarbon time scale: Radiocarbon, v. 24, p. 1–26.

Stuiver, M., and Polach, H. A., 1977, Reporting of ^{14}C data: Radiocarbon, v. 19, p. 355–363.

Thompson, J. M., Sims, J. D., Yadav, S., and Rymer, M. J., 1981, Chemical composition of water and gas from five nearshore subaqueous springs in Clear Lake: U.S. Geological Professional Paper 1141, p. 215–218.

Woillard, G. M., and Mook, W. G., 1982, Carbon-14 dates at Grande Pile; Correlation of land and sea chronologies: Science, v. 215, p. 159–161.

Printed in U.S.A.

Geological Society of America
Special Paper 214
1988

Amino-acid diagenesis and its implication for late Pleistocene lacustrine sediment, Clear Lake, California

*David J. Blunt**
Keith A. Kvenvolden
U.S. Geological Survey
345 Middlefield Road
Menlo Park, California 94025

ABSTRACT

The diagenesis of amino acids in sediments from Clear Lake core CL-80-1 is indicated by changes in amino acid concentrations, compositions, and stereochemistry. Concentrations of total amino acids decrease with depth, but the decrease is not systematic, possibly reflecting a nonuniformity in sedimentary and postdepositional processes affecting the amino acids. Ratios of neutral/acidic amino acids may indicate that the pH of interstitial water is slightly alkaline to slightly acidic and that the organic matter is well humified. Ratios of nonprotein/protein amino acids suggest that some changes in amino acids with depth result from microbial degradations. The extent of racemization of alanine increases with depth; the trends of these data may be explained, in part, by rapid sedimentation within the lake. Agreement between extents of alanine racemization for sediments from equivalent depths in two cores from the lake suggests that diagenetic temperatures are uniform within the sediments of the northern basin of Clear Lake.

INTRODUCTION

The diagenesis of a particular group of organic compounds—the α-amino carboxylic acids (amino acids)—can be used to study postdepositional chemical changes in lacustrine sediment. Amino acid geochemical reactions can provide information in three important areas: (1) relative rates of diagenesis of organic matter, (2) paleotemperature estimations, and (3) relative and absolute geochronology. Relative rates of organic matter diagenesis as reflected in changes of amino acid compositions have yielded information on the formation of humic acids and kerogen (Erdman, 1961; Abelson and Hare, 1970, 1971; Hoering, 1973). Diagenetic temperature estimations from amino acid reactions have provided limits on paleotemperature histories in lacustrine deposits from Lake Uinta (Jones and Vallentyne, 1960); Lake Bonneville (McCoy and Williams, 1983); Lake Ontario (Schroeder and Bada, 1978); and Clear Lake (Blunt and others, 1981). Relative and absolute geochronologic implications of using amino acid stereochemistry in lacustrine sediments have been reported for sapropelic clays (Dungworth and others, 1977; Schroeder and Bada, 1978; Tanaka and Handa, 1979; Dungworth, 1982; and Blunt and others, 1981, 1982), algal-tuff deposits (McCoy and Miller, 1979) and ostracodes (McCoy and Scott, 1980). This report represents continued research concerning amino acid diagenesis in lacustrine sapropelic sediment. Blunt and others (1981) reported preliminary findings on the geochemistry of amino acids in core CL-73-4 from Clear Lake, California, a core 115.2 m long and spanning about 130,000 yr. During the summer of 1980 a 177-m-long rotary-drilled core, designated CL-80-1, was taken 3 km northwest of core CL-73-4 in the northern basin of Clear Lake, California (Fig. 1). The lacustrine sediments in CL-80-1 record late Pleistocene sedimentary deposition over a time span of approximately 175,000 years, based on palynologic and paleomagnetic relationships (Sims and others, 1981b; Heusser and Sims, 1981; Liddicoat and others, 1981). This report examines the amino acid geochemistry in core CL-80-1 to interpret the processes of diagenesis and their application toward understanding paleotemperature and geochronology of the core.

*Present address: Exceltech, Inc., Fremont, California.

Analytical Procedure

Following recovery of core sections of CL-80-1, subsamples of sediment were placed in polyethylene bags and frozen according to the procedures described by Rymer and others (1981). At the U.S. Geological Survey Amino-Acid Biochemistry Laboratory in Palo Alto, approximately 2–3 g of thawed, wet sediment was placed in a 100-ml polypropylene tube and mixed with 40 ml of 6 N HCl. Low carbonate content in all samples was suggested by the lack of effervescence. The polypropylene tube with sediment and HCl mixture was placed in an oil bath at 110°C for 20 hours. The resultant hydrolysate was collected by filtering the sediment-hydrolysate mixture through glass-fiber filter paper into a 100-ml pear-shaped flask and dried by rotary vacuum-evaporation. The dried extract was taken up in pH 1 norleucine-containing solution and mixed into the top of an 80-ml column of freshly regenerated AG 50-X8 (H^+) cation-exchange resin. Amino acids were eluted with 2 N NH_4OH at a flow rate of 0.5 ml/min. The eluate was dried and split. One milliliter of pH 2.2 sodium citrate buffer was added to one portion for determination of amino-acid concentrations by ion-exchange chromatography (IEC). The other portion was acidified in 0.5 n HCl and dried to form an amino acid salt. The salt was esterfied with 0.2 ml (+)-2-butanol for 3 hours at 110°C, followed by acylation in a mixture of pentafluoropropionic anhydride and dichloromethane. The resulting amino-acid derivatives were resolved by gas chromatography (GC) on a 60-m × 0.08-cm stainless steel capillary column coated with UCON 75H-90,000. Amino acid D/L ratios were measured by peak heights on gas chromatograms.

The NH_4OH and HCl reagents were made by bubbling the respective gas (NH_3 or HCl) into water from a nonboiling quartz still. Analytic blanks taken through the entire procedure contain insignificant concentrations of amino acids.

RESULTS

Total concentrations of protein and nonprotein amino acids in whole sediment from core CL-80-1 range from 41.9 μM/g at 3.5 m to 32 nM/g at 155.50 m (Table 1), a change of about three orders of magnitude. Individual amino acids also show this trend (Table 1); for example, glycine decreases from a concentration of about 7.2 μM/g to 6 nM/g, and serine decreases from 2.9 μM/g to 2 nM/g. The trends of decreasing amino-acid concentration with depth are not systematic, as indicated by relatively larger concentrations at 74.60 and 123.00 m. The nonprotein amino acids β-alanine and γ-aminobutyric acid occur in highest concentration near the top of the core and together represent between 2 and 17 percent of the total amino acid concentration. Figure 2 shows the IEC record of amino acids from a sample taken at 61.35 m.

The amino-acid enantiomeric (D/L) ratios in the sediment from core CL-80-1 are reported in Table 2. Aspartic acid generally has the highest D/L ratios, and valine generally the lowest within each sample. At 3.5 m, the D/L ratio is 0.13 for aspartic acid and 0.03 for valine. At 155.50 m the D/L ratios of aspartic acid and valine increase to 0.40 and 0.06, respectively. The amino acid D/L ratios do not increase in a systematic manner with core depth, as evidenced by the relatively lower aspartic acid D/L ratios of 0.22 and 0.17 at sediment depths of 74.60 and 123.00 m. Other amino acid D/L ratios are variable to some extent downcore. Figure 3 is an example of the GC trace from which D/L ratios were obtained for amino acids in the sample from 61.35-m depth.

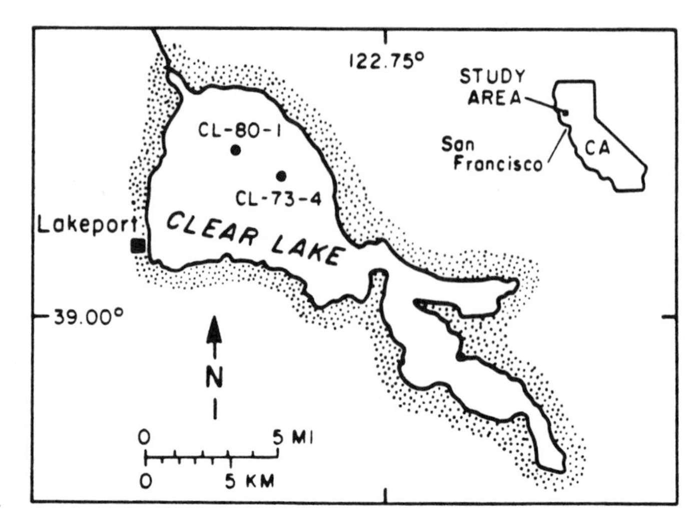

Figure 1. Location map of Clear Lake, California, showing site of cores CL-80-1 and CL-73-4.

The results for both amino acid concentrations and D/L ratios are generally in agreement with our previous investigation of amino acids in sediment at Clear Lake. Blunt and others (1981) reported that amino acid concentrations decrease rapidly as compared to slow increases for D/L ratios with depth in core CL-73-4. Absolute amino acid concentrations are greater in core CL-80-1, perhaps due in part to sampling bias and to changes in our analytic procedure. In this study, cations were collected on cation-exchange resin, whereas the previous method used precipitation and filtering to fractionate the amino acid residues from the hydrolysate mixture.

DISCUSSION

The amino acid data indicate that the postdepositional environmental conditions within Clear Lake sediments may be interpreted in terms of clay surface adsorption and processes of humification. Clay surfaces adsorb specific amino acids under certain pH conditions. For example, alkaline conditions tend to preserve acidic amino acids, whereas acidic conditions tend to preserve the basic amino acids. Stevenson and Cheng (1970) studied the adsorbance of specific amino acids on montmorillonite clay (Wyoming bentonite) and found the order of adsorption to be basic > neutral > acidic. Swain (1961) found that the

TABLE 1. AMINO ACID CONCENTRATIONS IN SEDIMENT FROM CORE CL-80-1, CLEAR LAKE, CALIFORNIA

Amino Acid	Depth (m)										
	3.50	19.66	37.00	61.35	74.60	83.25	103.25	123.00	133.00	142.00	155.50
Aspartic acid	41915	16382	3573	2043	4622	1433	1290	2820	733	527	32
Threonine	3290	1380	259	121	377	62	84	145	25	36	2
Serine	2860	871	176	50	187	56	30	157	19	16	2
Glutamic acid	3710	1490	241	96	238	87	50	180	42	31	3
Proline	2190	990	221	109	234	46	29	68	16	NR	NR
Glycine	7230	1610	404	107	174	157	60	352	89	55	6
Alanine	5560	2110	560	190	667	159	124	307	65	62	5
α-aminobutyric acid	65	17	6	3	14	3	3	9	2	2	NR
Valine	3300	1570	341	262	517	103	132	346	52	42	2
Isoleucine	2110	1670	208	306	705	176	220	214	92	66	1
Leucine	2560	1390	256	204	480	80	104	219	42	35	2
Tyrosine	721	349	57	39	107	11	19	32	3	3	NR
Phenylalanine	1000	528	105	108	206	34	53	115	15	13	NR
β-alanine	814	NR	109	33	60	82	48	93	36	25	NR
Ƴ-aminobutyric acid	172	128	177	83	89	97	130	141	86	51	1
Histidine	1200	355	142	83	133	75	56	104	44	37	NR
Lysine	869	788	188	190	304	106	125	147	68	51	7
Arginine	24	6	1	2	3	1	NR	NR	TR	NR	NR
Total nM/g	41915	16382	3573	2043	4622	1433	1290	2820	733	527	32

Notes:
NR = not reported; compound is below the limits of detection or masked by interfering compounds of unknown identity.
TR = trace amount; compound is within the limits of detection but less than 1 nM/g.

environmental conditions of amino acid preservation in lacustrine sediment could be measured by the abundance of neutral and acidic amino acids. He found that neutral/acidic amino acid ratios of about 6 represented well-humified lake deposits. In Clear Lake sediment, the neutral/acidic amino acid ratios range from about 4 to 12 (Table 3). This range of ratios suggests that the Clear Lake deposits are slightly alkaline to slightly acidic and well humified. Unfortunately, pH measurements of interstitial water were not conducted on the Clear Lake sediments. Weakly acidic conditions in Clear Lake sediments are also inferred by the general lack of preservation of calcareous ostracodes except in local cemented zones (Sims and others, 1981b).

Amino-acid Compositions

The relative abundance of individual amino acids can be used to characterize Clear Lake sediment. The relationship of relative molar percent for 10 amino acids in core CL-80-1 is shown in Figure 4. Relatively high percentages of aspartic acid occur at depths of 83.25 and 123.00 m. Lysine occurs in progressively greater relative abundances with depth. The relative abundance of glycine abruptly increases at 83.25 m. Blunt and others (1981) did not observe abrupt changes in relative amino acid abundances in core CL-73-4, probably because fewer samples were examined in core CL-73-4 and because core CL-80-1 extends to greater depth. Yet in the uppermost 70 m of both cores, the trends in the relative abundances of amino acids are similar. The specific factors responsible for the trends with depth of amino acid compositions in Clear Lake sediments are not known. However, the factors probably include those suggested by Kemp and Mudrochova (1973), Dungworth and others (1977), and Schroeder and Bada (1978) for amino acids in sediments of Lake Ontario. These authors attributed composition changes to adsorption, selective microbial usage, variable microbial abundances, and variable initial organic content.

Relationship of Glycine and Alanine

Systematic changes in the ratio of glycine (GLY) to alanine (ALA) have indicated diagenetic reactions in proteinaceous material in foraminifers (Bada and others, 1978), mollusks (Kvenvolden and others, 1980; Akiyama, 1980; Weiner and Lowenstein, 1980), and calcareous sediment (Kvenvolden and Blunt, 1981). The systematic decrease with depth of the GLY/ALA ratio in lacustrine sediment in Clear Lake Core CL-73-4 is attributed to the formation of alanine from serine during diagenesis (Blunt and others, 1981). In core CL-80-1, GLY/ALA ratios decrease from 1.4 at 3.50 m to 0.3 at 74.60 m, but increase in deeper samples to a GLY/ALA ratio of 1.4 to 133.00 m (Table 3). If the pH of the interstitial water affects the relative preservation of certain amino acids, then the variable abundances of glycine and alanine also tend to reflect chemical reactions occurring in the in situ environment. This is suggested by the decreasing glycine to alanine ratios as the neutral to acidic amino acid ratios increase with depth in CL-80-1.

Nonprotein Amino Acids

The extent of microbial diagenesis in core CL-80-1 can

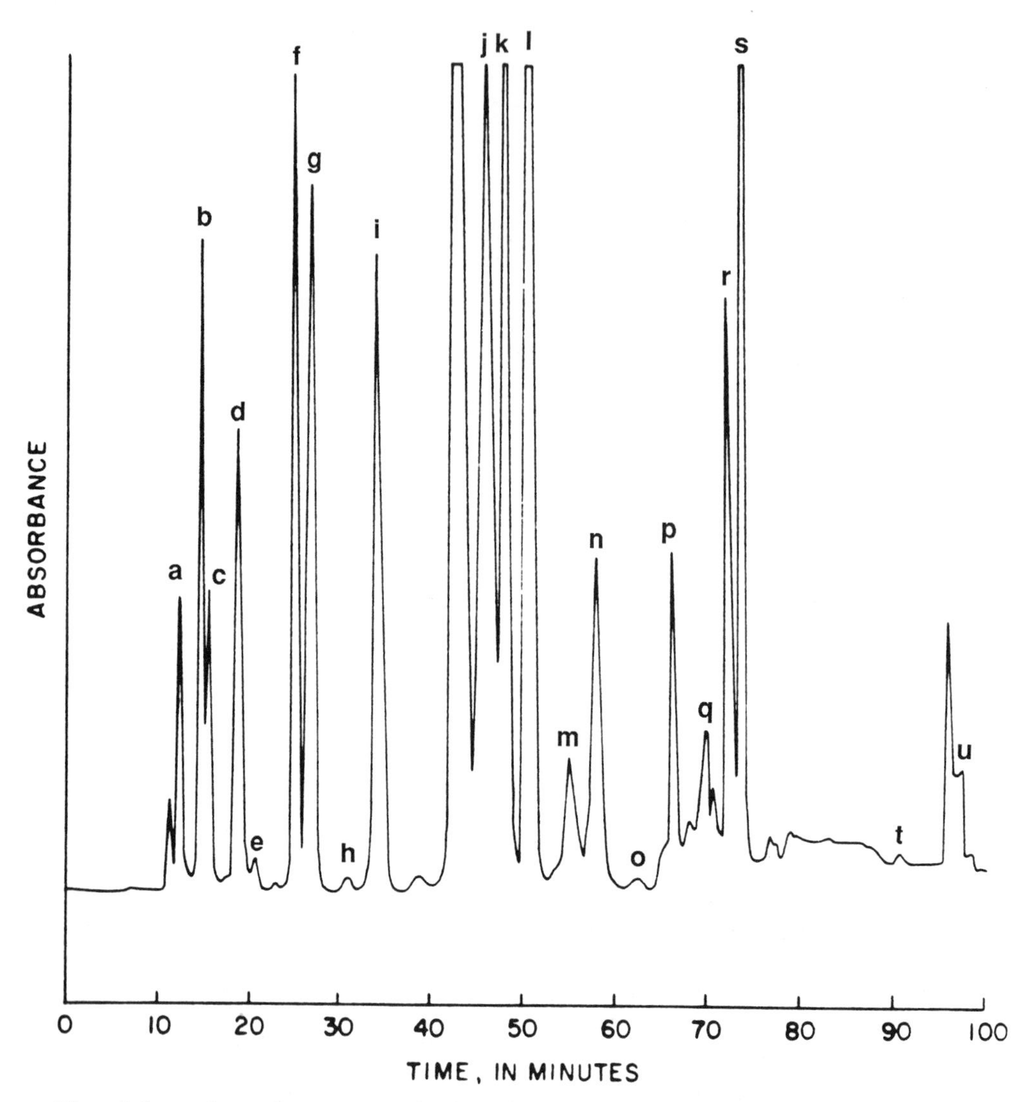

Figure 2. Ion-exchange chromatogram of amino acids in sediment at 61.35 m in core CL-80-1, showing 570-nm wavelength trace. This sample depth corresponds to a palynologic age of about 50,000 yr (Sims and others, 1981b). Unlabeled peaks may be unidentified amino acids, amino sugars, or dipeptides. Identification of known compounds are as follows: a, aspartic acid; b, threonine; c, serine; d, glutamic acid; e, proline; f, glycine; g, alanine; h, α-aminobutyric acid; i, valine; j, isoleucine; k, leucine; l, norleucine; m, tyrosine; n, phenylalanine; o, β-alanine; p, γ-aminobutyric acid; q, ammonia; r, histidine; s, lysine; t, arginine; u, NaOH regeneration.

perhaps be measured by the relative abundance of nonprotein amino acids. Nonprotein amino acids are those compounds not normally found in the proteins of living organisms. We report the relative abundance of α-aminobutyric acid, β-alanine and γ-aminobutyric acid—three nonprotein amino acids (Table 1). Most amino acids undergo natural oxidative deamination, but under anaerobic conditions microorganisms in sediments and soils may selectively decarboxylate glutamic acid and aspartic acid, yielding the nonprotein compounds γ-aminobutyric acid and β-alanine, respectively (Bada, 1971; Brown and others, 1972; Aizenshtat and others, 1973; Ithara, 1973; Kemp and Mudrochova, 1973). Nonprotein amino acids are also included in the N-acetyglucosamine polymer of bacterial cell walls and may point to bacteria as an important source of amino acids in Florida peat deposits (Casagrande and Given, 1980). Nonprotein amino acids might be useful to characterize bacterial effects in Clear Lake sediment.

The relationship of the ratio of γ-aminobutyric acid to glutamic acid (γ-ABA/GLU) and β-alanine to aspartic acid (β-ALA/ASP) in sediments of core CL-80-1 is shown in Figure 5. Although the ratios generally increase with depth, these changes are not systematic. Relatively low values in sediments from

TABLE 2. AMINO ACID D/L RATIOS IN SEDIMENT FROM CORE CL-80-1, CLEAR LAKE, CALIFORNIA

Amino Acid	Depth (m) 3.50	19.66	37.00	61.35	74.60	83.25	103.25	123.00	133.00	142.00	155.50
Aspartic acid	0.13	0.20	0.30	0.29	0.22	0.32	0.37	0.17	0.37	0.30	0.40
Alanine	0.08	0.09	0.13	0.17	0.13	0.19	0.18	0.19	0.23	0.20	0.19
Glutamic acid	0.07	0.10	0.13	0.14	0.14	0.16	0.16	0.19	0.22	0.19	0.13
Proline	0.03	0.05	0.08	0.11	0.14	0.17	.017	0.17	0.23	0.15	ND
Phenylalanine	0.05	0.05	0.09	0.09	0.09	0.14	0.14	0.10	0.16	0.15	ND
Leucine	0.04	0.06	0.05	0.07	0.05	0.08	0.09	0.10	ND	0.15	0.19
Valine	0.03	0.04	0.04	0.04	0.04	0.05	0.04	0.04	0.07	0.05	0.06

Note: ND = not determined; measurement of D- and L-amino acid was not determined due to interfering compound of unknown.

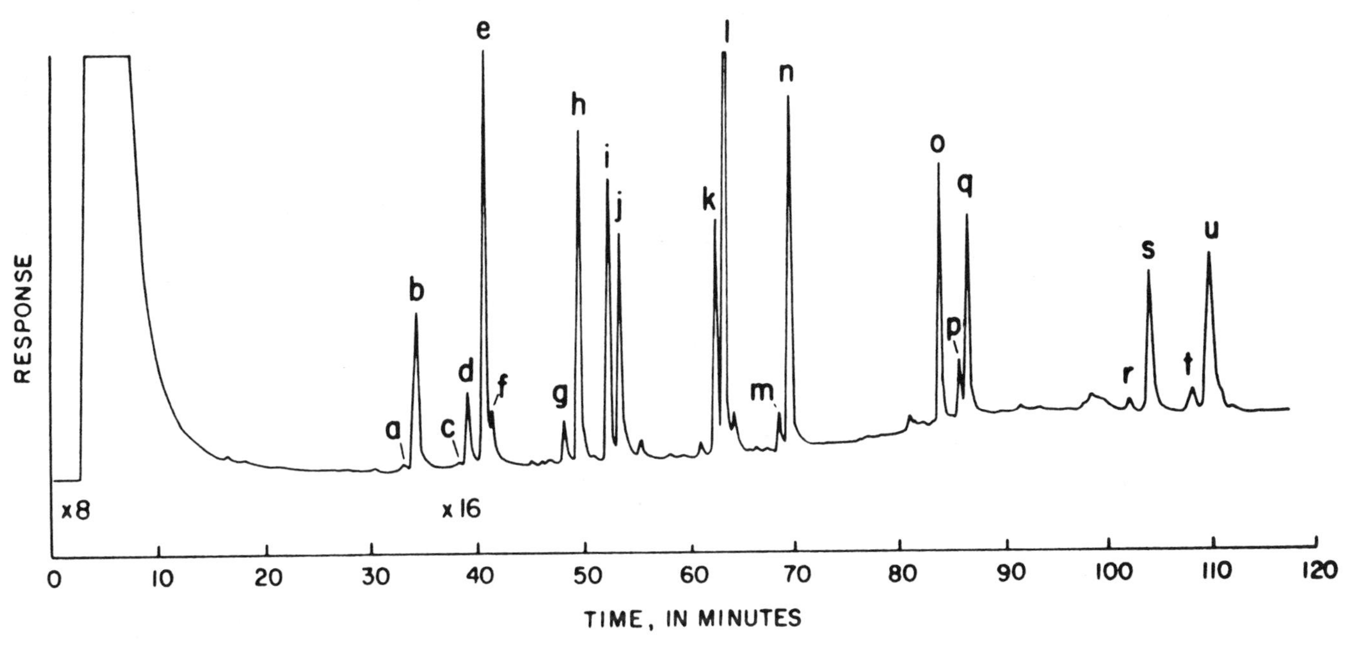

Figure 3. Gas chromatogram of amino acid enantiomeric relationships in sediment at 61.35 m in core CL-80-1. This sample depth approximates an age of about 50,000 yr (Sims and others, 1981b). Identification of known peaks are as follows: a, D-valine; b, L-valine; c, D-alloisoleucine; d, D-alanine; e, L-alanine; f, L-isoleucine; g, D-leucine; h, L-leucine; i, D-norleucine; j, L-norleucine; k, β-alanine; l, glycine; m, D-proline; n, L-proline; o, γ-aminobutyric acid; p, D-aspartic acid; q, L-aspartic acid; r, D-phenylalanine; s, L-phenylalanine; t, D-glutamic acid; u, L-glutamic acid.

depths of 74.60, 123.00, and 155.50 m suggest reduced microbial effects at these depths, whereas the relatively high values at 103.25 and 133.00 m suggest relatively greater microbial degradation at these depths.

Amino-acid Geochronology

The fundamental basis for amino-acid geochronologic studies relies on the fact that almost all living animals and plants have proteins which contain amino acids configured as L-stereoisomers. As diagenesis occurs in proteinaceous material the amino acids (excluding glycine) undergo a stereochemical interconversion by racemization

$$\text{L-amino acid} \underset{k_D}{\overset{k_L}{\rightleftarrows}} \text{D-amino acid,}$$

where the L-amino acid enantiomer interconverts to the D-amino enantiomer at a specific rate of racemization (k_L). The reaction is

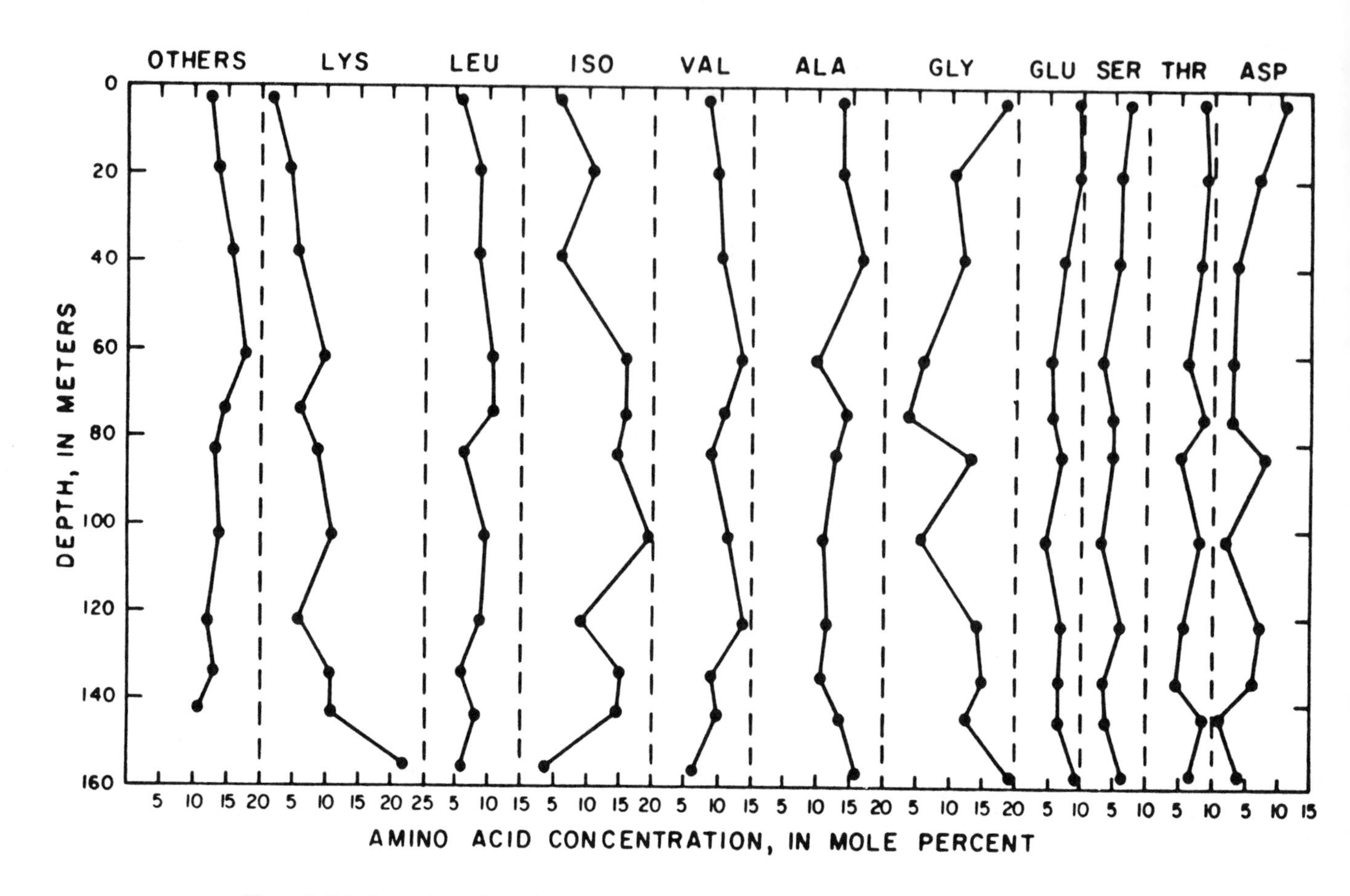

Figure 4. Relative molar amino acid concentrations in lacustrine sediment from core CL-80-1. Relationships for protein amino acids are shown. Abbreviations are as follows: ASP, aspartic acid; THR, threonine; SER, serine; GLU, glutamic acid; GLY, glycine; ALA, alanine; VAL, valine; ISO, isoleucine; LEU, leucine; LYS, lysine; and OTHERS, which includes proline, tyrosine, phenylalanine, histidine, and arginine.

TABLE 3. RELATIONSHIPS OF ANIMO ACID COMPOUNDS IN SEDIMENT FROM CORE CL-80-1, CLEAR LAKE, CALIFORNIA

Neutral/Acidic Ratio*	Depth (m)										
	3.50	19.66	37.00	61.35	74.60	83.25	103.25	123.00	133.00	142.00	155.50
Neutral/Acidic	3.9	4.7	7.1	9.3	10.1	4.9	11.8	5.3	5.3	10.0	5.0
Glycine/Alanine	1.4	0.8	0.7	0.6	0.3	1.0	0.5	1.2	1.4	0.9	1.2

*The neutral/acidic ratio of amino acid compounds are as follows:
Neutral--threonine, serine, proline, glycine, alanine, valine, isoleucine, leucine, tryosine, and phenylalanine;
Acidic--aspartic acid and glutamic acid.
Values of 6 (6n:1a) correspond to neutral peat deposits and well-humified lake deposits; values of approximately 3 correspond to alkaline bogs and well-humified organic marls; values of >6 ($75\text{-}95_n$:0-10a) correspond to conditions in acidic peat bogs (Swain, 1961).

reversible and may involve a carbanion intermediate stage (Neuberger, 1948). Over time a mixture of L- and D-amino acids develops until an equilibrium mixture is reached.

A rate expression for the reversible first-order racemization reaction has been derived by Bada and Schroeder (1972):

$$ln\left(\frac{1 + D/L}{1 - D/L}\right) - ln\left(\frac{1 + D/L}{1 - D/L}\right)_{t=0} = 2k_1t, \quad (1)$$

where D/L is the ratio of D- and L-amino acid enantiomers; k_1 is the rate of racemization; and t is time. The term at time-zero (t = 0) is a correction for D-amino acids in modern samples. Bacterial cell walls contain some amino acids configured as D-stereoisomers (Martin, 1966). The time-zero correction should also account for D-amino acid contributions from bacteria (Dungworth, 1982; Blunt and others, 1982). The rate constant k_1 can be calculated when time is known, for example, by radiocarbon dating. This method calibrates the apparent racemization rate constant for the average kinetic temperature of burial (Bada and Protsch, 1973).

Amino acid D/L ratios should increase systematically with time to be useful for geochronologic applications. Aspartic acid D/L ratios in sediments from CL-80-1 are quite variable with depth, as indicated by the relatively low D/L ratios at 74.60 and 123.00 m, and valine D/L ratios undergo little change in absolute values with depth (Table 2; Fig. 6). In Clear Lake sediment aspartic acid D/L ratios are generally highest, indicating the greatest extent of racemization, and valine D/L ratios are generally lowest (Blunt and others, 1981). These two amino acids have limited usefulness for geochronologic application in Clear Lake sediment.

In previous work with core CL-73-4 (Blunt and others, 1981), the racemization kinetics of alanine were used to determine ages in spite of the apparent complexity of the reactions affecting alanine. Calculations of absolute ages using alanine stereochemistry is limited, in part, by the equivocal nature of the time-zero correction constant in equation 1. This constant was used (see discussion by Dungworth [1982] and Blunt and others [1982] to try to correct for possible contributions to the sediment of D-alanine arising from the cell walls of bacteria that are present in the sediment, interstitial water, and overlying water column (Lee and Bada, 1977; Pollock and Kvenvolden, 1978; Dungworth and others, 1977; Bada and others, 1982; Henrichs and Farrington, 1980). In spite of the uncertainties, however, the apparent racemization of alanine appears to follow linear kinetics in some sediments (Blunt and others, 1977; Tanaka and Handa, 1979; Bada and Man, 1980; Warnke and others, 1980; Blunt and others, 1981).

Comparison with Lake Biwa

Comparison of alanine racemization kinetics in sediment from Lake Biwa, Japan (Tanaka and Handa, 1979), and in sediment from cores CL-73-4 and CL-80-1 from Clear Lake shows different trends with absolute age (Fig. 7). Ages are determined by fission-track dating at Lake Biwa and radiocarbon and palynologic relationships at Clear Lake. Clearly, the alanine D/L ratios in sediment at Lake Biwa increase with time at a rate greater than the alanine D/L ratios in sediment at Clear Lake. This difference may reflect, in part, differences in sedimentation rates and difference in diagenetic temperatures. The average sedimentation rate is about 40 cm/10^3 yr at Lake Biwa based on fission-track dating (Tanaka and Handa, 1979). At Clear Lake, the average sedimentation rate is about 100 cm/10^3 yr based on radiocarbon dating and palynologic relationships (Adam and others, 1981; Sims and others, 1981a, 1981b; Heusser and Sims, 1981). The apparent kinetics of alanine racemization in Clear Lake cores CL-73-4 and CL-80-1 are in agreement and may be useful for absolute geochronologic applications up to 4×10^5 yr.

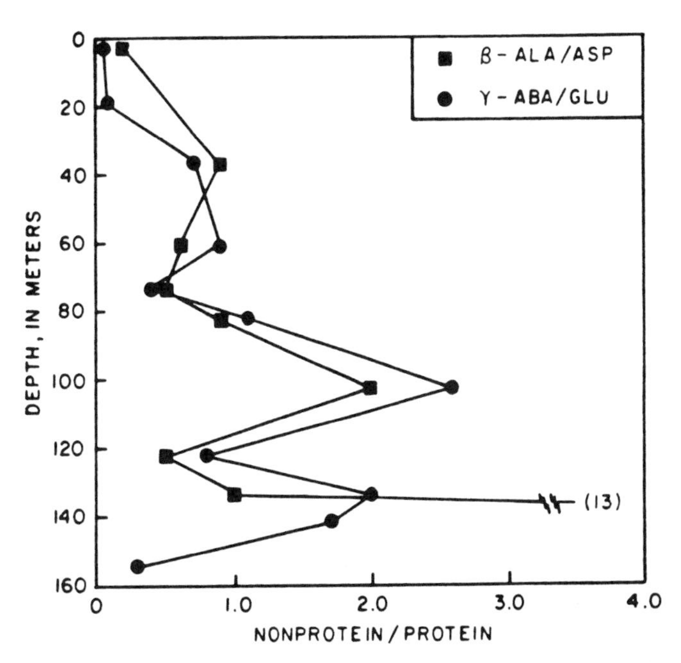

Figure 5. Relationship of nonprotein to protein amino acid precursors in sediment from core CL-80-1. Increasing ratios infer a greater degree of microbial diagenesis. Abbreviations are as follows: β-ALA, β-alanine; ASP, aspartic acid; γ-ABA, γ-aminobutyric acid; GLU, glutamic acid.

Relative Diagenetic Temperature Implications

Similar diagenetic temperature histories should result in similar apparent rates of alanine racemization if alanine geochronology is to be useful. The agreement of alanine D/L ratios in sediment from cores CL-73-4 and CL-80-1 suggests similar diagenetic temperatures in the two cores from the northern basin of Clear Lake. At a specific alanine D/L ratio, the age of sediment in Lake Biwa and Clear Lake should be the same if effective diagenetic temperatures are similar—a factor that is presently uncertain. We find that alanine D/L ratios of about 0.20 represent an age of about 100,000 yr by palynologic dating in Clear

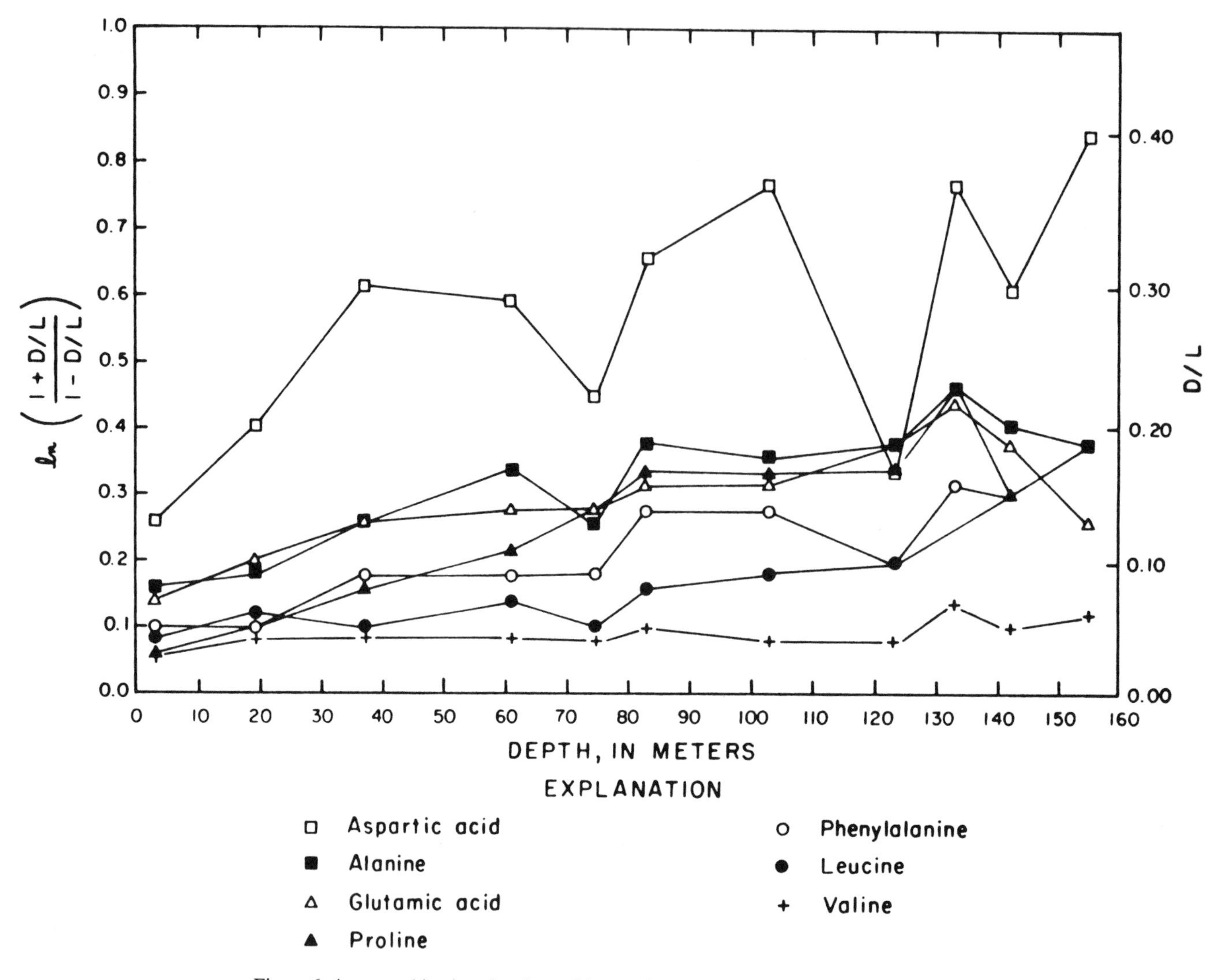

Figure 6. Apparent kinetics of amino-acid racemization in sediment from core CL-80-1.

Lake, whereas alanine D/L ratios of about 0.40 represent the same time interval by fission-track dating in Lake Biwa. The apparent alanine racemization rate constant from equation 1 for Lake Biwa sediment is 2.5×10^{-6} yr^{-1} and for Clear Lake sediment is 1.3×10^{-6} yr^{-1} by linear regression analysis. These apparent rate constants differ by a factor of about 2. The lower alanine apparent rate constant at Clear Lake implies cooler diagenetic temperatures. A difference of 3° to 4°C changes theoretical rates of amino-acid racemization in lacustrine sediment by a factor of 2 (Schroeder and Bada, 1978). This comparison indicates that the relative diagenetic temperatures of Lake Biwa sediment and Clear Lake sediment may differ by 3° to 4°C. It should be emphasized that the diagenetic temperature may not directly relate to climatic temperature due to the effect of geothermal heating. These data suggest a potential means to study relative temperature differences in lacustrine sediment.

SUMMARY

Amino acid geochemistry of sediments in core CL-80-1 from Clear Lake, California, is useful to clarify the diagenesis of organic matter and to estimate geochronology and relative diagenetic temperatures. The decrease with depth of absolute concentrations of amino acids from a total of 41.9 μM/g to 32 nM/g in core CL-80-1 occurs with changes in relative composition and stereochemistry. The ratio of neutral to acidic amino acids ranges between 4 and 12, which suggests that the sediments are slightly alkaline to slightly acidic and well humified. Nonuniform diagenetic processes are indicated by glycine to alanine ratios, which vary between 1.4 and 0.3, and nonregular changes with depth of the relative molar concentrations of individual amino acids in CL-80-1.

Evidence of bacterial decarboxylation is inferred from ratios of nonprotein amino acids to their possible protein amino-acid precursors. Ratios of β-alanine to aspartic acid and γ-aminobutyric acid to glutamic acid increase nonsytematically with depth.

The stereochemistry of alanine is useful for relative geochronologic and paleotemperature estimations. The relationships of alanine racemization kinetics to age in sediments from Lake Biwa, Japan, and in sediments from Clear Lake may in part be a function of different sedimentation rates and diagenetic temperatures. A diagenetic temperature difference of 3 to 4°C in sediment from the two lakes is required to account for apparent racemization rate constants that differ by a factor of 2. Agreement of alanine D/L ratios in sediment from cores CL-73-4 and CL-80-1 suggests similar diagenetic temperatures in the two cores from the northern basin of Clear Lake.

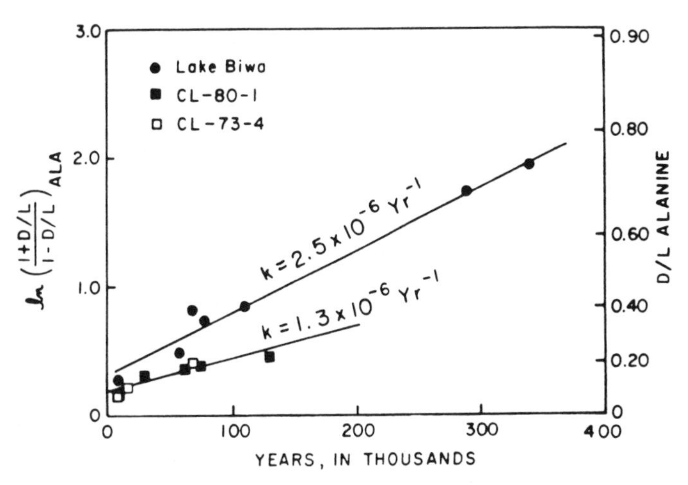

Figure 7. Apparent kinetics of alanine racemization in late Pleistocene lacustrine sediment from Lake Biwa, Japan, and Clear Lake, California. Sediment ages are determined by fission-track dating at Lake Biwa (Tanaka and Handa, 1979) and by radiocarbon and palynologic control at Clear Lake (Adam and others, 1981; Sims and others, 1981a, b; Heusser and Sims, 1981). Linear regression curve for Clear Lake sediment has equation $y = 1.29 \times 10^{-6}X + 0.18$, with correlation coefficient of 0.95.

REFERENCES CITED

Abelson, P. H., and Hare, P. E., 1970, Uptake of amino acids by kerogen: Carnegie Institute of Washington Yearbook, v. 68, p. 297–303.

—— , 1971, Reactions of amino acids with natural and artificial humus and kerogens: Carnegie Institute of Washington Yearbook, v. 69, p. 327–334.

Adam, D. P., Sims, J. D., and Throckmorton, C. K., 1981, 130,000-yr continuous pollen record from Clear Lake, Lake County, California: Geology, v. 9, p. 373–377.

Aizenshtat, Z., Baedecker, M. J., and Kaplan, I. R., 1973, Distribution and diagenesis of organic compounds in JOIDES sediment from Gulf of Mexico and western Atlantic: Geochimica et Cosmochimica Acta, v. 37, p. 1881–1898.

Akiyama, M., 1980, Diagenetic decomposition of peptide-linked serine residues in the fossil scallop shells, *in* Hare, P. E., Hoering, T. C., and King, K., Jr., eds., Biogeochemistry of amino acids: New York, John Wiley and Sons, p. 115–120.

Bada, J. L., 1971, Kinetics of the nonbiological decomposition and racemization of amino acids in natural waters, *in* Hem, J. D., and Gould, R. F., eds., Nonequilibrium systems in natural water chemistry: American Chemical Society, Advances in Chemistry Series, v. 106, p. 309–331.

Bada, J. L., and Man, E. H., 1980, Amino acid diagenesis in Deep Sea Drilling Project cores; Kinetics and mechanisms of some reactions and their applications in geochronology and in paleotemperature and heat flow determinations: Earth Science Reviews, v. 16, p. 21–25.

Bada, J. L., and Protsch, R., 1973, Racemization reaction of aspartic acid and its use in dating fossil bones: National Academy of Sciences Proceedings, v. 70, p. 1331–1334.

Bada, J. L., and Schroeder, R. A., 1972, Racemization of isoleucine in calcareous marine sediments; Kinetics and mechanism: Earth and Planetary Science Letters, v. 15, p. 1–11.

Bada, J. L., Hoopes, E., and Ming-shan Ho, 1982, Combined amino acids in Pacific Ocean waters: Earth and Planetary Science Letters, v. 58, p. 276–284.

Bada, J. L., Ming-Yung Shou, Man, E. H., and Schroeder, R. A., 1978, Decomposition of hydroxy amino acids in foraminiferal tests; Kinetics, mechanism, and geochronologic implications: Earth and Planetary Science Letters, v. 41, p. 67–76.

Blunt, D. J., Kvenvolden, K. A., and Sims, J. D., 1981, Geochemistry of amino acids in sediments from Clear Lake, California: Geology, v. 9, p. 378–382.

—— , 1982, Reply to comment on geochemistry of amino acids in sediments from Clear Lake, California: Geology, v. 10, p. 1f24–125.

Blunt, D. J., Warnke, D. A., and Pollock, G. E., 1977, Trends of amino acid racemization in a southern ocean core: Geological Society of America Abstracts with Programs, v. 9, p. 903.

Brown, F. S., Baedecker, M. J., Nissenbaum, A., and Kaplan, I. R., 1972, Early diagenesis in a reducing fjord, Saanich Inlet, British Columbia—III. Changes in organic constituents of sediments: Geochimica et Cosmochimica Acta, v. 36, p. 1185–1203.

Casagrande, D. J., and Given, P. H., 1980, Geochemistry of amino acids in some Florida peat accumulations—II. Amino acid distributions: Geochimica et Cosmochimica Acta, v. 44, p. 1493–1507.

Dungworth, G., 1982, Comment on geochemistry of amino acids in sediments from Clear Lake, California: Geology, v. 10, p. 124–125.

Dungworth, G., Thijssen, M., Zuurveld, J., Van der Velden, W., and Schwartz, A. W., 1977, Distribution of amino acids, amino sugars, purines, and pyrimidines in a Lake Ontario sediment core: Chemical Geology, v. 9, p. 295–308.

Erdman, 1961, Some chemical aspects of petroleum genesis as related to the problem of source bed recognition: Geochimica et Cosmochimica Acta, v. 22, p. 16–36.

Henrichs, S. M., and Farrington, J. W., 1980, Amino acids in interstitial waters of marine sediments; A comparison of results from varied sedimentary environments, *in* Douglas, A. G., and Maxwell, J. R., eds., Advances in Organic Geochemistry, 1979: Oxford, Pergamon Press, p. 435–443.

Heusser, L. E., and Sims, J. D., 1981, Palynology of core CL-80-1, Clear Lake, California: Geological Society of America Abstracts with Programs, v. 13, p. 61.

Hoering, T. C., 1973, A comparison of melanoidin and humic acid: Carnegie Institute of Washington Yearbook, v. 72, p. 682–690.

Ithara, Y., 1973, Amino acids in the Cenozoic sediments of Japan: Pacific Geology, v. 6, p. 51–63.

Jones, J. D., and Vallentyne, J. R., 1960, Biogeochemistry of organic matter—I. Polypeptides and amino acids in fossils and sediments in relation to geothermometry: Geochimica et Cosmochimica Acta, v. 21, p. 1–34.

Kemp, A.L.W., and Mudrochova, A., 1973, The distribution and nature of amino acids and other nitrogen-containing compounds in Lake Ontario surface sediments: Geochimica et Cosmochimica Acta, v. 37, p. 2191–2206.

Kvenvolden, K. A., and Blunt, D. J., 1981, Amino acids in sediments from Leg 68

Site 502, *in* Prell, W. L., Gardner, J. V., and others, eds., Initial Reports Deep Sea Drilling Project, v. 68: Washington, D.C., U.S. Government Printing Office (in press).

Kvenvolden, K. A., Blunt, D. J., McMenamin, M. A., and Strahan, S. E., 1980, Geochemistry of amino acids in shells of clam *Saxidomus, in* Douglas, A. G., and Maxwell, J. R., eds., Advances in organic geochemistry, 1979: Oxford, Pergamon Press, p. 321–332.

Lee, C., and Bada, J. L., 1977, Dissolved amino acids in the equatorial Pacific, the Sargasso Sea and Biscayne Bay: Limnology and Oceanography, v. 22, p. 502–510.

Liddicoat, J. C., Sims, J. D., and Bridge, W. D., 1981, Paleomagnetism of a 177-m-long core from Clear Lake, California [abs.]: Geological Society of America Abstracts with Programs, v. 13, p. 67.

Martin, H. H., 1966, Biochemistry of bacterial cell walls: Annual Reviews in Biochemistry, v. 35, p. 457–484.

McCoy, W. D., and Miller, G. H., 1979, Protein degradation in organically precipitated tufa; Reliable relative age indices: Geological Society of America Abstracts with Programs, v. 11, p. 474–475.

McCoy, W. D., and Scott, W. E., 1980, Amino-acid stereochemistry as an aid to differentiation and correlation of deposits of Quaternary Lake Bonneville: Geological Society of America Abstracts with Programs, v. 12, p. 280.

McCoy, W. D., and Williams, L. D., 1983, Temperature and precipitation estimates for the Bonneville Basin during the last lake cycle: Geological Society of America Abstracts with Programs, v. 15, p. 301.

Neuberger, A., 1948, Stereochemistry of amino acids, *in* Anson, M. L., and Edsall, J. T., eds., Advances in protein chemistry: New York, Academic Press, v. 4, p. 297–383.

Pollock, G. E., and Kvenvolden, K. A., 1978, Stereochemistry of amino acids in surface samples of a marine sediment: Geochimica et Cosmochimica Acta, v. 42, p. 1903–1905.

Rymer, M. J., Sims, J. D., Hedel, C. W., Bridge, W. D., Makdisi, R. S., and Mannshardt, G. A., 1981, Sample catalog for core CL-80-1, Clear Lake, California: U.S. Geological Survey Open-File Report 81-245, 53 p.

Schroeder, R. A., and Bada, J. L., 1978, Aspartic acid racemization in Late Wisconsin Lake Ontario sediments: Quaternary Research, v. 9, p. 193–204.

Sims, J. D., Adam, D. P., and Rymer, M. J., 1981a, Late Pleistocene stratigraphy and palynology of Clear Lake, Lake County, California, *in* McLaughlin, R. J., and Donnelly-Nolan, J. M., eds., Research in the Geysers–Clear Lake geothermal area, northern California: U.S. Geological Survey Professional Paper 1141, p. 219–230.

Sims, J. D., Rymer, M. J., and Perkins, J. A., 1981b, Description and preliminary interpretation of core CL-80-1, Clear Lake, Lake County, California: U.S. Geological Survey Open-File Report 81-751, 175 p.

Stevenson, F. J., and Cheng, C. N., 1970, Amino acids in sediments; Recovery by acid hydrolysis and quantitative estimation by a colorimetric procedure: Geochimica et Cosmochimica Acta, v. 34, p. 77–88.

Swain, F. M., 1961, Limnology and amino-acid content of some lake deposits in Minnesota, Montana, Nevada, and Louisiana: Geological Society of America Bulletin, v. 72, p. 519–546.

Tanaka, N., and Handa, N., 1979, Organogeochemical studies of a 200-meter core sample from Lake Biwa. V: Racemization of amino acids in the sediments determined by gas-liquid chromatography: Japan Academy Proceedings, v. 55, p. 169–174.

Warnke, D. A., Blunt, D. J., and Pollock, G. E., 1980, Enantiomeric ratios of amino acids in southern ocean siliceous oozes, *in* Hare, P. E., Hoering, T. C., and King, K., Jr., eds., Biogeochemistry of amino acids: New York, John Wiley and Sons, p. 183–189.

Weiner, S., and Lowenstein, H. A., 1980, Well-preserved fossil mollusk shells, Characterization of mild diagenetic processes, *in* Hare, P. E., Hoering, T. C., and King, K., Jr., eds., Biogeochemistry of amino acids: New York, John Wiley and Sons, p. 95–114.

MANUSCRIPT ACCEPTED BY THE SOCIETY SEPTEMBER 15, 1986

Printed in U.S.A.

Geological Society of America
Special Paper 214
1988

Clear Lake record vs. the adjacent marine record; A correlation of their past 20,000 years of paleoclimatic and paleoceanographic responses

James V. Gardner
U.S. Geological Survey
345 Middlefield Road
Menlo Park, California 94025

Linda E. Heusser
Lamont-Doherty Geological Observatory
of Columbia University
Palisades, New York 10964

Paula J. Quinterno
Sean M. Stone
John A. Barron
U.S. Geological Survey
345 Middlefield Road
Menlo Park, California 94025

Richard Z. Poore
U.S. Geological Survey
Reston, Virginia 22092

ABSTRACT

A deep-sea core collected on the continental slope off northern California contains a pollen stratigraphy for the past 20,000 yr that can be correlated to the pollen stratigraphy from the upper section of Clear Lake core CL-73-4. The occurrence in one sequence of pollen, reflecting the local continental paleoclimates, and marine microfossils reflecting the local paleoceanography, allows a comparison of concurrent responses of the local ocean and adjacent continental area to global climate changes. The interpretation of the two data sets gives a complex progression of changes that are probably interrelated, such as upwelling that produced coastal fogs. The changes in climatic and oceanographic environmental conditions that occurred in response to the switch from global glacial to interglacial conditions was not a smooth progression of increasingly moderate regimes; rather, the changes appear to be a complicated series of states that suggests a disequilibrium mode lasting from about 15,000 to 5,000 yr ago.

INTRODUCTION

The study of a sediment core 115 m long collected from Clear Lake, California (39°03′N; 122°50′W; 404 m in elevation) has provided a detailed pollen stratigraphy and paleoenvironmental history of the late Quaternary for this area (Adam and others, 1981). Microfaunal and microfloral analyses on deep-sea sediment from the continental slope off the coast of California at about the same latitude offer a unique opportunity to compare the regional terrestrial paleoenvironment with the regional paleoceanographic response to late Quaternary climatic changes. Earlier studies have shown that the highest pollen concentrations in surface sediments are in the vicinity of the mouths of rivers and reflect the vegetation along the course of the rivers (Heusser and Balsam, 1977). We anticipated that pollen sedimented in Clear Lake also would have fallen in the drainage basin of the nearby Russian River and would have been carried offshore onto the continental margin.

A series of piston and gravity cores was collected from the continental slope west of the mouth of the Russian River in the hope of obtaining a record of at least the last 150,000 yr. However, because of rapid accumulation rates and fairly high shear-strength sediment in this area, the longest cores are only about 4 m long and represent only the last 20,000 yr. After preliminary studies, one core, V1-80-P3, was selected for detailed analysis. It is a 3.85-m-long piston core collected at 38°25.51′N, 123°47.77′W in a water depth of 1,600 m (Fig. 1). The sediment facies is a diatom-bearing to nannofossil-bearing clayey silt of

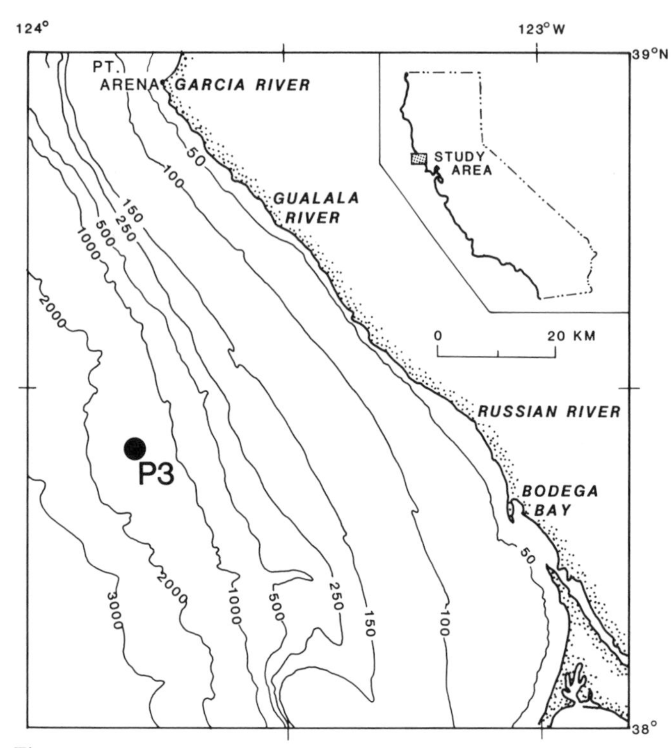

Figure 1. Location of core Vl-80-P3. Bathymetric contours in meters.

rather uniform grayish olive (10Y4/2) color. Although subtle, burrow mottling is present throughout the core.

The core was quantitatively analyzed to determine: (1) total pollen, (2) total benthic and planktonic foraminiferal faunas, (3) total diatom flora, (4) dominant nannofossil flora, (5) oxygen and carbon isotopes, (6) grain-size variations, (7) carbon budget, and (8) clay mineralogy. All of the raw data and descriptions of the techniques used for these analyses can be found in Gardner and others (1983). The following is a correlation of the upper part of the Clear Lake record with the adjacent deep-sea record and a synthesis of these data interpreted as changes in the paleoceanographic and paleoclimatic conditions that prevailed in this region during the time represented by the deep-sea core.

DATING

Low carbon content (<2%) in the sediment rendered reliable only two radiocarbon dates out of six submitted: 2370 ± 70 yr at 5 to 7 cm (USGS No. 1183) and 11,200 ± 200 yr at 135 to 142 cm (USGS No. 1185). The other four samples yielded so little gas that the uncertainties are quite large. However, the oxygen-isotope record, together with the two C-14 dates, gives a fairly good chronostratigraphy.

The oxygen-isotope record (Figs. 2A, 3) appears to span the range from the top of Isotope Stage 2 (<29,000 yr) through Isotope Stage 1 (Holocene) of Shackleton and Opdyke (1973). Our C-14 dates and their relationship to the oxygen-isotope curve show a remarkable resemblance to the oxygen-isotope data and C-14 dating of a core from the eastern North Atlantic discussed by Duplessy and others (1981) (Fig. 2B). Both termination I_A and I_B, the two phases of deglaciation at the Holocene-Pleistocene boundary as defined by Duplessy and others (1981), are evident in core V1-80-P3 (Fig. 2). In addition, our C-14 date of 11,210 yr falls on a section of the oxygen-isotope curve that closely correlates with a level dated at 11,680 yr in core CH73139C. The close correspondence of the C-14 and oxygen isotope data from these two cores suggests that core V1-80-P3 does not have any record of Isotope Stage 3. The isotope record does not show a plateau of large positive values at the bottom of the core (Fig. 2), as would be expected at the Isotope Stage 2/3 boundary, but the oxygen-isotope values approach +3 per mil at the bottom of core V1-80-P3, so it appears that the upper part of Isotope Stage 2 is represented.

The average accumulation rate between the two C-14 ages in core V1-80-P3 is 14.7 cm/1,000 yr and the average accumulation rate between our 11,210 yr C-14 date and the C-14-dated age of the middle of Termination I_A (14,800 yr from Duplessy and others, 1981), is 27.9 cm/1,000 yr. We used these accumulation rates to calculate an age for each sample so all data can be plotted versus age. We realize that accumulation rates can change with time, especially when the record spans changes in global climates, yet other than a correlation with our oxygen-isotope record, we have no other recourse. Figure 3 is a plot of the oxygen-isotope record of core V1-80-P3 vs. age using this age model.

MARINE STRATIGRAPHY

Sediment Parameters

We analyzed bulk sediment for grain size, clay mineralogy, and percentages of calcium carbonate and organic carbon. The calcium carbonate content is very low (<10%) throughout the core, which may be the result of dilution or dissolution of carbonate, or both. The core was recovered from a water depth of only 1,600 m; thus initially dissolution was discounted and we favored dilution as the principal reason for the low content of calcium carbonate. However, postdepositional dissolution of the carbonate may also have occurred in these sediments, especially in the top 20 cm of the core, because some planktonic foraminifers and nannofossils show evidence of corrosion. However, the floras and faunas discussed in detail later are similar to those found in presumably undissolved assemblages; this suggests that, if dissolution was a major postdepositional process, at least it did not considerably alter the floral and faunal compositions.

Analytical precision and accuracy for the percentage of calcium carbonate are both about ±1 percent; consequently, many of the smaller fluctuations probably are analytical variations. However, there is an increase in carbonate from less than 2 percent prior to about 18,000 yr to more than 3 percent between

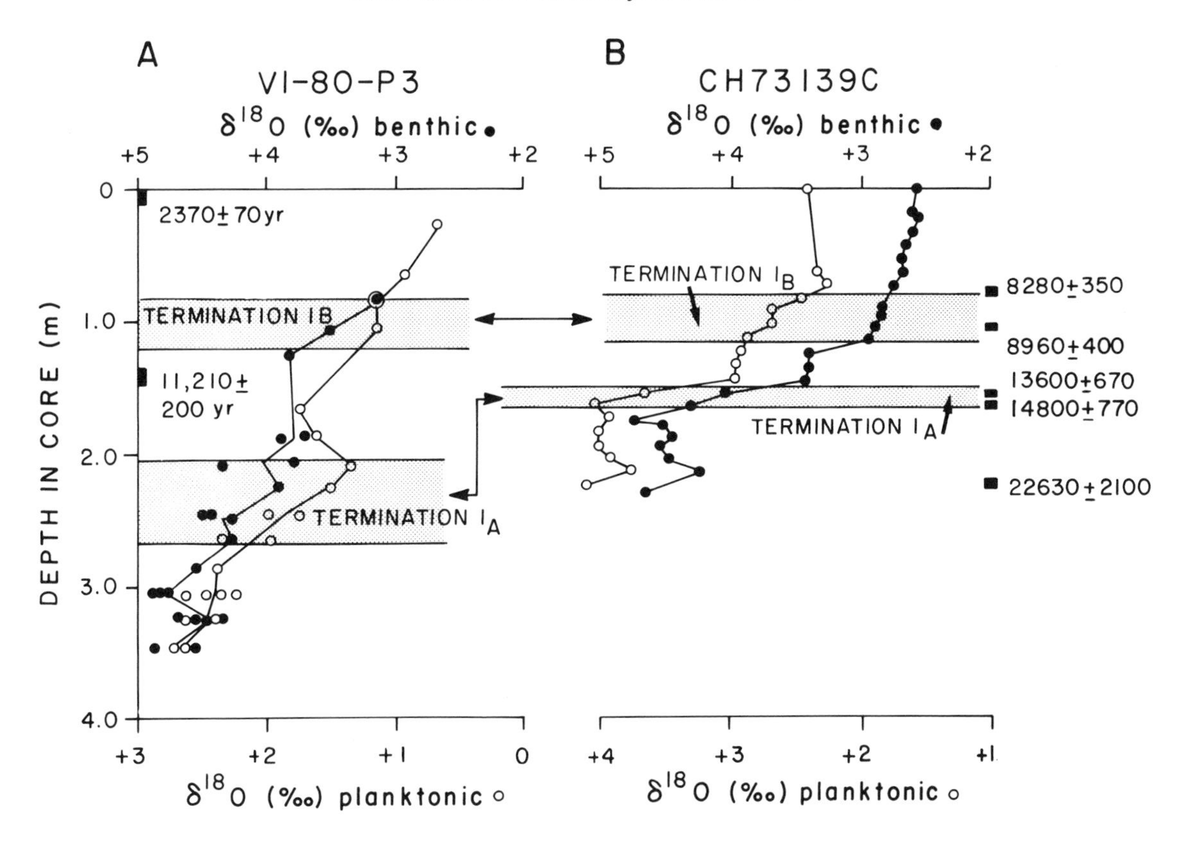

Figure 2. A, Oxygen-isotope record and C-14 ages for core V1-80-P3. Oxygen isotope values are δ O-18 relative to PDB standard. Locations of Terminations I_A and I_B are extrapolated from their positions in core CH73139C (Fig. 2B). B, Oxygen-isotope record for Core CH73139C from eastern North Atlantic with Terminations I_A and I_B and relevant C-14 ages. (From Duplessy and others, 1981.)

18,000 and about 12,000 yr. The carbonate values then drop to less than 2 percent, but at about 10,000 yr they increase to more than 5 percent and remain at these values until about 6,000 yr. The values drop to below 2 percent by about 5,000 yr and stay at this concentration to the top of the core. However, these carbonate concentrations are so low that any interpretation of them alone would be highly speculative.

Organic carbon values show a three-step increase from about 1 percent (by weight) between 20,000 and about 15,000 yr, to 1.5 percent between 15,000 and about 4,500 yr, to almost 3 percent in the surface sediment. This significant increase in organic carbon will be discussed below.

The data on the distribution of grain sizes show that the principal differences are in a reduced deposition of sand and increased silt accumulation at about 15,500 yr. Clay deposition remained fairly constant, at about 10 percent of the total sand-silt-clay facies. Clay mineralogy shows no significant changes in composition or abundances throughout the entire core. Average smectite is 52 percent ($\sigma = \pm 3$), average illite is 25 percent ($\sigma = \pm 2$), and average chlorite + kaolinite is 23 percent ($\sigma = \pm 3$).

Planktonic Foraminifers

Planktonic foraminifers were processed from a standard 10-cc sample of bulk sediment. Our intent was to count at least 300 specimens per sample, but many of the samples contained fewer than 300 specimens, making statistical treatment unwarranted. Despite the paucity of planktonic foraminifers in some samples, certain patterns are evident.

The highest abundances of planktonic foraminifers occur in species of *Neogloboquadrina pachyderma* (right- and left-coiling), *Globigerina bulloides,* P-D intergrade (an intergrade between *N. pachyderma* and *N. dutertrei*), *Globigerinita glutinata,* and *Turborotalia quinqueloba* (Fig. 4). The species appear to group into three stratigraphic successions. The first assemblage, PFI, reflects a cold subpolar fauna (Coulbourn and others, 1980)—dominated by *N. pachyderma* (left-coiling), *G. bulloides,* and, to a lesser extent, *G. glutinata*—that occurs from the bottom of the core to about 15,000 yr. We are uncertain whether this assemblage represents a regional water mass, the effects of upwelling, a shelf fauna, or some combination of these. This assemblage rapidly declined in importance at about 15,000 yr, and an assemblage, PFII—composed of *G. bulloides, T. quinqueloba* and right-coiling *N. pachyderma*—became important, but only for a short time. All three species dominate the transitional fauna (Bradshaw, 1959). This assemblage was replaced between about 13,000 and 14,000 yr by the third assemblage, PFIII, which is largely composed of *N. pachyderma* (right- and left-coiling) and PD intergrade, species of subpolar faunas (Bradshaw, 1959; Colbourn and others, 1980). The faunal assemblages then began a series of fluctuations, first back to a dominance by the cold subpolar assemblage PFI by about 12,000 yr, then a dominance by the transitional assemblage PFII by 10,500 yr, and then a dominance by the subpolar assemblage PFIII between 9,500 and 6,500 yr. About 5,000 yr ago the three stratigraphic assemblages disappeared and a fauna dominated by left-coiling *N. pachyderma* was established but this is probably a reflection of dissolution.

Benthic Foraminifers

Recent studies have questioned the traditional use of benthic foraminifers as paleodepth indicators. Streeter (1973), Schnitker (1974), Lohmann (1978), and Blake and Douglas (1980) have all shown that benthic foraminifers do not respond to water depth *sensu stricto* (i.e., pressure), but rather to water mass characteristics (such factors as temperature, oxygen content, and salinity). Spatial and temporal variations of benthic foraminiferal assemblages are responses to the fluctuations of the water mass preferred by those assemblages. The actual water mass preferences of different benthic faunas of the eastern Pacific margin have yet to be determined, but changes of assemblages can be interpreted as changes in water mass.

Benthic foraminifers were picked from the same >149-μm-size fraction that was processed for planktonic foraminifers. If benthic foraminifers were abundant, then a microsplitter was used to obtain approximately 300 specimens. If the sample contained fewer than 300 specimens, all specimens in the sample were counted.

Of the 85 species of benthic foraminifers identified from sediment from core VI-80-P3, only 13 species have significant abundances (>5 percent abundance; see Fig. 5). These 13 species show five distinct stratigraphic assemblages (Fig. 5) that we interpret as reflections of a progression of changing or evolving water masses. The stratigraphically oldest assemblage, BFI, is dominated by *Cassidulina translucens* but includes *Uvigerina peregrina, Bulimina spicata,* and *B. striata* var. *mexicana.* At about 15,000 yr, BFI was replaced by BFII, which was dominated by *Epistominella smithi, U. peregrina, Epistominella pacifica,* and *Bolivina spissa.* BFII prevailed until sometime around 11,000 to 10,000 yr, when BFIII appeared. BFIII is characterized by abundant *Uvigerina peregrina, Epistominella smithi, Bolivina spissa, Bulivina spicata,* and *Chilostomella sp.* BFIV, which replaced BFIII at about 8,000 yr, is composed of abundant *Uvigerina* sp. 1 and *Bolivina spissa.* BFV—composed of *Uvigerina proboscidea, Uvigerina* sp. 1, and *Bulimina striata* var. *mexicana*—appeared at about 5,000 yr ago and has predominated since then.

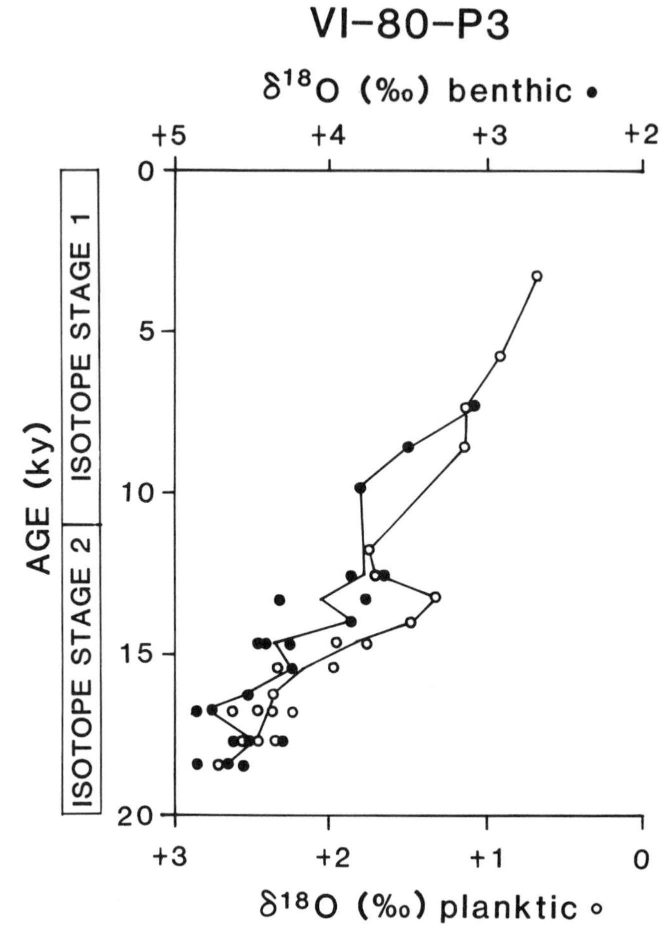

Figure 3. Oxygen-isotope record of core V1-80-P3 vs. age. See text for an explanation of age model. Age of isotope stage 1/2 boundary is from Pisias and Moore (1981).

Diatoms

Diatoms were tabulated by identifying up to 300 total specimens on smear slides and converting the counts of each species or group to percentages. The most common taxa or groups of taxa of diatoms found throughout core V1-80-P3 include *Pseudoeunotia doliolus, Thalassionema nitzschioides, Thalassiothrix longissima, Thalassiosira* spp. (mainly *T. decipiens, T. symbolophora,* and *T. pacifica*), tychopelagic diatoms (*Stephanopyxis* spp., *Actinoptychus* spp., and *Melosira sulcata*), benthic marine diatoms, and fresh-water diatoms (Fig. 6). *P. doliolus, T. nitzschioides,* and *T.* spp. are planktonic diatoms associated with subtropical to transitional conditions. *T. nitzschoides* and *T. longissima* are diatoms that prefer areas of upwelling (L. H. Burckle, personal communication, 1981; Sancetta, 1983; Barron and Keller, 1983). Tychopelagic diatoms are those that spend part of their life cycle attached to the sea floor and part as plankton.

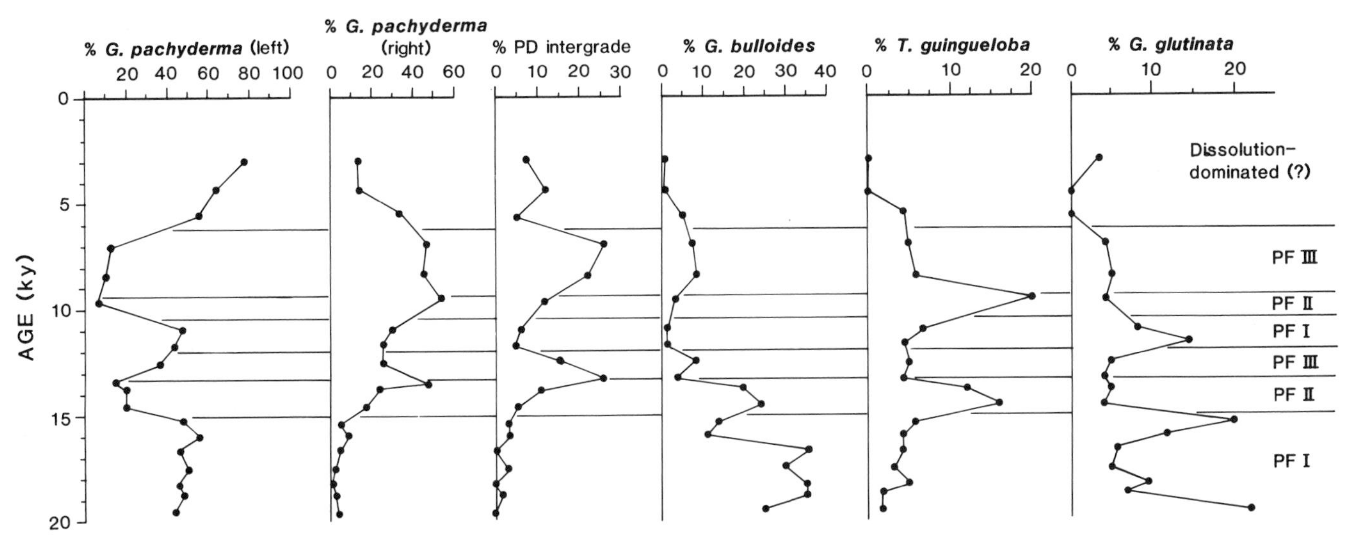

Figure 4. Plots of abundances of important species of planktonic foraminifers from assemblages in core V1-80-P3. PFI, PFII, PFIII refer to the different assemblages of planktonic foraminifers; see text for explanation.

Tychopelagic species are more common on the continental shelves and are generally common components of near-shore assemblages.

The diatom stratigraphy shows the progressive decrease in benthic marine and fresh-water species, presumably a reflection of the marine transgression that began about 18,000 yr ago. The distribution of *T. nitzschioides* and *T. longissima,* the upwelling indicators, suggests that upwelling in this area has, in general, increased in intensity with time, although episodically. Two distinct pulses of increased upwelling occurred: one began at about 15,000 yr and lasted until about 12,000 yr, and the second started about 9,000 yr and presumably continues today. The normal marine subtropical to transitional assemblage of diatoms found off northern California today, as typified by common *P. doliolus,* was virtually absent until about 8,000 yr.

Calcareous Nannofossils

Calcareous nannofossils are rare to common in samples from core V1-80-P3; however, counts of 100-specimen traverses with a light microscope did reveal variations in assemblages. Five recognizable species or groups appear in the samples: large *Gephyrocapsa,* small *Gephyrocapsa, Emiliania huxleyi, Calcidiscus leptoporus,* and *Coccolithus pelagicus.* Geitzenauer and others (1976) stated that *E. huxleyi* is an unreliable Quaternary paleoceanic indicator because it has been evolving throughout the late Pleistocene and Holocene. These authors also cast doubt on the usefulness of the *Gephyrocapsa* groups because at least one, *G. caribbeanica,* has undergone a marked decline in abundance that does not appear to be related to climatic changes. The two species that show significant variations in core V1-80-P3 are *C. leptoporus* and *C. pelagicus* (Fig. 7). The percentages of these two species in core V1-80-P3 are similar to the percentages discussed by Roth and Coulbourn (1982) for surface sediment along the northeastern Pacific margin. Consequently, we do not believe the floras in core V1-80-P3 are artifacts of varying degrees of dissolution; rather, they reflect actual assemblages. *C. leptoporus* gradually becomes dominant to *C. pelagicus* starting at about 16,000 yr. *C. leptoporus* disappears between 13,000 and 11,000 yr ago and then consistently occurs until 5,000 yr ago.

No calcareous nannofossils were found in sediment younger than 5,000 yr, probably because of a combination of clastic dilution and carbonate dissolution. The distribution of modern coccoliths in the Pacific (McIntyre and others, 1970; Geitzenauer and others, 1976; Roth and Coulbourn, 1982) show that *C. leptoporus* is an important species of transitional waters between the subpolar and subtropical water masses. *C. pelagicus* is found associated with present-day subpolar waters.

Pollen

Fifteen different pollen types were identified in samples from core V1-80-P3, using the techniques described by Heusser and Balsam, (1977). The only numerically important pollen are *Pinus, Quercus, Sequoia,* TCT (a combination of the families Taxodiaceae, Cupressacene, and Taxaceae), and *Alnus.* Heusser and Balsam (1977) discussed the distribution of *Pinus* in marine sediment of the northeast Pacific Ocean; they reviewed several studies suggesting high concentrations of *Pinus* pollen are due to the multiplicity of pine trees and the voluminous production of

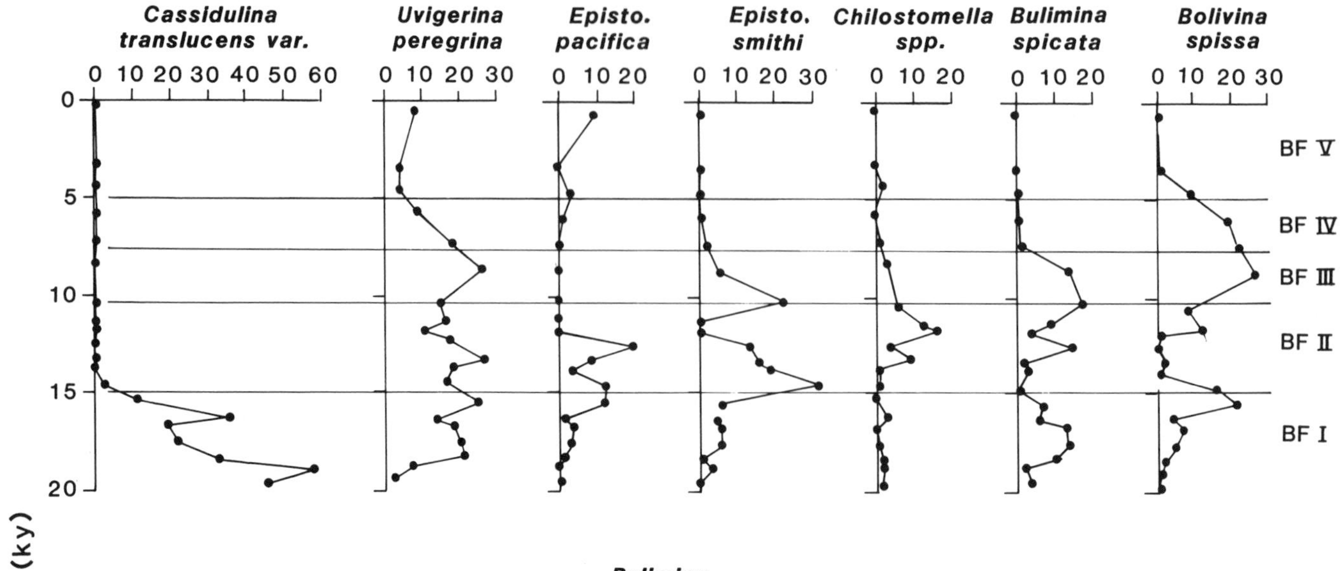

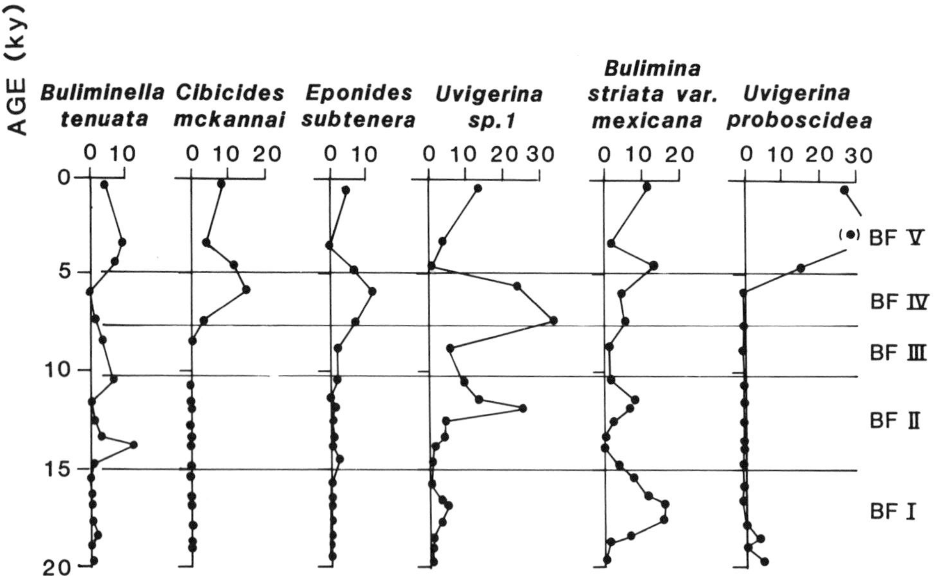

Figure 5. Plots of abundances of important species of benthic foraminifers from assemblages in core V1-80-P3. BFI-BFV refer to the different assemblages of benthic foraminifers; see text for explanation.

pine pollen . . ." along the Pacific Northwest, as well as to hydrodynamic efficiency (Heusser and Balsam, 1977, p. 58). However, a major factor that controls the presence of pine pollen in marine sediment is the distance of pine trees from shore. Distribution maps of pine pollen in marine sediment do correlate to the pine forests in North America (Heusser and Balsam, 1977), and the location of core V1-80-P3 is only about 100 km from shore and is in the central region of high *Pinus* concentrations.

Although we believe that the stratigraphy of *Pinus* pollen from V1-80-P3 is a record of pine in the overall flora of the California Coast Ranges, the possibility of selective concentration of pine pollen cannot be overlooked. Consequently, we have recalculated the pollen abundances excluding *Pinus. Quercus* is only locally important in continental pollen as at Clear Lake, but in general has relatively low abundances in the Pacific Northwest (Heusser and Balsam, 1977).

The *Pinus*-free pollen data, which are plotted in Figure 8, show *Quercus* (oak) increased in abundance starting about 15,000 yr ago and reached a stable, high, average abundance by about 14,000 yr that has continued through the Holocene. The abundance of *Sequoia sempervirens* (coastal redwood), a species found today in the coastal zone of northern California, correlates with that of *Querus,* but the increase in *Sequoia* abundance occurred more abruptly at about 12,000 yr (Fig. 8). The spectra of TCT shows a peak in abundance from 15,000 to about 12,500 yr and responds with a strong negative correlation with *Sequoia* (Fig. 8). However, TCT pollen is highly susceptible to alteration in the marine environment and we are reluctant to interpret this abundance. *Alnus,* principally alders that prefer cooler and wetter conditions than those that prevail at present in northern California, peaked in abundance from 16,000 to about 7,000 yr and then declined. Alder species appear to be opportunistic and estab-

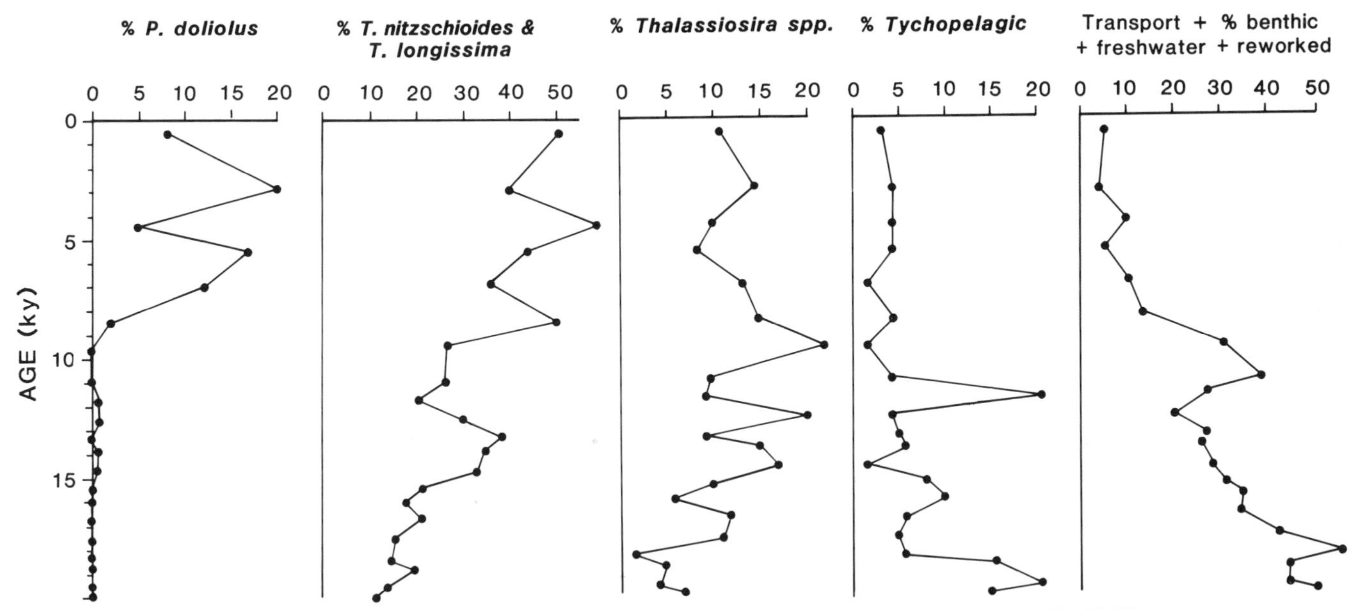

Figure 6. Plots of abundances of important diatom taxa or groups from assemblages in core V1-80-P3. Percentages for *P. doliolus, T. nitzschioides, T. Longissma, Thalassiosira,* and Tychopelagic are on a fresh-water and reworked diatom-free basis.

lish themselves soon after initial deglaciation (Heusser and Shackleton, 1979). Thus, this peak in abundance of *Alnus* may represent the period of changing conditions from a glacial to an interglacial climate.

The distribution of *Pinus* in the stratigraphic record of V1-80-P3 (Fig. 8) shows a general decline of importance with time but with some high-frequency fluctuations. The general decline in abundance of pine pollen reflects the negative correlation with *Quercus,* as would be predicted from Adam (1967) and Adam and others (1981). Therefore, we believe the decline in abundance of *Pinus* pollen does reflect the gross signal, but the high-frequency fluctuations could be artifacts of transport and postdepositional reworking.

CORRELATION OF CLEAR LAKE AND MARINE RECORDS

To compare our data directly to the record from Clear Lake core CL73-4, we first used the assigned ages of Adam and others (1981) to generate a time series. Unfortunately, the dating of this core is no better than our dating of core V1-80-P3. Although Adam and others (1981) placed 10 C-14 dates and one amino-acid date on their data curves for CL73-4 (their Fig. 3), they argued that 9 of the C-14 dates are questionable. Consequently, we have used their single accepted C-14 and amino-acid dates, interpolated an age for each data point, and plotted their pollen data for the upper section of CL73-4 vs. time.

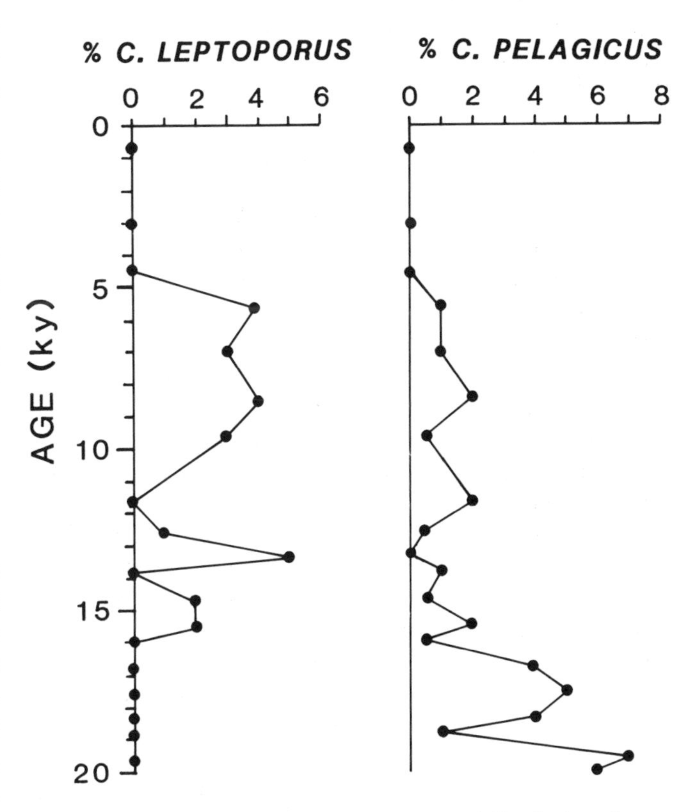

Figure 7. Plots of abundances of *C. leptoporus* and *C. pelagicus,* the two significant nannofossil species from assemblages in core V1-80-P3.

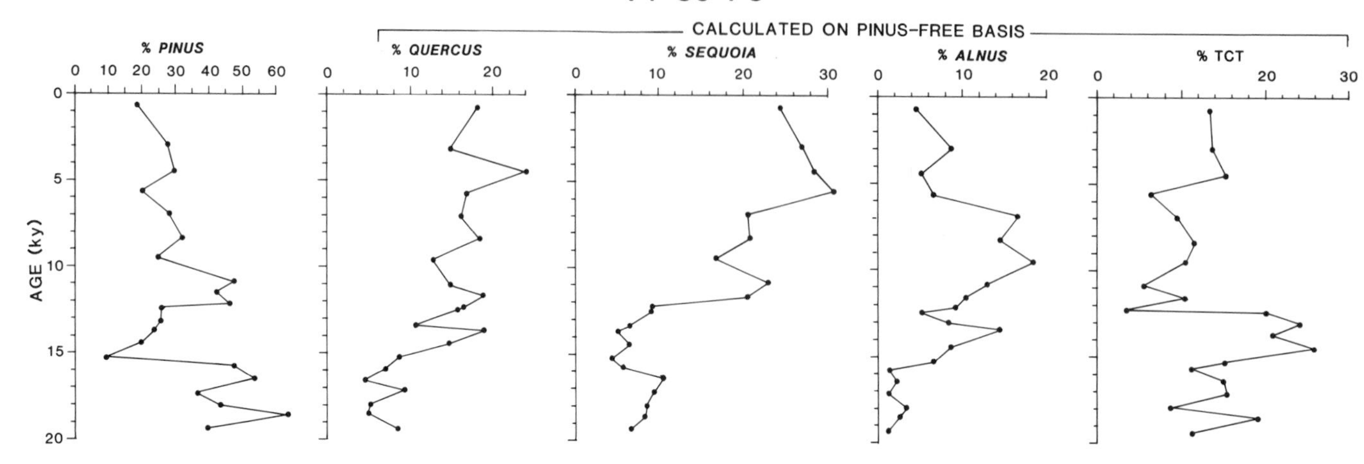

Figure 8. Plots of abundances of important pollen groups from core V1-80-P3. Abundances of *Quercus, Sequoia, Alnus,* and TCT were calculated on *Pinus*-free basis.

A comparison of the pollen data from the two cores, plotted on the same time scale, shows some offset in events that most certainly are coeval (Fig. 9A). There is a discrepancy of about 2,000 yr in the ages for the initial increases in *Quercus* and the decreases in *Pinus* at about 15,000 yr (on the marine time scale). We believe that the close correspondence of the record of oxygen-isotope vs. age to the data of Duplessy and others (1981), together with the uncertainty in the age model for CL73-4 as discussed by Adam and others (1981), suggests that our age model is appropriate. Therefore, the data for CL73-4 were adjusted for this offset and are plotted together with comparable data for V1-80-P3 (Fig. 9B).

The pollen records for CL73-4 and V1-80-P3 show some remarkable similarities. The overall trends of *Quercus* and *Pinus* are comparable. The abundance of *Pinus* are comparable between the two records, yet even though the trends are similar, the abundances of *Quercus* differ by more than a factor of 2. The difference in abundance of *Quercus,* as discussed above, is the result of oak being a dominant pollen producer in the immediate area of Clear Lake but a relatively minor producer in the California Coast Ranges. This correlation of marine and continental pollen stratigraphy is also similar to results of Heusser and Florer (1973) from the state of Washington and the northeastern Pacific area. Our pollen spectra show the transition of vegetation in the California Coast Ranges as global climates changes from a glacial mode (pre-18,000 yr), through a transition (18,000 to 13,500 yr), and eventually to an interglacial mode (post-13,500 yr). The occurrence of significant abundances of *Sequoia* pollen in the marine record but not in the continental record of CL73-4 adds an important paleoclimatic indicator. *Sequoia sempervirens* requires a mediterranean-type climate with coastal fogs and relatively high humidity. Today *S. sempervirens* is restricted to the western slopes of the California Coast Ranges in close proximity to the shoreline. The occurrence of *Sequoia* with *Quercus* and *Pinus* provide additional detailed information about the history of the present coastal climate.

The correlation of the two records allows us to construct a scenario of the paleoceanographic and paleoclimatic events that occurred in this region during the past 20,000 yr.

PALEOCLIMATIC AND PALEOCEANIC RECONSTRUCTION

Figure 10 is a columnar diagram that summarizes the changes we found. Approximately 18,000 yr ago, calcium carbonate increased from <2 to >3 percent. The low values of the carbonate throughout the entire core, as well as the preservation and composition of the floral and faunal assemblages, suggest that dissolution may have been moderately active. The increase in carbonate preservation at 18,000 yr coincides with the initial global deglaciation. By about 16,000 yr ago, the nannofossil flora appears to have shifted from a subpolar to a transitional assemblage. Because nannofossils are pelagic, photosynthetic algae that reflect environmental parameters of the surface waters, they should be among the first organisms to respond to changes in the ocean environment. The distribution of modern coccoliths in the Pacific show that *C. pelagicus* is a member of the subpolar flora (Geitzenauer and others, 1976).

C. leptoporus, the species that replaced *C. pelagicus* at about 16,000 yr, is an important species of the transitional flora. Maps from Roth and Coulbourn (1982) show that the modern distribution of *C. leptoporus* has maximum abundances north of 40°N, a boundary very close to the location of core V1-80-P3.

Thus, the nannofossil data, although low in abundances, suggest that subpolar water may have penetrated as far south as 38°N during the last maximum global glaciation conditions, al-

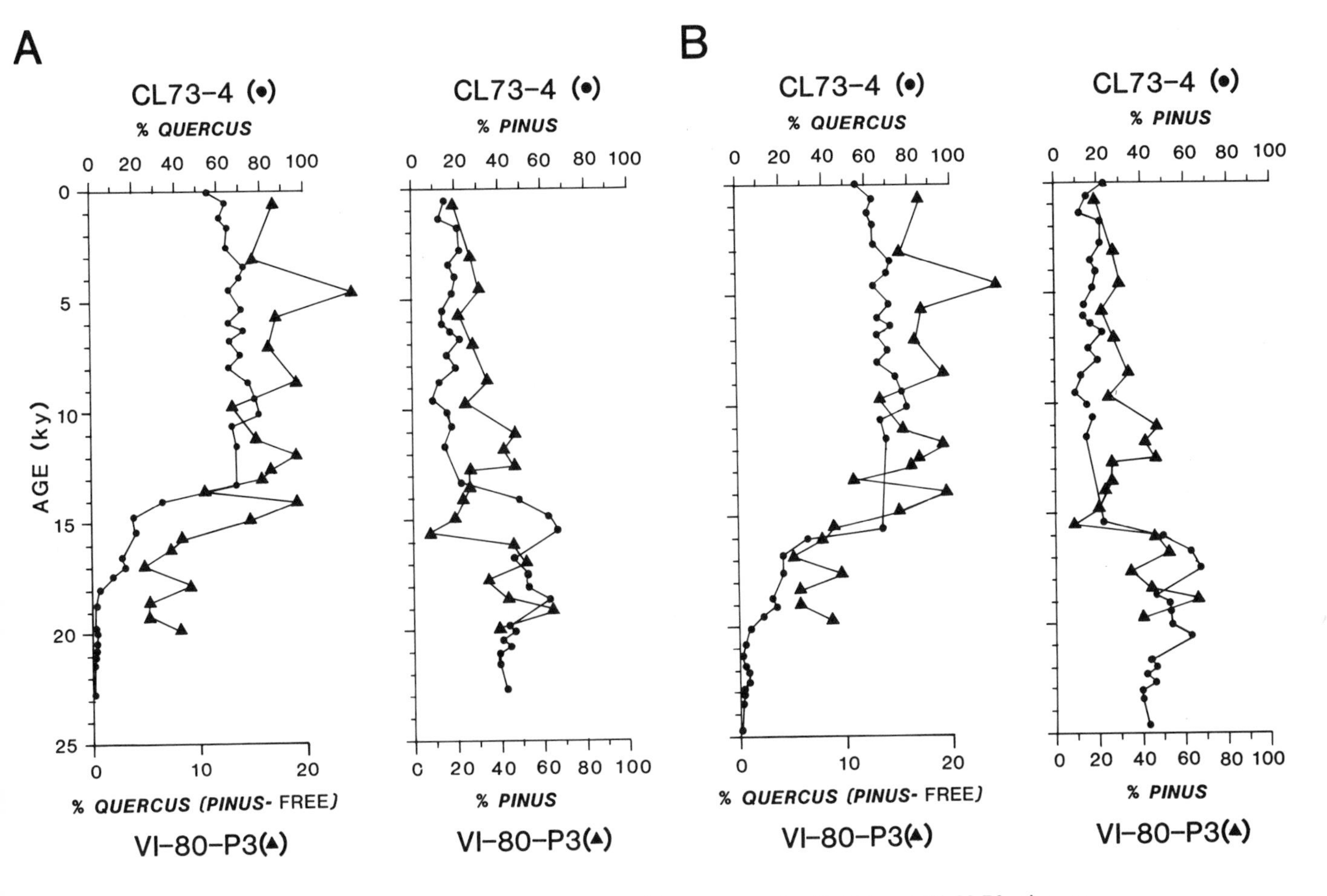

Figure 9. A, Plots of percentages of *Quercus* and *Pinus* vs. age for CL73-4 and V1-80-P3 using age model of Adam and others (1981) for CL73-4. B, Plots of percentages of *Quercus* and *Pinus* vs. age for CL73-4 and V1-80-P3 using adjusted age model for CL73-4 (see text for explanation). Data for CL73-4 are taken from Figure 2 of Adam and others (1981).

though probably not much farther. It appears that the first response to global deglaciation in this region was the retreat of subpolar waters, probably by reduced circulation, and the intrusion of transition waters. At the same time, *Alnus* (alders) began to increase in abundance, suggesting the advent of continental deglaciation. Conditions of temperature and precipitation combined to provide an environment for *Alnus* on the western slopes of the California Coast Ranges at this time. These conditions amenable to *Alnus* persisted until about 7,000 yr ago.

Pine abundance began to decline about 15,500 yr ago, about the same time that the sediment facies changed from a sandy silt to a silty clay. The facies change was probably the result of the initial marine transgression caused by the eustatic rise in sea level. The gradient of the continental shelf is quite gentle in this area, so that a small rise in sea level would have produced a relatively large transgression. This would have effectively moved the locus of sand input farther away from the deposition site. The reduced *Pinus* abundance may reflect a moderation of the coastal climate in response to the encroachment of the maritime environment or the increased distance of the input point.

Diatom assemblages, especially the increase toward dominance of *T. nitzschioides* and *T. longissima,* suggest that upwelling became significant in this region by about 15,000 yr ago. Organic carbon content of the bulk sediment reflects the increased upwelling with an increase from an average of about 1 percent (by weight) to an average value of about 1.5 percent. Planktonic foraminiferal assemblages show a replacement of a cold subpolar fauna by a transitional fauna at about this time. The appearance of warmer water at the same time as increased upwelling suggests the upwelling may have been seasonal.

As global deglaciation continued, planktonic foraminiferal faunas changed at 13,500 to 14,000 yr from a transitional to a cooler subpolar assemblage. This faunal change implies a continued cooling of surface waters, which is also suggested by the continued increase in dominance of *T. nitzschioides,* which is associated with cold water upwelling. *Quercus* had increased to

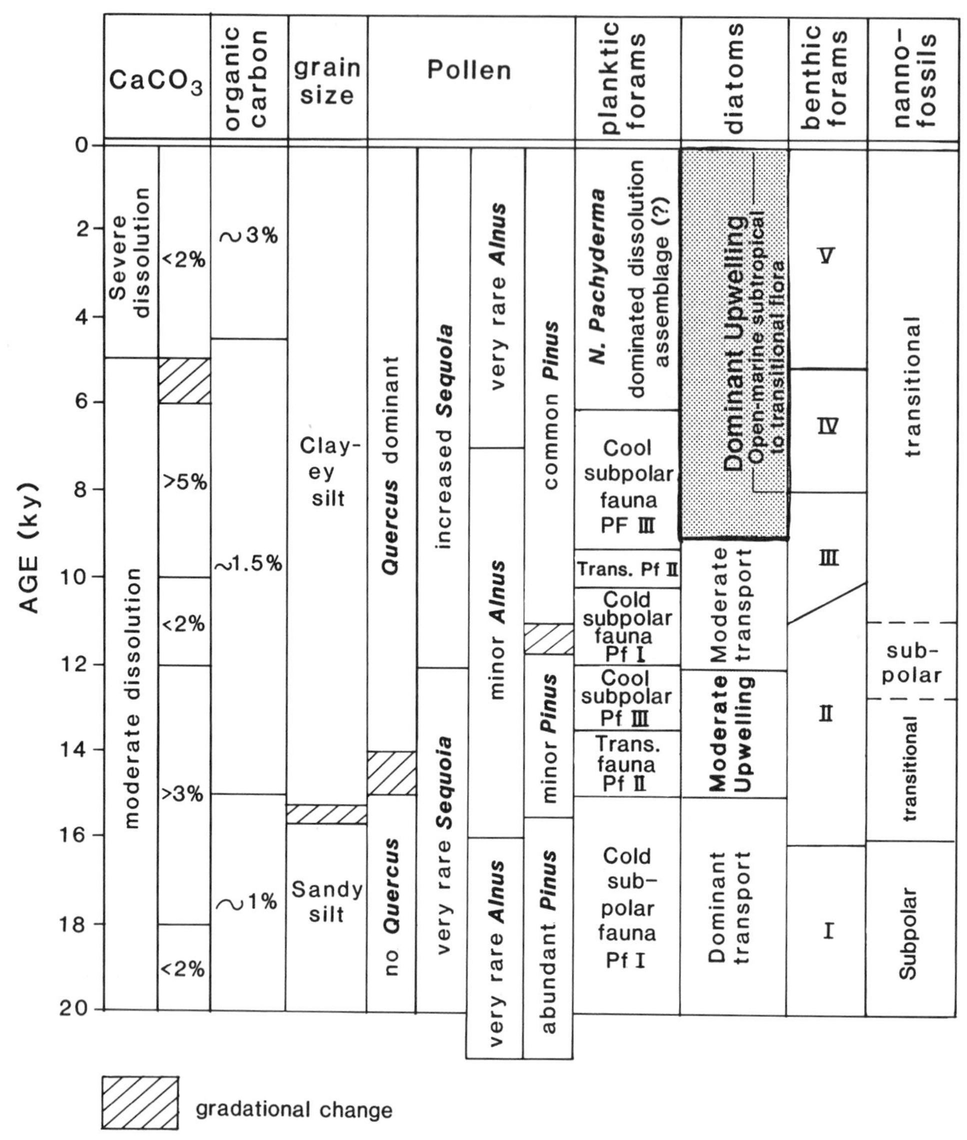

Figure 10. Summary diagram of faunal, floral, and sediment parameter changes during the last 20,000 yr, from records of V1-80-P3 and CL73-4.

modern abundances by about 14,000 yr. *Sequoia* requires temperatures and effective moisture in the form of coastal fog, and *Sequoia* pollen were not very abundant at this time. These observations suggest that upwelling did occur, but as a seasonal phenomenon, and the intensity of upwelling was not strong (i.e., the depth from which the upwelling waters rose was relatively shallow). Also, upwelling and temperature contrasts between the surface and land surface apparently were not great enough to produce strong seasonal coastal fogs.

The major portion of the marine transgression should have been completed by about 12,000 yr ago. Cool subpolar planktonic foraminiferal assemblages were replaced at this time by a colder subpolar assemblage, suggesting increased circulation and cooling of sea-surface temperatures. Upwelling, best represented by the diatoms *T. nitzschioides* and *T. longissma,* shows a reduced influence at this time. Curiously, this period also experienced an increase in *Sequoia,* which at first would suggest increased (not decreased) upwelling to produce a greater land-sea temperature contrast and coastal fogs. *Quercus* continued to dominate flora of the California Coastal Ranges, so that the increase in *Sequoia* may only indicate an increase in coastal precipitation. Benthic foraminiferal assemblages changed about 11,000 to 10,000 yr ago, probably reflecting the effects of the newly established high sea-level stand.

By about 10,500 yr ago, the subpolar planktonic foraminiferal faunas were replaced by transitional faunas, an indication of another period of reduced circulation and consequent sea-surface warming. Then, in succession, a cooling of sea-surface temperatures occurred at about 9,500 yr ago, indicated by the return of subpolar planktonic foraminiferal faunas; by about 9,000 yr ago significant upwelling was re-established. Benthic foraminiferal assemblages showed a change about 8,000 yr ago in water masses; by about 8,000 yr ago, a modern assemblage of diatoms, dominated by the upwelling-indicators *T. nitzschioides* and *T. longissma* (but with a common component of the open-marine subtropical-to-transitional flora associated with *P. doliolus*), was firmly established in the area. *Sequoia* pollen abundances suggest seasonal upwelling was developed, and by about 7,000 yr ago the region had essentially modern characteristics.

The latest changes recorded in our data sets, changes in benthic and planktonic foraminiferal faunas, occur together with a decrease in calcium carbonate and an increase in organic carbon content. We believe these changes are related to increased dissolution that occurred at about 5,000 yr ago, although the cause of this increased dissolution is still unknown.

CONCLUSION

This paper presents a scenario of the changing paleoclimatic and paleoceanographic conditions that prevailed in the California Coast Ranges and adjacent continental margin. We used a multivariate data set from only two reference points, and are aware of the limitations this represents; yet, we believe, there was much gained from such a limited study: (1) We successfully correlated the pollen record from the upper part of Clear Lake core CL73-4 with the pollen record of a deep-sea core. (2) We found that an interpretation of the two data sets gives a very complex progression of changes that are probably interrelated. (3) We found that the changes in climatic and oceanographic environmental conditions that occurred as responses to the change from global glacial to interglacial conditions did not represent a smooth progression of increasingly moderate regimes, but rather a complicated series of states suggesting a disequilibrium mode that lasted from about 15,000 to 5,000 yr ago.

Oceanographic conditions such as sea-surface temperatures and seasonal upwelling, together with local and regional geography, determine the climate of a region such as the California Coast Ranges. Consequently, the chance to compare the concurrent responses of the ocean and immediate continental area to global climate change may provide important information for global climate models and may eventually lead to unraveling the feedback systems that undoubtedly play important roles in the overall climate system.

ACKNOWLEDGMENTS

We thank the officers and crew of the research vessel *Velero IV* whose cooperation in this study was essential. We also acknowledge the work of David Klise and Christine Wilson, whose analytical work helped to provide the data base. The constructive reviews of Vera Markgraf, Phil Weaver, and Walter E. Dean are appreciated. This study was funded in part by the USGS Climate Program.

REFERENCES CITED

Adam, D. P., 1967, Late Pleistocene and Recent palynology in the central Sierra Nevada, California, *in* Cushing, E. J., and Wright, H. E., Jr., eds., Quaternary paleoecology: New Haven, Connecticut, Yale University Press, p. 275–301.

Adam, D. P., Sims, J. D., and Throckmorton, C. K., 1981, 130,000-yr continuous pollen record from Clear Lake, Lake County, California: Geology, v. 9, p. 373–377.

Barron, J. A., and Keller, G., 1983, Paleotemperature oscillations in the Middle and late Miocene of the northeastern Pacific: Micropaleontology, v. 29, p. 150–181.

Blake, G. H., and Douglas, R. G., 1980, Pleistocene occurrence of *Melonis pompilioides* in the California borderland and its implication for foraminiferal paleoecology: Cushman Foundation Special Publication 19, p. 59–67.

Bradshaw, J. S., 1959, Ecology of living planktonic foraminifera in the North and equatorial Pacific Ocean: Contributions to Cushman Foundation Foraminiferal Research, v. 10, p. 25–64.

Coulbourn, W. T., Parker, F. L., and Berger, W. H., 1980, Faunal and solution patterns of planktonic foraminifera in surface sediments of the North Pacific: Marine Micropaleontology, v. 5, p. 329–399.

Duplessy, J. C., Delibrias, G., Turon, J. L., Pujol, C., and Duprat, J., 1981, Deglacial warming of the northeastern Atlantic Ocean; Correlation with the paleoclimatic evolution of the European continent: Palaeogeography, Palaeoclimatology and Palaeoecology, v. 35, p. 121–144.

Gardner, J. V., and 7 others, 1983, Quantitative microfossil, sedimentological, and geochemical data on cores V1-80-P3, V1-80-G1, and V1-80-P8 from the continental slope off northern California: U.S. Geological Survey Open-File Report, 83-83, 51 p.

Geitzenauer, K. R., Roche, M. B., and McIntyre, A., 1976, Modern Pacific coccolith assemblages; Derivation and application to Late Pleistocene paleotemperature analysis, *in* Cline, R. M., and Hays, J. D., eds., Investigation of late Quaternary paleoceanography and paleoclimatology: Geological Society of America Memoir 145, p. 423–448.

Heusser, C. J., and Florer, L. E., 1973, Correlation of marine and continental Quaternary pollen records from the northeast Pacific and western Washington: Quaternary Research, v. 3, p. 661–670.

Heusser, L. E., and Balsam, W. L., 1977, Pollen distribution in the northeast Pacific Ocean: Quaternary Research, v. 7, p. 45–62.

Heusser, L. E., and Shackleton, N. J., 1979, Direct marine-continental correlation; 150,000-yr oxygen isotope–pollen record from the North Pacific: Science, v. 204, p. 837–839.

Lohman, G. P., 1978, Abyssal benthonic foraminifera as hydrographic indicators in the western South Atlantic Ocean: Journal of Foraminiferal Research, v. 8, p. 6–34.

McIntyre, A., Be, A.W.H., and Roche, M. B., 1970, Modern Pacific coccolithosphorida; A paleontological thermometer: Transactions of the New York Academy of Science, v. 32, p. 720–731.

Pisias, N. G., and Moore, T. C., Jr., 1981, The evolution of Pleistocene climate; A time-series approach: Earth and Planetary Science Letters, v. 52, p. 450–458.
Roth, P. H., and Coulbourn, W. T., 1982, Floral and solution patterns of coccoliths in surface sediments of the North Pacific: Marine Micropaleontology, v. 7, p. 1–52.
Sancetta, C., 1983, Biostratigraphic and paleoceanographic events in the eastern equatorial Pacific; Results of DSDP Leg 69, *in* Cann, J., White, S., and others, Initial Reports DSDP, v. 69, p. 311–319.
Schnitker, D., 1974, West Atlantic abyssal circulation during the past 120,000 years: Nature, v. 248, p. 385–387.
Shackleton, N. J., and Opdyke, N. D., 1973, Oxygen isotope and paleomagnetic stratigraphy of equatorial Pacific core V28-238; Oxygen isotope temperatures and ice volumes on a 10^5-year and 10^6-year record: Quaternary Research, v. 3, p. 39–55.
Streeter, S., 1973, Bottom water and benthonic foraminifera in the North Atlantic; Glacial-interglacial contrasts: Quaternary Research, v. 3, p. 131–141.

MANUSCRIPT ACCEPTED BY THE SOCIETY SEPTEMBER 15, 1986

Printed in U.S.A.

Geological Society of America
Special Paper 214
1988

Fish evolution and the late Pleistocene and Holocene history of Clear Lake, California

John D. Hopkirk
Department of Biology
Sonoma State University
Rohnert Park, California 94928

ABSTRACT

Clear Lake in Lake County, California, has an endemic fish fauna composed of five lake-adapted forms derived from lowland stream-adapted forms present in surrounding drainage basins. Two of the five endemic forms are extinct. The three remaining endemics maintain themselves despite the destruction of sloughs and tule beds surrounding Clear Lake that are used for spawning and nursery areas. Trophic specializations of the endemic fishes indicate past selection for feeding on small benthic and pelagic invertebrates. The presence of fine particles in the substrate and the reduced activity of tributary streams for at least the past 10,000 yr are major hydrographic features contributing to the evolution of these trophic adaptations.

Subfossil scales of the endemic Clear Lake tuleperch, (*Hysterocarpus traskii lagunae*) present in three U.S. Geological Survey cores (CL-73-7, -6, and -8), removed from the bottom of Clear Lake in 1973 were analyzed by Casteel and others (1975, 1977a, b, 1979) for age and growth rate. Periods of increased scale growth were inferred to represent warming of the lake. Comparison of the Casteel data with pollen data (Adam and others, 1981) indicate that maximum scale growth (core CL-75-8) occurred at about 19 ka (=15 ka, according to Robinson and others, this volume) during a cold interval. Fluctuations in scale density in cores CL-73-4 and CL-73-7, however, seem to follow fluctuations in oak pollen. It is therefore concluded that maximum-scale growth represents cool periods, whereas maximum-scale density represents warm periods in the history of the lake. During the period that maximum-scale growth occurred, Clear Lake basin may have also been closed off from surrounding basins and the lake enriched with nutrients.

INTRODUCTION

The evolution of Clear Lake basin played a major role in the evolution of California fresh-water fishes. The importance of the basin in fish evolution was first realized during an analysis of morphologic variation in the tuleperch (*Hysterocarpus traskii*) (Hopkirk, 1962). The study was prompted by an earlier reference by Tarp (1952) to the existence of a "lacustrine forma" of the tuleperch present in Clear Lake and possibly other lakes of California and a "fluviatile forma" of the same species in the rivers of California. Analysis of the variations in the tuleperch indicated the presence of three subspecies, rather than of two ecologic formae or ecophenotypes: one subspecies is confined to Clear Lake basin, one to the Russian River drainage basin, and one to the remainder of the species range (Sacramento–San Joaquin, San Francisco Bay, and Salinas-Pajaro drainage basins).

Subsequent analysis of other fish species native to Clear Lake basin revealed that at least 5 of the 14 forms then known from the basin were endemic or peculiar to it (Hopkirk, 1967, 1974). The endemic fishes of Clear Lake basin were lake-adapted derivatives of primarily lowland, river-adapted forms found in the Central Valley of California. The extent of differentiation in these fishes indicates that the lake is much older than the few hundred years theorized by Hinds (1952). Taylor (1966) concluded that Clear Lake and Klamath Lake were the only lakes in North America still containing large endemic molluscan faunas similar

Figure 1. Map of Clear Lake, Lake County, California, showing location of 1973 and 1980 core sites (after Rymer and others, 1981). Cores taken in 1973 shown by single-digit numbers. North end of lake shown as it was known to the Pomo Indians (after Kniffen, 1939).

to those that existed during late Pliocene and early Pleistocene time.

In 1973 our knowledge of the history of Clear Lake was vastly improved when the USGS removed eight 12- to 15-cm diameter cores ranging from 13.9 to 115.2 m in length from the lake bottom (Fig. 1). Sims and others (1981) have conclusively shown that the lake is at least 200,000 yr old. If one believes that Clear Lake is a remnant of an ancestral lake represented by deposits of the Cache Formation, then the lake is approximately 1.8 to 3.0 m.y. old (Casteel and Rymer, 1981).

This paper reexamines the fish evolution in Clear Lake basin in light of recent studies on the comparative morphology of Clear Lake fishes and preliminary analyses of material and literature on late Pleistocene and Holocene subfossil fish remains obtained through recent investigations of the paleolimnology of Clear Lake by Sims and colleagues of the U.S. Geological Survey.

NATIVE FISH FAUNA OF CLEAR LAKE BASIN

Clear Lake is presently dominated by introduced species that are characteristic of the low-elevation warm-water areas of the Central Valley of California. The community of fishes found in this type of habitat has been given various names: "fish association one" (Murphy, 1948); "hitch zone" (Hopkirk, 1967, 1974); "introduced fishes association" (Moyle and Nichols, 1973); "deep-bodied fishes zone" (Moyle, 1976); and "introduced warmwater fishes zone" (Moyle and Daniels, 1982).

Historical changes in the fish fauna of Clear Lake are described by Cook and others (1966) and Moyle (1976). Three primary factors caused the decline (or extinction) of native species in the lake: reclamation of peripheral sloughs and channelization of streams for agricultural purposes, removal of tules from around the edge of the lake, and introduction of exotic species.

The drainage of Tule Lake and peripheral sloughs and the destruction of the lower reaches of tributary streams had an adverse effect on species that spawned in those areas, namely the Sacramento perch, Clear Lake tuleperch, thicktail chub, Clear Lake roach, and Clear Lake splittail (the last three species are now extinct). The removal of tules removed the protective cover for young fish (verified by Ralph Holder, an Eastern Pomo Indian who fished the lake in the early 1900s, and by Wendy Jones and Larry Week, biologists for the California Department of Fish and Game). The introduction of exotic fishes provided competitors and predators not present before. Their influence on the native fish fauna is difficult to assess although it must have been substantial.

The following annotated list briefly updates our knowledge of the fishes native to the basin.

Native Fishes of Clear Lake Basin.

Family Petromyzontidae. Lampreys.

***Lampetra tridentata* (Gairdner, *in* Richardson, 1836).** Pacific lamprey. Jordan and Gilbert (1894) reported this species "occasionally taken" in Clear Lake. Ammocoetes recently collected from Kelsey Creek may belong to this species (Taylor and others, 1982); none were seen prior to that time.

***Lampetra pacifica* Vladykov, 1973.** Coast Range brook lamprey ("brook lamprey"). Taylor and others (1982) recorded larval lampreys from the lower sections of Kelsey Creek; Moyle and others (1982) referred to these as brook lampreys. If the Coast Range brook lamprey is native to Clear Lake basin, and there is no reason to doubt it zoogeographically, then the lamprey may possibly represent a new subspecies. Both Coast Range brook lamprey and the Pacific lamprey are present in drainages adjacent to Clear Lake basin.

Family Salmonidae. Trouts.

***Salmo gairdnerii* Richardson, 1836.** Rainbow trout. Rainbow trout occur in every stream and drainage of the basin (Taylor and others, 1982: Fig. 6). They are now rare in the lake. Stone (1874) recorded "trout" in tributaries of the lake and "salmon trout" in the lake itself. "Salmon trout" may have been steelhead from Cache Creek (see Jordan and Gilbert, 1894).

Family Cyprinidae. Minnows.

***Pogonichthys ciscoides* Hopkirk, (1974).** Clear Lake splittail. This endemic species was, until 1940, the most abundant species in the lake. Its abundance was documented by Lindquist and others (1943) and confirmed by Ralph Holder (Hopkirk and McLendon, MS in preparation). The Eastern Pomo name for this species was "hitch," and Kelsey Creek, the major spawning stream for it, was referred to as "hitchbidame" or literally, hitch creek. The Clear Lake splittail was last collected in the late 1960s (Shapovalov and others, 1981).

***Gila crassicauda* (Baird and Girard, *in* Girard, 1854).** Thicktail chub. The thicktail chub, endemic to the Sacramento fish province (see Hopkirk, 1974) and a member of the widespread Western genus *Gila,* was last collected in 1957 in the Sacramento River Delta (Shapovalov and others, 1981). Only two specimens, collected in 1873 and 1938, are known from Clear Lake (Miller, 1963). Based on information provided by Ralph Holder, we know that the thicktail chub or "Indian carp" spawned in sloughs. Along with the black chub, it was often seen in large numbers swimming along the edge of the lake (Hopkirk and McLendon, in preparation).

***Ptychocheilus grandis* (Ayres, 1854).** Sacramento squawfish. Although now rare in the lake, the Sacramento squawfish is still present in fair numbers in the larger exposed sections of Kelsey and Scott Creeks. It was present at 15 (13 percent) of the localities sampled by Taylor and others (1982). In times past, it was undoubtedly better represented in the lake and its tributaries.

***Mylopharodon conocephalus* (Baird and Girard, *in* Girard, 1854).** Hardhead. The hardhead was identified by Casteel, but not emphasized as a new locality record for the basin, in subfossil fish remains from core 8 made by the USGS (Casteel and others, 1979). A fragment of a pharyngeal bone and a tooth of the hardhead were present. Casteel and Rymer (1981) later officially recorded the hardhead from the Kelseyville Formation (Pleistocene) of Clear Lake basin. A pharyngeal bone with teeth intact was recently identified by Hopkirk in material removed from an archeological site north of the exit of the lake into Cache Creek. Because of its widespread occurrence elsewhere, its presence in Clear Lake basin was anticipated (Hopkirk, 1974). When it disappeared from the basin, however, is still unknown.

***Hesperoleucus symmetricus symmetricus* Baird and Girard, *in* Girard, 1854.** Sacramento roach. This species has either been introduced into the basin or represents a recent natural arrival into it: the latter hypothesis may be the correct one. Taylor and others (1982; Fig. 7) found roach at 46 percent of their stream collecting sites. Cook and others (1966) never collected this species from the lake, but did collect it from tributary streams.

***Hesperoleucus grandipinnis* (Hopkirk, 1974) new combination.** Clear Lake roach. This species, considered to be extinct, was formerly placed by Hopkirk (1974) in a new genus and species, *Endemichthys grandipinnis.* Because the name *Endemichthys* Hopkirk, (1974) is preoccupied by *Endemichthys* Forey and Gardiner, 1973, it can no longer be used. Because of certain features that seem to relate it to the genus *Hesperoleucus,* I have assigned it to that genus. The 12 specimens on which the taxon is based do not, as considered by Hubbs (1974) and Coad and Quadri (1978), represent hybrids; the type material consists of males and females with a sexual dimorphism in size (females larger) and in "ripe" condition.

***Lavinia exilicauda chi* Hopkirk, (1974).** Clear Lake hitch. The Clear Lake hitch is at present the most abundant native species in the lake. As mentioned earlier, this species was not the "hitch" of the Eastern Pomo. Their name for this species was "chigh" (Hopkirk and McLendon, in preparation). Recent workers (Avise and others, 1975; Moyle and Daniels, 1982) have, on the basis of electrophoretic data, decided to merge the genera *Hesperoleucus* and *Lavinia*. Because additional information is needed, my preference is to keep the preceding genera as separate taxa.

Orthodon microlepidotus **(Ayres, 1854).** Sacramento black chub ("Sacramento blackfish" of other authors). An abundant species in Clear Lake, this species spawns in the lake but not in its tributaries (Cook and others, 1966; Taylor and others, 1982). In other areas of its geographic range, *Orthodon* spawns and survives in low-gradient streams. It is relatively abundant in the Sacramento–San Joaquin Delta. Casteel and others (1977, 1979) found pharyngeal teeth of this species commonly represented in the USGS cores. Casteel and Rymer (1975) also found a pharyngeal tooth of *Orthodon* in the Cache Formation of Clear Lake basin west of the town of Lower Lake. Fossil mammals indicate a late Pliocene or early Pleistocene age for the Cache Formation of Anderson (1936). Pharyngeal teeth of this genus have been found at eight other fossil localities, ranging from Pliocene to late Pleistocene in age, in central California (Casteel and Hutchinson, 1973).

Family Catostomidae. Suckers.

Catostomus occidentalis occidentalis **Ayres, 1854.** Sacramento sucker. Taylor and others (1982) found suckers widely distributed at lower elevations and in the larger streams of the basin. They were present at 49 (41 percent) of the stream sites sampled. Suckers are at present uncommon in the lake. In times past, however, they were commonly present. In fact, two different spawning populations, stream and lake, may have existed within the basin, comparable to what occurs within the Tahoe sucker (*Catostomus tahoensis*) population of Lake Tahoe (Willsrud, 1971).

Family Gasterosteidae. Sticklebacks.

Gasterosteus aculeatus microcephalus **Girard, 1854.** Semiarmored threespine stickleback. Taylor and others (1982) found the stickleback only in Cole Creek. Cook and others (1966) reported the collection of two specimens from Clover Creek, near the town of Upper Lake. Based on a very small sample, Kevin Howe (personal communication) found the Clear Lake sticklebacks to have comparatively long dorsal spines. Casteel (1977, 1979) found the threespine stickleback to be the most abundant species, in terms of bone fragments, in USGS cores from Clear Lake. We may be witnessing the demise of an undescribed subspecies with a continuous fossil record of 130,000 yr.

Family Cottidae. Sculpins.

Cottus asper **Richardson, 1836, subspecies unnamed Hopkirk, (1974).** Clear Lake prickly sculpin. Krejsa (1965) described a number of "phenotypes," actually subspecies, within the species *asper*. Because he was misled by previous records of the riffle sculpin (*Cottus gulosus*) from Clear Lake, based on misidentifications, he failed to consider the prickly sculpin of Clear Lake basin. In 1974, I briefly described the Clear Lake prickly sculpin but did not name it. In their recent survey of the streams of the basin, Taylor and others (1982) did not find either *Cottus asper* or *C. gulosus* at any of their collecting sites. Cook and others (1966) never found *C. asper* outside of the lake. *Cottus gulosus* may be a recent extinction, similar to the hardhead, for the basin. Casteel and others (1977b) found a number of vertebrae and skull bones of *Cottus* (presumably *C. asper*) in core CL-73-6 from Clear Lake. The present population of the prickly sculpin is apparently maintaining itself even though other native species have either declined or disappeared.

Family Centrarchidae. Sunfishes.

Archoplites interruptus **(Girard, 1854).** Sacramento perch. Although widespread in the fossil record and widely introduced successfully in the West, the Sacramento perch has declined drastically within its original range. It survives in a few isolated farm ponds and localities that are unsuitable for introduced sunfishes. Although now rare in the lake (Peter Moyle, personal communication), it is commonly represented in material from the cores.

Family Embiotocidae. Viviparous Perches.

Hysterocarpus traskii lagunae **Hopkirk, (1974).** Clear Lake tuleperch. This endemic form is still common in the Blue Lakes and Clear Lake. Baltz and Moyle (1981) recently evaluated the systematics of the tuleperch and confirmed the earlier conclusions (Hopkirk, 1974). Casteel and others (1975, 1977b, 1979), in the analysis of cores from the lake, found the tuleperch to be the dominant species in terms of scales preserved in the bottom of the lake. Based on an age and growth study of the scales, Casteel and others (1977a), concluded that maximum-scale growth was correlated with warm periods in Clear Lake.

Endemic Fishes of Clear Lake Basin

Three of the five endemic forms are members of the minnow family (Cyprinidae): *Pogonichthys ciscoides, Hesperoleucus grandipinnis,* and *Lavinia exilicauda chi.* The first two forms are extinct; our future knowledge of them will therefore be based on museum material and evidence from ethnography, archeology, and paleontology. Because members of the family Cyprinidae are soft-rayed fishes with bones and scales that readily decompose, future knowledge of them from the fossil record will be based primarily on their pharyngeal bones and teeth. The remaining endemic forms are spiny-rayed fishes: *Hysterocarpus traskii lagunae* and *Cottus asper* subspecies. Additional endemic fishes may eventually be described from Clear Lake basin.

TABLE 1. GILL RAKER COUNTS OF LAKE AND RIVER POPULATIONS FROM THE CLEAR LAKE REGION*

Population (Lake/River)	Mean (Sample Size)	Index of Divergence
Gila pectinifer/G. bicolor obesa	32.2(30)/16.7(82) (Lake Tahoe)	1.93
Hesperoleucus grandipinnis/H. symmetricus	15.0(12)/9.0(31) (Central Valley)	1.67
Hesperoleucus grandipinnis/H. symmetricus	15.0(12)/10.2(40) (North Fork, Consumnes River)	1.47
Pogonichthys ciscoides/P. macrolepidotus	21.5(28)/15.8(67) (Central Valley)	1.36
Lavinia exilicauda chi/L. e. exilicauda	28.3(29)/21.4(13) (Russian River)	1.32
Hysterocarpus traskii lagunae/H. t. traskii	23.7(62)/19.0(264) (Central Valley)	1.25
Lavinia exilicauda chi/L. e. exilicauda	28.3(29)/23.5(40) (Central Valley)	1.20
Hysterocarpus traskii laguna/H. t. pomo	23.7(62)/21.9(78) (Russian River)	1.08
Ptychocheilus grandis/P. grandis	11.1(48)/10.9(35) (Central Valley)	1.01
Archoplites interruptus/A. interruptus	29.1(32)/29.0(61) (Central Valley)	1.00
Cottus asper/C. asper	6.4(5)/6.4(9) (Napa River)	1.00
Orthodon microlepidotus/O. microlepidotus	33.2(9)/34.9(33) (Central Valley)	0.95

*Except for Gila, all comparisons are between Clear Lake and surrounding river systems. Gill raker data based on counts made on right first gill arch of preserved material.

All of the endemic forms exhibited what I (Hopkirk, 1974) referred to as lacustrine adaptations. These adaptations were similar to those of other species adapted for living in lakes. They were at one end of the spectrum of adaptations that occurs between populations inhabiting small creeks and populations inhabiting large lakes. Lacustrine adaptations are comparable to pelagic adaptations of "off-shore" marine species. Fluviatile adaptations, those seen in "riverine" or stream fishes, are comparable to benthic adaptations of "in-shore" or bay species.

Endemism in the fishes of Clear Lake basin was attributed to fluviolacustrine speciation. This was one of five possible patterns of geographic isolation and speciation in fresh-water fishes (Hopkirk, 1974). Fluviolacustrine speciation results from the evolution, and subsequent geographic isolation, of a lake on a river system. In the new lake basin, the original river organisms are forced to adapt to a lake environment: more than one lake basin can be evolved, more than one river system can be involved in the process of fluviolacustrine speciation. Based on the extent of the lacustrine adaptations present, and the species that exhibit those adaptations, predictions can be made about how the lake evolved, the age of the lake, and the past ecology of the lake.

TROPHIC ADAPTATIONS OF CLEAR LAKE ENDEMIC FISHES

The morphologic adaptations of fishes can be divided into a number of categories, such as trophic or feeding related, natatory or locomotory, protective or defensive, respiratory, and reproductive. Of the many adaptations seen in the endemic fishes of Clear Lake, trophic adaptations are the most extreme and diagnostic.

Trophic Adaptations

The major adaptive difference between lake and river populations of the Clear Lake region was in a trophic character, the gill raker number (Table 1). Gill rakers sieve food particles from either the water or the substrate, protect the gill filaments, supplement the valvelike action of the gill filaments in separating buccal and opercular chambers, and taste food particles before they enter the esophagus.

The number, shape, length, and spacing of gill rakers relate to the diet of the fish. *Gila* (*Siphateles*) *pectinifer,* a Lahontan basin endemic with extremely fine gill rakers (greatest number of any minnow native to the United States: 29 to 42 counted on the first gill arch), presumably feeds on smaller foods than its river-lake adapted counterpart, *Gila* (*Siphateles*) *bicolor obesa* (Miller, 1951). *G. pectinifer* was confined to the boundaries of pluvial Lake Lahontan and evolved as a lake-adapted derivative of the widespread river species, *Gila* (*Siphateles*) *bicolor.* The origin of the species *pectinifer* antedates the origin of the subspecies *obesa* from *bicolor.* Both *pectinifer* and *obesa* apparently represent lake-adapted derivatives of *bicolor*: *pectinifer* may have originated with the first stage of Lake Lahontan, *obesa* with the third stage of Lake Lahontan.

The systematics of the two species (lake and lake-river) of *Gila* is confused by the occurrence of natural hydridization (sec-

TABLE 2. UPPER AND LOWER JAW LENGTHS OF LAKE AND RIVER POPULATIONS FROM THE CLEAR LAKE REGION

Population (Lake/River)	Mean (Sample Size)	Index of Divergence*
Upper Jaw Length:		
Pogonichthys ciscoides/P. macrolepidotus	63.2 (6)/55.3(10) (Central Valley)	1.14*
Lavinia exilicauda chi/L. e. exilicauda	54.1(8)/47.3(3) (Central Valley)	1.14*
Gila pectinifer/G. bicolor obesa	71.3(10)/70.9(7) (Pyramid Lake)	1.01*
Hysterocarpus traskii lagunae /H. t. traskii	88(56)/87(246) (Central Valley)	1.01
Archoplites interruptus/A. interruptus	152.5(8)/152.1(8) (Central Valley)	1.00
Hesperoleucus grandipinnis/H. symmetricus	68.1(12)/7508(15) (Central Valley)	0.90
Hysterocarpus traskii lagunae/H. t. pomo	88(65)/93(43) (Russian River)	0.95
Lower Jaw Length:		
Pogonichthys ciscoides/P. macrolepidotus	86.8(6)/71.1(12) (Central Valley)	1.22*
Lavinia exilicauda chi/L. e. exilicauda	77.1(11)/66.5(4) (Central Valley)	1.16*
Gila pectinifer/G. bicolor obesa	101.2(10)/97.0(7) (Pyramid Lake)	1.04*

*Lengths given in thousandths of standard length. Measurements are based on preserved material unless noted by an asterisk; those noted by an asterisk are based on dried skeletal material.

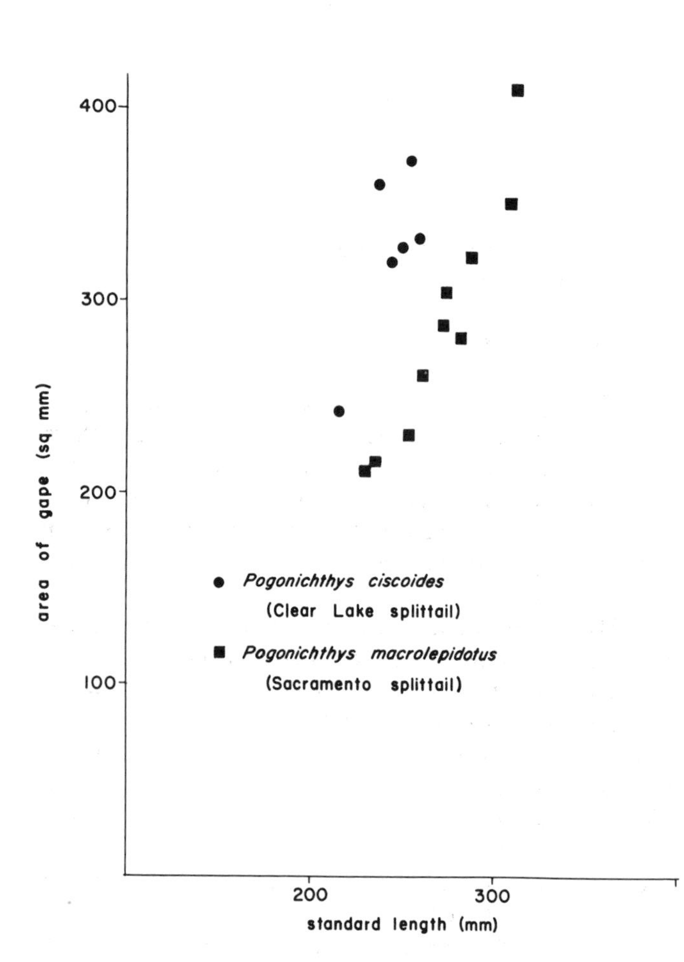

Figure 2. Estimated area of gape in square millimeters plotted against standard length for Clear Lake splittail (*Pogonichcthys ciscoides*) and Sacramento splittail (*P. macrolepitodus*).

ondary and introgressive) between the two forms in areas of sympatry. Gene flow from *pectinifer* to *obesa* through fertile F_1 hybrids is common in a number of lakes. However, as Carl Hubbs (1961) pointed out, the two forms are remarkable in maintaining their genetic integrity. Needless to say, reproductive mechanisms isolating the two forms are poorly understood. The great similarity in body shape and color often seen between the two forms is the result of introgressive hybridization, a beneficial process that allows the survival of a stream form (*bicolor obesa*) in an environment now deficient in streams.

Unlike the two forms of *Gila* in the Lahontan basin, the lake-adapted form (*ciscoides*) of *Pogonichthys* is essentially geographically isolated from its river congener (*macrolepidotus*) of the Sacramento River system. In terms of trophic adaptations, *ciscoides* parallels other lake-adapted minnows in possessing fine gill rakers (Table 1), long lower and upper jaws (Table 2) and therefore a larger gape (Fig. 2), and small pharyngeal arches (Table 3). Although advanced in the preceding features, the genus *Pogonichthys* still retains the primitive feature of two rows of teeth on each pharyngeal arch. The inner row of teeth is present in almost all cyprinid genera; the outer row, however, is reduced or absent in lake-adapted forms and in microphagous species (Table 4). A second evolutionary trend is for an increase in the number of inner row teeth, as seen in *"Endemichthys," Orthodon, Lavinia,* and *Gila* (*Siphateles*) (Table 4).

The increased jaw length in the lake minnows indicates that a greater volume of water (and plankton) can be taken in and passed over the gill rakers. The reduced pharyngeal arch implies that smaller foods (prey organisms) are represented in the diet of lake minnows. Even *Hysterocarpus,* a member of a completely different order of fishes, reveals the same evolutionary trend in the reduction of the pharyngeal arch. The Clear Lake tuleperch has a reduced surface area on the inferior pharyngeal arch for grinding up food (Fig. 3).

Hubbs and others (1974) emphasized the conservative nature of gill rakers. Once evolved, lacustrine types with many gill

TABLE 3. LENGTH, WIDTH, AND HEIGHT OF LOWER PHARYNGEAL ARCH IN LAKE AND RIVER POPULATIONS FROM THE CLEAR LAKE REGION*

Population (Lake/River)	Mean (Sample Size)	Index of Divergence*
Lower Pharyngeal Length:		
Pogonichthys ciscoides/P. macrolepidotus	65.3(38)/69.6(41) (Central Valley)	0.94
Gila pectinifer/G. bicolor obesa	81.3(20)/86.9(11) (Pyramid Lake)	0.94
Lavinia exilicauda chi/L. e. exilicauda	64.3(20)/68.8(40) (Central Valley)	0.93
Lower Pharyngeal Height:		
Pogonichthys ciscoides/P. macrolepidotus	51.0(20)/55.5(40) (Central Valley)	0.92
Gila pectinifer/G. bicolor obesa	42.8(35)/48.9(42) (Central Valley)	0.88
Lavinia exilicauda chi/L. e. exilicauda	54.2(20)/67.2(11) (Pyramid Lake)	0.81
Lower Pharyngeal Width (including teeth):		
Pogonichthys ciscoides/P. macrolepidotus	45.7(20)/48.3(40) (Central Valley)	0.95
Gila pectinifer/G. bicolor obesa	34.2(38)/38.3(42) (Central Valley)	0.89
Lavinia exilicauda chi/L. e. exilicauda	43.1(19)/53.3(10) (Pyramid Lake)	0.81

*Measurements given in thousandths of standard length (made with the aid of dial calipers). All measurements based on dried skeletal material.

TABLE 4. PHARYNGEAL TEETH OF LAKE AND RIVER POPULATIONS FROM THE CLEAR LAKE REGION

Population (Lake/River)	Mean (Sample Size)	Index of Divergence*
Outer Row (total no.):		
Pogonichthys ciscoides/P. macrolepidotus	1.36(44)/1.72(32) (Central Valley)	0.79
Both Arches (total no.):		
Hesperoleucus grandipinnis/H. s. symmetricus	10.2(12)/9.0 (Moyle, 1976) (Central Valley)	1.13
Lavinia exilicauda chi/L. e. exilicauda	10.0(13)/9.9(30) (Central Valley)	1.01
Gila pectinifer bicolor obesa	9.0(30)/9.1(29) (Lake Tahoe)	0.99
Pogonichthys ciscoides/P. macrolepidotus	12.64(22)/13.50(16) (Central Valley)	0.94

rakers are "slow" in undergoing an evolutionary reduction in number. The preceding authors cited (p. 146) as an example of this the Mohave tuichub, *Gila bicolor mohavensis* (Snyder, 1918), a lake derivative of *bicolor* adapted for pluvial Lake Mohave, which has persisted under postpluvial conditions that are conducive to fluviatile types (Hubbs and Miller, 1943).

Paleolimnologic Implications

Increased gill raker number and length, increased jaw length (hence increased gape), reduced dentigerous surface of the inferior pharyngeal bone, decreased height of the inferior pharyngeal bone, and a reduction in the number of pharyngeal teeth, as shown in the preceding tables and figures, suggest a prolonged period of selection for microphagy (feeding on small prey) in Clear Lake. The evolutionary progression toward microphagy means that a greater volume of water or substrate must be filtered. This requires a larger gape and a more active life style. A more active life style is indicated by an increase in the number of gill filaments (Hopkirk, 1967). Those species most influenced have been those feeding on zooplankton and small benthic insect larvae (mostly dipteran). Piscivores (Sacramento perch, Sacramento squawfish) and phytoplanktivores (Sacramento black chub) exhibit no great differences in trophic adaptations in Clear Lake.

The adaptations seen in the Clear Lake endemics thus lead to the assumption that Clear Lake has been completely or partially isolated from surrounding basins for long enough (late Pleistocene and Holocene time) for speciation to have occurred. The lake has been subjected to sedimentation (mostly clay, silt, and volcanic ash) and eutrophication for a long period of time; complete filling-in of the lake never occurred because of continued subsidence of the basin (Sims, 1976). Fine sediment has continually provided a substrate for dipteran larvae (Clear Lake is notorious for one of them, *Chaoborus astictopus,* the Clear Lake gnat), tules, and tule-associated organisms, and has created marshes and peripheral sloughs.

The endemic fish fauna of Clear Lake is not restricted to

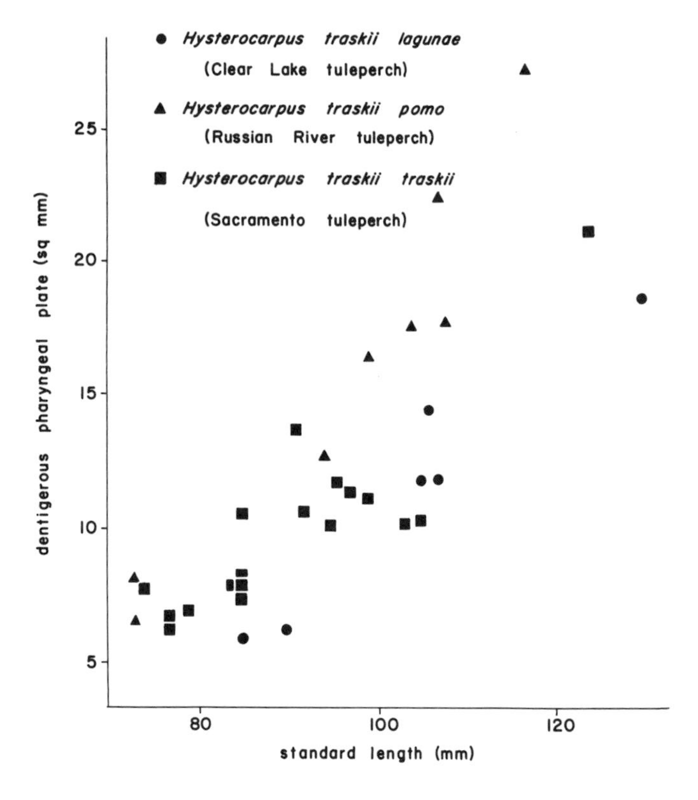

Figure 3. Surface area in square millimeters of dentigerous, or tooth-bearing, pharyngeal plate in the three subspecies of the tuleperch (*Hysterocarpus traskii*) (after Hopkirk, 1967).

Clear Lake but is also found in the Blue Lakes (Fig. 1) and other subbasins within the basin. Although fishes of the Blue Lakes diverge slightly from those of Clear Lake, they definitely belong to the Clear Lake population in terms of taxonomy. The amount of isolation and evolutionary divergence between the Blue Lakes and Clear Lake fishes is comparable to that seen between Lake Tahoe and Pyramid Lake fishes (Hopkirk, 1974).

Strong-flowing cold streams have not existed in the basin for thousands of years. Conspicuously absent from the basin are the riffle sculpin and the speckled dace. Both species are found in cold tributaries of Cache and Putah Creeks, immediately east of Clear Lake basin.

Fluviolacustrine speciation, considered previously as the type of geographic speciation responsible for the Clear Lake endemic fish fauna, is in the final analysis the product of substrate selection. In this example of substrate selection, however, fine particle size selects primarily for small interstitial organisms. Concealing coloration is of secondary importance. Benthic-feeding river fishes, when introduced into a lake environment, experience selection toward microphagy and planktiphagy, especially if tributary streams to the lake become inactive or intermittent. This type of selection has produced the endemic fishes of Clear Lake, Lake Biwa, and many other lakes characterized by a long history of fine-grained sedimentation.

SUBFOSSIL TULEPERCH SCALES AND THE PALEOLIMNOLOGY OF CLEAR LAKE

Fish bones and scales have been analyzed from a number of the 1973 Clear Lake cores, although fish bones and scales in the 1980 cores have yet to be analyzed. Casteel has analyzed fish subfossils from three of the cores removed in 1973: core CL-73-7 (Casteel and others, 1975, 1977a), core CL-73-6 (Casteel and others, 1977b), and core CL-73-8 (Casteel and others, 1979). I have analyzed scales, separated by the USGS, from the upper layers of core CL-73-4. Sims (1976) has described the physical features of the cores removed in 1973.

Preservation of Subfossil Fish Remains

Fish scales are represented in the last 25,000 yr of the cores examined so far. Unfortunately, the scales that are present represent only a limited portion of the total fish fauna. Most scales are those of the Sacramento perch and the Clear Lake tuleperch. These two species represent approximately one-eighth of the original fish fauna. Scales from the soft-rayed species, which composed the bulk of the biomass and the fauna, are rarely present and then appear only in the form of curled shreds. Conditions for preservation were adverse for scales with a low percentage of bone. Scales of the tuleperch are apparently ideal for preservation and often appear entire. Sacramento perch scales, on the other hand, are so bony and brittle that they usually appear as fragments.

Environmental conditions in shallow arms of the lake, as seen in cores CL-73-6 and -7, resulted in peat formation that may have destroyed many scales (see Casteel and others, 1977a). Preservation of tuleperch scales in core CL-73-7 was decidedly poor in comparison to preservation in cores CL-73-6 and -8. Core CL-73-7 has a preservation rate of 0.008 scales/yr, core CL-73-6 a rate of 0.016 scales/yr, and core CL-73-8 a rate of 0.025 scales/yr. The preceding rates are based on tuleperch scales present in the cores for roughly the last 11,000 yr.

Fish bones are usually fragmented and not especially common. Those seen are primarily those of the threespine stickleback. A few pharyngeal bones and teeth of tuleperch, minnows, and suckers are also seen.

Age and Growth Studies of Subfossil Tuleperch Scales

Casteel and others (1975, 1977b, 1979) aged subfossil scales of the tule perch, measured distances between year rings or annuli, and then estimated standard lengths in millimeters and weight in grams of the fish at the end of each year of growth or annulus. On the basis of age and growth data from core CL-73-7, the first core analyzed, Casteel and others (1975, 1977a) hypothesized that scale growth was essentially dependent on tempera-

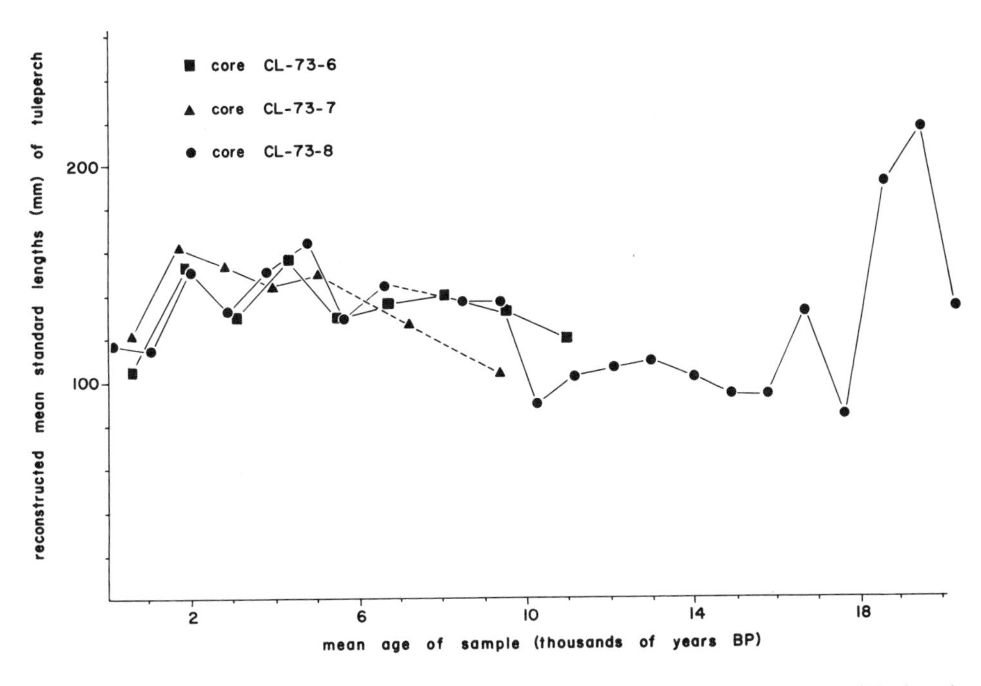

Figure 4. Reconstructed mean standard lengths of age class 3 tuleperch (*Hysterocarpus traskii*) plotted against mean age of core sample in thousands of years BP (based on data of Casteel and others, 1975, 1977b, 1979).

ture. It was inferred that warming of the lake began about 10 ka and ended about 2.8 ka. A marked period of warming occurred between about 4 ka and 2.8 ka. These findings appeared to be consistent with tree-line studies and palynology (Adam, 1967; La Marche, 1973).

Additional age and growth studies of subfossil tuleperch scales from cores CL-73-6 (Casteel and others, 1977b) and CL-73-8 (Casteel and others, 1979) augmented paleoclimatic trends reconstructed earlier from core CL-73-7. Based on the past 12,000 yr (8,000 years according to Robinson and others, this volume), scales from core CL-73-6 (Oaks Arm of the lake) exhibited the greatest average growth for fish in their first two years of life; core CL-73-8 (open-water area opposite the mouth of Rodman Slough), the least growth. Core 8 provided a longer history of tuleperch growth (and inferred temperatures) than did cores CL-73-6 and -7 (see Fig. 4). Age classes 1 and 2 exhibited their smallest estimated standard lengths during the latest Pleistocene (ca. 15 ka, or 11 ka of Robinson); age class 3 of core CL-73-8 had its smallest estimated standard length at about 17 ka (13 ka of Robinson). Core CL-73-8 had extraordinary growth at about 19 ka (15 ka of Robinson); the estimated standard length was 219 mm (Fig. 4). Even in the best environments (Sacramento Delta), the growth at the end of the third year in present populations is only about 150 mm in standard length.

A second possible explanation of growth rates in subfossil tuleperch scales is eutrophication. During the interval when growth was maximum in core CL-73-8, Kelsey Creek was actively bringing nutrients into the lake from the surrounding watershed. Clear Lake was ending a pluvial period comparable to that experienced by Lake Lahontan (see Benson, 1978). If we use the corrected radiocarbon datings of Robinson and others (this volume), Clear Lake was in the "late-glaciation transition" stage of Adam and others (1981). Fluctuations in scale growth seen in Figure 4 do not relate to fluctuations in oak pollen abundance (inferred warming) but instead to the fluctuations in abundance of pine and TCT (Taxodiaceae, Cupressaceae, and Taxaceae) pollen (inferred cooling).

The optimum water temperature for the tuleperch may well be much colder than previously thought. The tuleperch, as the only fresh-water member of a cold temperate marine family (Embiotocidae), would probably do well in moderately cold alkaline water. During the drought of 1976–1977 in California, the Russian River tuleperch became the dominant species in the alkaline Sonoma State University Pond (Rohnert Park, California), suppressing the introduced centrarchids. The Sacramento perch, an ecologic associate of the tuleperch and a native centrarchid, has been successfully introduced into a number of alkaline lakes in Nevada and elsewhere in the Great Basin (La Rivers, 1962; Moyle, 1976). Growth of the Sacramento perch is significant in these interior lakes.

Eutrophication of Clear Lake, if coupled with a closure of the drainage basin, would have resulted in a spectacular growth rate. Conversely, a dramatic opening of the basin to the Sacramento River system would have resulted in a drastic loss of nutrients and a sharp decline in growth. The greatest decline in growth occurred at either ca. 17.55 or 15.7 ka (13.35 and 11.5 ka, if corrected), depending on which age class is examined. A second explanation for a sharp drop in growth would be the rapid decline of a pluvial period. Pluvial Lake Lahontan lasted until 11.1 ka (Benson, 1978).

Besides influencing growth, opening of the basin would have allowed stream species to either repopulate basin streams or genetically merge with any stream population still present in the basin. A gradual reduction in the amount of water leaving the basin and a gradual increase in the nutrients coming into the lake would have resulted in a gradual increase in Holocene scale growth. Fluctuations within this gradually increasing growth trend, however, were brought about by temperature changes.

In conclusion, increasing growth rates seen in subfossil tuleperch scales may reflect declining water temperatures and increasing eutrophication. An increase in tuleperch scale density in the cores may indicate a warming trend that resulted in successful spawning and survival of young. The consequence of successful reproduction, however, would have been an increase in intraspecific competition and a decrease in growth rate. The small size of these tuleperch would have increased their vulnerability to the Sacramento perch and allowed that species to increase in size. A future age and growth study of the subfossil scales of the Sacramento perch from the USGS cores of Clear Lake is necessary to validate the latter hypothesis.

SUMMARY

Trophic adaptations in the endemic fishes of Clear Lake basin indicate a reduction in macroinvertebrates in the substrate and their replacement by microinvertebrates. The bottom of Clear Lake has been composed mostly of fine sediments for at least 130,000 yr. A coarse substrate developed around the mouth of Kelsey Creek when it was active during the last glacial period. Since then, tributary streams have been intermittent and have provided little in the way of a stream environment. An evolutionary trend from benthic macrophagy to benthic microphagy and planktiphagy has occurred.

Growth data derived from previous studies of subfossil tuleperch scales obtained from three USGS cores (CL-73-7, -6, and -8) of the bottom of Clear Lake suggest changes in temperature and in nutrients within the lake. Maximum estimated growth of scales occurred at about 19.4 ka (15.2 ka of Robinson) during a time period when the climate was cool and humid (indicated by the presence of pine and TCT pollen), and Kelsey Creek was actively bringing nutrients into the lake and providing a coarse stream-type substrate for macroinvertebrates. During the time period when maximum growth occurred, the basin may also have been isolated from surrounding basins.

ACKNOWLEDGMENTS

I am grateful to John D. Sims for inviting me to leave the classroom to again become active in studying the past history of Clear Lake and its fascinating endemic fishes. I leaned heavily on the creative endeavors of others, primarily Richard W. Casteel, David P. Adam, Michael J. Rymer, and John D. Sims, and thus I wish to acknowledge the contributions they have made to our understanding of the history of Clear Lake.

REFERENCES CITED

Adam,D. P., 1967, Late Pleistocene and Recent palynology in the central Sierra Nevada, California, *in* Cushing, E. J., and Wright, H. E., Jr., eds., Quaternary Paleoecology: Yale University Press, p. 275–301.

Adam, D. P., Sims, J. D., and Throckmorton, C. K., 1981, 130,000-yr continuous pollen record from Clear Lake, Lake County, California: Geology, v. 9, p. 373–377.

Anderson, C. A., 1936, Volcanic history of the Clear Lake area, California: Geological Society of America Bulletin, v. 47, p. 629–664.

Avise, J. C., Smith, J. J., and Ayala, F. J., 1975, Adaptive differentiation with the little genic change between two native California minnows: Evolution, v. 29, p. 411–426.

Ayres, W. O., 1854, New fishes: (San Francisco) Daily Placer Times and Transcript, 30 May.

Baltz, D. M., and Moyle, P. B., 1981, Morphometric analysis of tule perch (*Hysterocarpus traski*) populations in three isolated drainages: Copeia, v. 1981, p. 305–311.

Benson, L. V., 1978, Fluctuation in the level of pluvial Lake Lahontan during the last 40,000 years: Quaternary Research, v. 9, p. 300–318.

Casteel, R. W., and Hutchison, J. H., 1973, *Orthodon* (Actinopterygii, Cyprinidae) from the Pliocene and Pleistocene of California: Copeia, v. 1973, p. 358–361.

Casteel, R. W., and Rymer, M. J., 1975, Fossil fishes from the Pliocene or Pleistocene Cache Formation, Lake County, California: U.S. Geological Survey Journal of Research, v. 3, p. 619–622.

——, 1981, Pliocene and Pleistocene fishes from the Clear Lake area: U.S. Geological Survey Professional Paper 1141, p. 231–235.

Casteel, R. W., Adam, D. P., and Sims, J. D., 1975, Fish remains from core 7, Clear Lake, Lake County, California: U.S. Geological Survey; Open-File Report No. 75-173, 67 p.

——, 1977a, Late Pleistocene and Holocene remains of *Hysterocarpus traski* (tule perch) from Clear Lake, California, and inferred Holocene temperature fluctuations: Quaternary Research, v. 7, p. 133–143.

Casteel, R. W., Beaver, C. K., Adam, D. P., and Sims, J. D., 1977b, Fish remains from core 6, Clear Lake, Lake County, California: U.S. Geological Survey Open File Report No. 77-639, 154 p.

Casteel, R. W., Williams, J. H., Throckmorton, C. K., Sims, J. D., and Adam, D. P., 1979, Fish remains from core 8, Clear Lake, Lake County, California: U.S. Geological Survey Open File Report No. 79-1148, 98 p.

Coad, B. W., and Qadri, S. U., 1978, On the nomenclature of the genus name *Endemichthys* (Osteichthyes): Copeia, v. 1978, p. 330.

Cook, S. F., Jr., Moore, R. L., and Conners, J. D., 1966, The status of the native fishes of Clear Lake, Lake County, California: Wasmann Journal of Biology, v. 24, p. 141–160.

Forey, P., and Gardiner, B. G., 1973, A new dictyopyfid (*sic*) from the cave sandstone of Lesotho, South Africa: Palaeontology of Africa, v. 15, p. 29–31.

Girard, C., 1854, Descriptions of new fishes, collected by Dr. A. L. Heermann, naturalist attached to the Survey of the Pacific Railroad Route, under Lieut. R. S. Williamson, U.S.A.: Proceedings of the Academy of Natural Sciences

of Philadelphia, v. 7, p. 129–140.
Hinds, N.E.A., 1952, Evolution of the California landscape: California State Division of Mines Bulletin 158, 240 p.
Hopkirk, J. D., 1962, Morphological variation in the freshwater embiotocid *Hysterocarpus traskii* Gibbons [M.A. Thesis]: University of California at Berkeley, 159 p.
—— , 1967, Endemism in fishes of the Clear Lake region [Ph.D. thesis]: University of California at Berkeley, 356 p.
—— , 1973 [1974], Endemism in fishes of the Clear Lake region of Central California: University of California Publications in Zoology, v. 96, 136 p.
Hubbs, C., 1974, Review of: Hopkirk, J. D., 1973, Endemism in fishes of the Clear Lake region of Central California: University of California Publications in Zoology, v. 96, p. 1–135: Copeia, v. 1974, p. 808–809.
Hubbs, C. L., 1961, Isolating mechanisms in the speciation of fishes, *in* Vertebrate speciation, A University of Texas symposium: University of Texas Press, 642 p.
Hubbs, C. L., and Miller, R. R., 1943, Mass hybridization between two genera of cyprinid fishes in the Mohave Desert, California: Papers Mich. Acad. Sci., Arts, Letters, v. 28, p. 343–378.
Hubbs, C. L., Miller, R. R., and Hubbs, L. C., 1974, Hydrographic history and relict fishes of the North-Central Great Basin: Memoirs of the California Academy of Sciences, v. 7, 262 p.
Jordan, D. S., and Gilbert, C. H., 1894, List of the fishes inhabiting Clear Lake, California: Bulletin of the U.S. Fisheries Commission, v. 14, p. 139–140.
Kniffen, F. B., 1939, Pomo geography: University of California Publications in American Archeology and Ethnology, v. 36, p. 353–400.
Krejsa, R. J., 1965, The systematics of the prickly sculpin, *Cottus asper*; An investigation of genetic and non-genetic variation within a polytypic species [Ph.D. thesis]: Vancouver, University of British Columbia, 109 p.
La Marche, V. C., Jr., 1973, Holocene climatic variations inferred from tree line fluctuations in the White Mountains, California: Quaternary Research, v. 3, p. 632–660.
La Rivers, I., 1962, Fishes and fisheries of Nevada: Nevada Fish and Game Commission, 784 p.
Lindquist, A. W., Deonier, C. C., and Hancey, J. E., 1943, The relationship of fish to the Clear Lake gnat, in Clear Lake, California: California Fish and Game, v. 29, p. 221–227.
Miller, R. G., 1951, The natural history of Lake Tahoe fishes [Ph.D. thesis]: Stanford University, 160 p.
Miller, R. R., 1963, Synonymy, characters, and variation of *Gila crassicauda,* a rare Californian minnow, with an account of its hybridization with *Lavinia exilicauda*: California Fish and Game, v. 49, p. 20–29.
Moyle, P. B., 1976, Inland fishes of California: University of California Press, 406 p.
Moyle, P. B., and Daniels, R. A., 1982, Fishes of the Pit River system, McCloud River system, and Surprise Valley region: University of California Publications in Zoology, v. 115, p. 1–82.
Moyle, P. B., and Nichols, R. , 1973, Ecology of some native and introduced fishes of the Sierra Nevada foothills, central California: Copeia, v. 1973, p. 478–490.
Moyle, P. B., Smith, J. J., Daniels, R. A., and Baltz, D. M., 1982, IV. A review: University of California Publications in Zoology, v. 115, p. 225–256.
Murphy, G., 1948, Distribution and variation of the roach (*Hesperoleucus*) in the coastal region of California [M.A. thesis]: University of California at Berkeley, 55 p.
Richardson, J., 1836, Fauna Boreali-Americana, pt. 3, 327 p.
Rymer, M. J., Sims, J. D., Hedel, C. W., Bridge, W. D., Makdisi, R. S., and Mannshardt, G. A., 1981, Sample catalog for core CL-80-1, Clear Lake, Lake County, California: U.S. Geological Survey Open-File Report No. 81-245, 53 p.
Shapovalov, L., Cordone, A. J., and Dill, W. A., 1981, A list of the freshwater and anadromous fishes of California: California Fish and Game, v. 67, p. 4-38.
Sims, J. D., 1976, Paleolimnology of Clear Lake, California, U.S.A., *in* Horie, S., ed., Paleolimnology of Lake Biwa and the Japanese Pleistocene: Kyoto, Japan, Kyoto University, v. 4, p. 658–702.
Sims, J. D., Adam, D. P., and Rymer, M. J., 1981, Late Pleistocene stratigraphy and palynology of Clear Lake: U.S. Geological Survey Professional Paper 1141, p. 219–230.
Snyder, J. O., 1918, The fishes of the Lahontan system of Nevada and northeastern California: Bull. U.S. Bureau of Fishery, v. 35, p. 31–86.
Tarp, F. H., 1952, A revision of the family Embiotocidae (the surfperches): California Department of Fish and Game, Fisheries Bulletin 88, 99 p.
Taylor, D. W., 1966, Summary of North American Blancan nonmarine mollusks: Malacologia, v. 4, p. 1–172.
Taylor, T. L., Moyle, P. B., and Price, D. G., 1982, III. Fishes of the Clear Lake Basin: University of California Publications in Zoology, v. 115, p. 171–224.
Vladykov, V. D., 1973, *Lampetra pacifica,* a new nonparasitic species of lamprey (Petromyzontidae) from Oregon and California: Journal of the Fisheries Research Board of Canada, v. 30, 205–213.
Willsrud, T., 1971, A study of the Tahoe sucker, *Catostomus tahoensis* Gill and Jordan [M.S. thesis]: San Jose, California, San Jose State College, 96 p.

Manuscript Accepted by the Society September 15, 1986

Printed in U.S.A.

Geological Society of America
Special Paper 214
1988

Seismicity in the Clear Lake area, California, 1975-1983

Donna Eberhart-Phillips
U.S. Geological Survey
345 Middlefield Road
Menlo Park, California 94025

ABSTRACT

Earthquake locations for the time period March 1975 through March 1983 indicate diffuse seismicity in the Clear Lake area, with the area to the southeast of the lake having the highest level of activity. Swarms lasting 2 days, with events of magnitude ⩾3.5, have been observed in the Konocti Bay fault zone. In contrast, almost no microearthquake activity is associated with the Collayomi fault zone. A 10-km-wide northeast-southwest–trending zone of seismicity is apparent although there is no known corresponding geologic feature. Earthquake depths in the Clear Lake area are shallower than in the surrounding major right-lateral fault zones. Focal mechanism solutions show predominant strike-slip with significant normal dip-slip movement. The maximum and minimum compressive stress orientations, north-northeast and east-southeast, respectively, are consistent with the San Andreas right-lateral transform boundary region. However, the component of extensional stress illustrated by normal fault-plane solutions and the shallowness of seismicity may be related to an inferred nearby crustal partial melt body.

INTRODUCTION

Clear Lake, in northern California, is located approximately 60 km northeast of the San Andreas fault, within a zone of northwest-trending faults subparallel to the San Andreas fault (Fig. 1). The regional seismicity is predominantly clustered along mapped fault zones that have been active since late Pleistocene time (Hearn and others, 1981; McLaughlin and others, 1986). Large earthquakes have not been reported in the Clear Lake region, but following the great 1906 San Andreas earthquake, ground breakage was reported south of Clear Lake (Lawson, 1908). More recently two earthquakes of magnitude (m) 5.6 and 5.7 occurred in October 1969 on the Healdsburg–Rodgers Creek fault zone; a 4.9-m event occurred near Willits in November 1977, and a 4.4-m event took place near Ukiah in March 1978. Seismic, geologic (McLaughlin and Nilson, 1983), and geodetic (Prescott and others, 1979) observations suggest that the zone is currently undergoing dextral shear deformation parallel to the San Andreas fault that is consistent with inferred movement throughout northern and central California along the Pacific–North American plate boundary. Gravity and teleseismic P-wave delay studies indicate the presence of a low-density, low-velocity body as shallow as 6 km located between The Geysers and Clear Lake (Isherwood, 1976; Iyer and others, 1979; Fig. 1), which has been interpreted as a body of partial melt in the crust and upper mantle (Oppenheimer and Herkenhoff, 1981). Associated with this magma body is the Quaternary (2 m.y. to 1,000 yr BP) Clear Lake volcanic field that covers the south and east arms of Clear Lake (Hearn and others, 1981).

In this paper, the seismicity occurring in the Clear Lake area for the period March 1975 through March 1983 is examined in detail. The relationship of the spatial and temporal distribution of the seismicity to the mapped faults, inferred regional deformation, and inferred magma body is briefly discussed.

EARTHQUAKE DATA

The seismic data consist of P-wave arrival times and first motions recorded at stations of the U.S. Geological Survey microearthquake network shown in Figure 2. Prior to 1970, the network extended only as far north as Santa Rosa, 65 km south of Clear Lake; in August 1973 two stations were added in The Geysers area. From August 1973 through May 1975, seven 2- to 3.5-m earthquakes were located in the Clear Lake area, with hori-

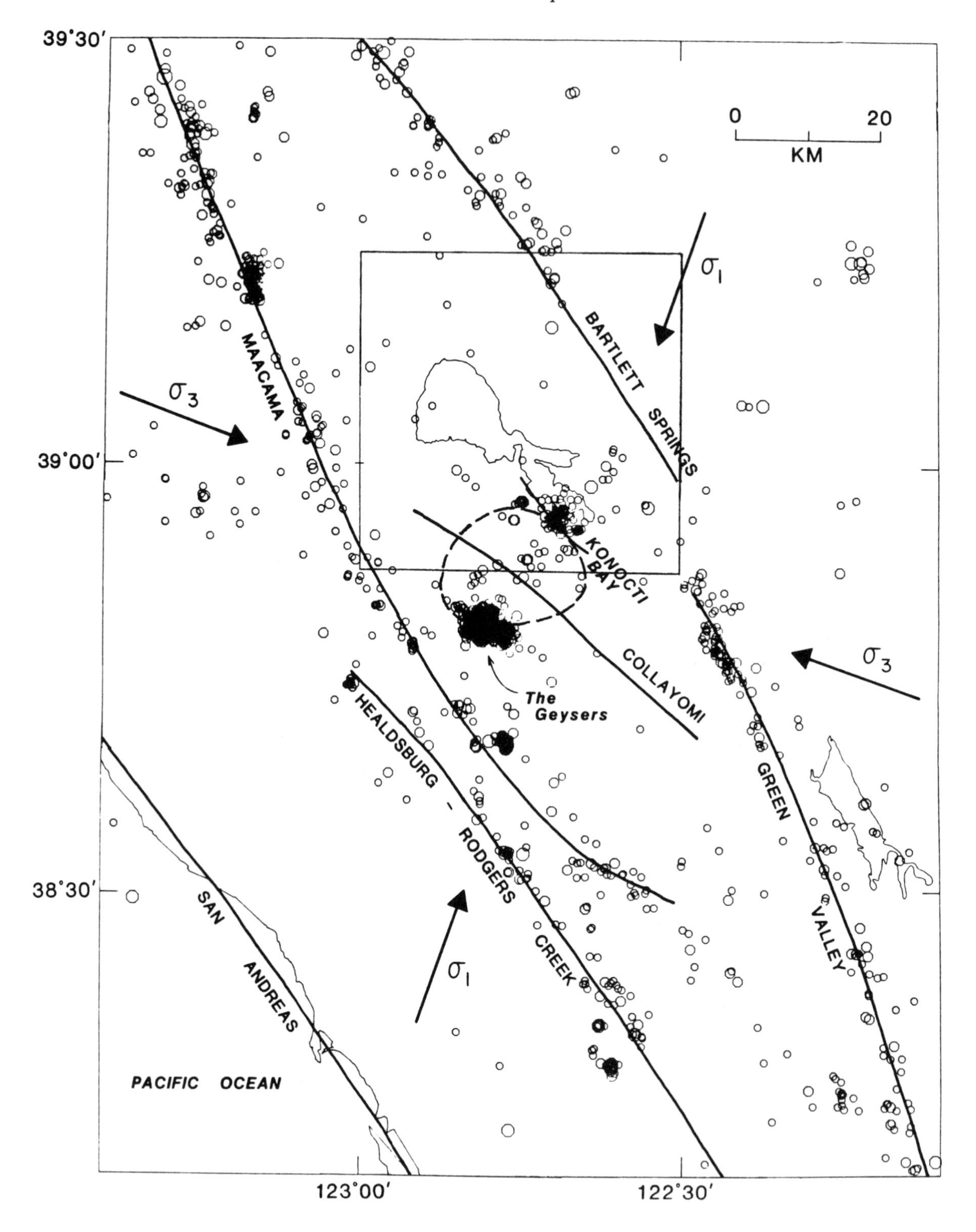

Figure 1. Map of Clear Lake area and surrounding region showing epicenters of earthquakes (M ⩾1.5; March 1972–December 1981) and locations of major fault systems (McLaughlin and Nilsen, 1983); σ_1 and σ_2 depict greatest and least compressive stress axes. Dashed line encloses 8 percent crustal velocity decrease modeled by Oppenheimer and Herkenhoff (1981). Clear Lake study area outlined on map.

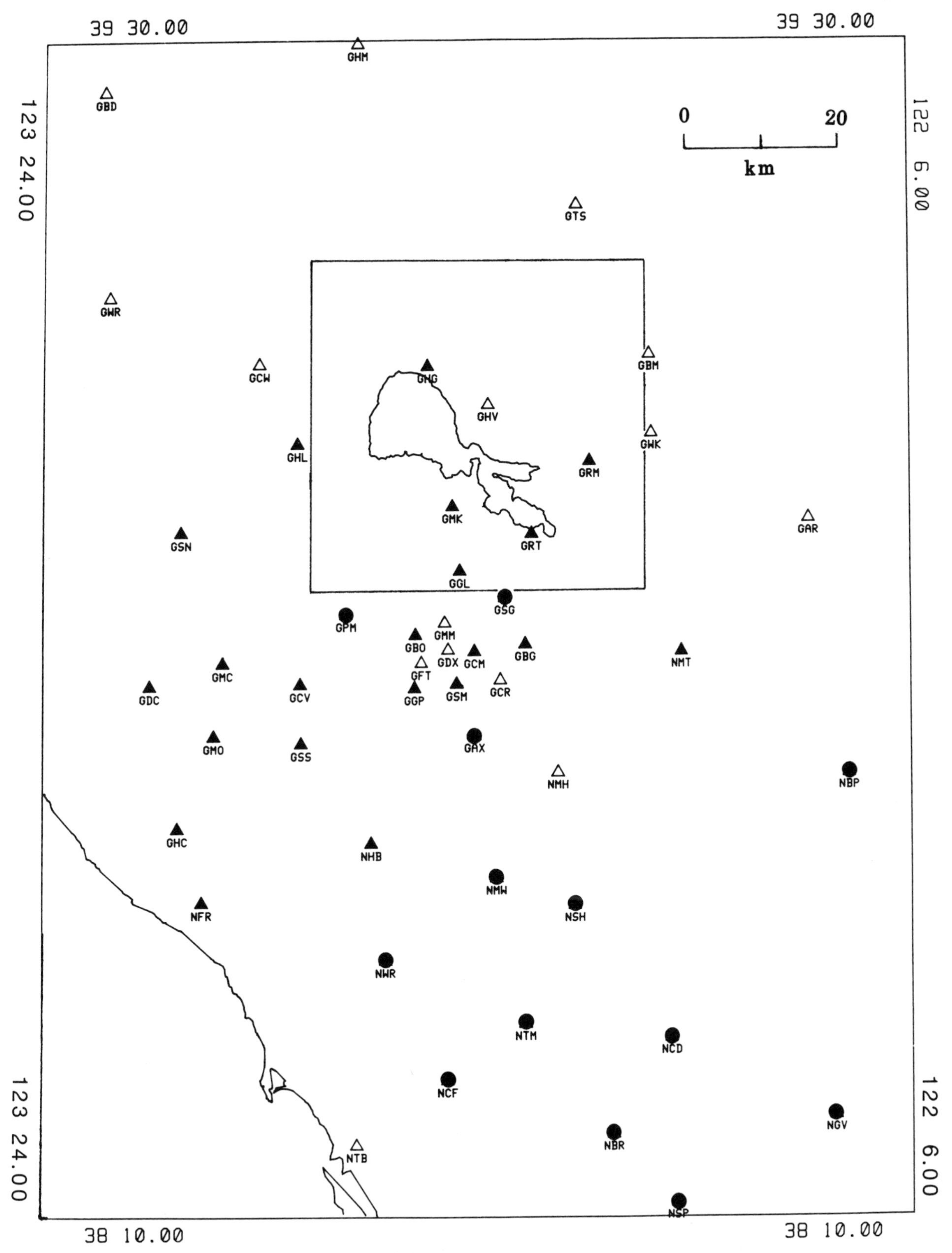

Figure 2. Map showing stations of the U.S. Geological Survey microearthquake network. Dots indicate stations operating prior to 1975. Solid triangles indicate stations added in 1975. Open triangles indicate stations added since 1977. Outlined area is Clear Lake study area.

TABLE 1. VELOCITY MODEL USED FOR RELOCATING EARTHQUAKES

Top of Layer (km)	Velocity (km/sec)
0.00	4.43
1.50	5.12
3.00	5.47
4.25	5.58
6.00	5.62
8.00	5.86

zontal location standard errors as large as 3.5 km. In spring 1975, stations GHG, GMK, GRM, and GRT were added to the network. These stations surround Clear Lake (Fig. 2) and provide adequate azimuthal coverage for accurate earthquake locations. After June 1975, horizontal location standard errors were reduced to approximately 1 km, and the data set is complete for magnitudes greater than 1.3.

The study area is defined by the rectangle in Figures 1 and 2. The velocity model (Table 1) and corresponding stations terms were obtained by simultaneous inversion for velocity and hypocenters from earthquakes throughout the region shown in Figure 1 (Crosson, 1976; Eberhart-Phillips and Oppenheimer, 1984). Although the velocity model was derived primarily for locating earthquakes in The Geysers area, the model is also appropriate for relocating Clear Lake seismicity since Clear Lake earthquakes composed about 20 percent of the earthquakes used in the velocity inversion. In this study, 488 earthquakes that occurred during the time period March 1975 through March 1983, and had a minimum of six stations recording arrivals, were relocated. The resulting locations have average horizontal and vertical standard mislocation errors of 0.75 and 1.5 km, respectively. However, the earthquake locations may be systematically biased due to lateral crustal variations not adequately accounted for by the model parameters. Confirmation of any mislocation bias was not possible because precise locations of known explosions were not available. The relocations are more tightly clustered and closer to mapped faults than the initial U.S. Geological Survey catalog locations.

The number of recorded earthquakes in the Clear Lake area is too small to allow accurate computation of the b value; however, the distribution of events with magnitude is plotted in Figure 3. The dashed line illustrates the fit to the data for a b value of 1.0, typical of California. It is apparent that the b value for the Clear Lake area is neither extremely small nor extremely large.

Figure 4 shows the time history of seismicity from March 1975 through March 1983. Steps in the cumulative number graph represent swarms of earthquakes. The three largest swarms during the study period occurred in the area immediately south of Clear Lake and account for 21 percent of the total number of earthquakes. One occurred during September 1975; its largest magnitude event was only 2.3. Another occurred during December 1980, and included four ⩾3-m events; and another occurred during August 1982 and included one ⩾3-m event.

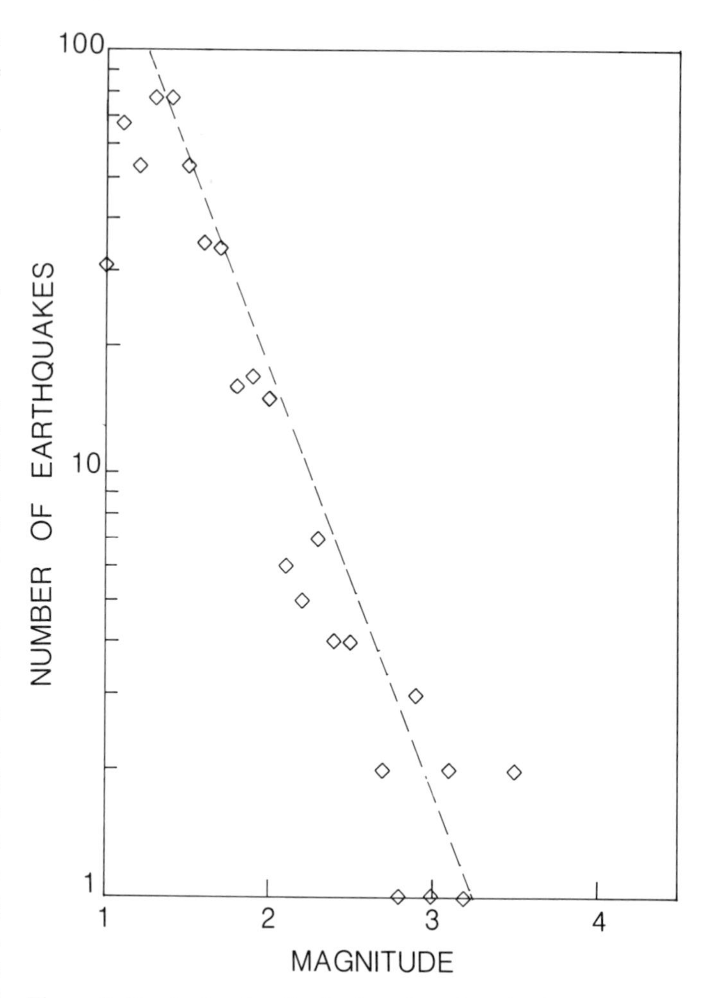

Figure 3. Graph showing distribution of earthquakes by magnitude. While data set contains too few events to determine accurately b value, a line representing a b value of 1.0 is shown for comparison.

Steps in the cumulative moment plot (Fig. 4b) generally contain one or more strong earthquakes. The 1980 swarm is the most outstanding feature in the moment plot, whereas the other two swarms have much smaller associated moments. Two other large steps in the moment plot correspond to solitary 3.5-m events: in March 1975 along the Bartlett Springs fault zone and in February 1980 6 km east of the southern end of Clear Lake. The three swarms account for 50 percent of the total seismic moment, whereas the two 3.5-m earthquakes account for 30 percent of the moment.

SPATIAL DISTRIBUTION OF EARTHQUAKES

On a regional scale, the most apparent seismicity patterns are fairly continuous linear zones of seismicity trending northwest-southeast and extending approximately 50 to 150 km in length. These zones are subparallel to the San Andreas fault and spatially correspond to the Bartlett Springs and Green Valley

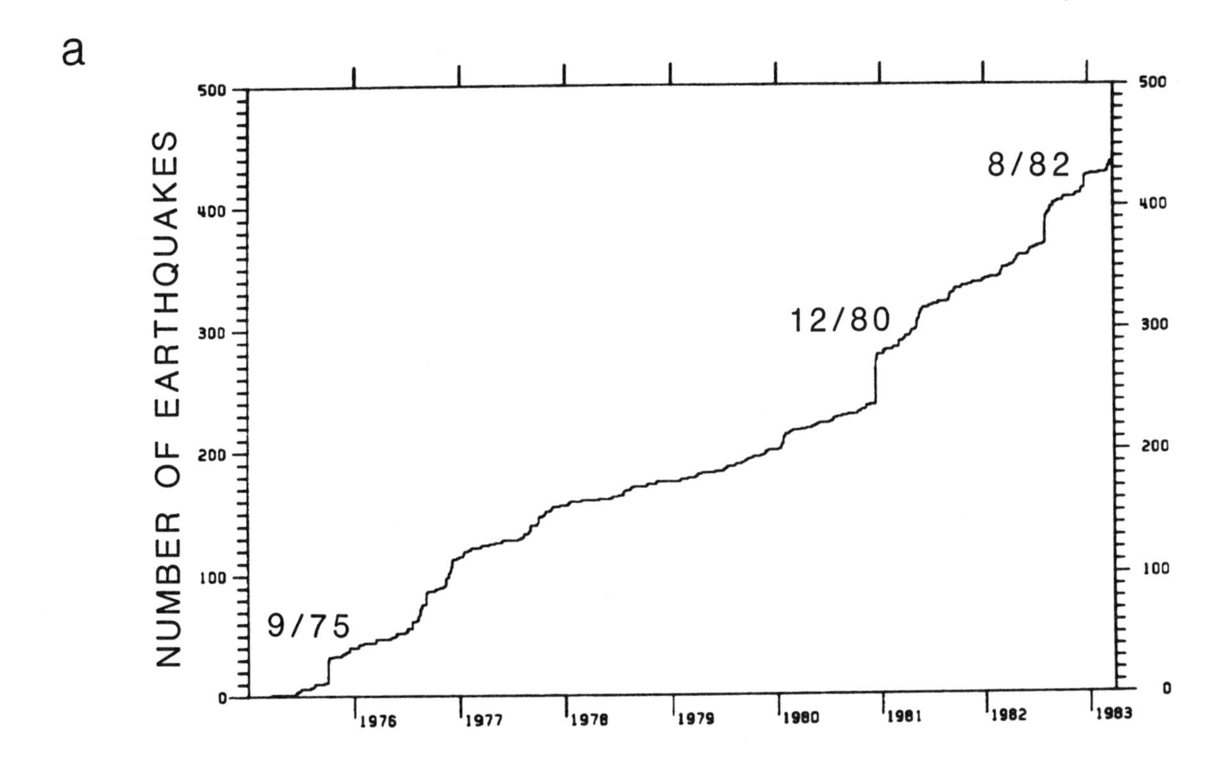

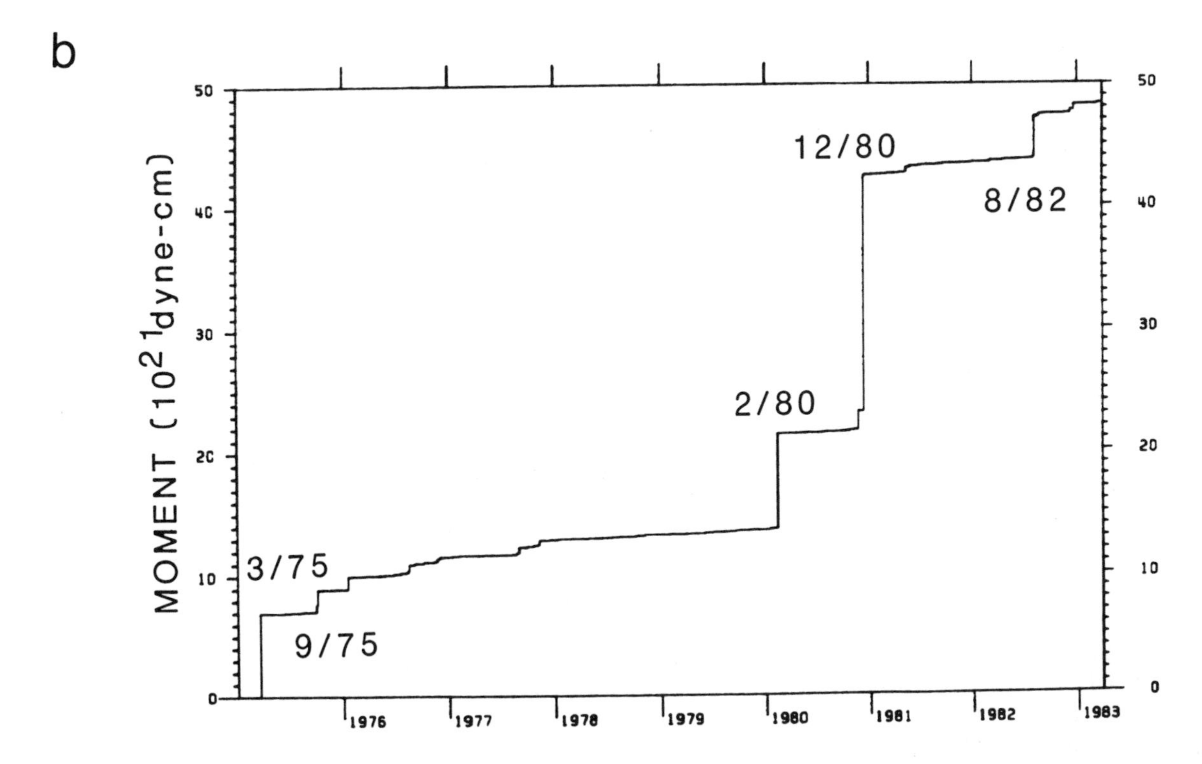

Figure 4. a, Plot of cumulative number of earthquakes in Clear Lake area. b, Plot of cumulative seismic moment, computed using M_o = 10 exp (17 + 1.5 · magnitude).

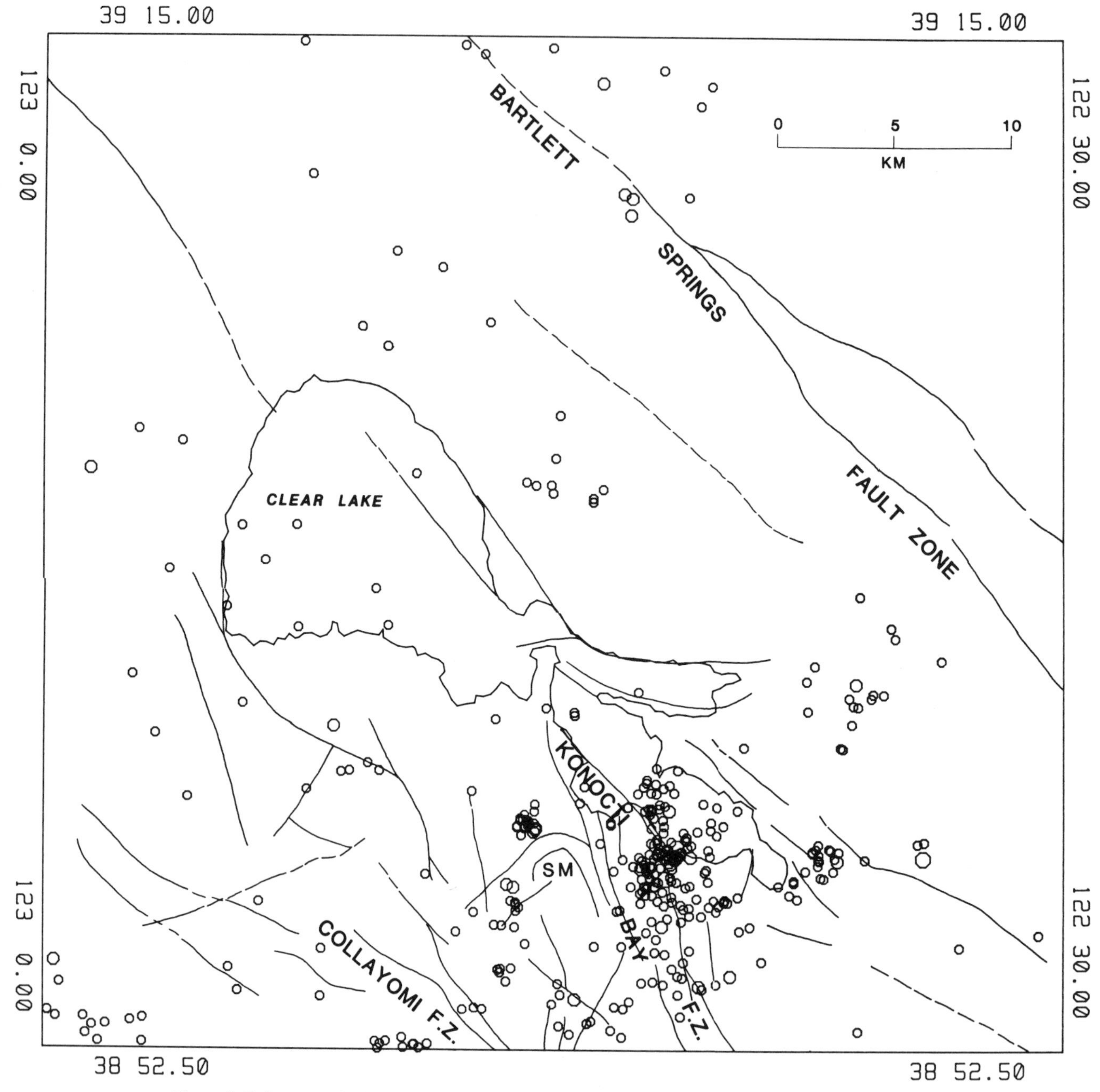

Figure 5. Epicenters of relocated earthquakes in Clear Lake study area (M ⩾1.0; March 1975–March 1983; horizontal standard error, ⩽2 km) shown on generalized fault map (after Hearn and others, 1981; McLaughlin and Nilsen, 1983; McLaughlin, unpublished data). SM indicates Sugarloaf Mountain.

fault zones east of Clear Lake, and the Maacama and Healdsburg–Rodgers Creek fault zones to the west (Fig. 1). These recently active dextral strike-slip fault zones are composed of branching and en-echelon faults with numerous small pull-apart basins (Pampeyan and others, 1981; McLaughlin and others, 1986; McLaughlin and Nilsen, 1982). Seismic activity is conspicously absent along the San Andreas fault, while earthquakes are scattered throughout the region between the other major fault zones. A linear trend of epicenters southeast of the Maacama fault zone suggests that it may connect with the Green Valley fault zone. The most intense cluster of earthquakes in Figure 1 is located at The Geysers, where the majority of seismicity is induced by geothermal power production activities (Eberhart-Phillips and Oppenheimer, 1984). Another large cluster of earthquakes is located at the southern end of Clear Lake along the Konocti Bay fault zone.

The Clear Lake area is illustrated in more detail in Figure 5. Mapped faults (Hearn and others, 1981; McLaughlin and Nilsen,

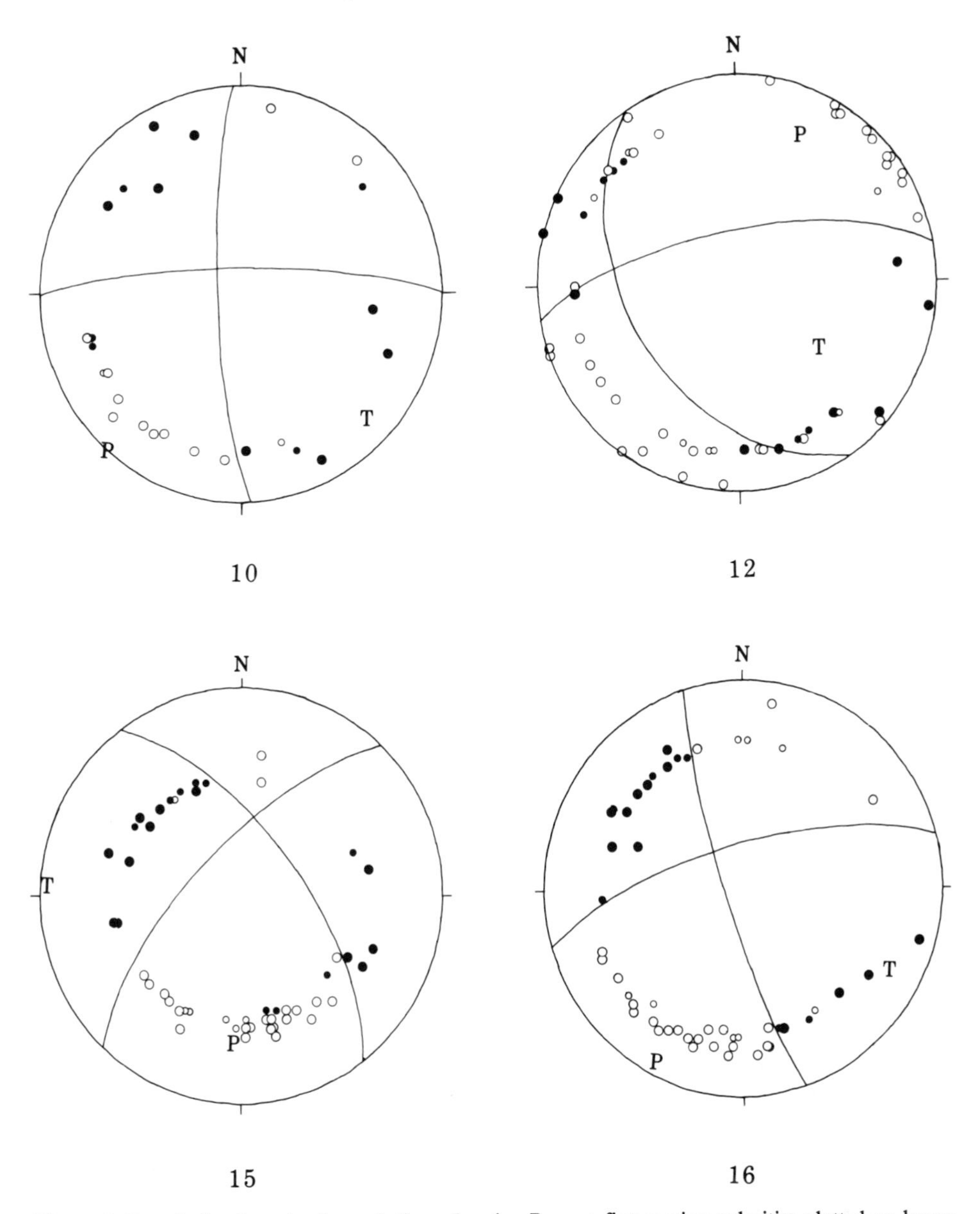

Figure 6. Sample focal mechanism solutions showing P-wave first motion polarities plotted on lower hemisphere equal area projection. Solid circles indicate compression; open circles, dilatation. Larger symbols indicate better quality observations. P and T indicate inferred axes of maximum and minimum compressive stress. Numbers correspond to event locations indicated in Figure 8 and listed in Table 2.

1983; and R. J. McLaughlin, unpublished data) of Quaternary age in the Clear Lake area are shown with the relocated earthquakes. Faulting on the Bartlett Springs fault zone is younger than 2.0 m.y. and possibly younger than 0.7 m.y. (McLaughlin and others, 1986). The faults that constitute the Konocti Bay fault zone and the faults between the Konocti Bay and Collayomi fault zones displace volcanic rocks with ages from 0.1 to 2.1 m.y., while faulting in the Collayomi fault zone is younger than 1.1 m.y. (Hearn and others, 1981).

While diffuse seismicity occurs throughout the Clear Lake area, most of the seismicity occurs at the northeast end of a 10-km-wide band of activity that extends southwest from Konocti Bay to The Geysers. This southwest-northeast trend of seismicity is not correlated with any known fault structure. The numerous mapped fault traces generally trend southeast-northwest or south-southeast–north-northwest (Hearn and others, 1976; McLaughlin, 1978 and unpublished data). The few fault traces trending northeast-southwest are minor faults and cannot be shown to be more recently active than adjacent faults with different orientations.

Several events locate near mapped faults in the Bartlett Springs fault zone, while the Collayomi fault zone does not ap-

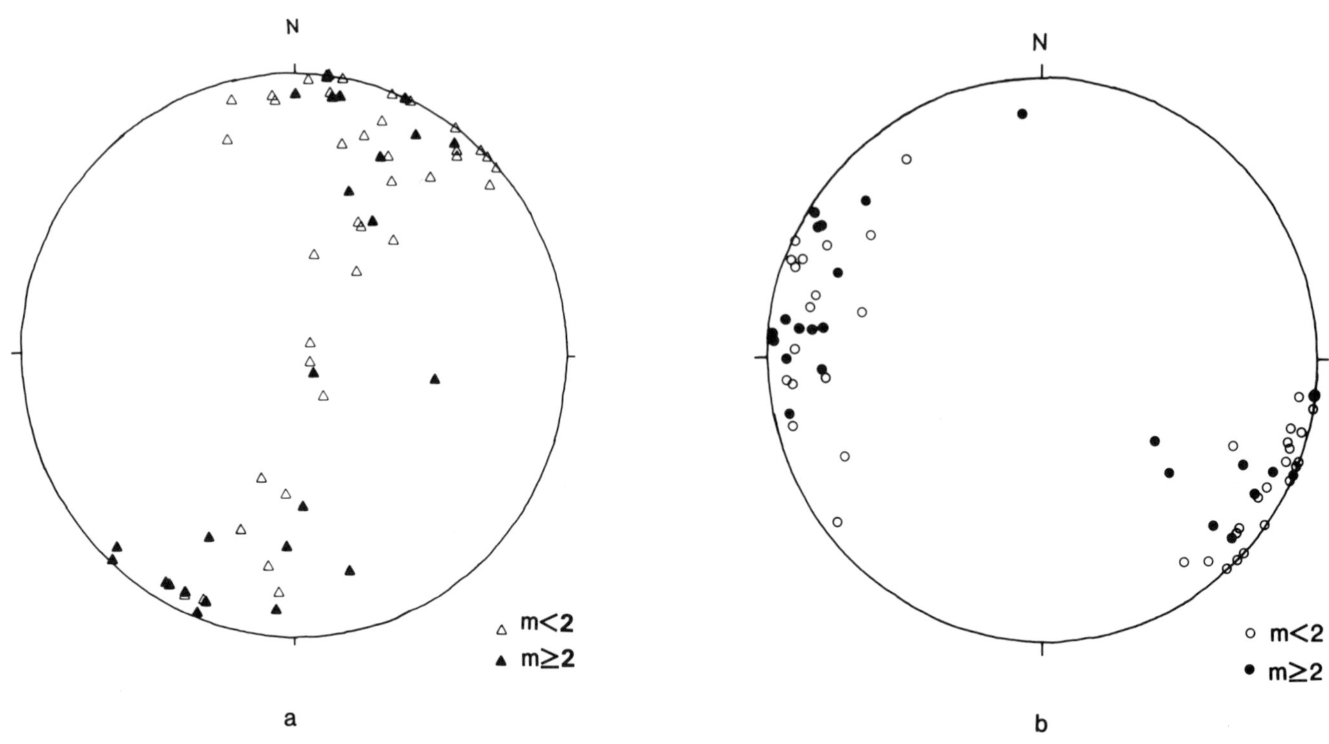

Figure 7. a, P-axes; b, T-axes plotted on lower hemisphere equal-area projection for focal mechanisms of 60 Clear Lake earthquakes.

pear to have an associated linear zone of epicenters. A sparse, northwest-southeast trending group of epicenters occurs between Clear Lake and the Bartlett Springs fault zone, but is not associated with any mapped fault trace. The largest concentration of activity in the Clear Lake area is located in the Konocti Bay fault zone immediately south of Clear Lake. The two largest swarms are located here and the third swarm is located just 5 km west near Sugarloaf Mountain.

FAULT-PLANE SOLUTIONS AND INFERRED STRESSES

Focal mechanisms were determined for 63 earthquakes in the Clear Lake area. From 10 to 61 P-wave first motions were used for each solution. The strike and dip of the nodal planes are constrained within $\pm 10°$ and $\pm 15°$, respectively, although in many cases at least one of the nodal planes is constrained within 5°. Four representative focal mechanism determinations are shown in Figure 6 to illustrate the azimuthal distribution of stations and the quality of solutions.

Most fault-plane solutions are strike-slip with a component of normal dip-slip as depicted by the pressure (P) and tension (T) axes plotted in Figure 7. The larger magnitude and better constrained solutions are listed in Table 2 and shown on a map as balloon diagrams with compressional quadrants shaded (Fig. 8). Although earthquakes are not precisely enough located to be associated with individual fault segments, the focal mechanisms of earthquakes along the Konocti Bay and Bartlett Springs fault zones are compatible with dextral strike-slip motion on north-northwest–trending fault planes subparallel to mapped faults. This inferred orientation and sense of motion is consistent with the pattern of regional deformation in the Pacific–North American plate boundary.

Focal mechanisms were determined for some events that are not spatially associated with mapped fault zones. Here the choice of which nodal plane corresponds to the fault plane is more ambiguous since left-lateral faults in areas of pull-apart basins may occur (McLaughlin and Nilsen, 1983; Hill, 1982). While the northeast-southwest–trending belt of seismicity (Figs. 1, 5) suggests the presence of a sinistral fault zone, there is no corresponding major northeast-southwest–trending mapped fault trace.

The only exception to the pattern of strike-slip and normal dip-slip movement is event 12, which has a large component of reverse dip-slip. This 3.5-m earthquake is located 10 km east of the Konocti Bay fault zone; in contrast to other events of comparable size, it has almost no other earthquakes spatially or tem-

TABLE 2. DESCRIPTION OF EARTHQUAKES SHOWN IN FIGURE 8

	Date (yr/mo/da)	Time (GMT) (HrMin)	Latitude (°)	Latitude (')	Longitude (°)	Longitude (')	Magnitude	No. of First Motions	P-axis Azimuth (°)	Dip (°)	T-axis Azimuth (°)	Dip (°)
1*	75/09/29	1708	38	57.40	122	44.69	2.00	--	24	0	114	0
2	75/10/03	2145	38	56.32	122	45.57	2.87	27	209	7	300	7
3	75/10/03	2210	38	56.24	122	45.96	2.27	20	223	6	132	7
4	76/08/10	2028	38	53.26	122	46.16	1.69	15	30	29	122	8
5*	76/08/16	2248	38	56.90	122	40.88	2.53	--	184	10	275	2
6	76/10/09	0311	38	54.89	122	41.15	2.07	24	205	7	115	7
7	77/09/03	1153	39	12.50	122	42.18	2.30	30	0	7	267	21
8	77/09/03	1213	39	10.96	122	42.54	2.19	31	10	6	278	21
9	78/10/10	0812	38	56.51	122	42.17	1.92	16	43	0	133	0
10	79/06/01	1041	38	53.52	122	44.11	2.00	26	222	1	133	15
11	79/09/01	0910	38	55.55	122	43.05	1.64	16	201	7	292	7
12	80/02/13	0745	38	56.84	122	34.24	3.49	61	24	23	125	48
13	80/10/23	0608	39	0.74	122	36.18	2.10	45	31	43	302	3
14	80/11/07	0510	39	11.12	122	51.40	1.61	15	3	2	272	11
15	80/11/24	1533	38	59.75	122	51.07	2.97	52	182	32	274	3
16	80/12/12	1424	38	56.80	122	41.61	3.11	57	210	7	117	19
17	80/12/12	1427	38	56.49	122	41.54	2.93	56	30	10	121	10
18	81/03/07	0630	38	57.55	122	40.33	1.67	30	39	6	130	7
19	81/09/22	0237	38	58.87	122	49.75	1.77	28	18	19	109	4
20	81/11/22	0044	38	53.11	122	58.73	1.63	28	21	11	291	3
21	82/03/06	1042	38	56.74	122	41.01	2.06	28	221	2	311	16

*These focal mechanisms were determined by Bufe and others (1981). Dashes indicate that the number of first motions is unknown.

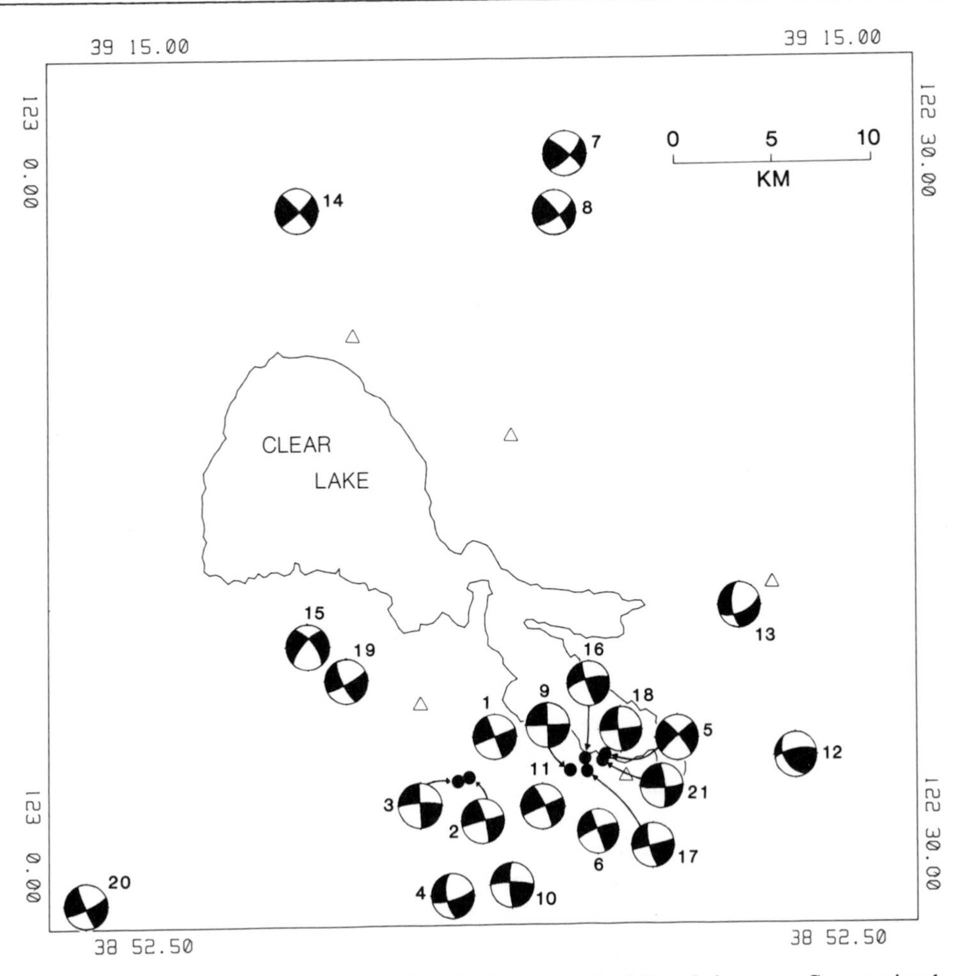

Figure 8. Balloon diagrams of focal mechanisms for best-constrained Clear Lake events. Compressional quadrants are shaded. Numbers correspond to list of events in Table 2. Stations indicated by triangles.

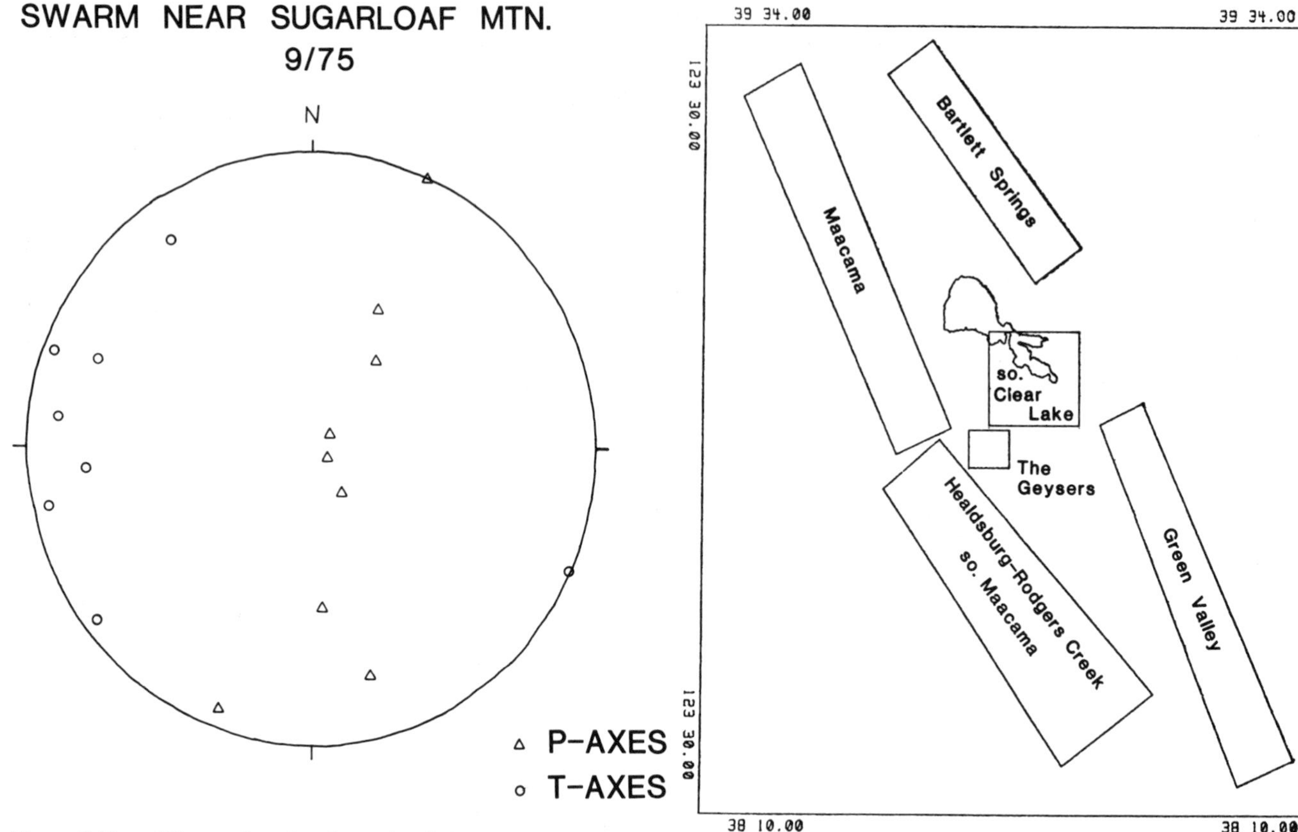

Figure 9. P- and T-axes plotted on lower hemisphere equal-area projection for focal mechanisms of the swarm of September 29-30, 1975, south of Clear Lake.

Figure 10. Map showing areas used for subdivision of earthquakes for depth calculations shown in Table 3.

porally associated with it. Reverse dip-slip is consistent with the regional stress field when it occurs on east-west trending faults (Zoback and Zoback, 1980). In this case, one of the nodal planes is nearly east-west, and the nearest mapped fault bends sharply to a more easterly trend near the epicenter. Some reverse dip-slip solutions are also found in the eastern San Francisco Bay region (Ellsworth and others, 1982).

The local orientations of the maximum and minimum compressive stress axes in the Clear Lake area are inferred from the P- and T-axes. The orientation of the minimum stress axis determined from the Clear Lake focal mechanisms is nearly horizontal and trends west-northwest–east-southeast (Fig. 7b). The orientation of the maximum stress axis is more variable; the trend of the P-axes is consistently south-southwest–north-northeast, but the plunge ranges over 180° from horizontal to vertical (Fig. 7a). While on average the P-axis is subhorizontal, this variation indicates that the maximum and intermediate stresses in this region may be of nearly equal magnitude.

Of particular interest is the distinct swarm of seismic activity located 5 km west of the Konocti Bay fault zone (Fig. 5). The cluster is 2 km northwest of a large circular basin surrounding Sugarloaf Mountain, which may be related to caldera collapse after rhyolite eruptions (Hearn and others, 1981). Fault-plane solutions for events in this swarm are compatible with east-west extension; six indicate normal faulting while three indicate strike-slip faulting (Fig. 9). The predominant normal mode of faulting, and the proximity of the locations to the inferred location of the collapse feature associated with extension tectonics, suggest that these events are related to the underlying magma body.

EARTHQUAKE DEPTHS

The mean depths of earthquakes subdivided into seismic provinces (Fig. 10) are compared in Table 3. The lower depth limit of seismic activity in the Clear Lake and Geysers zones is near 6 km and is at least 2 km shallower than that of the surrounding northwest-southeast–trending fault zones. Geothermal production activities induce most seismicity at The Geysers (Eberhart-Phillips and Oppenheimer, 1984) and may be responsible for the shallowest hypocenters in the region, since the steam extraction zone extends from 0 to 2.5 km in depth. The partial melt body, thought to be the heat source for the geothermal area

TABLE 3. COMPARISON OF EARTHQUAKE* DEPTHS IN FAULT ZONES NEAR CLEAR LAKE (Fig. 10)

Fault Zone or Area	Number of Events	Maximum Magnitude	Mean Depth (km)	Maximum Depth (km)
Bartlett Springs	20	2.81	4.27 ± 0.28	8.56 ± 0.7
Clear Lake (southern part)	88	3.49	2.61 ± 0.15	5.80 ± 1.0
Green Valley	93	2.93	5.82 ± 0.34	13.10 ± 1.0
Healdsburg-Rodgers Creek, southern Maacama	202	3.78	4.79 ± 0.18	12.14 ± 0.5
Maacama	178	4.87	3.57 ± 0.17	12.81 ± 0.5
The Geysers	924	3.52	1.83 ± 0.03	6.49 ± 0.4

*For earthquakes with magnitudes ≥ 1.5 and vertical standard error of ≤ 2.0 km.

(Fig. 1), (Isherwood, 1976; Iyer and others, 1979; Oppenheimer and Herkenhoff, 1981) would also dictate a shallower limit to the seismogenic zone at The Geysers (Marks and others, 1978; Ludwin and Bufe, 1980). Similarly, in the Clear Lake area, the proximity of the shallow seismicity to the inferred magma body and the recent volcanic rocks surrounding southeast Clear Lake suggests that geothermal processes may also explain the abnormally shallow seismicity in southern Clear Lake.

CONCLUSIONS

For the period March 1975 through March 1983, diffuse seismicity is apparent throughout the Clear Lake area. However, the largest magnitude earthquake during this period is only 3.5. A southwest-northeast zone extending from The Geysers to the southern end of Clear Lake encompasses most seismic activity, although this zone is not correlated with any known geologic structure. The largest cluster of earthquakes locates along the Konocti Bay fault zone, indicating that the fault zone is presently active. There is little activity associated with the Collayomi fault and it may be inactive.

Most fault-plane solutions for events in the Clear Lake region indicate strike-slip movement with a component of normal dip-slip movement. The solutions are compatible with right-lateral movement along north-northwest–trending faults, which is the sense of motion expected from the regional tectonics. The orientations of the maximum and minimum compressive stresses, inferred from P- and T-axes of fault plane solutions, are consistent with the Pacific–North American plate boundary, which is the primary influence on Clear Lake tectonics. Variation in the P-axes, however, indicates that the maximum and intermediate stresses locally may be of nearly equal magnitude. Hypocenters with extensional fault mechanisms spatially associated with recent volcanism in the Clear Lake basin and abnormally shallow seismicity distinguish Clear Lake from the surrounding major fault zones. Hence the inferred crustal magma chamber is likely exerting a secondary influence on Clear Lake tectonics.

ACKNOWLEDGMENTS

I am grateful to Paul Reasenberg, Chris Stephens, and David Oppenheimer for many comments and suggestions helpful to this paper.

REFERENCES CITED

Bufe, C. G., Marks, S. M., Lester, F. W., Ludwin, R. S., and Stickney, M. C., 1981, Seismicity of the Geysers–Clear Lake region: U.S. Geological Survey Professional Paper 1141, p. 129–133.

Crosson, R. S., 1976, Crustal structure modeling of earthquake data, 1. Simultaneous least squares estimation of hypocenter and velocity parameters: Journal of Geophysical Research, v. 81, no. 17, p. 3036–3046.

Eberhart-Phillips, D., and Oppenheimer, D. H., 1984, Induced seismicity in The Geysers geothermal area, California: Journal of Geophysical Research, v. 89, no. B2, p. 1191–1207.

Ellsworth, W. L., Olson, J. A., Shijo, L. N., and Marks, S. M., 1982, Seismicity and active faults in the eastern San Francisco Bay region: Proceedings, Conference on Earthquake Hazards in the Eastern San Francisco Bay Area, Special Pub. 62, California Division of Mines and Geology, p. 83–91.

Hearn, B. C., Jr., Donnelly, J. M., and Goff, F. E., 1976, Preliminary geologic map and cross-section of the Clear Lake volcanic field, Lake County, California: U.S. Geological Survey Open-File Report 76-751, scale 1:24,000.

Hearn, B. C., Jr., Donnelly-Nolan, J. M., and Goff, F. E., 1981, The Clear Lake volcanics; Tectonic setting and magma sources: U.S. Geological Survey Professional Paper 1141, p. 25–45.

Hill, D. P., 1982, Contemporary block tectonics; California and Nevada: Journal of Geophysical Research, v. 87, no. B7, p. 5433–5450.

Isherwood, W. F., 1976, Gravity and magnetic studies of The Geysers–Clear Lake geothermal region, California: Proceedings of the 2nd United Nations Symposium on Development and Use of Geothermal Resources, v. 2, p. 1065–1073.

Iyer, H. M., Oppenheimer, D. H., and Hitchcock, T., 1979, Abnormal P-wave delays in The Geysers–Clear Lake geothermal area, California: Science, v. 204, p. 495–497.

Lawson, A. C., 1908, The California earthquake of April 18, 1906; Report of the State Earthquake Investigation Commission: Carnegie Institute of Washington Publication 87, v. 1 and atlas, 451 p.

Ludwin, R. S., and Bufe, C. G., 1980, continued seismic monitoring of The Geysers, California geothermal area: U.S. Geological Survey Open-File Report 80-1060, 50 p.

Marks, S. M., Ludwin, R. S., Louie, K. B., and Bufe, C. G., 1978, Seismic monitoring at The Geysers geothermal field, California: U.S. Geological Survey Open-File Report 78-798, 26 p.

McLaughlin, R. J., 1978, Preliminary geologic map and structural sections of the central Mayacamas Mountains and The Geysers steam field, Sonoma, Lake, and Mendocino Counties, California: U.S. Geological Survey Open-File Map 78-389, scale 1:24,000.

McLaughlin, R. J., and Nilsen, T. H., 1983, Neogene nonmarine sedimentation and tectonics in small pull-apart basins of the San Andreas fault system, Sonoma County, California: Sedimentology, v. 29, p. 865–876.

McLaughlin, R. J., Ohlin, H. N., and Thormalhlen, D. J., 1986, Geologic map and structure sections of the Little Indian Valley–Wilbur Springs geothermal area, northern Coast Ranges, California: U.S. Geological Survey I-Series Map (in press).

Oppenheimer, D. H., and Herkenhoff, K. E., 1981, Velocity-density properties of the lithosphere from three-dimensional modeling at The Geysers–Clear Lake region, California: Journal of Geophysical Research, v. 86, no. B7, p. 6057–6065.

Pampeyan, E. H., Harsh, P. W., and Coakley, J. M., 1981, Preliminary map showing active breaks along the Maacama fault zone between Laytonville and Hopland, Mendocino County, California: U.S. Geological Survey Miscellaneous Field Studies Map MF-1217, scale 1:24,000.

Prescott, W. H., Savage, J. C., and Konoshita, W. T., 1979, Strain accumulation rates in the western United States between 1970 and 1978: Journal of Geophysical Research, v. 84, no. B10, p. 5423–5435.

Zoback, M. L., and Zoback, M., 1980, State of stress in the conterminous United States: Journal of Geophysical Research, v. 85, no. B11, p. 6113–6156.

MANUSCRIPT ACCEPTED BY THE SOCIETY SEPTEMBER 15, 1986.

Printed in U.S.A.

Geological Society of America
Special Paper 214
1988

The thermal regime of the shallow sediments of Clear Lake, California

T. C. Urban
W. H. Diment
U.S. Geological Survey
Box 25046, Mail Stop 966
Denver Federal Center
Denver, Colorado 80225

ABSTRACT

During 1973 and 1974, precision temperatures (±0.02°C absolute) were measured in four drill holes, ranging in depth from 42 to 122 m, in Clear Lake, Lake County, California. The departure of the measured temperatures from their predrilling values was found to vary considerably among the holes. It is a function of the length of time after drilling that the buoy system supporting the plastic casing survived the storms on the lake, as well as the duration of the drilling disturbance. With one exception, CL-73-7, most of the holes were lost within a month after drilling.

Thermal-conductivity measurements on the sediment cores from the holes were measured using the needle-probe technique. These measurements indicate that the sapropelic muds that underlie most of the main body of the lake and the peat of the southeast, or Highlands Arm, of the lake have thermal conductivities only slightly greater than that of water. Conductivities of the coarse-grained sediments associated with the deltaic deposits of Kelsey Creek are two to three times that of water and are not grouped as tightly as the fine-grained muds and peats.

Heat flows calculated from the above measurements are 1.5 to 1.6 heat-flow units (HFU) in the main basin of Clear Lake and about 2.4 HFU in the Highlands Arm of the lake. These values are considerably lower than expected, based on heat-flow measurements in The Geysers 25 km south of Clear Lake. A correction for an average sedimentation rate in the lake of 0.68 mm/yr would raise the observed heat flows about 13 percent. Although the exact thickness of sediment is unknown, a correction for the refraction effect caused by the thermal conductivity contrast between the low-conductivity lake sediments and the higher conductivity surrounding rock would tend to increase the heat flow at depth in the main basin only slightly for any reasonable thickness of sediment. In the Highlands Arm of the lake, the geometry is different, and the correction could raise the observed heat flux at depth as much as three times that observed, again for reasonable sediment thicknesses. However interesting these numbers are, it should be cautioned that sediment composition and thickness below about 200 m in the lake are unknown, and large differences in conductivity could drastically change these corrections. The possibility of such high heat flows, though, is encouraging enough that future measurements at greater depths should be contemplated.

Aside from the refraction effects there are other possible causes of the low heat flows: (1) Clear Lake is bounded by faults and at least one fault is inferred to pass beneath the main body of the lake—downward cold water movement along such faults could absorb heat and decrease the heat flux; and (2) water may be moving down

through the sediments, although generally the permeability of lake sediments is rather low. Depending on the velocity of the movement, a substantial reduction in heat flow could result. The downward movement of water over such a large area (114 km^2) could be a source of recharge for The Geysers–Clear Lake geothermal system.

INTRODUCTION

The economic geology of the Clear Lake region of northern California has long been of interest. Sulfur was mined in the Sulfur Bank Mercury Mine from 1865 to 1868 at the eastern edge of the Oaks Arm of Clear Lake. In 1873, the mine was reopened for mercury and mined intermittently until 1957 (Sims and White, 1981). The "mining" activity in recent years has been centered about 25 km south of Clear Lake in The Geysers, where steam is produced from the vapor-dominated hydrothermal system for electric power. The source of the heat and fluids for the system and the associated geologic environment have been investigated for a number of years; the collection of papers edited by McLaughlin and Donnelly-Nolan (1981) summarizes the current knowledge of the region.

There have been few published heat-flow measurements in The Geysers–Clear Lake area (Lachenbruch and Sass, 1980; Fig. 1). Most of the temperature-gradient measurements were obtained by industry and remain proprietary. Urban and others (1976) measured temperatures in The Geysers steam field away from active steam seeps and found that the plot of temperature versus depth was nearly linear to the maximum depth logged (0.8 km). After correction for effects of terrain and extrapolation of the corrected temperatures to the depth of "first steam" (1.5 km at this locality in The Geysers), it was found that the temperature was close to that of the steam reservoir (~240°C). Depth to first steam is simply the depth to the first obvious manifestation of steam in an air-drilled hole and might be thought of as the depth to the first large crack in the steam reservoir. This suggests that a similar method might indicate anomalous heat sources in other parts of The Geysers–Clear Lake area. At Cloverdale (Urban and others, 1976), 13 km west of The Geysers, the heat flow is high (~4 HFU, where 1 HFU = 1 microcal/cm^2·sec = 41.84 milliwatts/m^2), suggesting that anomalous conditions extend far beyond the area of the known steam field.

The steam reservoir appears to be confined to the northeast limb of the Mayacmas antiform (Fig. 1). The boundaries roughly coincide with the northwest-trending Mercuryville fault zone on the southwest and the Collayomi fault zone on the northeast (McLaughlin, 1981). Farther to the north and east into the Clear Lake area, the system is considered to be largely composed of hot water, although to what extent is unknown. Hearn and others (1981) noted that "The young volcanic activity and thermal anomalies around the east arm of Clear Lake, and east-northeast of Clear Lake as far as the Wilbur Springs area, indicate geothermal potential, but the distribution of thermal springs shows the zone to be narrow and possibly not continuous." The question remains, what role does Clear Lake play in the system?

The Clear Lake basin is defined by north-, northwest-, and west-trending faults (Hearn and others, 1981; Sims and Rymer, 1976). The main basin of the lake is almost circular, with the Franciscan assemblage on the northeast, alluvial valleys and low hills underlain by the Franciscan on the west and south, and the Clear Lake volcanics to the south. The southeast or Highlands Arm of the lake is a trough-shaped depression bounded on the northeast by Franciscan rocks and valleys filled with Quaternary fluvial and lacustrine sediments, and by the Clear Lake Volcanics on the southwest. The sedimentation rate at various depths in the lake ranges from 0.37 to 1.44 mm/yr, and subsidence apparently has matched deposition (Sims, 1976). Indeed, the pollen record from the 115-m sediment core of CL-73-4 indicates fairly continuous sedimentation for more than 130,000 yr (Adam and others, 1981). Although the thickness of the sediments in Clear Lake is unknown, lake beds associated with basaltic eruptions have been dated at approximately 0.5 m.y. If the sedimentation rates calculated for the upper 100 m have remained about the same for the entire history of the lake, then the total sediment thickness would be about 300 to 1,000 m. Similar thicknesses have been noted for other valleys in the region (Berkland, 1972).

Drill Holes

During 1973, eight core holes 12 to 15 cm in diameter were drilled for J. D. Sims by the Pitcher Drilling Company into the sediments in Clear Lake in order to investigate earthquake-induced deformational structures in sediments. Clear Lake was chosen due to its location near the seismically active San Andreas fault and the probability of obtaining an uninterrupted sedimentary record back to the late Pleistocene (Sims, 1976). Six of the eight holes were completed with 4-cm (outside diameter) polyvinylchloride (PVC) plastic pipe. The water-filled pipe was capped at the bottom and supported at the lake surface with a buoy. Although this method proved satisfactory during calm weather, unanticipated high waves of a meter or two occurred, which led to the loss of most of the buoys and casings after a short period. In two of the six holes, no temperature measurements were ever obtained. The remaining four survived for periods ranging from a few days to 6 months.

Measurements

Precision temperature measurements were obtained in four drill holes (Fig. 2) on one to four occasions (Table 1), using thermistor probes with a nominal resistance of 10,000 ohms at

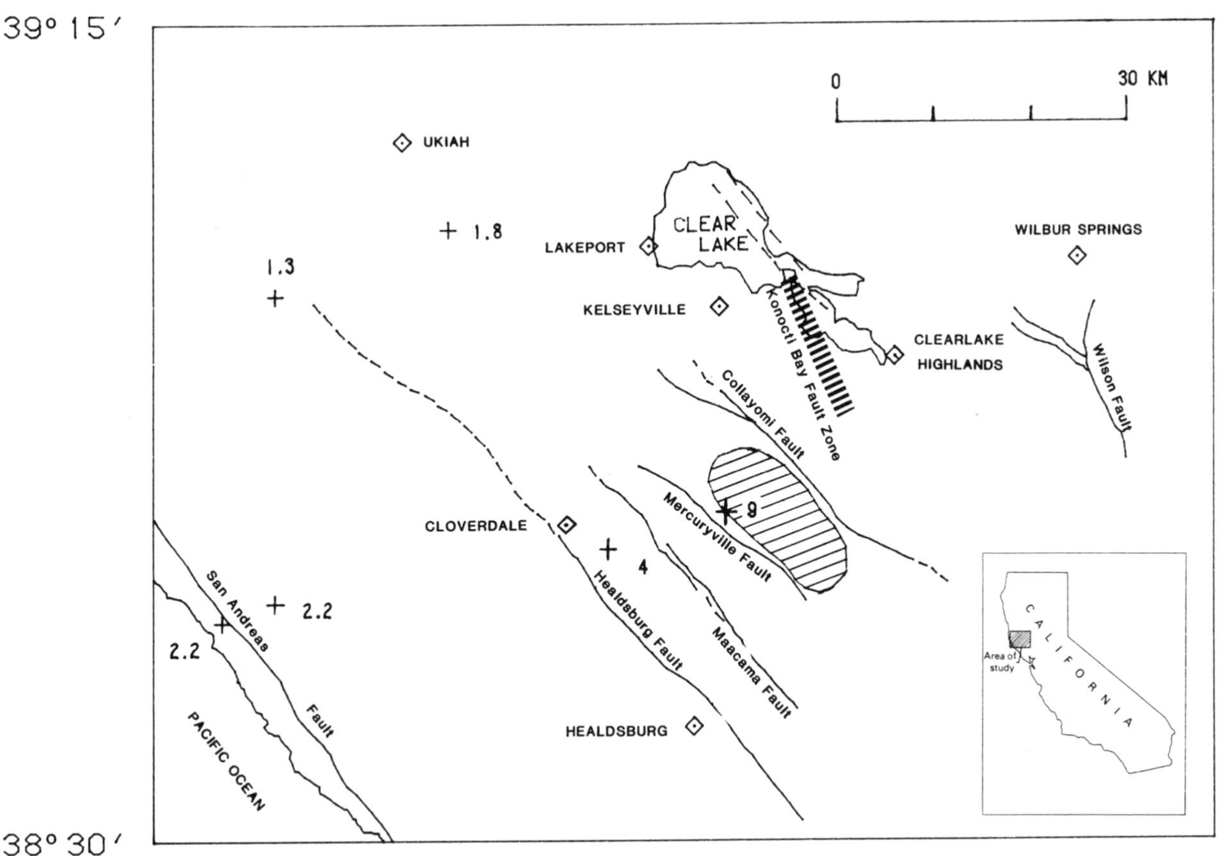

Figure 1. Map showing regional setting of Clear Lake. Faults are dashed where inferred. Heat-flow values (+) from Urban and others (1976) and Lachenbruch and Sass (1980). Cross-hatched area between Mercuryville and Collayomi fault zones is approximate extent of The Geysers steam field.

25°C. The absolute precision is ±0.02°C; the relative accuracy, ±0.002°C. The probe was mounted on the end of a four-conductor armored cable having a 0.3-cm diameter. Thermistor resistances were measured at approximately 30-cm intervals with a battery-powered Data Precision model 2530 digital multimeter in a four-wire configuration that eliminated the lead resistance. Temperatures were calculated from these resistances using the method of Sass and others (1971).

Thermal-conductivity measurements were made on the sediment cores using the needle-probe method described by Von Herzen and Maxwell (1959). The quality of the needle-probe data varies considerably (Figs. 3 through 6). For the sapropelic muds and peats, the results are fairly good, whereas for the coarser grained materials of the Kelsey Creek deltaic deposits, the variability is much greater. This variability is due to two causes: the thermal properties of these sediments have a much greater range, and the measurement error increases significantly for coarse-grained materials, especially gravels, due to the occurrence of pebbles and void spaces near the needle probe. In addition, several months had elapsed between the coring operation and the needle-probe measurements. An unknown amount of water had been lost from the CL-73-1 and CL-73-3 cores. Sealing of the plastic containers (PVC pipe) was improved for the later cores and the subsequent water loss was considerably less. Replacement of water by air would tend to lower the measured thermal conductivities, and thus the values shown should be considered minimums. It is estimated that the uncertainty in the values is 10 percent and possibly as much as 20 percent for some of the coarse-grained sediments and gravels.

TEMPERATURES AND THERMAL CONDUCTIVITY

Composite temperature plots for each hole logged in Clear Lake are shown in Figures 3 through 6. In the main basin of Clear Lake, CL-73-3, CL-73-1, and CL-73-4 form a southwest-northeast line across the lake in the direction of decreasing thickness of the deltaic deposits of Kelsey Creek (Fig. 2; Sims, 1976). The section labeled B in Figures 3 and 4 corresponds to these coarse deltaic sediments. No corresponding sediments that could be correlated with the Kelsey Creek deposits were found in CL-

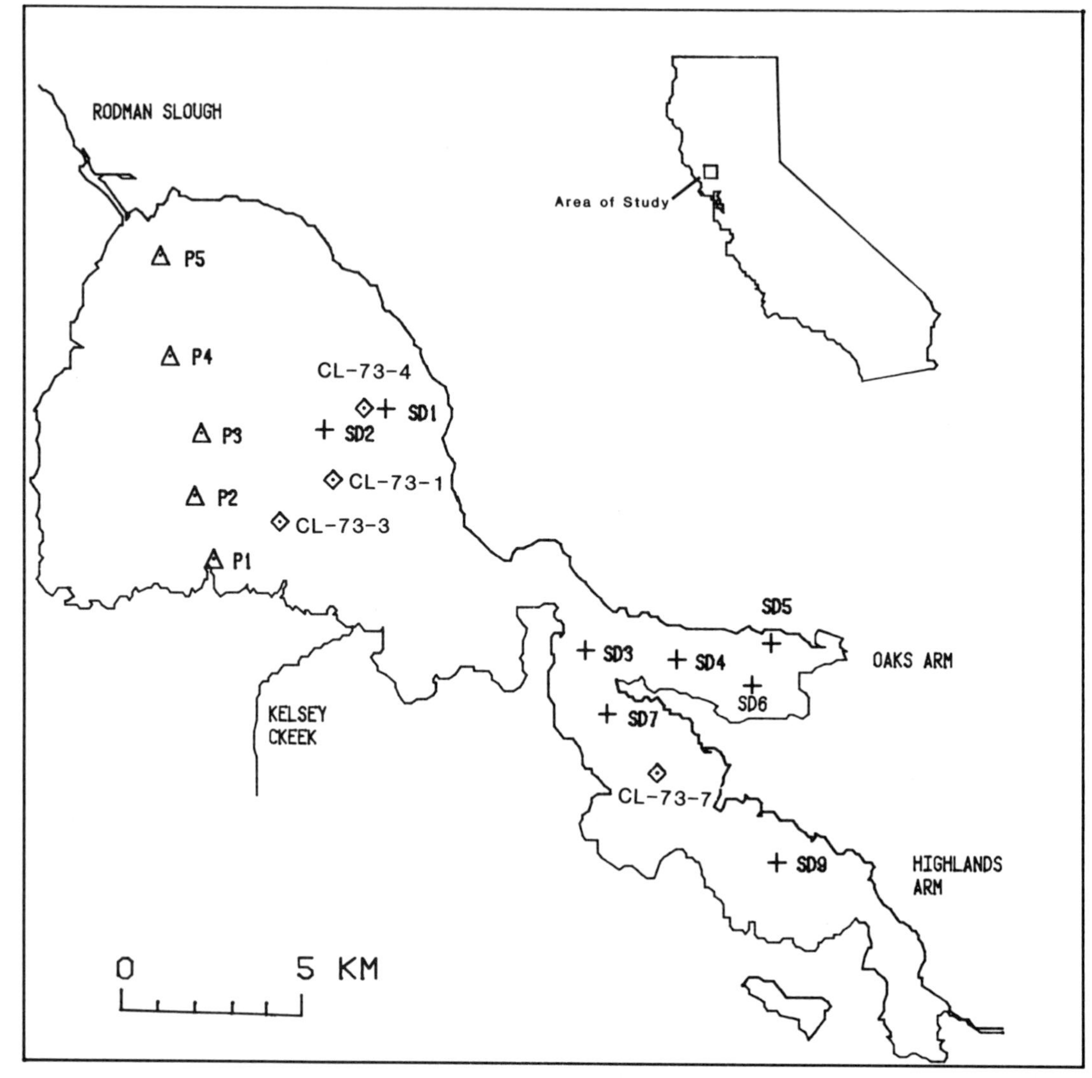

Figure 2. Map showing location of Clear Lake and measurement sites: diamond symbols indicate drill holes; triangles, shallow temperature probes; plus signs, Secchi disk transparency measurements.

TABLE 1. CLEAR LAKE DRILLING HISTORY AND DATES OF TEMPERATURE LOGS*

Hole	Start Date	Finish Date	Drilled Depth (m)	Casing Depth (m)	Date Logged	Curve Label
CL-73-1	8/28/73	8/30/73	61	42	8/30/73	1
					9/ 4/73	2
					9/26/73	3
					10/10/73	4
CL-73-3	9/12/73	9/21/73	77	48	9/26/73	1
					10/10/73	2
CL-73-4	9/25/73	10/17/73	124	122	10/18/73	
CL-73-7	11/ 2/73	11/ 4/73	42	41	11/28/73	1
					1/10/74	2
					5/ 7/74	3
					5/ 8/74	

*Depths listed are depths in meters below lake surface in PVC casing; in all cases they are less than total depth drilled due to caving of lower sections.

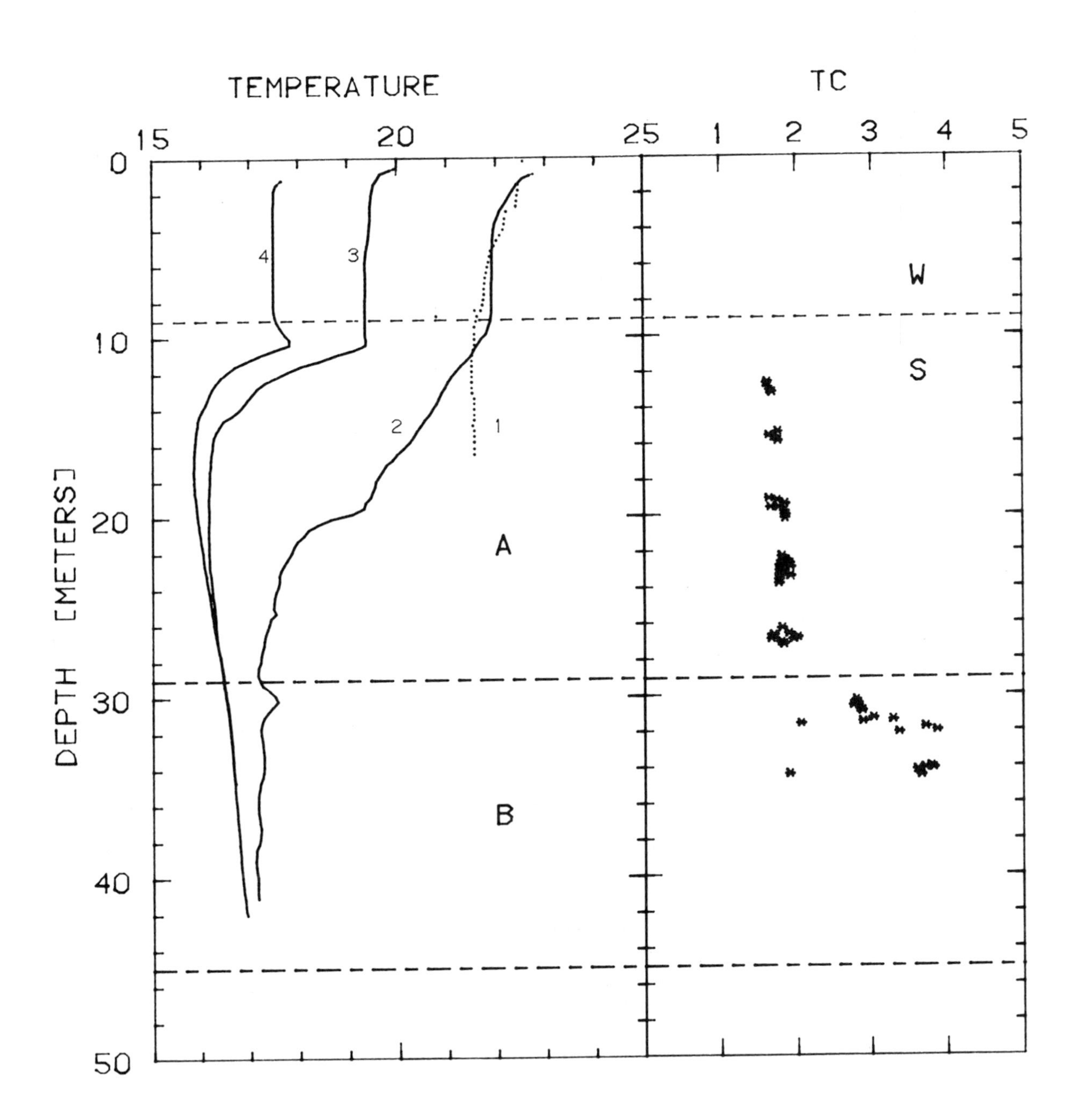

Figure 3. Temperatures (°C) and thermal conductivities (millical/cm·sec·°C) from Clear Lake CL-73-1 site. Dates for four temperature logs listed in Table 2. Section A corresponds to fine-grained sapropelic mud; section B to deltaic deposits of Kelsey Creek (Sims, 1976). Dashed line at 8.7 m, labeled W/S, is bottom of Clear Lake at this site. Temperatures illustrate decay of drilling disturbance and mixing of lake to bottom. Diffusivity of sediments in section A is estimated to be about 0.001 cm^2/sec, based on penetration of annual temperature wave (Carslaw and Jaeger, 1959, p. 66).

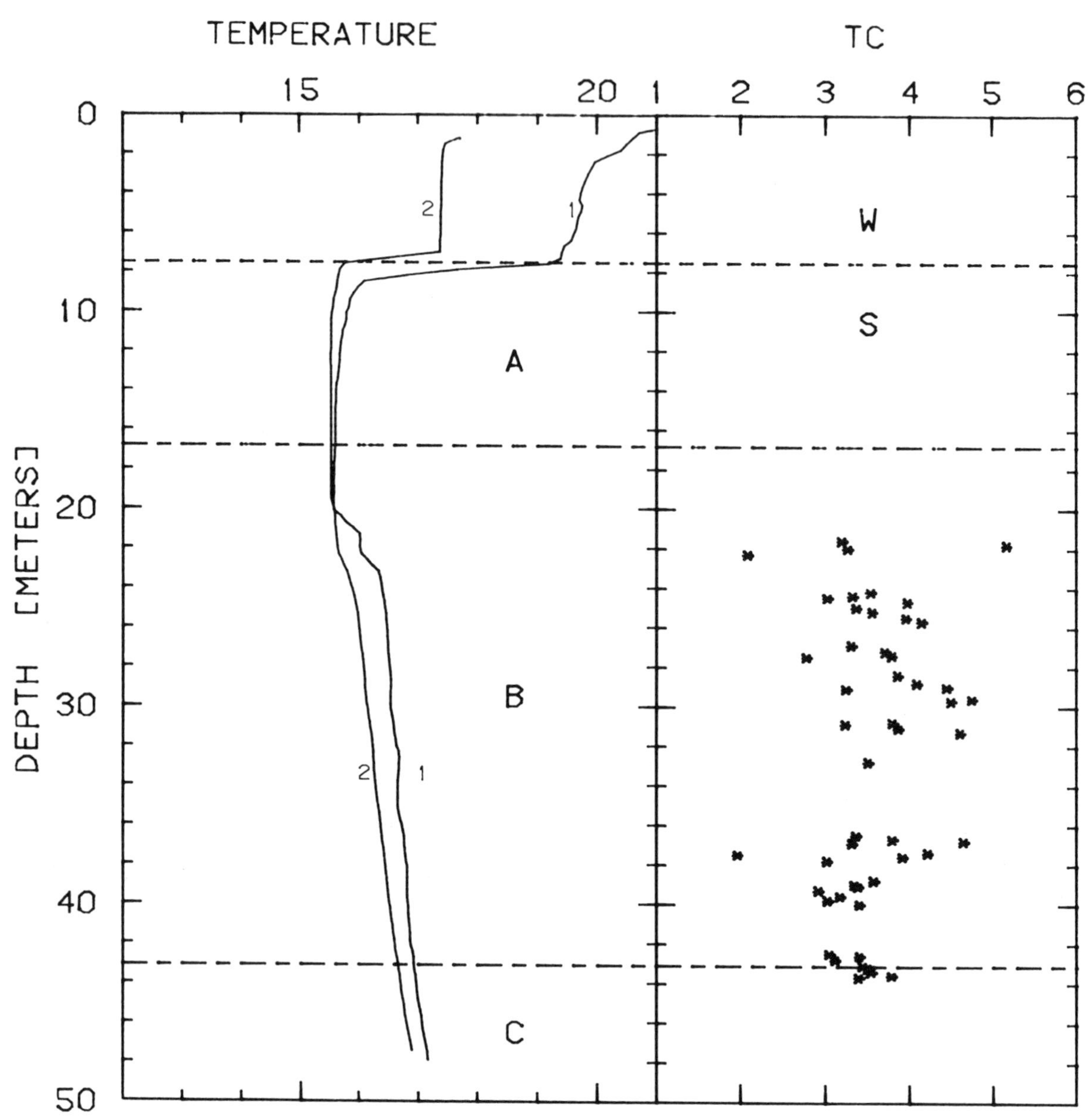

Figure 4. Temperatures (°C) and thermal conductivities (millical/cm·sec·°C) from Clear Lake drill hole CL-73-3. Dates for two temperature logs listed in Table 2. Section A corresponds to fine-grained muds; section B, to deltaic deposits of Kelsey Creek; section C, to silty sands, sands, and clays (Sims, 1976). Dashed line at 7.5 m, labeled W/S, is bottom of Clear Lake at this location. Temperatures illustrate decay of drilling disturbance and weak stratification of lake (curve 1).

73-4 (Sims, 1976). The remainder of the deposits, section A of Figures 3 and 4 and all of Figure 5, consisted primarily of fine-grained mud that is similar to that presently accumulating in the lake.

Drill Hole CL-73-1

Temperature logs were obtained in CL-73-1 for almost 2 months after the hole was drilled (Fig. 3). Curve 1 in Figure 3 (dotted line) was obtained immediately after the completion of drilling and the installation of the PVC casing, and illustrates the near-isothermal character of the drilling fluid in the hole. Measurements were terminated at 17 m due to the failure of the batteries in the multimeter. Subsequent logs—labeled 2 through 4 in Figure 3—show the decay of the drilling disturbance. The last log obtained, curve 4, is very close to equilibrium, and temperatures below 30 m are approximately the same for curves 3 and 4, within the error of each measurement.

There is no obvious change in lithology or thermal conduc-

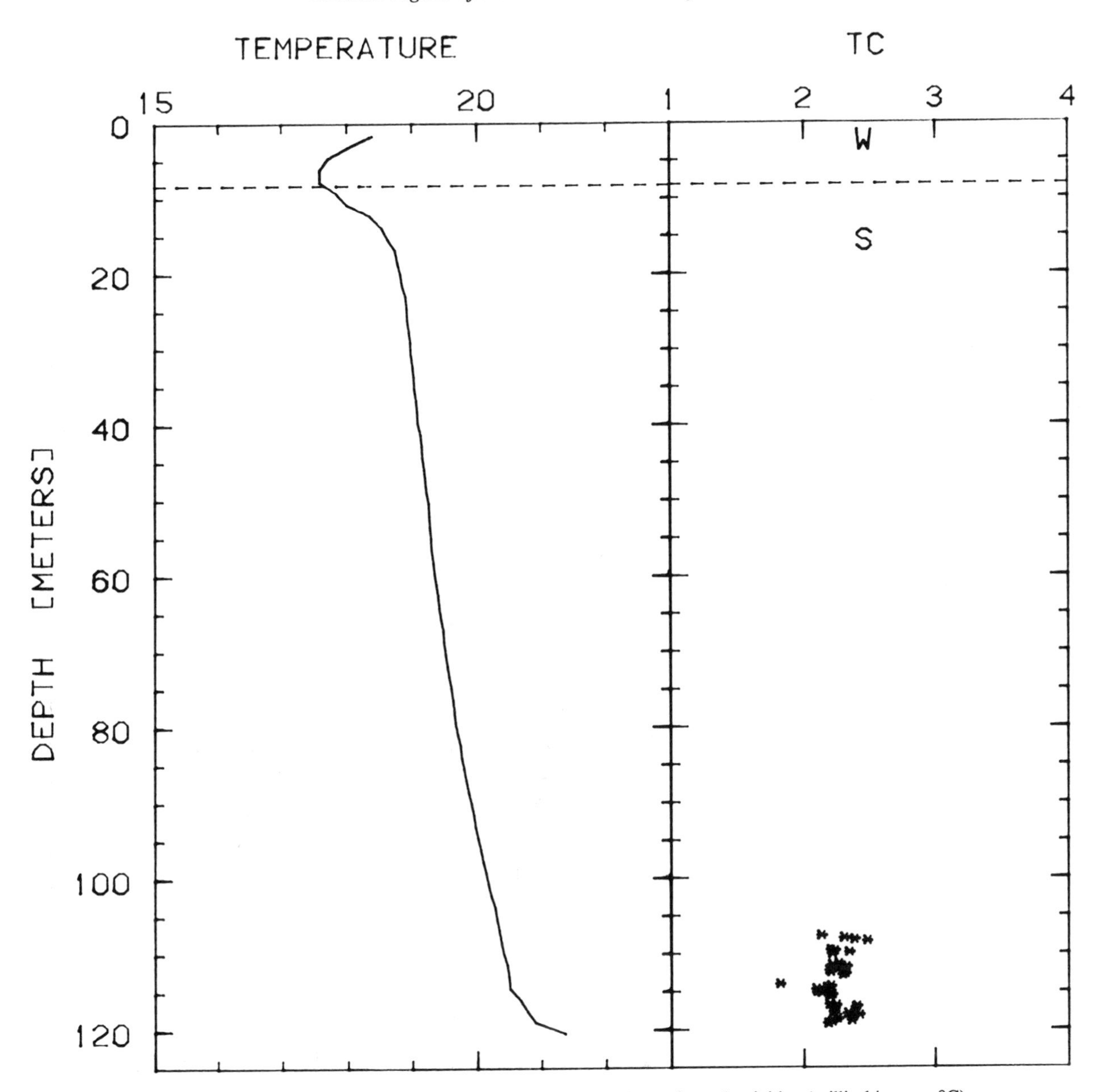

Figure 5. Temperatures (°C) for October 18, 1973, and thermal conductivities (millical/cm·sec·°C) from Clear Lake drill hole CL-73-4. Thermal conductivities were only measured in deepest part of hole where hole is least affected by drilling disturbance. Dashed line at 8.2 m, labeled W/S, is bottom of Clear Lake at this site. Temperatures are severely affected by drilling disturbance and are not considered representative of this location.

tivity to account for the sharp break in the temperatures at about 20 m in curve 2. Examination of the drilling record indicates that this is approximately the depth at which drilling ceased after the first day, and it may be that the larger positive offset can be attributed to the additional circulation time above the 20-m depth in the hole. A similar break at about 29 m also corresponds to a break in drilling, but here the situation is complicated by a change in lithology. An alternate explanation is that the 20-m depth may correspond to a decrease in the permeability or porosity due to compaction. If this is the case, then considerably less drilling fluid may have permeated the formation between 20 and 29 m, and thus the decay of the drilling disturbance would have been much more rapid. In section B, the Kelsey Creek deltaic deposits are coarser and exhibit a high permeability, as illustrated by the high irregularity of curve 2 below 29 m. This irregularity probably would have been much more pronounced had the drilling lasted longer than a day in this part of the hole. The deltaic sediments also exhibit a much higher mean thermal conductivity than do the fine-grained muds (Figs. 3, 4) although the scatter is much greater.

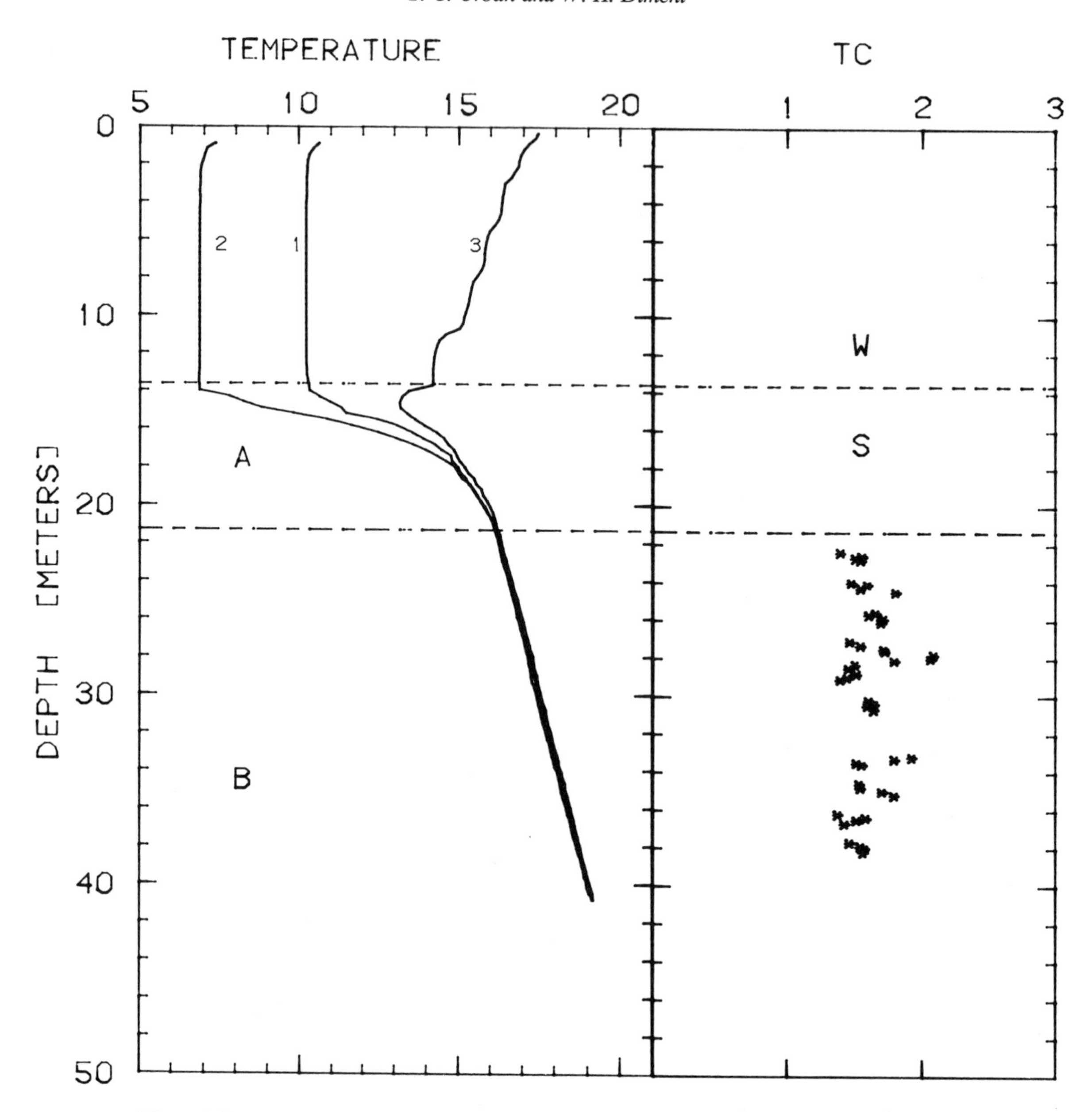

Figure 6. Temperatures (°C) and thermal conductivities (millical/cm·sec·°C) for Clear Lake drill hole CL-73-7. Dates for temperature logs are listed in Table 1. Lithology for site from Sims (1976); section A is sapropelic mud, and section B is peat, mud, and volcanic ash. Temperatures illustrate effect of annual temperature wave in sediments, lake's weak stratification in early May (curve 3), and lake's well-mixed character during winter (curves 1 and 2). Diffusivity of sediments in section A is estimated to be about 0.002 cm^2/sec, based on penetration of annual temperature wave (Carslaw and Jaeger, 1959, p. 66).

Drill Hole CL-73-3

Only two temperature logs were obtained in CL-73-3 before it was lost. With a long 9-day drilling period and such a short recovery time (Table 1), the temperatures are only beginning to approach equilibrium in curve 2 (Fig. 4) but are on the order of 0.1° to 0.2°C too warm compared with extrapolated equilibrium temperatures. The break at 22 m corresponds to a break in drilling as noted for CL-73-1. However, it seems unlikely that this is the cause of such a lingering effect, considering that this depth marks only the first of 9 days of drilling. Curiously, this is the same zone in which the largest range of thermal conductivities occurs in the hole and may reflect a slight change in lithology or permeability. Section C, below 43 m, represents a slight reduction in grain size, and thus a corresponding reduction in the error and apparent variability in the thermal conductivity as the sediments become more uniform. Compared with section A, Figure 3, the higher mean thermal conductivity reflects the sand content of this mud. Some of the higher conductivities appear to correlate with the numerous volcanic ash layers that were identified in the cores (Sims, 1976).

Drill Hole CL-73-4

Only a single temperature log was obtained in CL-73-4 (Fig. 5). Although the profile is fairly linear over a significant portion of the hole, the temperatures were measured only a day after the completion of drilling (Table 1) and are severely affected by almost 3 weeks of drilling circulation. The least-affected temperatures would be those near the bottom. An examination of the turnover point—the point about which the shallow temperatures appear to pivot in response to the annual temperature wave—indicates that the maximum depth of penetration is about 8 m below the bottom of the lake (Figs. 3, 6) with a temperature of about 16°C. Eight meters below the bottom of the lake corresponds to a depth of about 16 m below the surface of the lake for CL-73-4 (Fig. 5) and is about the same depth as the turnover point in Figure 5. If the thermal conductivity is fairly uniform throughout the hole, then a straight line drawn from the temperature of the deepest point in the hole (120.5 m) through the turnover point at 16 m and 16°C and extrapolated to the lake bottom yields a mean annual temperature of about 15.5°C. The slope of this line is the gradient, 51.3°C/km; the average thermal conductivity from Figure 5 is 2.24 millical/cm·sec·°C; the resulting heat flow is 1.15 HFU. If the depth point at the water-sediment interface is moved back to 15.2°C, similar to that for CL-73-1 and CL-73-3, then the heat flow can be increased to about 1.22 HFU. This would move the 16°C turnover point down to about 22 m; this is not too unreasonable, considering the apparent higher thermal conductivity and, therefore, diffusivity of these shallow sediments as compared with those of CL-73-1 and CL-73-3. At best, with the disequilibrium in the temperature profile and the scant knowledge of the thermal properties at this site, it can be concluded that the heat flow appears to be low and even a 30 percent correction would not dramatically alter this conclusion.

Drill Hole CL-73-7

The only hole to survive for a long time was CL-73-7, with the last data set being collected 6 months after drilling was completed. This hole is outside the main basin, in the southeast or Highlands Arm of the lake. The sediments, starting at a depth 21 m below the surface of the lake, consisted almost entirely of peat (section B in Fig. 6). There was little drilling disturbance in this hole as there was no circulation of drilling fluid; the core barrel on successive runs was pushed deeper into the hole. This proved to be a highly successful method, with almost 95 percent core recovery (Sims, 1976). With little or no circulation, the hole recovered to its predrilling temperatures very rapidly, and the first log obtained (curve 1 in Fig. 6), about 3 weeks after drilling, showed little departure from subsequent logs in the lower part of the hole. What variations that exist could possibly be attributed to some circulation in the annulus between the casing and the hole. It was observed that, from the time the hole was completed until the first log, the lake level rose almost 1 m but did not submerge the buoy. By January 1974, when the second log was made, the lake had risen an additional meter, but this time the buoy was partially submerged and had only pulled the casing out about half a meter. Such motions indicate a very slow collapse of the hole around the casing. Although adjustments were made in the depths of the temperatures, it is possible that some vertical fluid motion did occur in the annulus. The size of this motion, if it occurred, is deemed either to be small or to have encompassed the entire length of the hole, since the temperatures are quite linear in section B (Fig. 6) and the thermal properties are rather uniform.

HEAT FLOWS AND CORRECTIONS

Equilibrium Temperatures

During the coring operation, fluid was pumped down through the drill pipe to the bottom of the hole, circulated around the bit, and returned to the surface through the annulus between the drill pipe and the sediment wall. In this process, heat is exchanged between the drilling fluid and the sediment, resulting in the sediment wall becoming warmed or cooled. When the drilling operation ceased, the water or drilling mud stagnated in the hole and the temperature at each depth asymptotically approached the value it had before drilling commenced. If a long enough period of time has elapsed, then the observed temperatures are nearly equal to their predrilling values. As a measure of the amount of time involved, Diment and Weaver (1964, p. 80) noted that, for a core hole near Mayaguez, Puerto Rico, the gradient reached its equilibrium value in less than 10 times the length of the drilling disturbance and was within 93 percent of its equilibrium value in a time 1.5 times the length of the drilling disturbance. In a hole drilled in permafrost, Lachenbruch and Brewer (1959, p. 107) found that the gradient was within 95 percent of its equilibrium value in 1.3 times the duration of the drilling disturbance. For CL-73-1 the last temperature log was taken about 10 times the length of the drilling disturbance (Table 1) and should be near equilibrium. CL-73-3, however, had a 9-day drilling disturbance, which would imply about 3 months to equilibrium. The last temperature log was obtained less than 3 weeks after drilling ceased, but should be roughly within 95 percent of the equilibrium gradient. CL-73-4, with 17 days of disturbance and with only 1 day after drilling, is significantly out of equilibrium, and the temperatures reflect more the temperature of the drilling fluid than that of the surrounding rock.

Because the primary quantity of interest in a heat-flow determination is the thermal gradient and not the actual temperature, some estimate of the relative departure of the last gradients—obtained from actual temperature measurements—to the equilibrium gradients is required. In order to do this, an estimate was made of the predrilling, equilibrium temperatures.

Equilibrium temperatures were determined using the method of Lachenbruch and Brewer (1959): temperatures for each date at a specific depth were plotted against $E = ln(t/t\text{-}s)$, where t is the time elapsed since the drill bit first reached the depth in question, s is the time elapsed since the drill bit first reached the depth in question until the drilling operation ceased and ln is the natural log. The plot of temperature versus $ln(t/t\text{-}s)$ should show that the return of the temperatures to their equilibrium value is nearly a linear function of $ln(t/t\text{-}s)$. By extrapolating to $E = 0$ (i.e., to infinite time), the equilibrium temperatures were determined, assuming that the sole cause of the disturbance was the conductive warming or cooling caused by circulation during the drilling process.

This was done for a number of depths for CL-73-1 and CL-73-3. For the intervals in Table 2, the gradients for CL-73-1 were at their equilibrium values for the last temperature log on October 10, 1973 (Table 1). However, for CL-73-3 the gradient error was as great as 20 percent for the last log, and the values shown in Table 2 are for the calculated equilibrium gradients. This is not unexpected, since the porosity and the permeability of these sediments are probably higher than most rocks, and, in addition to heating or cooling by conduction, fluid loss probably was a significant factor in the drilling disturbance. In the case of CL-73-7, there was little variation among the temperatures in all the logs; because of the short drilling time and no circulation, the gradient from the May 7, 1974, temperature log was used in Table 2.

Terrain Corrections

Because there is significant topography in the vicinity of CL-73-7 (Mt. Konocti) that could possibly affect the observed heat flow, a terrain correction was calculated. The geometry in the vicinity of CL-73-7 is roughly that of an infinite trough surrounded by mountains on the northeast and southwest; thus, a two-dimensional steady-state terrain correction was calculated (Birch, 1950, p. 587). The unknown factor for this calculation was the thermal gradient in the rock surrounding the lake. If the thermal conductivity is approximately 5 millical/cm·sec·°C, and if one uses the heat flow from CL-73-7, then the regional gradient is approximately 46°C/km. The resulting terrain correction is about −2°C/km and reduces the observed heat flow for CL-73-7 about 0.03 HFU. With little topography near the drill holes in the main basin and given the small correction for CL-73-7, no terrain corrections were calculated for those holes.

Sedimentation

Clear Lake has accumulated sediment for at least 130,000 yr (Adam and others, 1981) and possibly for as long as half a million years (Donnelly-Nolan and others, 1981, p. 55). Sedimentation rates at various depths in the lake have varied from

TABLE 2. GRADIENTS, THERMAL CONDUCTIVITIES, AND HEAT FLOWS FOR CLEAR LAKE DRILL HOLES

Hole	Interval (m)	Gradient (°C/km)	Thermal Conductivity*	Heat Flow (HFU)
CL-73-1	27-32	51.6	3.03 (12)	1.56
	32-35	43.3	3.42 (6)	1.48
CL-73-3	38-41	48.9	3.26 (7)	1.59
	43-46	46.3	3.40 (8)	1.57
CL-73-7	24-41	149.6	1.60 (44)	2.39

*Thermal conductivity is expressed in (millical/cm·sec·°C). Number in parentheses indicates number of values used to determine average thermal conductivity.

0.37 to 1.44 mm/yr, averaging approximately 0.68 mm/yr. For sedimentation with constant surface temperature, the ratio of the observed gradient (G') to the steady-state gradient (G) is (Carslaw and Jaeger, 1959, p. 388; Jaeger, 1965, p. 14)

$$G'/G = 1 + F(p)$$

and

$$F(p) = 1/2\, p^2 - (1 + 1/2\, p^2)\mathrm{erf}(1/2\, p) - (1/\pi^{1/2})p \exp(-p^2/4),$$

$$p = Ut/(at)^{1/2}$$

where U is the sedimentation rate, t is the elapsed time since sedimentation commenced, a is the diffusivity (0.001 cm^2/sec), and erf is the error function. For the average sedimentation rate observed in Clear Lake

$$G'/G = 0.8882,$$

which would increase the observed heat flow for CL-73-1, CL-73-3, and CL-73-7 to 1.7, 1.8, and 2.7 HFU, respectively, or roughly 13 percent.

Lake Effect

Deep lakes in temperate continental climates tend to be stratified and the temperatures of the deep waters fairly constant. Even in lakes that do circulate and are stratified only during part of the year, the average bottom temperature can vary significantly from that of the surrounding land. This lake effect can depress the isotherms in the vicinity of the lake, resulting in a higher apparent gradient and heat flow in the lake.

In addition to the sediment temperatures in Clear Lake, water temperatures were also measured in the lake (Figs. 3 through 6). They show that the lake in general tends to be well

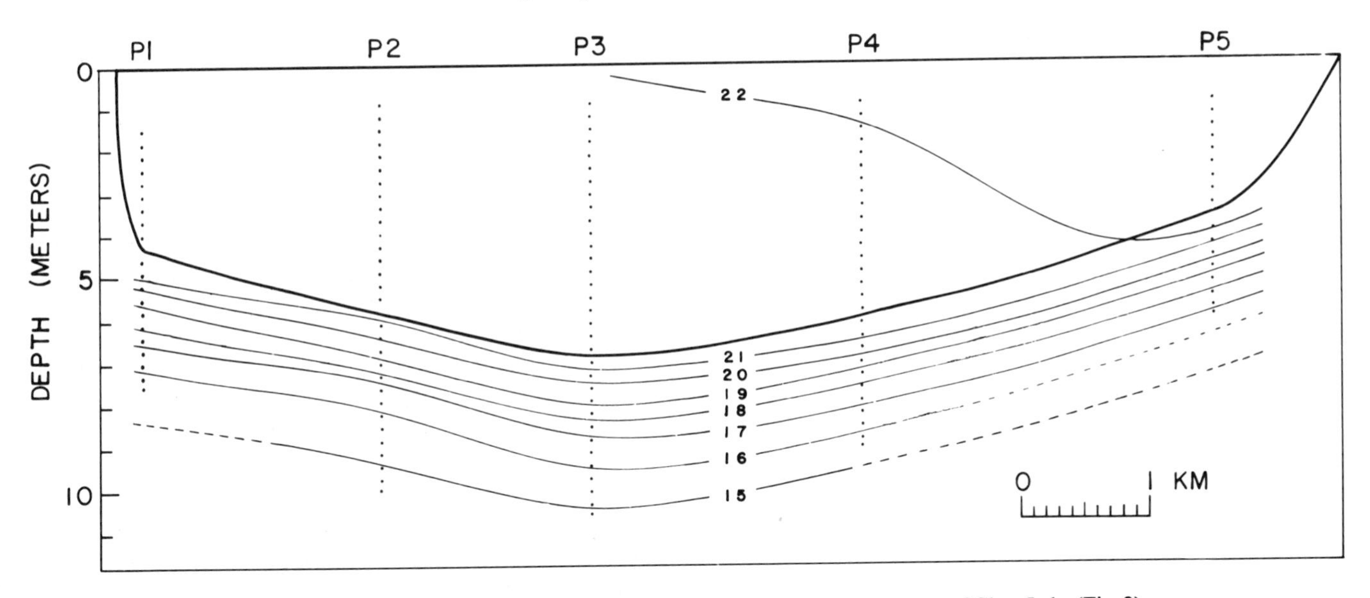

Figure 7. Isotherms for P-series of temperature measurements in the western part of Clear Lake (Fig. 2). Dots indicate actual measurement points; heavy solid line, sediment-water interface. Numbers are isotherms in degrees Celsius. Configuration of 22°C isotherm in vicinity of P4 and P5 is influenced by influx of warm water into lake from Rodman Slough.

mixed. At the same time, a series of temperature measurements were also made at shallow depths (5 m) in the sediments (September 11, 1973) to examine the western part of the main basin (P series in Figs. 2, 7). The method employed was to push a 3.3-cm (outside diameter) aluminum pipe, capped on the bottom to exclude any sediment, into the sediments of the lake. The tubing was filled with water and allowed to come to equilibrium. After initial experimentation it was determined that there was little temperature change after 30 min and therefore all subsequent measurements were made 30 min after insertion. Although the tubing could be pushed deeper than about 5 m into the sediments, the insertion depth was restricted to approximately this amount to facilitate removal of the pipe. A cross section of the lake and the sediments penetrated by the temperature measurements and the isotherms at 1°C intervals are plotted in Figure 7. The isotherms generally tend to follow the contour of the lake bottom, indicating little or no effect of any stratification. The anomalous configuration of the 22°C isotherm reflects the infusion of warm water from nearby Rodman Slough.

On three occasions, Secchi disk transparency measurements were obtained in the lake to determine if direct heating of the bottom sediments could occur. The measurement is crude but easy to make; it consists of lowering a 20-cm diameter white disk until it no longer can be seen. This depth corresponds to a light penetration of 5 to 10 percent (Hutchinson, 1957, pp. 399–403). A synoptic series of measurements was conducted in May 1974 (Table 3; Fig. 1). They showed that the Secchi disk transparency for Clear Lake was only about 150 to 200 cm for that date. Comparison of three measurements at CL-73-7 over 2 days indicates that there is considerable variability in the transparency but that the lake is fairly opaque. Hence, it is unlikely that there is any significant direct heating of the sediments.

Extrapolation of the temperatures in CL-73-1 (20 to 30 m), CL-73-3 (below 25 m), and CL-73-7 (below 22 m) to the bottom of the lake indicates a long-term mean annual temperature of 15.2°C. Although slightly warmer, this number is close to the recorded mean annual air temperatures for Lakeport (14.0°C) and Clearlake Highlands (13.5°C) for the period from 1964 to 1974 (U.S. Department of Commerce, 1964 through 1974).

Conductivity Contrast

The observed heat flows (Table 2) in Clear Lake are low in comparison to those found in The Geysers steam field (~9 HFU) (Urban and others, 1976) but are similar to other heat-flow values in the Coast Ranges of northern California (Lachenbruch and Sass, 1980). The corrections previously discussed would increase the heat flow only about 0.2 to 0.3 HFU. This would tend to indicate that Clear Lake possesses a normal background heat flow. This may indeed be the case, assuming that the sediment in the lake is relatively thin and that the thermal conductivity contrast between the sediments of the lake and the bounding rocks is not great enough to cause any significant refraction effect.

Although there are no conductivity values available from rocks adjacent to Clear Lake, thermal conductivities have been measured for the Franciscan graywacke from a core hole near Cloverdale and from outcrops at The Geysers (Urban and others, 1976). These rocks have thermal conductivities ranging from 5 to 7 millical/cm·sec·°C and can be as high as 9. Other rock types that may be present in the area are serpentinite or metabasalts,

TABLE 3. SECCHI DISK TRANSPARENCY MEASUREMENTS IN CLEAR LAKE

Location	Date	Depth (cm)	Time (PST)
CL-73-1	9/4/73	97	13:50
CL-73-7	5/7/74	122	09:45
	5/8/75	79	10:50
SD1	5/7/74	229	13:35
SD2	5/7/74	213	13:45
SD3	5/7/74	144	14:53
SD4	5/7/74	152	15:02
SD5	5/7/74	233	15:07
SD6	5/7/74	144	15:11
SD7	5/7/74	229	15:21
SD8 (CL-73-7)	5/7/74	144	15:25
SD9	5/7/74	165	15:33

neither of which have thermal conductivities greatly different from graywacke (Diment, 1964; Clark, 1966). Robertson and Peck (1974) measured thermal conductivities of a variety of vesicular basalts from Hawaii and found that the conductivity of water-saturated samples ranged from 2 to 5.8 millical/cm·sec·°C. The lower values appear to have been the result of the insulating effect of micropores and microfractures that were not completely filled with water.

In general, it appears that the bulk of the rocks in the vicinity of Clear Lake have thermal conductivities in the range of 5 to 7 millical/cm·sec·°C, which is significantly greater than that of the sediments in the lake (Figs. 3 through 6). The ratio of the conductivity of the rocks to the sediments in the main basin would be approximately 2 to 4, whereas for the Highlands Arm of Clear Lake the ratio would be 3 to 5. Such a contrast might have a significant effect on the observed heat flow. Unfortunately, the exact sediment thickness in the lake is unknown.

Subbottom seismic profiling was done by Sims (1976, p. 662) to determine the approximate thickness of the sediments and the presence of internal structures. However, the acoustic penetration was only 80 m at a maximum and did not reveal any bedrock. Subsequent drilling in 1973 and 1980 has shown that the sediment is more than 177 m thick (Sims and others, 1981). Depths of alluvium in the valleys in the Central Coast Ranges are as thin as 20 m in Bachelor Valley on the northwest edge of Clear Lake and as thick as 1.2 km in the Willits Basin about 50 km northwest of Clear Lake (Berkland, 1972, p. 11). Sedimentation rates, discussed previously, have varied from 0.34 to 1.44 mm/yr. On the basis of the oldest sediments dated around Clear Lake, this would imply a total accumulation of sediments on the order of 300 to 1,000 m, which is consistent with the observations in other basins.

Several models (Fig. 8) were calculated based on various conductivity contrasts and sediment thicknesses for the main basin and the Highlands Arm of Clear Lake. The models are based on an ellipsoid in an infinite medium (Carslaw and Jaeger, 1959, p. 427), with temperatures at a great distance from the ellipsoid approaching Gz, where G is the steady-state geothermal gradient, z is the depth, and the rectangular coordinate system is centered in the ellipsoid. Solutions for the circular basin and trough were given by Von Herzen and Uyeda (1963, p. 4238) and were used to generate the curves of Figure 8. The observed heat flow is assumed to be measured at zero depth with no sediment in the lake. The solution to this problem shows that the heat flow in the ellipsoid is constant, and that the ratio of the gradient observed in the ellipsoid to the regional gradient far removed from the ellipsoid is determined by the thermal conductivity contrast for a given basin geometry.

The main basin of Clear Lake can be approximated by an ellipsoid, which is circular in plan view and about 11 km in diameter. The corrected heat flows for various conductivity contrasts are shown as the solid lines in Figure 8. The heat flow is increased for all the models but not significantly enough to distinguish it from the normal regional heat flow of 1.5 to 2 HFU. In this case the model has a geometry in which the sediment thickness is less than 10 percent of the diameter of the lake, which leads to a very small correction.

For the southeast or Highlands Arm of Clear Lake, the geometry is different and may be treated as a semi-elliptical trough with a width of about 2.6 km. In this case (dashed lines, Fig. 8) not only is the conductivity contrast and heat flux greater than for the main basin, but the sediment thickness is significantly greater in proportion to the width of the trough, resulting in a larger correction. Indeed, for the largest conductivity contrast shown and the greatest sediment thickness, the corrected heat flow approaches that observed in The Geysers. Whether such large conductivity contrasts or thick accumulations of sediment exist is subject to speculation. What is clear, however, is that these results tend to indicate the possibility of some geothermal potential and that further investigation is warranted.

DISCUSSION

If the sediments in Clear Lake are only on the order of 100 to 200 m, then even a refraction correction would not significantly alter the observed heat flow, and other possible causes of the low heat flux should be examined. Sims and Rymer (1976) noted numerous gas seeps in the lake, one trail of which passed near CL-73-4 along an inferred fault. As the gas percolates its way up along the fault, quite possibly it could have an effect on the observed temperatures and the resulting heat-flow calculation. However, the temperatures in CL-73-4 are severely out of equilibrium due to the drilling disturbance. In addition, the precise amount of gas produced from the vents in the lake is unknown but has been estimated to range from a few liters/minute to milliliters/minute (Sims and Rymer, 1976). Thus, given the small quantity of gas being produced, its limited areal extent, and the disequilibrium temperatures, no satisfactory assessment of the effect on the observed temperatures was possible. It does not,

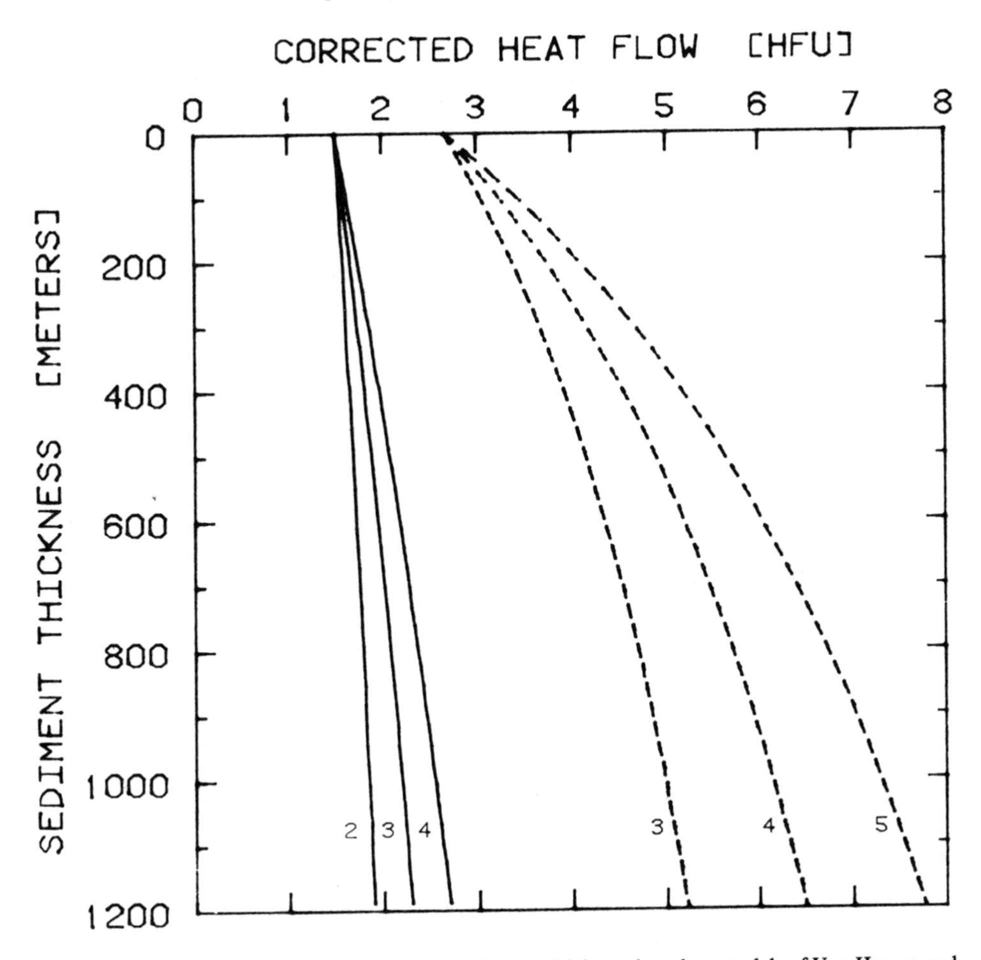

Figure 8. Corrected heat flows as a function of sediment thickness based on models of Von Herzen and Uyeda (1963). Curves represent corrected heat flow for refraction effect caused by thermal-conductivity contrast between surrounding rock and lake sediment. Solid curves are for circular hemiellipsoid of main basin, with observed heat flow of 1.5 HFU at sediment surface. Dashed lines are for trough representing Highlands Arm of Clear Lake, with observed heat flow of 2.7 HFU at surface of sediment, corrected for sedimentation rate of 0.68 mm/yr. Numbers on each curve indicate ratio between assumed thermal conductivity of surrounding rock to that of lake sediment and represent possible range of values for each basin. Actual sediment thickness is unknown, but range of depths shown reflects accumulation of sediments for several of nearby basins (Berkland, 1972).

however, appear that this mechanism would be a significant contributor to alteration of the observed heat flux.

Although most of the gas produced in Clear Lake is thought to be associated with faults, there are other faults—including those bounding the lake—that do not produce gas. Downward movement of cold water along such faults could cool the boundaries of the lake and reduce the observed heat flux. Thermal springs in the Clear Lake area are warm but generally are not hot. For example, of the 113 samples from springs and wells listed by Thompson and others (1981), 50 have temperatures less than 20°C, 58 are in the range of 20° to 50°C, and only 5 are higher than 50°C. Similarly, Waring (1965, pp. 21–22) listed 33 locations, but only 9 with temperatures higher than 50°C. The flow rates also tend to be low: of 113 sites, 85 flowed less than 20 l/min, only 17 were between 21 and 100 l/min, and less than 10 percent or 11 sites had flow rates greater than 100 l/min (Thompson and others, 1981). Again, Waring (1965) noted 33 locations but only 8 with flow rates greater than 100 l/min. This suggests either some mixing of the rising hot water with downward moving cold water, the source of which most likely is the faults in the region, or the circulation of the water that produces the springs is not very deep.

The low heat-flow values observed in the Clear Lake sediments, as compared with those to the south in The Geysers steam field, may have their origins in a general downward movement of water through the sediments, although the vertical permeability of lake sediments generally is rather low. With the exception of the deltaic deposits of Kelsey Creek, most of the sediments in the lake have thermal conductivities not too different from that of water. Depending on the velocity of such a vertical water movement, a substantial reduction in the observed heat flow could occur. Such large-scale vertical fluid motion has been utilized to explain the Eureka heat-flow low of southern Nevada (Sass and others, 1971). If a large-scale downward movement of water is occurring in Clear Lake, then conceivably it could be a source of the recharge for The Geysers–Clear Lake geothermal system. McLaughlin (1981, p. 14) suggested that, in the area north and east of the Collayomi Fault Zone, the vapor-dominated steam system passes into a hot- then cold-water dominated system. The occurrence of the low heat flow in the main basin of Clear Lake may indicate the transition into such a cold-water system.

However, probably not all the area around Clear Lake is part of such a cold-water system. The heat flow in the Highlands Arm (CL-73-7) is significantly higher than that measured in the main basin. At the east end of the Oaks Arm of Clear Lake in the Sulfur Bank mine, mercury as metacinnabar is currently being deposited in active spring vents in the pond that occupies the abandoned mine pit (Sims and White, 1981, p. 238). Temperatures as high as 80°C were encountered during the underground mining phase that was terminated in 1897. The hydrothermal mineralization at Sulphur Bank mine is entirely of late Quaternary age, with most of the deposits younger than 44,000 B.P. (Sims and White, 1981, p. 238). Northward or northeastward tilting of the lake basin has apparently continued to recent times, based on the northward tilting of lake and fluvial deposits south of Kelseyville and on the presence of steeper slopes and deeper water along the northeastern shores of Clear Lake (Hearn and others, 1981, p. 29). This recent activity indicates that the region is still active and the heat sources are most likely being masked by downward moving water.

CONCLUSIONS

Temperature measurements from four core holes in Clear Lake and thermal-conductivity measurements using the needle-probe method on the cores have yielded heat-flow values of 1.5 to 2.4 HFU, which are similar to measurements obtained in The Geysers–Clear Lake region but lower than those values measured in and near The Geysers steam field (4 to 9 HFU). Corrections to the heat flow for terrain, sedimentation, and topography do not substantially change this conclusion. A correction due to the refraction effect caused by the thermal-conductivity contrast between the lake sediments and the surrounding rock is relatively small for the main basin since the ratio of the sediment thickness to the diameter of the lake is small. A similar correction for the Highlands Arm of Clear Lake yields a significantly higher heat flow due to the larger ratio of sediment thickness to width. Indeed, the largest correction increases the heat flux at depth to that approaching The Geysers steam field for reasonable thermal-conductivity contrasts and for maximum sediment thicknesses observed in similar basins (~1,200 m). Whether such thick accumulations of sediment and large heat flows exist is subject to speculation at this time. Such corrections do indicate that if there is an anomalous heat source associated with Clear Lake, then it is partially masked. One mechanism for reducing the heat flux would be a general downward movement of water through the sediments. However, a downward water flow over the entire area of the lake seems unlikely, as the permeability of most lake sediments is rather low. An alternate explanation centers on the numerous faults that bound Clear Lake, at least one of which is inferred to lie entirely beneath the main body of the lake. Some of these faults are known to produce small quantities of gas and fluids. The downward movement of cold water along these faults could cool the boundaries of the lake and reduce the observed heat flux. Hot springs in the Clear Lake area generally tend to have low flow rates and are more warm than hot, although there are exceptions. This would tend to suggest either some mixing of the rising hot water with downward-moving cold water, the source most likely being the faults in the region, or that the circulation of water that produces the springs is not very deep. In either case the cooling effect of such downward-moving cold water would decrease the observed heat flux and could possibly provide a source of recharge for The Geysers–Clear Lake hydrothermal system.

ACKNOWLEDGMENTS

We are indebted to John D. Sims for providing the opportunity to case the Clear Lake holes for temperature measurements, and to the numerous individuals in the Clear Lake area who provided support and information about Clear Lake for this study. We especially thank John P. Kennelly and John P. Ziagos, who aided in the temperature measurements; Eugene P. Smith, who measured the thermal conductivities; and Robert J. Munroe, who calibrated the thermistors used in this study. The manuscript was read and improved by Arthur H. Lachenbruch and John H. Sass.

REFERENCES CITED

Adam, D. P., Sims, J. D. and Throckmorton, C. K., 1981, 130,000-Yr continuous pollen record from Clear Lake, Lake County, California: Geology, v. 9, p. 373–377.

Berkland, J. O., 1972, Clear Lake basin; A deformed Quarternary caldera, *in* Moores, E. M., and Matthews, R. A., eds., Geology guide to the northern Coast Ranges Lake, Mendocino and Sonoma Counties, California: Davis, California, Geological Society of Sacramento, p. 6–26.

Birch, Francis, 1950, Flow of heat in the Front Range, Colorado: Geological Society of America Bulletin, v. 61, p. 567–630.

Carslaw, H. S., and Jaeger, J. C., 1959, Conduction of heat in solids: London, Oxford University Press, 2nd ed., 510 p.

Clark, S. P., Jr., 1966, Thermal conductivity, *in* Clark, S. P., Jr., Handbook of physical constants: Geological Society of America Memoir 97, p. 459–482.

Diment, W. H., 1964, Thermal conductivity of serpentinite from Mayaguez, Puerto Rico, and other localities, *in* Burk, C. A., ed., A study of serpentinite: NAS-NRC Publication 1188, p. 92–106.

Diment, W. H., and Weaver, J. D., 1964, Subsurface temperatures and heat flow in the AMSOC core hole near Mayaguez, Puerto Rico, *in* Burk, C. A., ed., A study of serpentinite: NAS-NRC Publication 1188, p. 75–91.

Donnelly-Nolan, J. M., Hearn, B. C., Jr., Curtis, G. H., and Drake, R. E., 1981, Geochronology and evolution of the Clear Lake Volcanics, *in* McLaughlin, R. J., and Donnelly-Nolan, J. M., eds., The Geysers–Clear Lake geothermal area, northern California: U.S. Geological Survey Professional Paper 1141, p. 47–60.

Hearn, B. C., Jr., Donnelly-Nolan, J. M. and Goff, F. E., 1981, The Clear Lake volcanics: Tectonic setting and magma sources, *in* McLaughlin, R. J., and Donnelly-Nolan, J. M., eds., Research in The Geysers–Clear Lake geothermal area, northern California: U.S. Geological Survey Professional Paper 1141, p. 25–45.

Hutchinson, G. E., 1957, A treatise on limnology: New York, John Wiley & Sons, v. 1, 1015 p.

Jaeger, J. C., 1965, Application of the theory of heat conduction to geothermal measurements, *in* Lee, W.H.K., ed., Terrestrial heat flow: American Geophysical Union Monograph Series, no. 8, p. 7–23.

Lachenbruch, A. H., and Brewer, M. C., 1959, Dissipation of the temperature effect of drilling a well in Arctic Alaska: U.S. Geological Survey Bulletin 1083-C, p. 73–109.

Lachenbruch, A. H., and Sass, J. H., 1980, Heat flow and energetics of the San Andreas Fault Zone: Journal of Geophysical Research, v. 85, p. 6185–6222.

McLaughlin, R. J., 1981, Tectonic setting of pre-Tertiary rocks and its relation to geothermal resources in The Geysers–Clear Lake area, *in* McLaughlin, R. J., and Donnelly-Nolan, J. M., eds., Research in The Geysers–Clear Lake geothermal area, northern California: U.S. Geological Survey Professional Paper 1141, p. 3–23.

McLaughlin, R. J., and Donnelly-Nolan, J. M., ed., 1981, Research in The Geysers–Clear Lake geothermal area, northern California: U.S. Geological Survey Professional Paper 1141, 259 p.

Robertson, E. C., and Peck, D. L., 1974, Thermal conductivity of vesicular basalts from Hawaii: Journal of Geophysical Research, v. 79, no. 32, p. 4875–4888.

Sass, J. H., Lachenbruch, A. H., Munroe, R. J., Greene, G. W., and Moses, T. H., Jr., 1971, Heat flow in the western United States: Journal of Geophysical Research, v. 76, no. 26, p. 6376–6413.

Sims, J. D., 1976, Paleolimnology of Clear Lake, California, U.S.A., *in* Horie, S., ed., Paleolimnology of Lake Biwa and the Japanese Pleistocene: Kyoto, Japan, Kyoto University, v. 4, p. 658–702.

Sims, J. D., and Rymer, M. J., 1976, Map of gaseous springs and associated faults in Clear Lake, California: U.S. Geological Survey Miscellaneous Field Studies, Map MF-721.

Sims, J. D., and White, D. E., 1981, Mercury in the sediments of Clear Lake, *in* McLaughlin, R. J., and Donnelly-Nolan, J. M., eds., Research in The Geysers–Clear Lake geothermal area, northern California: U.S. Geological Survey Professional Paper 1141, p. 237–241.

Sims, J. D., Adam, D. P., and Rymer, M. J., 1981, Late Pleistocene stratigraphy and palynology of Clear Lake, *in* McLaughlin, R. J., and Donnelly-Nolan, J. M., eds., Research in The Geysers–Clear Lake geothermal area, northern California: U.S. Geological Survey Professional Paper 1141, p. 219–230.

Thompson, J. M., Goff, F. E., and Donnelly-Nolan, J. M., 1981, Chemical analyses of waters from springs and wells in the Clear Lake volcanic area, *in* McLaughlin, J. M., and Donnelly-Nolan, J. M., eds., Research in The Geysers–Clear Lake geothermal area, northern California: U.S. Geological Survey Professional Paper 1141, p. 215–218.

U.S. Department of Commerce, For years 1964 through 1974, Climatological data, California: Asheville, North Carolina, National Oceanic and Atmospheric Administration Environmental Service.

Urban, T. C., Diment, W. H., Sass, J. H., and Jamieson, I. M., 1976, Heat flow at The Geysers, California, U.S.A., *in* Second United Nations Symposium on the Development and Use of Geothermal Resources, San Francisco: University of California, Lawrence Berkeley Lab., v. 2, p. 1241–1245.

Von Herzen, R., and Maxwell, A. E., 1959, The measurement of thermal conductivity of deep-sea sediments by a needle-probe method: Journal of Geophysical Research, v. 64, p. 1557–1563.

Von Herzen, R. P., and Uyeda, S., 1963, Heat flow through the eastern Pacific Ocean floor: Journal of Geophysical Research, v. 68, no. 14, p. 4219–4250.

Waring, G. A., 1965, Thermal springs of the United States and other countries of the world; A summary, rev. by Blankenship, R. R., and Bentall, R.: U.S. Geological Survey Professional Paper 492, 390 p.

Manuscript Accepted by the Society September 15, 1986.

Printed in U.S.A.

Index

[Italic page numbers indicate major references]

Typeset by WESType Publishing Services, Inc., Boulder, Colorado
Printed in U.S.A. by Malloy Lithographing, Inc., Ann Arbor, Michigan

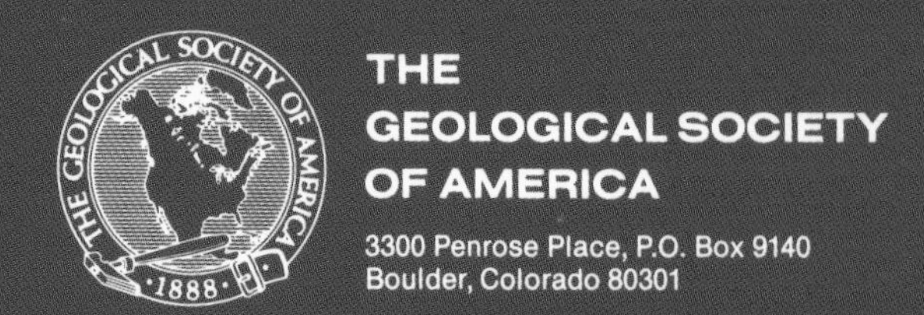

THE
GEOLOGICAL SOCIETY
OF AMERICA

3300 Penrose Place, P.O. Box 9140
Boulder, Colorado 80301

CONTENTS

ISBN 0-8137-2214-4